H. Kikuchi (Ed.)

Environmental and Space Electromagnetics

With 304 Figures, Including 2 in Color

Springer-Verlag
Tokyo Berlin Heidelberg
New York London Paris
Hong Kong Barcelona

Hiroshi Kikuchi
Professor, URSI Commission E Chairman, Nihon University, College of Science & Technology,
1-8-14, Kanda Surugadai, Chiyoda-ku, Tokyo, 101 Japan

ISBN-13:978-4-431-68164-9 e-ISBN-13:978-4-431-68162-5
DOI: 10.1007/978-4-431-68162-5

Library of Congress Cataloging-in-Publication Data
Environmental and space electromagnetics/H. Kikuchi, ed. p. cm. Based on the papers presented at the International URSI Symposium on Environmental and space electromagnetics held in Tokyo on 08/04-06/1989. Includes bibliographical references. ISBN-13:978-4-431-68164-9
1. Radio meteorology–Congresses. 2. Electromagnetism–Congresses. 3. Atmospheric electricity–Congresses. 4. Radio noise–Congresses. I. Kikuchi, H. (Hiroshi), 1926- . Electromagnetics (1989; Tokyo, Japan), QC972.6.E68 1991, 537.5'34–dc20 91-31912

Preface

This book is based on the updated versions of a lively mixture of tutorials, topical papers, and scientific and technological contributions collected from the International URSI Symposium on Environmental and Space Electromagnetics held in Tokyo on 4-6 September, 1989. It was sponsored by the International Union of Radio Science (URSI) as an activity of Commission E (Chairman: present editor) preceding the URSI General Assembly in Prague, Czechoslovakia in 1990.

The aim was an exchange of information and views to highlight the state of the art in radio science and interdisciplinary areas. Along this line, the editor has attempted to cover quite new, novel or unconventional subjects besides more traditional or conventional ones.
Although a great many subjects have apparently been covered, this book has been edited so the reader can find some common concepts or views presented. On this basis, a group of many subjects can be treated in a unified fashion, and new ideas and views can be gained as a most valuable addition to current knowledge. This is one of the major features of this volume that cannot be found in any of the monographs or proceedings that cover a narrow range of limited topics. Its broad scope does not stand for presentation in a superficial and shallow manner, but stands for a strong focus on the search for a common nature in basic concepts or views in apparently diverse subjects, and a focus on the advanced or innovatory nature of each contribution.

The symposium consisted of seventeen sessions with seventy papers, 38 from abroad and 32 from Japan, and there were more than one hundred participants from fifteen countries: Canada, China (CIE; SRS), Finland, France, Federal Republic of Germany, Ireland, Italy, Japan, Korea, Netherlands, Sweden, United Kingdom, United States, and USSR.
The topics covered are:

(1) spacecraft EM environment;
(2) lightning discharges;
(3) winter thunderstorms and triggered lightning;
(4) atmospherics, whistlers and emissions;
(5) natural VLF/ELF radio noise;
(6) terrestrial and extraterrestrial noise environment;
(7) planetary lightning and cosmic plasma noise;
(8) man-made noise environment and EMC;
(9) noise and communication statistics;
(10) space communication and measurement;
(11) high power electromagnetics;
(12) seismoelectric emissions;
(13) ball lightning;
(14) meteorologico-electric phenomena and EMHD vortices.

Of the topics mentioned above, (1), (3), (6), (7), and (12)-(14) provide the reader with quite new, novel or unconventional subjects that have never been discussed in an open forum or in a book. Topic (1) discusses spacecraft active experiments and environment, and (3) presents recent results of winter thunderstorms and rocket-triggered lightning. Chapter (7) provides its readers with the updated results of observations from the major spacecraft in deep space, Pioneer and Voyager, in particular reference to planetary lightning and plasma noise, and the X-ray astronomy satellite Ginga, as well as a new approach to gravitational dusty space plasmas. Chapters (12)-(14) are special topics that have come up rather recently. In particular, an example of EHD or EMHD vortex phenomena covered in Chaper (14) is the so-called crop circles emerging in England, which have intrigued the public worldwide. The rest of the topics, (2), (4), (5), and (8)-(11), are concerned with more traditional and conventional subjects, but present scientific updates of each topic.

As in the last URSI workshop and two other international plasma workshops organized by the present editor and held in Tokyo in 1984, 1980 and 1986, the present symposium also endeavored to assign as much time as possible to speakers for lectures and discussions, and to maintain a free, enjoyable, and stimulating atmosphere in a rather informal style. The editor hopes that this volume will convey this spirit, as it contains a mixture of short and long originals and reviews, even including unfinished work that is in progress, and personal rather than established views, which may not find their proper place in the literature of the field. This may enable us to exchange information, perhaps broader in scope than in a journal article, and this is considered another major feature of this volume.

Many thanks are due to the contributors for their prompt preparation and excellent presentations. Special thanks are due to the staff of Springer-Verlag, Tokyo, for their conscientious and patient collaboration in editorial work. It has been greatly appreciated indeed.

The editor is grateful to Professor F.L.H.M. Stumpers, Honorary President and past Vice President of URSI and my predecessor as Chairman of URSI Commission E, for his encouragement and support during all phases of this activity; to Professor J. Van Bladel, Secretary General of URSI, for his support as one of the URSI Symposia; to Professor A. Kimpara, past President of National URSI, for his opening address of welcome and for his keynote address at the reception party; to Professor H. Ishikawa, past Chairman of National URSI Comission E, for his suggestions and support during this activity; to Professor T. Ogawa, past President of the International Atmospheric Electricity Commission, for his closing address; and to Professors and Drs. L.R.O. Storey, J. Hamelin, V. Scuka, Ja. I. Likhter, H. Volland, A.C. Fraser-Smith, R.L. Gardner, P.V. Bliokh, M.B. Gokhberg, J.W. Warwick, G.C. Dijkhuis, and G.T. Meaden, who served as session chairmen during the symposium and contributed greatly to its success by conducting their sessions in a most engaging and stimulating manner.

In addition, Dr. M. Hayakawa, Chairman of National URSI Commission E was most helpful in organizing the symposium and preparing the technical programme. It is a pleasure to thank the staff of Springer-Verlag Tokyo for their editorial assistance and Ms. I. Satoyoshi for her secretarial work.

Direct financial assistance was offered the symposium by Research Grants from the Ministry of Education, Inoue Science Foundation, International Communications Foundation, Telecommunications Advancement Foundation, NTT, Toshiba, Fuji, Sumitomo, NEC, Tokyo and Tohoku Power Corporations under the support of the URSI, and this is gratefully acknowledged.

Tokyo

H. Kikuchi
Chairman
URSI Commission E.

Table of Contents

8 Man-Made Noise Environment and EMC

9 Noise and Communication Statistics

10 Space Communication and Measurement

11 High Power Electromagnetics

15 Assembled Abstracts

1 Introduction

1 Introduction

H. KIKUCHI
Nihon University, College of Science and Technology, Tokyo, 101 Japan

As stated in the Preface, this book offers the updated version of contributions collected from the URSI International Symposium on Environmental and Space Electromagnetics held in Tokyo on 4-6th September, 1989. It was sponsored by the International Union of Radio Science (URSI) as an activity preceding the URSI General Assembly in Prague, Czechoslovakia in 1990. The topics covers the state of the art in radio science and interdisciplinary areas consisting of fifteen chapters on the subjects of natural and man-made radio noise in the broadest sense. Among them, one final chapter consists of assembled abstracts amounting to seventeen, supplementary to the preceding fourteen chapters containing full papers. Many of topics have never been discussed in an open forum, and offer novel unconventional problems such as spacecraft EM environment; terrestrial and extraterrestrial noise and plasma environment; planetary lightning and emissions including the results from Voyager-Neptune flyby; triggered and ball lightning; seismoelectric emissions; meteorologicoelectric phenomena and EHD (electrohydrodynamics) including atmospheric dusty charged vortex phenomena which have drawn a world-wide public interest quite recently, besides more conventional or traditional subjects such as man-made radio noise and EMC; space communication, measurement and frequency management; noise and communication statistics; lightning; atmospherics, whistlers and emissions.

While a great many topics have apparently been collected in it, this book has been edited for its readers to be able to find some common nature of basic concepts or views on which a group of many papers can be grasped in a unified scope. In addition, the summarized remarks that follow should help readers with different backgrounds to better understand the purpose, discussions, relations among various articles, and the result of the Symposium, namely the contents of this volume.

1. Spacecraft EM Environment

Spacecraft encounters various plasma environments: deep space and solar probes experience tenuous energetic plasmas in interplanetary space and dusty plasmas in planetary and cometary atmospheres; geostationary orbits encounter energetic plasmas in the magnetosphere; polar orbits traverse cold dense plasmas as well as varied auroral and polar plasmas; low earth orbits encounter cold dense plasmas. As a result, spacecraft charging takes place, occasionally producing a potential so high with respect to the ambient plasma as to cause an electrostatic discharge (ESD) leading to related EMC problems for both passive and active spacecraft.

While space exploration is thought to be reaching maturity from some aspects, further efforts for new kinds of missions are still actively being continued particularly for low earth orbits with large structures such as Space Shuttle and Space Station. In this connection, space tether experiments are being planned and are actively under investigation with a variety of new ideas and proposals such as: 1. Tethered Satellite System (TSS); 2. tether momentum transfer system to space station operations; 3. tethered elevator; 4. tether assisted micro-gravity laboratory. These space tether techniques will be applied to a space power station, a space power generator, a space power line, an attitude/orbit control system, a VLF wave antenna in space and so on.

Storey gives an overview of the Shuttle Electrodynamic Tether Mission being developed by NASA and the Italian Space Agency under a cooperative agreement and describes the functions of instruments, the planned investigations, and the expected scientific results of the mission. The first TSS mission is currently manifested in the fall of 1991. Its main objective is to verify the capability of the facility to deploy, control, and retrieve a tethered satellite; to demonstrate the electrodynamic science capability of the system; to perform various scientific investigations.

Sasaki reports the result of a tethered mother-daughter rocket experiment on spacecraft charging and associated wave emissions by an electron beam injection (up to 48 mA at 1 keV) into the ionosphere from the mother paylord. The charging effect was measured by floating probes onboard the mother paylord and associated wave emissions were detected by the tethered daughter paylord at distances of up to 426 m. The spacecraft potential exceeded 500 V during a high current beam injection and featureless broad band emission was detected during the strong charging. The emission in the HF range (0.2 ~ 10 MHz) is thought to be generated by the beam-plasma interaction, while that in the VLF range up to 30 kHz was directly related to the charging effect.

Sasaki and Oyama describe space tether experiments by a space flyer unit (SFU) and by a large satellite (1500 ~ 2000 kg for LEO) under investigation. A tethered diagnostic package containing plasma, neutral gas and wave detectors will be deployed up to 10 km in various directions from the space flyer unit orbiting at 500 km altitude. The goal of the experiment is to study the electrodynamic effect of tether system and spacecraft plasma environment. With a 100 km-scale tether connecting subsattelites deployed from a large satellite orbiting at 250 km altitude, the system will be used to sound the ionosphere down to the lower limit (around 120 km). Besides the tethered plasma diagnostics, the study of dynamics of the large-scale tether is another important subject from the standpoint of space technology.

Nagai surveys a future space power technology that is emerging and becoming more and more important for the space station era now just ahead. Although the present rating of electric power for the Space Station under planning is the order of 100 kW, the electric capacity in the near future is expected to be increased 1 MW to 10 MW when a space factory or a free flyer is needed. While the current space power generating systems are mainly depending on a fuel cell or solar cell supplying the direct current, it is needed to use an alternating current generator with a rotating machine for the purpose of introducing a power transformer and a higher voltage power line in space for power transport. With increasing electric capacity, a number of problems on electrical insulation in space are coming up because of its special space environment with very sparse air and micro-gravity and are outlined in view of electric power demand for future space exploration and utilization. Further, from the scientific point of view, high voltage phenomena in such a space environment are discussed mainly from the arc and partial discharge viewpoints in comparison with those on the earth, since they are hardly surveyed so far.

Hamelin outlines research conducted at the European Space Agency (ESA) on spacecraft/plasma interactions and their electromagnetic effects, specifically for geostationary spacecraft in operation since 1982 which has led to a comprehensive understanding of the phenomena together with prediction and analysis codes, and design and test methodology. For low-earth orbits and particularly for large structures such as Space Station and Shuttle, similar work has been in progress for the last few years (15. ASSEMBLED ABSTRACTS).

2. Natural and Triggered Lightning

Natural cloud-to-ground lightning is usually initiated by a stepped leader moving downward toward the ground. In contrast, in many strikes to towers and rockets, the leader is observed to travel upward into the cloud from the top of the tower or rocket. When the leader is initiated from the structure or spacecraft, the lightning is termed to be triggered by them (H. Kikuchi (ed.), Laboratory and Space Plasmas, Springer-Verlag, New York, 1989, p.331).

It has recently been observed from lightning experiments in a coastal region of the Sea of Japan that the cloud shape and charge distribution in winter thunderstorms are extended more horizontally at lower altitudes due to a wind shear in contrast to its vertical extent at higher altitudes in summer. This leads to significant differences in lightning characteristics between summer and winter thunderstorms, which will be a main subject of the present chapter to be discussed in the following six articles.

Tuomi, Summanen and Pirjola present an extensive report on lightning observations made in Finland by lightning flash counters for thirty years and by a lightning location system (LLP) since 1984 and on a theoretical calculation of the lightning electromagnetic pulse but still for a perfectly conducting earth. The counters seem to record more flashes than the LLP system does, possibly due to the low conductivity of the earth in Finland.

Grigoryev and Shiryaeva propose a model for the preliminary stage of a lightning flash based on the concept of the charge drop instability in an electric field that is the superposition of a quasi-stationary intracloud field and an impulsive time-dependent field produced by large, strongly charged, water-covered flying pellets of hail. A plasma region is formed in the vicinity of the pellet with a typical time-scale 0.1 ms and its polarization and growth into the intracloud field provides initiation of a lightning channel.

Goto and Narita present unique observations of winter lightning strokes in a coastal region of the Sea of Japan with a new video camera system driven automatically. One hundred twenty events of winter lightning flashes were photographed by using this system. The results reveal the following findings: 1. The winter thundercloud is as low as 200 $\sim$ 800 m high above sea level; 2. Most of winter lightning flashes triggered by a tower are upward developing ones that branch toward the base of thunderclouds and the flashes often spread out horizontally to a considerable distance; 3. The median flash duration was 270 ± 33 ms much longer than that in summer, indicating that the continuing current was flowing for very long time in luminous channels; 4. The angle of incidence of winter lightning to the tower was widely distributed and its median value was approximately 30 degrees much larger than that in summer; 5. The cumulative frequency distribution of the angle of incidence, $F(\theta)$ can be approximated as $(\cos\theta)^4$.

Yamamoto et al. report anomalous features of electromagnetic pulses caused by positive lightning return strokes during winter thunderstorms in a coastal region of the Sea of Japan that are quite different from those of the summer negative return strokes. This is related to the differences between the results detected by N-S and E-W magnetic loops due to the different configuration of the channel of positive return strokes in winter from that of negative return strokes in summer.

Nakano et al. describe VHF/UHF radiation from positive and negative lightning, based on measurements made for close lightning in summer and winter together with the electric field change measurements. More than 150 negative and 10 positive lightning discharges were observed during summer and winter thunderstorms. Strong radiations are associated with return strokes of a ground flash in summer. The radiation intensity associated with the positive return stroke quite differs from that in the negative stroke. The intensity from positive return strokes in winter is several times as small as that from negative ones in summer. The time delay of VHF/UHF bursts from the onset of the electric field change caused by a return stroke is 17 μs for UHF radiations and 27 μs for VHF in average for 80 negative first return strokes, and is roughly 28 μs in the UHF range for four positive lightning strokes.

Nakamura and Horii summarize rocket triggered lightning experiments carried out with electrical, optical and acoustic measurements for winter thunderclouds in Japan. Eighty-four strikes have been obtained successfully from 1977 to 1987 by using four triggering methods: 1. zero impedance grounding; 2. finite impedance grounding; 3. no grounding with a short or long gap; 4. insulating with a nylon string. The main results are as follows: 1. For zero impedance grounding, thirty-six strokes were obtained with a nearly equal success ratio for both positive and negative clouds and the predischarge process lasted for 1 ~ 2 s before the main discharge; 2. For finite impedance grounding and for negative clouds, an abnormal discharge occurred frequently in such a way that the main discharge did not propagate along the vaporized wire but through the free air apart from the wire; 3. For the experiment using a long nylon string, seven strikes occurred on the test power line and tower, indicating that this triggering method is quite feasible for striking the power line to study the striking distance; 4. The lightning discharge test for various equipments can be performed by changing a gap length; 5. The discharge current and the luminosity duration were 10 kA and 20 ms in average, respectively.

Dijkhuis presents a new electrohydrodynamic model for horizontal thunderclouds in the atmospheric boundary layer combining turbulent shear flow with space charge from fair-weather electricity. Horizontal electrification is obtained from circulation of charged eddies carrying a spin current through wind-sheared air layers. Upstream deflection of downward field lines by eddy circulation in stratified space charge is found to be compatible with available fair-weather data. In weather conditions with reduced ohmic conductivity, the spin current accumulates positive charge on leading cloud sections as seen in winter thunderstorms above the Sea of Japan. Poisson's equation is solved on a square mesh for dipolar thundercloud electrifications approached by a wire-trailing rocket. Equipotential plots show electric field reversal at ground level in bifurcation points of the zero potential line. Field strength plots identify dipolar clouds as a saddle point, and show regions of field reversal as electric cusps. These calculations provide a first numerical simulation for a model of triggered lightning with a horizontal thundercloud and charge distribution described in editor's two articles (Kikuchi, in Proc. 6th Symp. Electromag. compact. Zürich, 1985, p.47; H. Kikuchi (ed.), Laboratory and Space Plasmas, Springer-Verlag, New York, 1989, p.331).

Berger discussed the formation of a lightning upward discharge originated from a protrusion on the earth surface by a relatively slow rise of the electric field induced at ground level when a negative downward leader approaches the ground (15. ASSEMBLED ABSTRACTS).

Scuka reviews the state of the art in lightning research from a global aspect and then outlines a study initiated by the ESA Meteosat program for the second generation meteorological satellites instrumentation with the availability for including an advanced optical lightning flash detector into its instrumentation load for geophysical research (15. ASSEMBLED ABSTRACTS).

Kawamata et al. report simultaneous lightning flashes from multiple points observed in a coastal region of the Sea of Japan for several winter seasons, indicating a rather horizontal cloud and charge distribution at a lower height in winter. The average distance between striking points was 1.2 km and the average time interval was 12 ms for those simultaneous flashes (15. ASSEMBLED ABSTRACTS).

3. Atmospherics, Whistlers, and Emissions

Atmospherics or sferics are radio noise emitted from electrical discharges in the atmosphere, accompanied by thunderstorms, showers, snow storms, developed cumulonimbi, or even sandstorms, and are caused by a variety of origins such as:

thunderstorms; cold fronts; polar fronts; jet streams; typhoons; hurricanes; cyclones; tornadoes; whirlwinds; volcanic eruptions; nuclear explosions. They can naturally been extended to include radio noise emitted from discharges in the planetary or extraterrestrial atmospheres such as emitted from planetary lightning and volcanic eruption.

Volland gives an overview on the excitation of sferics by lightning strokes and the propagation of these electromagnetic pulses within the atmospheric waveguide between ground and ionosphere. Their maximum spectral energy is within VLF-band and their waveforms are modified by dispersive guiding within the atmospheric waveguide or within the magnetosphere when they propagate along the geomagnetic field lines as whistlers. Several methods are reviewed for locating lightning strokes from the observation of sferics.

Ondoh presents an extensive study of polar emissions observed by ISIS satellites, based on VLF electric field data (50 Hz ~ 30 kHz) received at Syowa station, Antarctica from 1976 to 1983. Polar VLF emissions include polar cusp hiss, auroral hiss, VLF saucer and ELF hiss. The broad-band auroral hiss is defined by a criterion of simultaneous intensity enhancements greater than 5 dB above the quiet level at 5, 8, 16 and 20 kHz bands observed at invariant latitudes above 60°. Its high-latitude limit lies at about 82° in invariant latitude for all geomagnetic local times with two high occurrence rate regions from 76° to 78° in the afternoon sector and also from 78° to 81° in the midnight-morning sector, while its low-latitude contour is approximately symmetric with respect to the 22-10 geomagnetic local time meridian, lying at 74° on the dayside higher than at 67° on the nightside in geomagnetic latitude. An apparent lack of the broad-band auroral hiss in the 10-11 geomagnetic local time sector is due to the polar cusp hiss with very irregular spectrum in frequency and time. The occurrence map for broad-band auroral hiss is on the whole similar to that for the inverted-V electron precipitations, indicating that the broad-band auroral hiss could be generated through the process of Cerenkov radiation by inverted-V electrons precipitating from the boundary plasma sheet.

The following three papers are concerned with low latitude whistlers and VLF/ELF emissions, based on ground and spacecraft direction finding (DF) measurements. First, Hayakawa presents an overview of recent studies on the generation and propagation mechanism of magnetospheric VLF/ELF emissions, based on the DF measurements onboard the spacecraft, specifically for the following plasma waves: 1. half-gyrofrequency VLF emissions; 2. ELF hiss in detached plasma regions; 3. plasmaspheric ELF hiss; 4. VLF/ELF chorus. Half-gyrofrequency VLF emissions are generated at a frequency above one half the local electron gyrofrequency near the equator with wave normal directions close to the oblique resonance cone by an electrostatic Harris-type instability. This mechanism is supported by the DF measurements on GEOS-1 in the off-equatorial region. ELF hiss has distinctly been observed in detached plasma (DP) regions outside the plasmapause by GEOS-2 measurements, with wave normals nearly aligned with the geomagnetic field. This DP hiss is thought to be generated by a quasi-linear electron cyclotron instability. Plasmaspheric ELF hiss is a broadband and structureless whistler-mode noise within the plasmasphere and is sorted into two classes in terms of its wave normal. One is a group of field-aligned waves generated just inside the plasmapause by the electron cyclotron instability due to medium energy (10-100 keV) electrons, and the other is a group of oblique waves with two different subgroups of wave normal directions: one is a medium wave normal angle ranging from 20° to 50°, and the other is a large wave normal angle between 70° and 80° close to the oblique resonance angle. The possible generation mechanism for oblique waves might be the Cerenkov instability. Chorus is an intense naturally occurring VLF emissions in the magnetosphere characterized by its discrete spectra and is observed mainly at L values between the plasmapause and magnetopause at all local times with two types of structures: rising tone and impulsive chorus. While rising tone chorus seems to have two groups in terms of wave normals: one is field-aligned (5° $\sim$ 20°) and the other is close to the oblique resonance angle (30° $\sim$ 55°), impulsive chorus has a large scatter in wave normal angle from 5° to 60° and its generation mechanism is not clear, requiring detailed study. Finally, the technical problems of spacecraft DF measurements are summarized for the future satellite experiments.

Hayakawa, Ohta and Shimakura report spaced direction finding measurements of very-low-latitude whistlers carried out in January, 1988 at three stations nearly in the same magnetic meridian in South China: Zhanjiang (10.1° geomag. lat.); Guilin (14.1°); Wuchang (19.4°). Two horizontal and one vertical electric field components were simultaneously recorded over a wide frequency band from 2 to 6 kHz for comprehensive direction finding. The analyses of whistler data on several days indicate that the ionospheric transmisson latitude of equatorial latitude whistlers is restricted to a range of the geomagnetic latitude, 10°-14°, and the high occurrence of echo-train whistlers is a strong indication of their field-aligned propagation. Their propagation mechanism is under investigation.

Ohta, Hayakawa and Shimakura specifically discuss the frequency dependence of ionospheric exit points and polarization of daytime whistlers at low latitudes, based on DF measurements mentioned above over a broad range of frequencies up to 10 kHz, particularly using the digital recorders and a fast Fourier transform analyzer at Yamaoka station (25°N geomag. lat.). No appreciable frequency dependence has been found on both ionospheric exit points and polarization for daytime whistlers, supporting their field-aligned ducted propagation.

Hagino, Ogawa and Yano present a new approach to waveform simulation of whistlers, calculating an integral form of the electric field due to a δ-function sheet current source with the aid of the suddle point method. The calculated whistler waveforms are compared with observations on the ground and can explain the observed characteristics such as polarized modulation or fading.

Likhter, Larkina and Gdalevich report small scale variations of ELF/VLF magnetic and electric field components in subauroral zones, based on observations from "Intercosmos-14 and -19" satellites. Across the region of inhomogeneities of concentration in the topside ionosphere close to the high latitude troughs, simultaneous and synchronous fluctuations were found among plasma concentration, plasma temperature, magnetic and electric field components of ELF/VLF emissions. Concerning the probability characteristics of magnetic and electric field fluctuations, its correlation coefficient turned out to be smaller than 0.5 below the lower hybrid resonance (LHR) frequency of 4.65 kHz, but it increased up to 0.8 ~ 0.9 above the LHR frequency. This indicates that locally-excited electrostatic waves play a noticeable role in fluctuations below the LHR frequency, while the electromagnetic waves undergoing inhomogeneities of concentration are recorded above the LHR frequency (15. ASSEMBLED ABSTRACTS).

4. Natural VLF/ELF Radio Noise

Natural VLF/ELF radio noise is not only atmospherics emitted from electric discharges but also covers plasmaspheric and magnetospheric noise often generated from the plasmapause, high latitude troughs or holes, magnetopause and other boundary discontinuities or unstable regions.

A global survey of ELF/VLF radio noise (frequency in the range 10 Hz - 32 kHz) has been conducted by STAR Laboratory, Stanford University with three high latitude stations: Thule (TH: 76.5°N, 68.8°W) and Søndrestrømfjord (SS: 67.0°N, 50.1°W) in Greenland; Arrival Heights (AH: 77.8°S, 193.3°W) in the Antarctic and five lower latitude stations: Grafton, New Hampshire (43.6°N, 72.0°W); L'Aquila, Italy (43.4°N, 13.3°E); Stanford, California (37.4°N, 122.2°W); Kochi, Japan (33.3°N, 226.5°W); Dunedin, New Zealand (45.8°S, 189.5°W).

Fraser-Smith et al. describe a comprehensive review of this survey with some of the results obtained recently. Comparison is made, in particular between the noise measurements at high (polar) latitudes and at lower latitudes. Although most of the natural ELF/VLF noise observed everywhere in the world is lightning-generated, the high-latitude noise often contains additional components that are of magnetospheric origin. This noise consists predominantly of polar chorus, which is concentrated in the range 300 Hz - 2 kHz, and produces a characteristic signature in the noise statistics. Less frequent occurrences of broad-band (auroral) hiss can occasionally mask most or all of the lightning-generated noise in the ELF/VLF range.

Fraser-Smith et al. report measurements of radio noise by a Stanford University ELF/VLF radiometer located near Kochi, Japan since 1985. The measurements provide information on the average and root-mean-square (rms) amplitudes in 16 narrow frequency bands (5% bandwidth) distributed through the range 10 Hz to 32 kHz. The noise amplitudes tend to decline with increasing frequency in agreement with earlier measurements at other locations. The decline is roughly proportional to 1/frequency. The voltage deviation V_d that is the ratio in dB of the rms to average amplitudes, on the other hand, tends to increase with frequency, thus indicating increasing 'spikiness' of the noise with increasing frequency with an exception of a remarkable drop in V_d around 1 kHz. The noise amplitudes at Kochi are roughly 3-10 times larger during the summer than during the winter, and, when taken as a whole, they are at all times substantially larger than those measured at high latitude locations. This difference is attributed to its close proximity to thunderstorms that are the primary sources of the noise.

Tanaka and Nishino report propagation characteristics of LF whistler mode waves, based on observations of Decca navigation signals at Birdsville (L = 1.55; 25.83°S, 139.33°E geographic coordinate), Australia nearly conjugate to Biei (L = 1.54; 43.60°N, 142.45°E geographic coordinate), Japan, where Decca transmitter is situated. Results are as follows: 1. The whistler-mode signals were detected at 85.725 kHz but not at 114.300 kHz at Birdsville at sunset and sunrise, and occasionally at night; 2. They usually revealed the frequency shift: a dusk positive shift (< 0.5 Hz) and a dawn negative shift (> -0.5 Hz), being independent of magnetic activity; 3. Their EW intensity remarkably increases in the nighttime in association with magnetic storms; 4. Storm-associated energetic electrons and magnetospheric electric fields penetrating into the low-latitude magnetosphere may generate the intensity increase and the frequency shift; 5. For the reception of the relevant signals on the ground, whistler ducts seem to be required around the field line intersecting the transmitter and/or at somewhat lower latitudes than the field line.

Ishikawa et al. investigate the ray focussing of whistler-mode waves in terms of the critical frequency which results in an enhanced wave-particle interaction. The critical frequency in a slightly inhomogeneous magnetoplasma is defined as a frequency with zero curvature of the refractive index surface at a longitudial wave normal angle, and approaches one half the electron gyrofrequency when $f_H/f_p \rightarrow 0$ (f_H: electron gyrofrequency; f_p: electron plasma frequency). This critical frequency is also found to be consistent with the zero diffraction coefficient in the full-wave theory and proves important for wave-particle interactions near the plasmapause and in the inhomogeneous outer magnetosphere outside the plasmapause.

Yano et al. propose a calculation method for obtaining waveforms of tweek atmospherics which is thought to propagate between the ionosphere and the ground under multiple reflections. Lightning current is assumed to be a short dipole antenna situated vertically on the ground with an infinite conductivity, while the assumed sharp boundaries of both the ionosphere and the ground could be replaced by the electric images. Thus, the problem is reduced to evaluating the radiation electric field from the Fourier-expanded mode currents. Even for the first mode only in consideration, calculated waveforms are in good agreement with observed ones for both the initial rising part and the guitar-shaped main part. The effect of the Earth's curvature appeared at the beginning of a tweek in the frequency-time spectrogram.

Kimura et al. outline the objective, instrumentation, and preliminary results of the Akebono satellite (EXOS-D) launched on 22 February, 1989 for the exploration of aurora, its energy source and relating electromagnetic environment in the magnetosphere. One of unique VLF instruments is a propagation vector analyser by two components of electric field and three components of magnetic field for swept frequency or a fixed frequency in the frequency range from 100 Hz to 12.7 kHz (PFX) by which one can determine the wave normal direction of signals for a band width of 50 Hz, the center frequency being swept or fixed by command (15. ASSEMBLED ABSTRACTS).

5. Terrestrial and Extraterrestrial Noise Environment

Natural radio noise described so far is of terrestrial origin and mainly for low frequencies. This chapter, however, is extended to include noise of extraterrestrial origin and for higher frequencies, for example in interplanetary space.

Smith and Flock offer an overview of natural radio noise in terms of brightness temperature as observed from two vantage points. The first is from an Earth station located at 40 degrees north latitude and observing at elevation angles from 0 to 90 degrees with an ideal antenna (no sidelobes but finite beamwidth). The second is a satellite in geostationary orbit communicating with the Earth. Earth station noise at VHF and UHF is dominated by galactic and solar noise. Emission from atmospheric gases and hydrometeors are dominant at EHF and SHF and Oxygen actually has a minor peak around 1 GHz. The sun is a strong emitter between 50 MHz and 1 GHz and is extensively studied in this frequency band where emission occurs from well beyond the limb of the Sun. At EHF and SHF (3 to 300 GHz) emission as seen by the Earth station will come largely from the atmosphere irrespective of whether it is clear, cloudy or rainy. Radiative transfer theory is invoked in the calculation of brightness temperature from the atmosphere, while multifrequency radiometry can resolve the contributions from the liquid and gaseous components of water. Emission by molecular oxygen and water vapor are the primary contributors on clear days particularly at resonant frequencies. The situation is not vastly different as viewed from the geostationary satellite, particularly when looking at the Earth. The principal difference is that emission from land or sea can no longer be denied, even under idealized conditions. Land is a much more effective radiator than sea water, but at frequencies above 30 GHz the differential becomes much less due to the increasing significance of atmospheric emission. Observing the planets with a few exceptions, such as Venus and Jupiter, is a simpler proposition as planets do not intercept a significant portion of any radio beam from Earth orbit.

It is now generally accepted that short period geomagnetic pulsations in the Pc-1 range are generated in a region of the equatorial plasmapause rather than the magnetopause during the recovery phase of a magnetic storm as originally indicated by the present editor (Kikuchi, Nature Phys. Sci., 229, 79, 1971; Kikuchi and Taylor, J. Geophys. Res., 77, 131, 1972; Kikuchi, J. Atmos. Terr. Phys., 38, 1055; 1127, 1976). Higuchi discusses amplification and propagation of Pc-1 pulsations in terms of an electromagnetic ion cyclotron instability caused by temperature anisotropy of ring current protons, specifically on the role of heavier ions in the amplification mechanism for Pc-1 pulsations in the plasmapause region. While the unstable frequency bands are mainly below the helium ion gyrofrequency, there is an optimum value of the cold helium ion density to hot proton density for the maximum growth rate. The total wave amplification values are calculated along a single propagation path through the unstable equatorial region. The maximum amplification occurs just inside the plasmapause with a gain of 13 dB.

The simultaneous occurrence of hiss and chorus is frequently observed onboard satellites in the outer magnetosphere, and is the phenomenon that chorus is accompanied by a background hiss. As for the triggering of chorus, two possibilities have been indicated. One is a triggering by power line harmonic radiation (PLHR) which is considered to be a coherent signal, while the other is the excitation from the background hiss which is incoherent and turbulent. Hattori et al. discuss how a chorus is generated and show that a chorus is triggered by the background hiss, based on the detailed spectral analyses and direction finding measurements for the data from the GEOS-1 satellite near the magnetic equator in the outer magnetosphere. The following findings are noteworthy: 1. Each chorus element is very likely to be originated from the underlying hiss band; some of chorus elements are asymptotic to the hiss band and the df/dt at the foot of them are nearly zero; 2. The intensity and occurrence of chorus emissions are found to be closely correlated with the intensity of the underlying hiss band; 3. On some occasions, wavelets at the foot of chorus elements are considered to be capable for chorus generation; 4. The direction finding indicates that both hiss and chorus come from the same source region.

Meanwhile, it is well-known that geomagnetic disturbances produce large electric fields in Earth's environment in particular at auroral latitudes where disturbances are most intense. They also induce large quasi-direct currents, the so-called geomagnetically induced currents (GIC) in power transmission grids and other man-made conductors, occasionally disturbing power systems by saturating transformers. The following two papers are concerned with geomagnetic induction caused by geomagnetic disturbances and an auroral electrojet and are inserted among the topics of VLF/ELF radio noise. Pirjola and Viljanen report studies of GIC in both measurements and theoretical calculations for the Finnish 400 kV power grid. The highest GIC recorded so far is 165 A as a ten-second mean value. Theoretical estimates indicate that GIC of some hundreds of amperes may flow through Finnish 400 kV transformers during large geomagnetic storms.

Pirjola and Häkkinen propose a theoretical model of the auroral electrojet current system, assuming a horizontal sheet electrojet of a finite length with a Gaussian transverse distribution of the current and additional field-aligned current with a harmonic dependence in time and in space along the electrojet. The electromagnetic field thus produced on the earth's surface by such an electrojet current system has been calculated by using a homogeneous or four-layer earth model in expectation of a future comparison with EISCAT magnetic data.

Going back to the problem of VLF/ELF natural radio noise, the following two articles are specifically concerned with the so-called Wave Distribution Function (WDF) Analysis that is a generalized form of radiowave direction-finding, applicable to natural radio waves in space plasmas and that was created by L.R.O. Storey. In this connection, it should be noted that another wave distribution function approach based on quasi-particle concept was originated by N. Marcuvitz for waves in most general media, inhomogeneous, anisotropic, nonlinear, or turbulent, and was applied by H. Kikuchi and M. Hirota specifically to waves in nonlinear, dispersive, and dissipative media (Marcuvitz, Proc. IEEE, 68, 1380, 1980; Kikuchi and Hirota, in Nonlinear and Environmental Electromagnetics, edited by H. Kikuchi, Elsevier, 1985, p.83). Shimakura and Hayakawa propose a new ground-based direction finding technique to estimate the ionospheric exit regions of magnetospheric VLF waves on the basis of wave energy distribution of those waves in wave number space which is evaluated by maximum entropy method which is, however, model-dependent on the inversion or wave polarization. Attempts are made to obtain a most reliable solution by using a suitable model for the inversion and polarization toward the study of the ionospheric transmission and magnetospheric propagation mechanisms of magnetospheric VLF waves.

Storey offers a comprehensive review of the state of the art in natural ELF/VLF waves with particular reference to plasmaspheric hiss, based on WDF analysis as a procedure for making sky maps of the sources of natural radio waves in space plasmas, given local measurements of some or all of the magnetic and electric field components. The prescription for future work on WDF analysis includes solving basic methodological problems, translating the solutions into efficient algorithms, and embodying the algorithms in computer software. One important scientific use of WDF analysis is to identify the mode of origin of plasmaspheric hiss, for example hopefully by analyzing wave data from the satellite Akebono (EXOS-D).

Bashkuev et al. summarize measurements on global resonances of the earth-ionosphere cavity made in middle latitudes of Asia (Transbaikalia) over the years 1979-1988. The resonances from 3 to 6 were observed in the most spectra at 7.7, 13.7, 20.1, 26.3, and 32.3 Hz (average value), Q-factors were varied from 3 to 8, and resonance frequencies were changed from 0.5 to 1 Hz. Maximum spectral intensity levels exceed minimum levels from 1.6 to 4 times. The spectral properties of the global resonances reveal large variability of the level spectral densities, their frequency content, and the Q-factors. The main cause of it is the periodical change of the low ionosphere parameters. The propagation parameters were obtained from the Schumann spectra. The phase velocity is in the range from 0.71c to 0.79c and increases with increasing frequency. The average values of the attenuation rate are 0.16, 0.24, and 0.34 dB/Mm at day and 0.22, 0.35, and 0.56 dB/Mm at night at the first, second, and third resonance frequencies, respectively (15. ASSEMBLED ABSTRACTS).

Kikuchi reports a new type of interplanetary radio bursts (IRB) with electron plasma oscillations (EPO), associated with a small scale of sporadic electrostatic shocks that are characterized by superthermal electrons just like type II IRB but well behind an interplanetary hydromagnetic shock front in a highly turbulent downstream region near a HELIOS perihelion. Such a typical IRB-EPO event was obtained from HELIOS 2 on 22 April, 1977 and this IRB has tentatively been termed as a modified type II. In situ HELIOS measurements indicate that as close as to 0.3 AU a downstream region behind large-scale flare-associated hydromagnetic shocks are often highly variable and turbulent, sometimes producing magnetic clouds as reported and possibly electric clouds as well which could generate local electrostatic shocks with superthermal electrons responsible for the EPO-IRB (15. ASSEMBLED ABSTRACTS).

6. Planetary Lightning and Cosmic Plasma Noise

While the preceding chapter already covered the topics of extraterrestrial noise, the present chapter is exclusively devoted to planetary, solar, and galactic noise from astrophysical or cosmical rather than geophysical aspects, based on observations from spacecraft such as Pioneer, Voyager and Ginga and from ground as well. A topic of planetary lightning is considered as a continuing issue from an open session on Planetary Lightning, Discharges, and Emissions at the Second International Workshop on the Relation between Laboratory and Space Plasmas held in Tokyo on 25-26 November, 1986, and is found in Chapter V (pp. 273-344) of its Proceedings entitled Laboratory and Space Plasmas, edited by H. Kikuchi, Springer-Verlag, 1989.

Warwick offers a comprehensive overview of the Voyager program for the exploration of the outer solar system culminating in the flyby of Neptune on the background of his association over eighteen years. Particular attention is focussed on results of Voyager at Neptune, emphasizing Voyager radio astronomy data. Both Voyagers have now passed through the entire solar system, from Earth to beyond Neptune and are now on trajectories that will intersect the heliopause at some unknown time in the next several decades. After that the spacecraft will move within interstellar space. This final phase of the Voyager program beginning now is known as "VIM", the Voyager Interstellar Mission. Given the excellent condition of both spacecraft, it will continue for at least the next twenty to thirty years. An early image of Neptune already shows a large dark spot associated with white clouds at one side. The center of the image lies at about 30° south latitude. The Great Dark Spot (GDS) is about the size of Earth and the material within it circulates counterclockwise as it is observed from space. This means that the circulation is like that around a high pressure center in Earth's southern hemisphere. The rotation of Neptune brings GDS into view about once every eighteen hours. Further imaging revealed a small white spot at high southern temperate latitudes which was named "scooter" by the imaging team. On time lapse movies of Neptune with GDS fixed in the middle of the frame, scooter whips by rapidly to the east, passing GDS at a speed corresponding to a rotation period of only about 16.7 hours. The relative velocity of GDS and scooter is about 1,000 km/hr. This enormous shear corresponds to winds of about 300 m/s which must somehow be rationalized between scooter and GDS. Maps of thermal emission from Neptune obtained from an Imaging Infrared Spectrometer (IRIS) demonstrated that the equatorial regions of Neptune are hot, that temperate regions are cool, and that polar regions were hot, like the equator. These were interpreted by the IRIS team in terms of a Hadley cell model of Neptune's atmospheric circulation, in which gas decends at the equator and poles, but rises in temperate latitudes. They showed that this temperature and circulation are consistent with winds corresponding to the relative motion of GDS and scooter, where scooter is at high levels (white and bright) and GDS is at lower levels. While the Radio Astronomy Experiment (PRA) detected Neptune's radio emissions as early as about 12 days before closest approach on 25 August 1989, the importance of these early observations lies in the fact that PRA has been able to provide the program with accurate information on the rotation of Neptune observed over an interval of several weeks duration. These radio emissions were detected first in a dynamic spectral mode, in which the bursts lay as much as 18 dB above the noise background. For the most part, a given burst appeared in only one frequency channel of the PRA, which has channels spaced at intervals of 19.2 kHz. These bursts drifted upwards in frequency from burst to burst in some kind of average sense, from 700 kHz to 850 kHz. In addition, these bursts are extremely short and are strongly left-hand polarized. They may be produced in regions near the south pole of Neptune and at frequencies close to the electron cyclotron frequency in that region. In that case, Neptune's magnetic pole nearest the southern rotational pole has polarity opposite that of Earth's magnetic pole near Earth's southern rotational pole. The sense of Neptune's magnetic field therefore has the same relation to Neptune's rotation axis as do the magnetic fields of Jupiter, Saturn, and Uranus, but opposite that of Earth. The remarkable feature of these bursts (other than their bandwidth and polarization) is the precision with which they define a period of 16 hours, to within +/- 0.1 or 0.2 hours. From about four days before closest approach, "continuum" radio emission continued to be detected through and after closest approach as well, and is decidedly non-bursty, lasting for several hours during each episode of occurrence. Its radio frequency lies in the range generally about 300 to 600 kHz, distinctly below that of most of the bursts. The power in the continuum emission versus longitude of the sub-spacecraft position on Neptune, a diagram, shows that the continuum is emitted in a rotating searchlight pattern; this pattern is observed from a spacecraft which is itself moving around the planet. The period of rotation is 16.11 +/- 0.05 hours. The bursts do not follow the same pattern as the continuum. They occur instead at strictly repetitive intervals throughout the times before and after closest approach; they are fixed in terms of Sun. In conclusion, Neptune's rotation is very unusual in relation to the rotation of his atmosphere. Apparently, the interior rotates more rapidly than, or as rapidly as, any point in the atmosphere, being equalled only by scooter.

Taylor gives a comprehensive survey on the controversial problem of Venusian lightning, providing additional PVO (Pioneer Venus Orbiter) data, although this problem was already discussed extensively in the preceding book mentioned at the beginning of this chapter. He firmly concludes that: 1. The persistent correlation between 100 Hz noise and nightside ion troughs strongly suggests that the noise is generated near the spacecraft, and is related to the interaction of the solar wind with the nightside ionosphere; 2. Some of reported characteristics of the electric field noise are incorrectly based on telemetry interference, and not actual electric field signals; 3. The plasma noise detected by the PVO is unrelated to either lightning or volcanism, and provides no such evidence.

As one of topics on unconventional plasmas, the problem of dusty plasmas is receiving increasing attention from various aspects, laboratory, fusion, meteorological, astrophysical, and even industrial, and is already found in

Chapter II (pp. 49-107) of the preceding book mentioned above. Such plasmas exist in planetary rings, planetary and cometary atmospheres, and interstellar dust clouds in space. Bliokh and Yaroshenko consider collective processes in a plasma with charged dust particles, taking into account gravitational interaction between the particles. Formulae for the dielectric constant are derived in an isotropic, magnetized and selfgravitational plasma. As an application of this theory, they analyze some plasma processes in planetary rings that may be considered as a system interpenetrating streams of charged particles described by a modal equation. For the Saturn rings, the characteristic time for the principal ring to become subdivided into smaller ringlets are estimated as 10 years, and the lifetime of the radial irregularities known as "spokes" is estimated as 10 minutes.

Solar activity of the cycle 22nd is rising sharply aiming at the maximum around the years 1990 and 1991. During the maximum solar activity period, Nobeyama Radio Observatory plans to study solar radio bursts using the 17 GHz Radio Heliograph in close cooperation with the solar-A X-ray satellite and ground based optical telescopes. Shibasaki et al. describe the functions and instrumentations of the 17 GHz Radio Heliograph for studies of solar radio bursts with emphasis on structure of burst sources. It consists of T-shaped array of 490 m in E-W direction and 220 m in N-S direction with 84 antenna elements. Burst sources can be mapped in every 50 milliseconds with 10 arc second spatial resolution for the study of the behavior of high energy electrons accelerated in solar flares.

The X-ray astronomy satellite "Ginga" that stands for "galaxy" in Japanese, is the third Japanese X-ray astronomy satellite launched on 5 February, 1987, following Hakucho (Kondo et al., Space Sci. Instr., 5, 211, 1981) and Tenma (Tanaka et al., Pub. Astron. Soc. Japan, 36, 641, 1984) which have already made significant contributions in the field of X-ray astronomy. Makino et al. present a comprehensive overview of scientific activities in Ginga on its functions, instrumentations, and achievements. Three kinds of the following instruments are onboard: Large Area Proportional Counter (LAC); All Sky X-ray Monitor (ASM); Gamma-ray Burst Detector (GBD). The LAC is the main instrument with an effective area of 4,000 cm^2 and with the highest sensitivity to hard X-rays so far obtained. Ginga observed about 250 targets up to the end of 1989. Main results are as follows: 1. power law spectra with resonant absorption structures in the hard X-ray region detected from five X-ray pulsars: Her X-1; 4U1538-52; V0332+53; 1E2259+58; 4U0115+634, possibly due to cyclotron absorption; 2. five new transient X-ray pulsars: Cep X-4; GS1843+00; GPS1722-362;

GS1843-02; Sct X-1, in the region of 5 kpc ring of the Galaxy; 3. intense diffuse iron line emission from a high temperature plasma at the Galactic center; 4. X-ray scattering by interstellar dust grains; 5. X-rays from SN 1987A; 6. spectral structures and variations of Seyfert galaxies; 7. red-shifted iron line emission from quasars; 8. two new bright X-ray novae: GS2000+25; GS2023+338; 9. large amplitude quasi-periodic oscillations (QPO) from X-ray source GX339-4 and the rapid burster for the first time.

Borucki reports some new results for lightning activity on Venus and Jupiter, and, based on this, gives his predictions of lightning activity on Neptune. He introduces the recent papers of Russel et al. showing that the source of signals observed by the PVO electric field detector on Venus must be below 200 km and therefore the signals must be due to lightning rather than magnetospheric processes. However, Venusian iron troughs as a signature of the solar wind-ionosphere boundaries can partly be below 200 km and become the source origin of those signals. In addition, preliminary results of a new search for optical pulses from Venusian lightning indicate that lightning flashes must be weak, infrequent, or both. A new analysis of Voyager 2 images of Jupiter confirms the existence of a band of thunderstorms, 100,000 km long, at 49 degrees North latitude and of a second band at 17 degrees North latitude. No lightning activity is observed in the southern hemisphere. A determination of the latitude and longitude of each storm center and the association of these positions with cloud patterns show that the northern line of storms has its origin below the ammonia clouds, possibly in the water clouds (15. ASSEMBLED ABSTRACTS).

7. Man-Made Noise Environment and EMC

While we have so far been concerned with natural radio noise, several chapters hereafter deal with man-made radio noise sources and interference in our environment as well as natural electromagnetic pulse impact on man-made systems. Betsky gives a comprehensive overview of man-made noise sources and environments interference, starting with a historical review of interaction between live organisms and a variety of electromagnetic fields. With increasing radio and electric installations radiating both useful signals and spurious noises, two major problems have been coming up: EMC of radio and power equipments or installations; effects of electromagnetic interference in various frequency bands on live organisms. Specifically, spurious oscillations in various electronic devices are emphasized as a major, universal noise source due to their nonlinear nature of interaction between electric charges and the electromagnetic field, and are discussed in detail from the points of view of sources, devices, and suppression measures. Despite all these efforts, electromagnetic pollution of the environment is continuously growing, calling forth the notion of electromagnetic ecology. As one of current and future problems on electromagnetic interaction with live organisms, the use of millimeter waves of low intensity is receiving increasing attention, based upon experiments carried out for the last two decades, which indicate that millimeter waves for power intensities ranging from 0.5 mw/cm^2 to 10 mw/cm^2 can be harmless, and even medically beneficial, having a therapeutic effect on the human body. This radiation can also be used in biotechnology. For example, experiments on blue-green algae during the last three years, carried out at the Institute of Radio Engineering and Electronics of the USSR Academy of Sciences jointly with the Moscow State University, indicate a faster growth of the biomass under irradiation of millimeter waves, especially by greater secretion of biologically active substances into their environment. In summary, it can be stated that: 1. The problem of man-made radio interference is largely an ecology problem, rather than the EMC problem of radio and electronic devices; 2. The appearance of the problem of electromagnetic ecology and electromagnetobiology is being evidenced; 3. In the area of electromagnetobiology, a finding of medically useful frequency and power ranges of electromagnetic waves is an urgent task. It might be stated that one should seek such a pollution of the electromagnetic environment where the electromagnetic background will be not only harmless or neutral, but even would produce medicinal effects for human beings; 4. A positive example of this approach has been reached with electromagnetic waves in the millimeter waveband.

As mentioned in the previous paper, electromagnetic interaction with an environment often goes basically to the problem of nonlinear electromagnetics where the interaction of strong and short electromagnetic pulses with nonlinear or resonant media is presently the subject of active research. Along this line, <u>Kislov</u> and <u>Taranov</u> attempt to answer the following two questions on the basic notion of electromagnetic ecology on the mathematical and physical basis: 1. Whether strong and short pulses should be rapidly diminishing in environment, or may travel even in dissipative media to long distances?; 2. If they are propagating still with a significant amplitude, whether they could affect living systems and biologically essential processes there? Concerning Question 1, there may physically occur, under certain conditions, the so-called self-induced transparency (SIT) phenomena for intensive and short pulses in a dielectric medium even with dissipation at a frequency close to the resonant frequency of the medium. Mathematically, this is described by the Maxwell-Bloch equations with new dissipative terms which are solved exactly by means of the inverse scattering transform (IST) method. The result indicates possible propagation of electromagnetic pulses over substantial distances, preserving there possibilities of affecting organized systems (including electronics, bioelectronics, etc.), which are illustrated in response to Question 2. Thus, one can evaluate the EM pulse intensity necessary to produce biologically essential effect on macromolecules, leading to change in their conformation. For microwave pulse action on macromolecules and their conformation, for example, one needs the EM pulse intensity greater than 3 Mega Watt/cm^2 (with their relaxation time $\tau \simeq 10^{-9}$ sec). Laboratory experiments on microwave pulse action on macromolecules have been carried out up to $W \simeq 10^8$ W/cm^2 in pulse intensity with pulse duration of $\tau \simeq 10^9$ sec. Short pulses used in experiments tend to avoid direct heating effects, but initiating conformational changes directly by EM pulses.

Gao and Lu present an alternative formula for the mutual impedance between two infinite parallel wires above ground by using Struve's function. On this basis, a formula is also obtained for the mutual inductance between three phase power lines and a communication line. In this connection, it is noted that the editor gave a formula for the mutual inductance between oblique conductors above ground in 1954 in particular reference to an accurate estimate of magnetic induction for actual configurations of power and communication lines (Kikuchi and Iguchi, Bull. Electrotech. Lab. Japan, 18, 527, 1954).

8. Noise and Communication Statistics

Stumpers offers a comprehensive overview from radio interference to EMC with classification of the electromagnetic environment including activities of CISPR, CCIR, CCIF, IEC and related Committees for the past several decades. The contents are so exhaustive collecting and summarizing all the documents on EMC issued by the above-mentioned Committees and should be a most valuable source for those who are concerned.

Nitta et al. report experiments and their analyses on RF noise induced on DC power supply test lines which exhibits damped oscillation waveforms due to lossy parallel or series resonance under certain lumped circuit conditions inserted at resonance points on the test line. RF noise can be suppressed by inserting certain lumped elements, resistor and capacitor at resonance points to make impedance characteristics flat.

Arata and Yamazaki propose a method for eliminating narrow band interference waves caused by sporadic-E ionospheric propagation. The method uses a quadrature synchronous detection for television signals and a Hilbert transformation for extracted FM interference waves on the basis of theoretical analysis. An experimental system developed automatically eliminates a maximum of five interference waves, and suppresses an interference signal of sine wave by more than 50 dB and an FM interference wave by more than 40 dB in a TV signal band.

Geivandov summarizes his general approach to multimode measurements of EMC-parameters and statistical theory of multimode systems. For example, the method can be applied to multimode antennas on collateral radiation frequencies to evaluate their inner and outer parameters by using multimode measurement devices such as mode selective directional couplers and multimode power type of directional couplers. In addition, statistical recalculation of measurement results makes it possible to analyze multimode antennas in detail.

9. Space Communication and Measurement

Stumpers gives an extensive review for the future satellite communications and their EMC problems, looking back over the last two decades. Integrated Service Digital Networks are expected to be in the domain of optical fibres. Recently INTELSAT stressed that the Intermediate Data Rate system, derived from the FDMA access technique, will play a major role in providing cost-effective satellite path on a global basis. The provision of communication links to land mobile users is one of the most challenging tasks of satellite communications and is discussed in detail. There are frequency allocations for ships, aircraft, and land mobile satellite service in the range around 1.6 GHz. Frequencies above 10 GHz have advantages, possibilities for smaller antennas, and no terrestrial interference. Taking into account rain attenuation and depolarization, spatial diversity with two earth stations at 10 km distance is a possible solution, giving 10 dB gain at 30 GHz. Adaptive forward error correction, or variable rate coding schemes are another possibility. Space in the geostationary orbit is limited. A high tensile strength suspension cable might allow a suspended communications satellite to be maintained in orbit thousands of kilometers below the anchor satellite. Another approach is SPIDERSAT, a very low mass, gossimer-like communication satellite, the power for the satellite's operation and for its stabilization in geosynchronous orbit some 800 km above the earth, being provided by ground-based energy transmission. Direct Broadcast Satellites are already functioning in several countries. In most cases, they are received via cable (CATV). There is a potential demand of large European road freighters for a private wide area network. Originally USSR was using the Molnya and Tundra satellites in elliptic orbit, but later they reverted to the geostationary orbit. Recently, U.K. has also suggested a highly elliptic orbit satellite option for land mobiles.

Ito and Yamada consider an efficient use of the geostationary satellite orbit (GSO) for allotment planning and its practical use. Two methods are proposed to practically use the allotment plan. First is to simply relax the required C/I value which only degrade the C/N by 0.4 dB. Second is to introduce a new concept of "c/e (contraction/expansion) ratio" which is able to measure the impact of new entering systems. Then, the concept of the predetermined arc is fully utilized to flexibly adjust the locations of already alloted systems. These methods will be useful to improve the use of the allotment plan.

Sonoda et al. propose the AROW (Audio, Radio, and Optical Wave Ranging) system as a new means to measure the atmospheric parameters averaged along the propagation path between two observation points, which can be used to improve the measurement accuracy of EDM (Electro Optical Distance Measurement). The accuracy of order of 1×10^{-7} of the EDM is expected by correcting the atmospheric parameters with the aid of this new technique. To achieve this, however, the phase measurement accuracy of a few tens degrees for 400 MHz is required for the AROW system. A preliminary experiment indicates this possibility by adopting the DMTD (dual mixing time difference) signal processing method.

10. High Power Electromagnetics

This chapter concerns high power electromagnetics (HPE) including lightning and nuclear electromagnetic pulse (LEMP and NEMP) and high power microwaves. Kawamata et al. evaluate the current induced on a rectangular loop by the magnetic flux through it due to a heavy lightning current, paying attention to its waveform. Based on this result, a magnetic shielding practice is proposed for an application of rectangular conductive loop coils to the building. Many loop coils made of copper bar or tube are assembled as a framework, and are set besides the structural steel or the steel bars to enhance magnetic shielding. Thus, the walls of the building act as a shield to diminish electromagnetic noise travelling to both directions, from open air to the building and from the room in the building to open air. On the contrary, such assemblies may be useful for small electrical signals such as pocket bell and cordless telephone signals as a communication medium among different floors and rooms in the building.

Next three papers by Baum are all concerned with fundamentals of high power electromagnetics in highly mathematical terms with direct applications to LEMP and NEMP problems. The first paper discusses simulation of electromagnetic aspects of lightning. The problem is separated into two classes; one is simulation of distant lightning and the other is that of direct-strike lightning. For the case of distant lightning, commonly used EMP simulation techniques are applicable with some modification in the sources. For direct-strike lightning, one has to elucidate the physical processes involved to develop the simulation. Given the state of lightning understanding at present, however, one can design a simulator which is consistent with this understanding. At low frequencies, quasi-static considerations lead to the importance of both current and charge on the system. At resonant frequencies, the frequency-spectral contents of the excitation and properties of the lightning arcs are significant. Potential simulation concepts are presented and their relative merits are discussed. However, there is still the fundamental need of obtaining an adequate understanding of the physical processes involved in lightning discharge.

The second paper presents some features of waveguide/horn design with reference to a NEMP simulator. A conducting waveguide operating in the lowest order H mode can be optimized for its power handling capabilities. As a common rectangular waveguide, H_{10} mode has geometric properties which allow E-plane subdivision into some number of rectangular guides. This in turn allows convenient division or recombination of electromagnetic waves. This concept is extended to pyramidal horn which can launch the waves from one or more waveguides. The E-plane subdivision can be used for designing a metallic grating which allows passage of the wave while mechanically supporting a dielectric sheet separating vacuum from gas regions.

The third paper defines an energy norm, based on the enrgy delivered to some port in a system. It can be expressed as a weighted 2-norm. Criteria are developed for the frequency spectrum of time-domain excitation waveforms to bound the 2-norms and energy norm at the port. Canonical incident waveforms are introduced to give convenient bounds.

Baum and Gardner begin with an overview of HPE and lead into EMP (NEMP) hardening issues. The paper is divided into singularity expansion method (SEM), topology, and transmission line theory. SEM divides the system response into a series of decaying exponentials so that response from different paths of a scattering body can be obtained. Topology provides a practical means of designing a system to be hard to incident fields and currents (15. ASSEMBLED ABSTRACTS).

Gardner reviews progress in lightning electromagnetics, based on a book by Hemisphere which, in turn, is an expansion of the special issue on the electromagnetics of lightning in Journal of Electromagnetics. The subject is divided into four categories: 1. Models of Fundamental Lightning Process; 2. Propagation; 3. Measurement of Lightning Parameters; 4. Lightning Interaction with Systems. There is a concentration of the quantitative physical development of streamers and their transmission line description. The line elements are then used in nonlinear, nonuniform transmission line models which predict the electromagnetic observables from a set of initial conditions. Chapter 3 is a compendium of data presentation from a number of the active experimental groups. Of interest are fast measurements of electromagnetic parameters such as currents and fields. Chapter 4 considers application of the theory and data presented and empirical data to particular systems (15. ASSEMBLED ABSTRACTS).

URSI contribution in a broad interdisciplinary scientific program to the SCOPE-ENUWAR project (Scientific Committee on Problems of the Environment, Environmental Consequences of Nuclear War) has previously been presented and published (H. Kikuchi (ed.), Nonlinear and Environmental Electromagnetics, Elsevier, 1985, p.341). Wik summarizes the SCOPE-ENUWAR overall results. The project was initiated by ICSU (International Council of Scientific Unions) in 1982 and involved approximately 300 scientists from more than 30 countries. The scientific results have been published in two volumes by John Wiley & Sons. The first volume describes the physical and atmospheric effects of a nuclear war, and the second volume describes the ecological and agricultural effects of a nuclear war. The findings by SCOPE-ENUWAR and URSI have also supported a report to the United Nations (15. ASSEMBLED ABSTRACTS).

11. Seismoelectric Emissions

There is a growing interest in electromagnetic phenomena associated with earthquakes, and a number of articles on this topic have appeared for the last decade. Gershenzon and Gokhberg present an overview of seismoelectromagnetic phenomena, specifically on seismoelectrotelluric anomalies, electromagnetic emission, and ionosphere and magnetosphere disturbances. According to publications, there are several hundreds of cases of the electrotelluric field (ETF) anomalies appearance before earthquakes. Amplitudes of disturbances are usually of a few mV to a few tens mV, the duration is of few minutes to few weeks, and the period of disturbances before earthquake is from an hour to some weeks. As a matter of fact, electromagnetic fields at the earth's surface can only appear near inhomogeneities of the crust electrophysical properties. This feature of the electrokinetic ETF origin allows to explain "mosaic" anomalies field and weak dependence of signal magnitude. Estimations of the field magnitude that depend on the earthquake magnitude and epicenter distance show that the electrokinetic effect can explain observational data. Disturbances have been registered in a frequency range from some kHz to MHz. Earthquakes with magnitude 5 and more, have been felt for a distance of about some hundreds of km. The signal intensity has been more than background value from 2 to 10 times. There are some indications for the possibility of electromagnetic emission of the lithosphere origin. For instance, it may be generation of electromagnetic impulses during rock destruction that were registered in laboratory conditions and before rock strikes. The abnormal crust deformation exists in the zone of earthquake preparation. Rapid changes in the state of crustal stress and strain, assuming a hierarchical block structure in the crust, will necessarily lead to the local concentration of stresses and micro crack density increases in this area. In spite of increasing publications on the relation between earthquake preparation processes and ionosphere disturbances or on the lithosphere-ionosphere coupling, no positive indications have so far been obtained from the point of view of mechanoelectric conversion mechanisms. However, the possibility of the atmospheric electromagnetic field influence is not excluded as long as the relation between earthquake preparation processes and atmospheric electricity involving the ionosphere is not clear.

Ogawa et al. describe laboratory experiments on radio emissions from rock, iron, and wood by hitting those samples with either an iron hammer or the same kind of matter and their relevance to seismoelectromagnetic emissions, referring to an example of enhanced ELF/VLF noise observed from the Stanford University radiometer in association with the earthquake that occurred 200 km NW of the observatory. Results reveal the following findings: 1. The electric (E) and magnetic (M) fields of ELF/VLF radio noise are observed for any kinds of samples; 2. The E field is dominant for rock and wood, in general for dielectric materials, while the H field is dominant for iron, in general for magnetic materials; 3. The waveforms observed from rock and iron have large positive or negative initial pulses followed by damping oscillations; 4. Emission frequencies range from DC to some 20 MHz; 5. The frequencies of the peak field are found to be 5.7 kHz - 5.95 kHz for granite, 1.15 kHz and 9.55 kHz for iron, and

350 Hz - 400 Hz for wood. From these laboratory results, it is inferred that similar radiations may occur on the occasion of earthquakes with dominant E or H field when the geological structure is dielectric or magnetic, respectively.

There exists an extensive world literature concerned with radio emissions from earthquakes. Most of reports pertain to emissions at very low frequencies below 100 kHz. Warwick describes a different kind of emission that appears to have been associated with the great Chilean earthquake of 22 May 1960. On 16 May 1960, a strange emission was observed at no less than four stations distributed on the dimensions of a continent or larger. The network comprised Michigan, Colorado, New Mexico, and Hawaii, at each of which was recorded 18 MHz signals that were noisy in essentially identical patterns on this continental scale. Radio observations few days before the actual earthquake can be interpreted as follows: microfractures might be primarily involved and not necessarily full-scale breaking of rocks, and the stresses within competent rocks lead to such "Griffith" cracks before the rock actually fractures. He provisionally accepts the validity, demonstrated in the laboratory and exhibited in Nature, of the hypothesis that radio emissions can be produced, and are likely to be observed, in earthquake zones. Aside from ground observations of faults, he strongly suggests that an event like this might be observed from spacecraft orbiting Earth, because of the much greater efficiency and global coverage possible from that perspective.

Yoshino reports magnetic impulsive noise bursts at 82 kHz observed before the major volcanic eruptions at Mt. Mihara in Izu-Ohshima Island, Japan on 15 and 21 November, 1986, and also touches upon a new volcanic eruption at the city of Itoh very close to it that occurred only one month before this presentation (15. ASSEMBLED ABSTRACTS).

12. Ball Lightning

Ball lightning is a fireball that has been observed in the atmosphere or inside a plane, a submarine, or a house. Because of a rare phenomenon, attempts to place instruments over extended periods of observation have failed to encounter the desired object. Thus, the information available has come almost from eyewitnesses except from laboratory experiments. The typical lightning ball is a luminous sphere as bright as a strong household fluorescent lamp. The sphere may range in diameter from several to several tens cm. Most seem to shine steadily, but some pulsate. The ball can be, white, yellow, orange, red, blue, although green and violet are rare. It can last from a second or less up to several minutes. The ball usually materializes immediately after an ordinary lightning stroke. Some fade out calmly; others disappear abruptly, occasionally with an explosion or breaking up into pieces. Normally the ball moves horizontally, sometimes vertically through the air with a speed of one tenth to several m/s. The ball would contain hundreds kJ to a couple of MJ of energy. Its study requires interdisciplinary efforts from many aspects: atmospheric aspect/lightning; plasma/fusion and superconductor; aerosol or dusty plasma; chemical reaction; statistical survey from measurements and from eyewitnesses; laboratory simulation or modelling. The problem is basically twofold: (1) how the lightning ball is formed; (2) how it is stably sustained so long. A number of models are sorted into 4 classes from energy source aspects: (1) DC discharge of electrostatic energy (2) AC discharge of electromagnetic energy; (3) chemical reactions; (4) nuclear reactions. Some new concepts to be introduced in the next chapter (14.1) may also be relevant to a ball lightning model, i.e. electric reconnection, critical velocity, ponderomotive force, and associated plasmoid/fireball production.

Dijkhuis offers a new approach to ball lightning plasmas, based upon a state equation and its phase diagram for their fractal growth, after a comprehensive review of the state of the art. A state equation for electron pressure is derived by interpolating classical and quantum limits for pressure of free electrons with attractive exchange interaction. Van der Waals-like isotherms with a critical temperature 1844°K are shown in phase diagrams of this state equation. The locus of electron fractions coexisting at a different density is obtained with Maxwell's lever rule. A subregion of the phase diagram is obtained where universal phase separation dynamics creates structures with fractal dimension. A many-body theory for conduction electrons with exchange interactions at intermediate degeneracy spells out a phase transition in discharge plasmas and evaporating metals with critical parameters fixed by fundamental constants. The exchange instability grows fractal structures in cold, dense plasma regimes relevant to lightning discharges, high-voltage breakdown facilities, current interruption and laser ablation experiments. Fractal condensate in ball lightning has exchange interaction parameters similar to conduction electrons in ceramic compounds with high-T_c superconductivity. Its critical state parameters match the palladium-deuterium system known for cold nuclear fusion.

Grigoryev et al. present a correlative relationship between the radiation intensity of ball lightning in the optical range and its typical linear dimension, based on a statistical computer analysis of 3115 descriptions of ball lightnings. It turns out that the radiation intensity is proportional to the 2.06 power of ball lightning diameter, pointing to a surface character of radiation rather than volume radiation.

Ohtsuki and Ofuruton describe laboratory experiments on ball lightning or fire balls made with electric discharge in fuel gas and/or aerosol. Fire balls with lifetimes as long as 1 - 2 sec. were observed in 2.7% ethane and cotton fibers (100 cm^3). In another experiment, a microwave oscillator with a frequency of 2.45 GHz and power ranging from 1 to 5 kW was operated in the open air and microwaves were guided through a waveguide to a cylindrical cavity made of copper. The one side of the cavity was closed by an immovable aluminum foil and the other side was terminated by a movable meshed aluminum plate. Two types of plasma fires were observed. One was flames located at the end of the waveguide, their colour changing from white red to blue and orange. The size was from 1 to 3 cm, and the lifetime was few minutes. The other type was a very bright homogeneous flash over a whole of semicavity. The colour was white, and the lifetime was as short as 0.2 sec. Just after this bright flash, there appeared a separated plasma ball, moving up and down, horizontally, and rounding along the cavity wall. The colour was orange and its lifetime was 2 - 3 sec.

Grigoryev summarizes statistically processed results by computers from 3200 descriptions of ball lightning (BL) events collected from eyewitnesses on its appearance, disappearance, movement, heat radiation etc. The relation of BL observation frequency with the intensity of thunder activity turned out to be quite rigid. One of distinct results obtained in these processes is that radiation proceeds from the BL surface layer the thickness of which is less than the radius as mentioned in his another paper (15. ASSEMBLED ABSTRACTS).

13. Meteorologico-Electric Phenomena and EHD

There are a number of meteorologico-electric phenomena involving or accompanying electric discharge, ionization or electrification such as thunderstorms, cold and polar fronts, jet streams, typhoons, hurricanes, cyclones and tornadoes in the terrestrial environment, and planetary lightnings, rings, cometary tails, dark interstellar clouds in the extraterrestrial environment. For these phenomena, conventional hydrodynamics (HD) or magnetohydrodynamics (HMD) breaks down and even conventioal plasma physics is also not relevant, because their environments are considered neither fully ionized gases nor collisionless plasmas as often assumed for conventional plasmas, but may be termed unconventional plasmas as a whole. They may be partially ionized and collisional gases, dusty or dirty plasmas containing charged dust grains or aerosols, which can be an environment, for example, for natural, triggered and ball lightning, sandstorms or even whirlwinds. Consequently, conventional HD or MHD should be extended to a new area of electrohydrodynamics (EHD) and/or electromagnetohydrodynamics (EMHD). To effect this, the editor has introduced three basic physical concepts: 1. electric reconnection; 2. ponderomotive force; 3. critical ionization velocity, useful for understanding the meteorologico-electric phenomena in a unified manner at least qualitatively, and thus for establishing a new EHD or EMHD. The next problem is how to treat random noise fluctuations and chaos accompanied by meteorologico-electirc phenomena on the basis of a statistical approach to nonlinear and turbulent processes, incorporating two more concepts of self-organization and dynamic chaos in a new EHD or EMHD regime. Kikuchi describes the former three and latter two concepts briefly, and then attempts to apply them to a couple of real phenomena, traditional and novel, such as natural, triggered, ball lightning, sandstorms and whirlwinds. Finally, a new set of EHD and/or EMHD equations with additional terms of ponderomotive forces are presented, incorporating Maxwell's equations or Poisson's equation. In this connection, it should be noted that a Working Group, "Extraterrestrial and Terrestrial Meteorologico-Electric Environment with Noise and Chaos", has been established in URSI Commission E to promote a joint work in such an interdisciplinary, new field (Chairman, the editor).

Meaden describes a new meteorologico-electric phenomenon in the atmosphere that is an electrified vortex system laying down spiral-centered circular patterns in crops and upon the ground. Much information is now available about the processes that create the vortices which are caused by the breakdown of a columnar eddy vortex downwind of hills. Besides eyewitness reports of the vortices in action, there is proof of simultaneous acoustic, luminous and electromagnetic effects. Not only does the assembled evidence point to an electrical source of natural electromagnetic radiation in the atmosphere, with its necessary consequencies for radio-communications interference and radar ghost-images, but the work explains reports of previously unidentified luminous phenomena seen close to the ground or high in the air. Indeed, the new charged vortex phenomenon is possibly related to ball lightning if one can judge by the inferred character of its internal discharge processes, although it is distinguishable from ball lightning by its much larger size and considerably greater lifetimes. A wide class of vortex groups are shown to be possible. These are illustrated with reference to the circular-symmetric ground patterns and have been surveyed and photographed over a period of decade. The effects range from a relatively 'careful-laying' of sometimes complex patterns of a neat and attractive type, to violent downblasts that scour the ground and empty ponds. However, the editor might suggest that the latter may be different from the former type in characters, closer to ball lightning or natural lightning. In fact, a radial circular-blast event occurred in Japan in 1986, accompanied by light and sound emissions which caused simultaneous television interference.

Kikuchi finds a new type of large-scale vortex formation theoretically in helical turbulence in terms of hydrodynamic, electric, magnetic, and space charge fields in an external electric (and magnetic) field. It is called "Electromagnetohydrodynamic Vortices" and is generated as a result of self-organizational processes in nonequilibrium media by the transfer of energy from small-scale to large-scale sizes. The theory is based upon a new set of electromagnetohydrodynamic (EMHD) equations and Maxwell's equations, and a new equation of EMHD vortices is derived. The corn circles, which regularly crop up in southern England, are tentatively identified as a kind of electromagnetohydrodynamic (EMHD) vortices, and some of circle-effect phenomena are explained on the basis of a new theory of EMHD.

2 *Spacecraft EM Environment*

2.1 The Shuttle Electrodynamic Tether Mission

L.R.O. Storey

National Space Science Data Center (Code 630.2), NASA Goddard Space Flight Center, Greenbelt, MD 20771, USA

The Tethered Satellite System (TSS) is a Shuttle-based facility being developed by NASA and the Italian Space Agency under a cooperative agreement. The joint development work began in 1984, and the first TSS mission is currently manifested in the fall of 1991. Its main objective is to verify the capability of the facility to deploy, control, and retrieve a tethered satellite. Additionally, the mission will demonstrate the electrodynamic science capability of the system and perform various scientific experiments. The instrumentation, the scientific objectives, and the planned investigations are summarized; references are given to more detailed accounts.

1. INTRODUCTION

The Tethered Satellite System (TSS) is a cooperative program between NASA and the Italian Space Agency (ASI) to conduct unique electrodynamic and atmospheric missions in low Earth orbit, using the Space Shuttle as the base from which to deploy the system. NASA is responsible for the TSS "fishing reel" deployer and for systems integration. ASI is building the spherical satellite, 1.6 m in diameter, that will be attached to the tether, while both NASA and ASI are building scientific instruments to go on the tethered satellite and the Shuttle orbiter. All of the hardware for the TSS is unique.

In the TSS, a satellite is deployed from the Shuttle on the end of a long cable, which is the tether (Fig. 1). The two linked spacecraft are kept apart, and the tether kept taut, by the gradient of the Earth's gravitational field. The system is stable when either of the two spacecraft is directly above the other, with the cable vertical; thus the satellite can be either above or below the Shuttle. On the first mission (TSS-1), concerned with electrodynamics, the satellite will be deployed upwards on a 20 km tether in the form of a thin multi-strand copper cable with teflon and kevlar insulation. The second mission (TSS-2), however, will be concerned with atmospheric science, and for this purpose the satellite will be deployed downwards on a 100 km non-conducting tether. The present paper describes only the first mission, which, at the time of writing, is scheduled to take place in the fall of 1991; the nominal orbit is circular, at 300 km altitude, with 28° inclination.

The electrodynamic aspects of the mission arise because, as the TSS moves at its orbital velocity of 7–8 km/s through the Earth's magnetic field, the resulting dynamo action sets up a substantial potential difference between the satellite and the Shuttle, the satellite becoming positive and the Shuttle negative [1–3]. For a 20 km tether, this potential difference is of the order of 3000 V; the exact value varies around the orbit. With the conducting ionospheric plasma present to close the circuit, an electric current is driven along the tether. This current is limited mainly by the tether resistance (about 1600 ohms), and, on both spacecraft, by the small areas of exposed conducting surface available for collecting charged particles from the plasma; it is not expected to go much above 0.5 A on the TSS-1 mission.

Figure 1. Artist's impression of the first Tethered Satellite System.

2. OBJECTIVES

Since the TSS-1 mission is the first of its kind, its prime objectives are operational and technical, namely to demonstrate the feasibility of deploying, stabilizing, and retrieving a satellite on the end of a long tether, and to overcome any difficulties that may arise. Out of the seven-day Shuttle mission, about 36 h are being reserved for TSS operations, during about 10 h of which the tether will be fully extended. After some initial tests lasting 2-3 h, the rest of this 36 h period will be used mainly for scientific experiments, the objectives of which are: (1) to understand the electrodynamics of operating a tethered satellite in the quasi-steady state; (2) to understand time-varying electrodynamic operations; (3) to investigate fundamental processes in space plasmas; and (4) to investigate the mechanics of the tether.

2. INSTRUMENTATION

The scientific instruments for the TSS-1 mission have been selected to meet the technical as well as the scientific objectives. Some of them will be mounted on the Shuttle orbiter and some on the tethered satellite.

The instruments on the orbiter include: equipment for measuring the tether current and voltage; electrostatic probes for measuring the densities and energies of the ionospheric electrons and ions;

electron generators to close the circuit between the orbiter and the ionosphere, and for modulating the tether current; electrostatic analyzers for measuring the potential of the orbiter with respect to the ambient plasma; a hollow-cathode plasma bridge to help with discharging the orbiter; and an optical imaging device.

The instruments on the satellite include: triaxial fluxgate and search-coil magnetometers; a triaxial AC electric field meter; an electrostatic probe; a differential ion flux probe; and a soft particle energy spectrometer to measure the fluxes of electrons and ions in the energy range from 10 eV to 10 keV. Most of these instruments will be held away from the body by means of booms; there will be two 2.5 m extendable booms and one 1 m fixed boom (see fig.1). The satellite will spin at up to 0.7 revolutions per minute, varying the directions of the booms with respect to the satellite velocity vector, and thus enabling the instruments mounted on them to explore the perturbed environment of the satellite. Meanwhile, within the satellite itself, an ammeter will measure the tether current and a triaxial accelerometer will provide data on tether dynamics; these instruments are located in the Payload Module in the upper half of the satellite, while the supporting equipment (power supplies, telemetry, etc.) is in the Service Module in the lower half. Between the two, in the equatorial plane of the satellite, is the Auxiliary Propulsion Module containing cold-gas thrusters that serve to orient and stabilize the satellite, to make it spin, and to help in the initial phase of its separation from the orbiter. The scientific payload weighs 66 kg, while the total weight of the satellite is 518 kg.

During the operations in space, receivers on the ground will be used to search for the fields and waves generated by the TSS.

3. INVESTIGATIONS

3.1. Steady-state Electrodynamics

A major scientific investigation concerns the current/voltage relationship for the TSS acting as a generator of steady current, and this is governed mainly by the physics of charge collection on the spherical satellite. The current flowing through the tether can be controlled by varying the emission from the electron generators on the Shuttle. At the larger currents, the satellite is expected to be at a large positive potential with respect to its ambient ionosphere, and it will be surrounded by a region of strongly perturbed plasma, which is known as the *plasma sheath*. Under these conditions, if neutral gas is released from the thrusters, a discharge may occur and reduce the satellite potential for a given current, or, more significantly, increase the current collected for a given potential; such a discharge is likely to produce optical emissions observable by the imager on the Shurttle. Instabilities in the sheath also may assist current collection, by enabling ionospheric electrons to diffuse towards the satellite across the magnetic field lines.

Another important issue is how the current closes through the ionosphere between the satellite and the orbiter. At the altitude of the Shuttle orbit, the Earth's magnetic field has a strong influence on the electrons of the ionospheric plasma and on those emitted from the electron generators, preventing them from moving in directions other than up or down the field lines. It would seem, therefore, that the current must flow in a large loop: the current-carrying electrons must first travel from the Shuttle down the field lines to the ionosphere at about 100 km altitude, where collisions with the neutral gas allow them to move across the field, then they move up another of set of field lines to the satellite. This is the generally accepted picture of how the current closes, but recently it has been questioned in the light of results from some laboratory experiments [4]. The data from TSS-1 should help to decide the issue.

A related subject of investigation is the steady-state wave emission from the TSS. The pattern of current flow is steady in a coordinate frame moving with the Shuttle, but in a frame fixed relative to the ionosphere the pattern is time-varying, resulting in the excitation of waves that, in fact, carry the current. The customary view is that these are shear Alfvén waves [5,6], but, on the alternative view just mentioned [4], they are waves in the whistler mode.

3.2. Unsteady-state Electrodynamics

If the current flowing in the tether is modulated by varying the current emitted by the electron generators, the entire TSS will act as an antenna 20 km long and will radiate waves. This process is seen as a promising method for launching waves into the ionosphere in the ULF, ELF, and VLF bands. At such low frequencies, transmitting antennas on the ground are inefficient because their dimensions are small compared with the free-space wavelength, and much of the radiated energy is lost during propagation up through the lower ionosphere. A long antenna in the ionosphere should be more efficient because the wavelengths are much smaller there, and because the antenna is in direct electric contact with the plasma; measurement of the electrodynamic properties of the TSS antenna (*i.e.* tether impedance as a function of frequency and of the plasma characteristics) and ground-based reception of the radiated waves are among the investigations planned.

If this expectation is fulfilled, then the TSS could be used as a spaceborne transmitter for a variety of experiments in magnetospheric physics, including stimulation of natural VLF radio wave emissions and precipitation of energetic electrons from the radiation belts.

3.3. Plasma Physics

Even when the satellite is close to space potential, so that there is little or no sheath around it, it perturbs the plasma by sweeping up all the ions that hit its forward surface, thus creating a partial vacuum that stretches out as a wake behind it. The physics of this phenomenon has been extensively studied in laboratory plasmas and already to some degree in space, but the TSS-1 mission offers the best opportunity so far to study it in space with an appropriately instrumented satellite.

A phenomenon of broader interest is the the so-called critical ionization velocity (CIV) effect, whereby a neutral gas moving through a plasma in the presence of a magnetic field, with a velocity directed more or less perpendicular to the field and exceeding a certain threshold (the CIV), becomes ionized much faster than would be expected from the simple action of collisions [7]. This phenomenon also has been observed in laboratory plasmas but only episodically in space. It is believed to involve a collective plasma process that is still only partially understood. The TSS investigators will study it by releasing gas from the satellite's thrusters and detecting the resulting ionization with the instruments on board. The CIV effect may be important in the physics of comets [8].

3.4. Tether Mechanics

The planned investigations of tether dynamics concern such topics as the dynamic drag on the tether due to its motion through the tenuous atmosphere, and the additional electrodynamic drag that arises whenever an electric current is allowed to flow along it. The natural modes of oscillation of the tether and of the tethered satellite, and how these modes are excited, are among the other topics of interest.

4. CONCLUSION

The first TSS mission will study the generation of an electric current in the conductive tether as it cuts the Earth's magnetic field lines. Instruments will measure the effects on the space plasma around the satellite and the orbiter while the current is varied. The results from this mission, besides leading to better understanding of how electric currents flow in the upper atmosphere, could lead to an alternative method of generating electricity for future space structures, such as the Space Station "Freedom." Later missions will investigate the composition and density of the Earth's upper atmosphere in regions where data are currently scarce. Additional applications of tether technology, in such areas as momentum exchange, artificial gravity, waste disposal, crew rescue, geology, and long-wave antennas are also being studied.

5. ACKNOWLEDGEMENTS

At the time of the URSI Symposium, the author was employed by the Jet Propulsion Laboratory of the California Institute of Technology, in Pasadena, as a Visiting Senior Scientist, and was assigned to NASA Headquarters in Washington, D.C., in the capacity, among others, of Program Scientist for the TSS-1 mission; his present successor in this role is D.S. Evans. The Program Manager, T.D. Stuart, and Project Scientist, N.D. Stone, kindly gave him advance copies of two unpublished papers on the TSS-1 mission, and supplied fig. 1 which was drawn for NASA by the Martin Marietta Astronautics Group.

5. REFERENCES

1. Grossi, M.D., and Columbo, G. (1978). Interactions of a tethered satellite system with the ionosphere. In: Wu, S.T. (ed), Proceedings of the University of Alabama/NASA Workshop on the Uses of a Tethered Satellite System, University of Alabama, Huntsville.

2. Williamson, P.R., Banks, P.M., and Oyama, K. (1978). The electrodynamic tether. In: Wu, S.T. (ed), Proceedings of the University of Alabama/NASA Workshop on the Uses of a Tethered Satellite System, University of Alabama, Huntsville.

3. Banks, P.M., Williamson, P.R., and Oyama, K.-I. (1981). Electrical behavior of a Shuttle electrodynamic tether system (SETS), Planet. Space Sci. 29: 139–147.

4. Stenzel, R.L., and Urrutia, J.M., (1990). Currents between tethered electrodes in a magnetized laboratory plasma, J. Geophys. Res. 95: 6209–6226.

5. Barnett, A., and Olbert, S. (1986). Radiation of plasma waves by a conducting body moving through a magnetized plasma, J. Geophys. Res. 91: 10,117–10,135.

6. Estes, R.D. (1988). Alfvén waves from an electrodynamic tethered satellite system, J. Geophys. Res. 93: 945–956.

7. Newell, P.T. (1985). Review of the critical ionization velocity effect in space, Rev. Geophys. 23: 93–104.

8. Formisano,V., Galeev, A.A., and Sagdeev, R.Z. (1982). The role of the critical ionization velocity phenomena in the production of inner coma cometary plasma, Planet. Space Sci. 30: 491–497.

2.2 High Frequency Wave Generation by Electron Beam Injection in Space

S. SASAKI

The Institute of Space and Astronautical Science, Kanagawa, 229 Japan

ABSTRACT

Wave generation by an electron beam in space has been studied in a tethered mother-daughter rocket experiment(CHARGE-2). An electron beam up to 48mA at 1keV was injected from the mother payload and waves generated by the beam were observed by the tethered daughter payload at distances of up to 426 meters. This paper presents four kinds of wave emission in a high frequency range detected in the electron beam experiment and discusses the mechanism of the wave generation.

1. INTRODUCTION

During the tethered rocket experiment(CHARGE-2), electron beams were injected while the tethered payloads were separating. The purposes of the experiment were to study the electrodynamic effects of a space tether and to investigate the beam plasma interaction and spacecraft charging utilizing the tethered payload system. The experiment was conducted as a US-Japan joint space program. Utah State University provided the core instruments for the active experiments and photometers. The Institute of Space and Astronautical Science prepared plasma probes, film cameras, wave receivers, and a tether deployment system. Stanford University provided charge probes, tether biassing and i-v sensors, and University of Michigan a particle energy analyzer. For a description of the instruments and general results, see Raitt et al.[1]. For a description of the results on the electrodynamic effects without beam injection, see Sasaki et al.[2].

2. EXPERIMENTAL

Figure 1 shows the construction of the payload instruments and the pitch angles of the beam injection. The electron gun onboard the mother payload injected an electron beam in perpendicular to the tethered mother/daughter system which lay in a magnetic east-west direction. The hf(0.2-10MHz) wave receiver equipped with a 2.8m-dipole antenna(tip-tip) was onboard the daughter payload. Hf wave signals were analyzed as a frequency spectrum every 250msec with a band width of 200kHz. The measurable range of the receiver

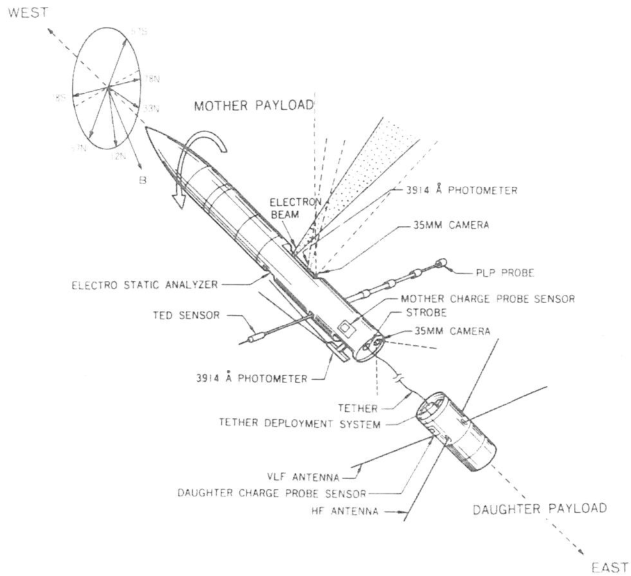

Fig.1 Configuration of CHARGE-2 payload instruments. Pitch angles of the beam injection are also illustrated.

was from 150μV to 5mV in low gain mode and from 5μV to 150μV in high gain mode. The analyzed signal was digitized into 9 bit-words and transmitted to ground. The payload was launched at 00:16:00:420 on 14th December 1985 from White Sands Missile Range, New Mexico, using Black-Brant IV (NASA 36.009UE). It reached an apogee of 262km at 252sec. At 68.0sec, the rocket was despun and then oriented in a magnetic east-west direction. At 115.8sec(161km), the preprogrammed sequence of the active experiment started. 9 active sequences were executed during the flight. The daughter payload was separated into the east direction at 141.9sec(193km) with an initial velocity of 1.05m/sec. Both payloads were connected with an insulated conductive wire until the separation reached 426m at 442sec(118km downleg). A rate control system onboard the daughter payload was intermittently operated to keep the separation velocity. During the separation, the mother payload was rolled by an attitude control system so that the electron beam was injected at different pitch angles with respect to the geomagnetic field in 45 degree increment. The electron beam at 1kV was injected in DC mode up to 31mA in the first half of each programmed sequence and in pulsed mode with current up to 48mA in the latter half.

3. RESULTS AND DISCUSSION

Four kinds of wave emissions were observed, discrete emissions near the electron cycltron harmonics and near the upper hybrid frequency, a broadband emission extending beyond the electron cycltron frequency, and a broadband emission in the whistler wave region (several hundreds of kHz). The typical wave spectrum for each type of emission is shown in Fig.2. The background wave spectrum (just before or after the beam emission) is also shown

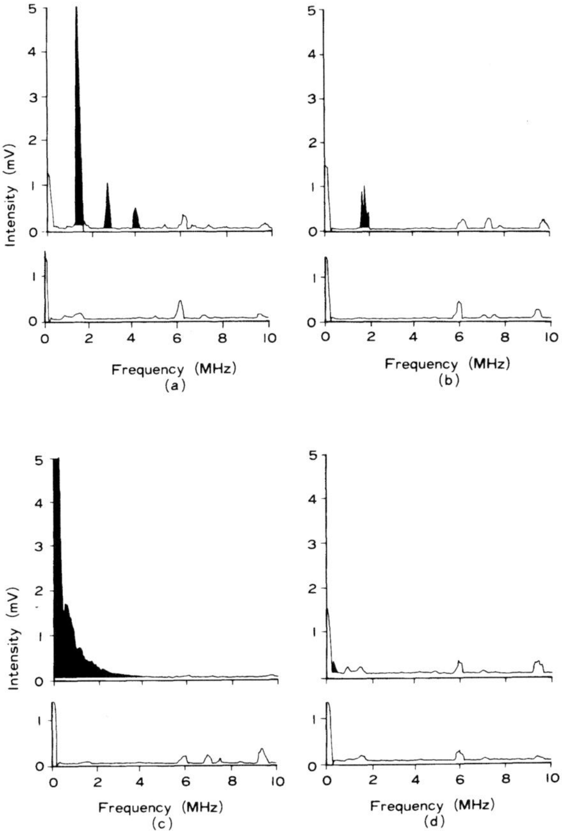

Fig.2 Typical examples of wave spectrum during beam injection (upper panel), together with background spectrum(lower panel). They are from the low gain channel.

below each panel for comparison. Shaded spectra are resulted from the beam injection. The papameters varied in the experiment were the beam current (1-31mA), pitch angle (12, 33, 57, 78 degrees), altitude (120-262km; background plasma density $5x10^2$-$1x10^5$/cc), and the distance between the two payloads (0-426m). The beam current and background plasma density were the major factors which dominate the type of wave emission. The dependence of the pitch angle was not evident if any. Figure 3 shows the parameter region (beam current in x-axis and background plasma density in y-axis) for each type of wave emission. The discrete emission near the upper hybrid frequency was detected only when the beam current was relatively low (less than 5mA). The emissions near the electron cyclotron harmonics were not detected after the

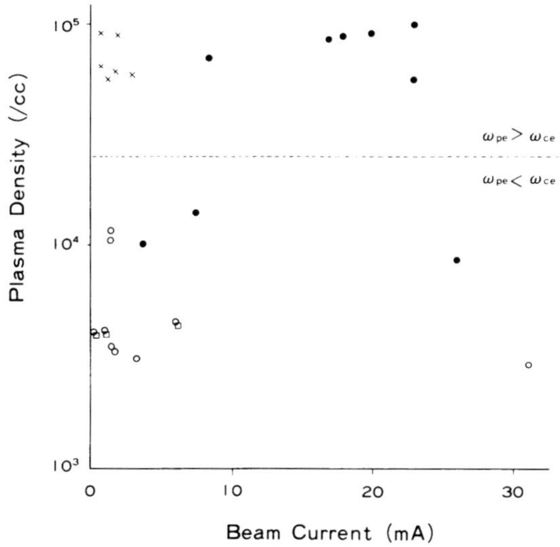

Fig.3 Parameter region for each type of wave spectrum;
X at upper hybrid frequency
○ in the whistler region
□ at cyclotron harmonics
● broadband emission extending beyond the cyclotron frequency

mother/daughter separation, which indicates that this type of emission did not propagate away. Up to 4th harmonics were detected in high gain channel. The discrete emission near the upper hybrid frequency was excited in the region of high plasma density($\omega_{pe} > \omega_{ce}$) and was detected as far as 270m. The broadband emission extending beyond the electron cyclotron frequency was also excited in the region of comparatively high plasma density and was detected as far as 340m. The broadband emission in the whistler wave region was excited in the region of low plasma density($\omega_{pe} < \omega_{ce}$) and was detected even at the final stage (distance 426 m). Since we did not measure the magnetic component of the wave emissions, it is very difficult to identify the wave mode decisively. However, the configuration in which the wave receiver was separated in the direction perpendicular to the magnetic field and to the beam propagation suggests the observed waves must have been more electrostatic rather than electromagnetic. In the electron beam experiments on Spacelab-2 [3] and by NVB-06 rocket [4], they detected a whistler-mode emission with a magnetic component by free-flying wave receivers, but it was measured near the magnetic conjunction (nearly parallel propagation). Both the wave emissions near the upper hybrid frequency and in the whistler region are likely explained by the Landau-type interaction between electrostatic waves and the electron beam. According to the linear theory by Christiansen et al.[5], the largest growth rate of the electrostatic waves occurs on the upper hybrid branch when $\omega_{pe} > \omega_{ce}$ and on the plasma branch when $\omega_{pe} < \omega_{ce}$. These types of emission were also observed in the ECHO-1 electron beam experiment [6]. Although

ECHO-1 was carried out under the condition of $\omega_{pe} < \omega_{ce}$, the plasma radiation near the upper hybrid frequency increased as the density of the background plasma increased while the radiation less than the electron cyclotron frequency ("whistler mode" in their paper) decreased as the plasma density. The wave emission near the cyclotron harmonics is possibly related to the cyclotron coupling with a large k in the beam-hot plasma system, which does not propagate long distance. The broad band emission extending beyond cyclotron frequency was detected when the high current beam was injected in a high dense plasma. This type of emission was observed in NVB-06 [4] and CHARGE-1 [7] when the

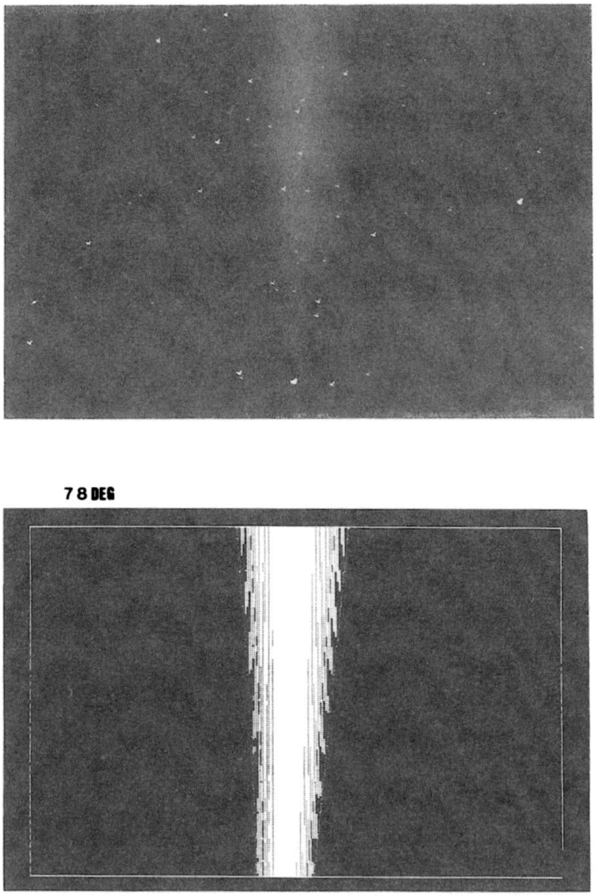

Fig.4 Beam appearance observed by the film camera(upper panel). Lower panel is a result of model calculation based on a single particle motion of electrons in a magnetic field without any scattering effect.

beam current was sufficiently large. Very similar emission with the featureless spectrum has been reported in the laboratory BPD (Beam Plasma Discharge) experiments [8,9]. It is observed when the beam current exceeds a certain threshold level. The generation of the broadband emission suggests that the beam-plasma interaction entered into a non-linear region in which the linear processes are saturated and an equilibrium turbulence is achieved. In the laboratory experiments, the broadband emission is usually accompanied with a scattering of the electron beam and with an abrupt energy transfer to the waves from the electron beam [9], resulting in an rf discharge (BPD). However, in the CHARGE-2 experiment, no evidence for the beam scattering nor discharge was observed by the film camera which monitored the beam trajectory from the rocket skin up to 1.5 m. Figure 4 shows that the beam appearance (upper panel) is almost the same as a model calculation (lower panel) in which the beam electrons are presumed to be propagating without any scattering (effects of the geomagnetic field and defocusing of the lens system are considered in the calculation). This indicates that the beam-plasma interaction with the featureless spectrum in space did not enter a stronger stage of the beam plasma discharge, different from the laboratory BPD experiments.

4. CONCLUSION

The wave generation by an electron beam in space was studied by a tethered diagnostic package which was separated in perpendicular to the magnetic field. Four types of wave emission were detected depending on the background plasma density and beam current. The emissions near the upper hybrid frequency, in the whistler wave region, and near the cyclotron harmonics are believed to be generated by the electrostatic instabilities, and the broadband emission extending beyond the electron cyclotron frequency to be attributed to a strong coupling between the beam and waves in a non-linear region.

REFERENCES

[1] Raitt,W.J., J.V.Eccles, N.B.Myers, D.C.Thompson, P.M.Banks, P.R.Williamson, R.I.Bush, J.Hawkins, S.Sasaki, K.I.Oyama, N.Kawashima, and W.F.Sharp, Active vehicle charging measurements in sounding rocket and Space Shuttle Orbiter environments at low earth orbit(LEO) altitude, AGARD, 9, 1-16, 1986.

[2] Sasaki,S., K.I.Oyama, N.Kawashima, T.Obayashi, K.Hirao, W.J.Raitt, N.B.Myers, P.R.Williamson, P.M.Banks, W.F.Sharp, Tethered rocket experiment(Charge 2):Initial results on electrodynamics, Radio Science, 23, 975-988, 1988.

[3] Gurnett,D.A., W.S.Kurth, J.T.Steinberg, P.M.Banks, R.I.Bush, and W.J.Raitt, Whistler-mode radiation from the Spacelab 2

electron beam, Geophys.Res.Lett., 13, 225-228, 1986.

[4] Kellogg,P.J., S.J.Monson, W.Bernstein, and B.A.Whalen, Observations of waves generated by electron beams in the ionosphere, J.Geophys.Res., 91, 12065-12077, 1986.

[5] Christiansen,P.J., V.K.Jain, and J.W.Bond, Laboratory beam-plasma interactions-linear and nonlinear, in Artificial Particle Beams in Space Plasma Studies edited by B.Grandal, Plenum Press., 439-469, 1982.

[6] Cartwright,D.G. and P.J.Kellogg, Observations of radiation from an electron beam artificially injected into the ionosphere, J.Geophys.Res., 79, 1439-1457, 1974.

[7] Sasaki,S., K.I.Oyama, N.Kawashima, Y.Watanabe, T.Obayashi, W.J.Raitt, A.B.White, P.M.Banks, P.R.Williamson, W.F.Sharp, T.Yokota, and K.Hirao, Results from a series of tethered rocket experiments, JSR 24, 444-453, 1987.

[8] Boswell, R.W. and P.J.Kellogg, Characteristics of two types of beam plasma discharge in a laboratory experiment, Geophys.Res.Lett., 10, 565-568, 1983.

[9] Liobet,X., W.Bernstein, and A.Konradi, The spatial evolution of energetic electrons and plasma waves during the steady state beam plasma discharge, J.Geophys.Res., 90, 5187-5196, 1985.

2.3 Space Tether Experiment

S. SASAKI and K. OYAMA
The Institute of Space and Astronautical Science, Kanagawa, 229 Japan

ABSTRACT

Space tether experiments studied in the Institute of Space and Astronautical Science(ISAS) are described. One is a 10km-tether experiment on a reusable space platform, Space Flyer Unit(SFU), to establish the technologies of space tether. It is also aimed to study plasma environment surrounding a spacecraft by a tethered plasma diagnostics. Another is a 100km-tether system deployed from a large satellite to be launched by our new rocket(M-V), to measure the geophysical parameters of the lower ionosphere by multiple-tethered subsatellites.

1. INTRODUCTION

Tethers in space promise to provide various kinds of scientific and technology applications as a unique space structure for the Space Station era now just ahead. A tethered subsatellite system to study the lower ionosphere[1], a momentum transfer system for space transportation[2], a tether elevator[3], and a tether-assisted micro-gravity laboratory[4] are feasible examples of space tether utilizing its structural characteristic. With a conductive wire, the system will be applied to a space power generator[5], an attitude/orbit control system[6], and a vlf wave radiator in space[7].

In spite of promising futures in the application, the technologies to build a tether system in space have not been established yet. The theoretical studies have been done quite actively since NASA entered into the investigation in 1970's. However, the demonstrations in space have not been done yet and the theories to control the space tether have not been verified. A space tether of 20km scale is going to be demonstrated in the US-Italy Tethered Satellite System-1 by the Space Shuttle in the fall of 1991[8], but the experiment time will be limited only a few days, which is not sufficient to study the dynamics of space tether systematically.

In order to establish tether technologies for the future application, we propose an engineering study of space tether on the Space Flyer Unit(SFU) which is an experimental platform dedicated to science and technology research. A 10km tether with a small subsatellite will be deployed and retracted repeatedly. The subsatellite will carry diagnostic instruments to study plasma environment surrounding the SFU. Once the tether technologies are well established by the SFU tether experiment, we will apply them to a longer space tether system for geophysical research. A 100km tether will be extended downward from a large satellite orbiting at 200km height. 4 subsatellites connected to the tether wire will be used to measure the geophysical parameters in the lower ionosphere which are hardly measured by a free-flying satellite due to a strong effect of air drag.

2. SFU TETHER EXPERIMENT(STEX)

A free-flying space platform, Space Flyer Unit(SFU), is now under development jointly by ISAS, the Ministry of International Trading and Industry(MITI), and the National Space Development Agency(NASDA). SFU is an unmanned, multi-purpose reusable platform for science and technology experiments, space observation, and flight tests of space technologies and industrial technologies. The weight is about 4000kg including the payloads of 1000kg. The scale is 4.55 m octagonal in diameter by 1.4m height. The SFU Mission-1 will be launched by H-2 rocket early in 1994 and will be retrieved by the Space Shuttle after 6 months mission operation. The mission orbit will be circular at the height of 500 km.

The SFU will provide a good opportunity in the research of space tether. The experiment time is long enough to study the dynamics of the tether system in various conditions. Furthermore, there is no operational constraints peculiar to the manned spacecrafts, such as the Space Shuttle and the Space Station. In the tether experiment on the SFU(STEX;Space Tether EXperiment) shown in Fig.1, the deployment, station-keeping, and retrieval of a

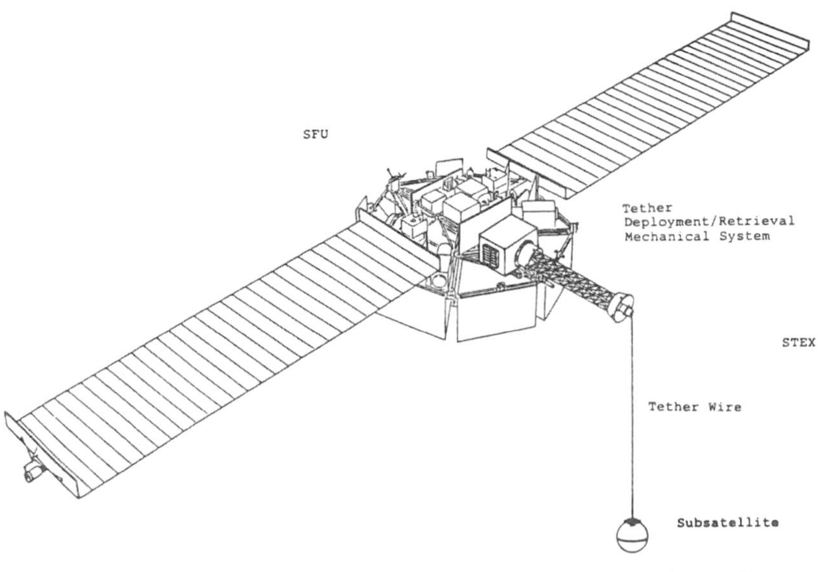

Fig.1 Space Tether Experiment on SFU(STEX)

tethered subsatellite are tested repeatedly and the theories to control the tether system are examined. The STEX system consists of a tether deployment/retrieval subsystem, an experiment controller, and a tethered subsatellite. The conceptual design of the deployment/retrieval mechanism is shown in Fig.2. The tether wire will be made of Kevlar of 1mm diameter. The total weight of the system is estimated as 170kg including the subsatellite. The subsatellite is equipped with the sensors to detect the dynamics of the tether system, as well as plasma diagnostics. The dynamics of the tether system is also observed by TV cameras from the SFU. In the STEX, two types of tether experiment are to be carried out with a short tether and a long tether. In the former experiment, the tethered subsatellite is deployed 1 km in various directions and retrieved repeatedly. This experiment will enable us to select an optimum control logic of tether sysem out of several theoretical predictions. In the latter experiment, the

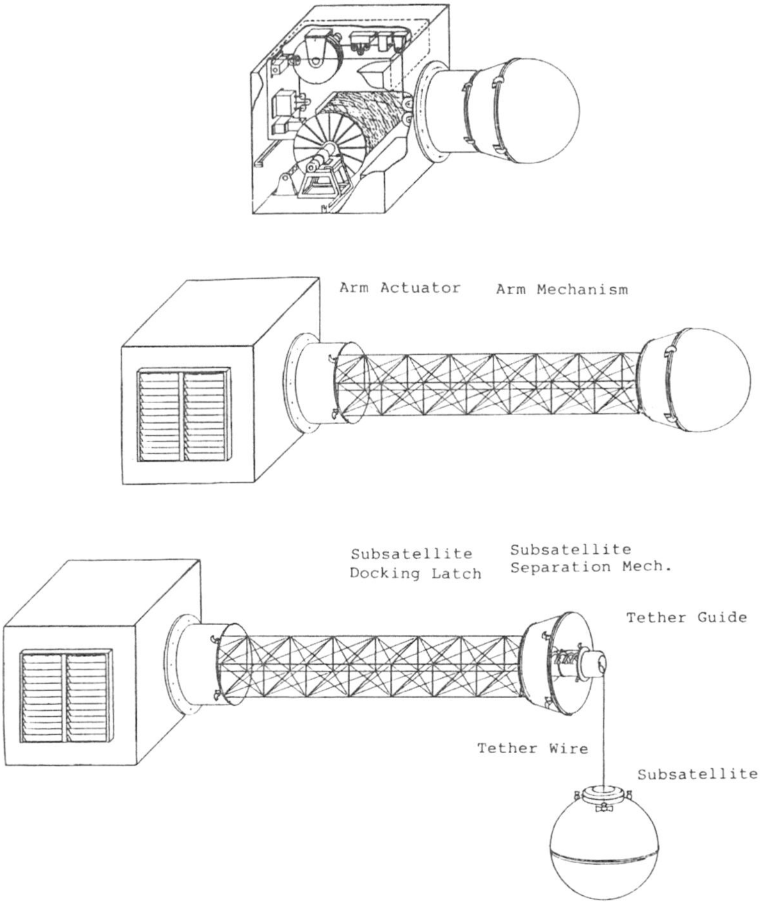

Fig.2 Tether deployment/retrieval mechanism in STEX

stability of a long tether (10km) deployed upward and downward under gravity gradient will be studied.

In the STEX, plasma environment surrounding the SFU is studied by the subsatellite. SFU moving with a supersonic velocity will generate a large scale disturbances in space plasma as shown in Fig.3. In the ram side, ion beams and non-collisional shock structure will be generated. In the wake side, plasma is greatly depleted, generating a steep gradient of plasma density. The density gradient results in plasma diffusion, associated wave generation and plasma heating. The tethered subsatellite will contain a plasma probe, a wave receiver and a pressure gauge to study these plasma effects. With the tethered plasma diagnostics, the plasma environment up to 1 km surrounding the SFU will be studied both in the ram and wake sides. The research of the spacecraft environment will provide scientific knowledges regarding to the interaction of the planets and comets with the solar plasma.

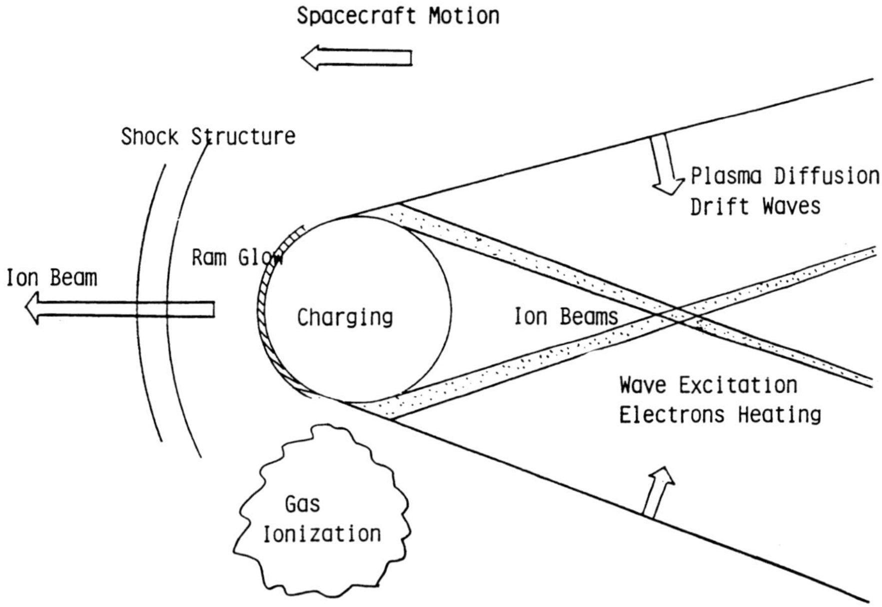

Fig.3 Plasma environment surrounding a spacecraft

After the fundamental technologies for space tether are established, the SFU tethered system will be used for the other science and technology experiments as an SFU experimental facility. The tethered subsatellite can be used as a receiving antenna for the Microwave Transmission Experiment(METS)(Fig.4) and Laser Transmission Experiment.

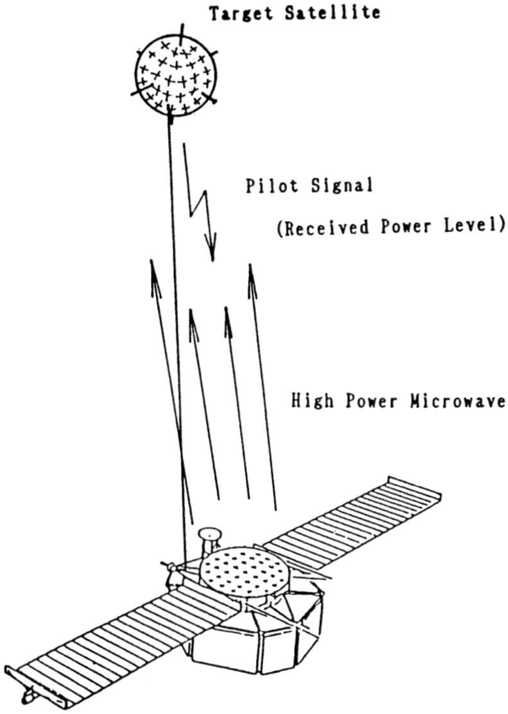

Fig.4 Microwave transmission experiment on SFU

3. IONOSPHERE TAPPING TETHER SUBSATELLITES(TSS-J)

ISAS new powerful rocket(MV-5), now under development, is capable to put a satellite of 1.8 ton into LEO. In the initial phase, the satellite is put into an orbit of 400 km. Then 4 tethered subsatellites 25 km apart from each other are deployed downward(Fig.5). The lowest one reaches 300 km height. The system is stabilized by the gravity gradient. The mother satellite will have a tether deployment system, a telemetry system with ground stations and subsatellites, a thruster system to keep its orbit, a TV camera to observe the tether system, particle injectors for active experiments and a geophysical observation system. Each subsatellite will be identical, carrying a power system, an attitude control/monitoring system, a tension monitor, a telemetry system and an observation system. All data obtaied by the subsatellites are relayed to the mother satellite and then transmitted to the ground station. The observation system consists of a magnetometer, wave receivers, particle detectors, plasma detectors, an electric field meter, a pressure gauge, a mass spectrometer, and visible/UV imagers. The lowest satellite also carries the instruments to study its aerodynamics. The weight of each subsatellite will be approximately 100kg. In the initial phase, technology experiments of the long tether will be carried out, as well as geophysical observation by the five

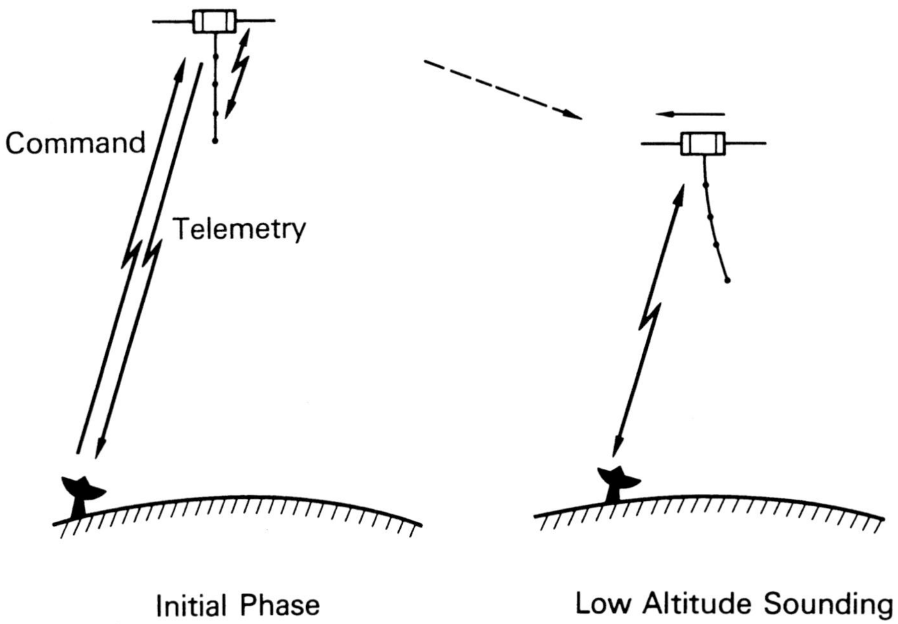

Fig.5 Geophysical observation by tethered multiple subsatellites(TSS-J)

satellites. The control logics for tether deployment and station-keeping under gravity gradient are investigated. Dynamic stabilization of the long tether will be studied by applying a thruster impulse to the tethered subsatellites. The retraction of the tether system will not be tested in this experiment.

After the technology experiments are completed, the mother satellite is transferred to a circular orbit at 200 km. In this configuration, the tethered system is subject to a strong aerodynamic force. The lowest subsatellite is positioned around 120 km height. During this phase, the vertical profile of geophysical parameters in the lower ionosphere are obtained by the 5 satellites. By operating the particle beam accelerators on the mother satellite, the beam/plasma interaction related to the aurora phenomena will also be studied by the subsatellites. The lowest subsatellite will be used to study the aerodynamics at the height of 120km. The mother satellite will carry the fuel enough to keep this configuration more than 3 months.

4. SUMMARY

Two programs of space tether proposed at ISAS are presented. The first one is an engineering test on the Space Flyer Unit to study the tether technology, and the second one is an application of the space tether to the geophysical research. Both have not yet been accepted officially, but they are studied actively by a working group of ISAS to promote these programs.

REFERENCES

[1] Colombo G., E.M.Gaposchkin, M.D.Grossi, and G.C.Weiffenbach, Shuttle borne "Skyhook": A new tool for low orbital-altitude research, proposal, Smithson.Inst.Astrophys.Obs., Cambridge, Mass., 1974.
[2] Woodies W.R., and J.M.VanPelt, Benefits of tether momentum transfer to space station operations, Adv.Astronaut.Sci., 62, 513-531, 1987.
[3] Merlina,P., W.Bogo, and S.Ciardo, Tethered elevator and pointing platform demonstrations: A shuttle flight test of scaled engineering models, Adv.Astronaut.Sci., 62, 147-171, 1987.
[4] Kroll,K.R., Tether implications on space station gravity level, Adv.Astronaut.Sci., 62, 533-544, 1987.
[5] Martinez-Sanchez,M and D.E.Hastings, A systems study of a 100kW electrodynamic tether, J.Astronaut.Sci, 35, 75-96, 1987.
[6] Lemke,L.G., J.D.Powell, and X.He, Attitude control of tethered spacecraft, J.Astronaut.Sci., 35, 41-55, 1987.
[7] Estes,R.D., and M.Grossi, Self-powered, drag-compensated, tethered satellite system as an orbiting transmitter at ULF/ELF, Adv.Astronaut.Sci., 62, 483-495, 1987.
[8] Bonifazi,C., Tethered satellite system(TSS) core science equipment, Adv.Astronaut.Sci., 62, 173-191, 1987.

2.4 High Voltage Technology in Space

K. NAGAI
Electrotechnical Laboratory of Ministry of International Trade and Industry, Ibaragi, Japan

Abstract

This paper discusses high voltage phenomena in space mainly from the arc and the partial discharge viewpoints in comparison with those on the earth.
For the future space exploration and utilization the electric power demand must be increased and for this purpose rising voltage used in space is the most important subject at present even for the future space solar generation program.

1. Introduction

For space usage of energy, electric power will be the most suitable in convenient and safe respects. Though the present rating of electric power for the space station under planning is in the order of 100kW (75kW at first stage, 150kW or so later), the electric capacity in the near future, when a space factory or a free flyer is necessary, is expected to be increased 1MW to 10MW taking a figure up one or two places.
In the case that this order of capacity is distributed at a working voltage to the load terminal, for instance at 150V, the bus current becomes a large value of about 3kA to 40kA for both sigle-phase three wire system and three-phase three wire system. This will cause unnecessary problems of switching phenomena resulting in higher level of surge impulse voltage, power loss and increase of conductor weight leading to a significant burden of payload, which must be avoided as far as possible. Consequently the adoption of higher voltage transformation is indispensable in space electric power system, the same as on the earth.
The various physical space factors which have been surveyed untill now show pessimistic influence on making the voltage higher because each of factors has a triggering action tending to initiate discharges.
This paper discusses the voltage situations used in space from the space features viewpoints.

2. Electric power demand in space

The electric power ratings in space started at about 100W for an artificial satellite in which the voltage used was as low as 10 to 30V and terminals at generating and load were connected with the same conductors directly. But beyond the year of 2000, MW order could be expected to one space station for factory or experimental usage. To meet the requirement, the type of voltage is of course alternating current, but the present space power generating systems mainly depend on fuel cell or solar cell supplying only direct current.
Necessary is to use an alternating generator with a solar thermal rotating machine or an invertor for the purpose of introducing a power transformer.

A higher voltage should be preferably selected taking account of supression of the current level to as low as 100A order. The required level of voltage will be about 6kV for 1MW, 60kV for 10MW irrespective of transmission systems with single-phase three wire or three-phase three wire.
The important matter is that up to what level of voltage can be utilized in space without serious discharges leading to any electrical failure.

3. High voltage consideration in space

In space, most electric power system so far is composed as one unit of generation, distribution and load sites not having each separated long division. But in a bigger space generation system for example in more extended system than in the space station the power system should be separated into each section. The matter is to know how to design an electric power line system in special space environment of very sparse air and micro order of gravity, including generation,substation at the sending side, transmission section, substation and distribution and load at the receiving side either for the purpose of space consumption or sending to the earth.
As with a transmission section there were not given so many considerations in comparison with generation method up to now. The reason for this may be that no opportunity existed to need to use a bigger electric power rating in space. The option of transmission section has to leave among insulation covered type of wire, cable type and bare conductor which is most convenient to use but not clarified whether it is usable in space that is, because of having no space withstand voltage data. The solution which type we should select is at present too premature because it is necessary for us to understand the electric insulation characteristics of space itself.
The most of sections from the substation at the receiving side toward the sending side would be exposed to the space itself and only the load section including it's distribution section would be set up isolated from space itself in for instance a space ship or an airtight type of space factory or free flyer. These inside rooms would have some functions for protecting electric equipment from electrically hazardous due to strong solar radiations, cosmic particles radiations, plasma and debris particles so on, and in many cases, surely they would be filled with air for astronaut workers to live.
The electrical situation becomes very much different between the space itself and the inside room. In the former space electrical discharges would be prompted and insulation materials be aged but in the latter space the situation would be rather like terrestrial one except only being micro gravity. When considering the high voltage phenomena in space, cares must be taken that the site in space is shielded from many cosmic factors or has air or experimental gases sealed. If air exists and under micro gravity the electric arc would show a unique trace. The arc trace on the earth, due to buoyance mostly goes up with the help of air convection. But in space both functions by buoyance and convection do not work upon the arc column then the trace would become very whimsical and importantly be controlled by the initial condition. The consequence may be that the estimation to which direction the arc proceeds is difficult and the probability to deviate to the opposite electrode or conductor would be very high resulting in ground or phase faults. This may mean that in space using cut-out types of switches or relays for rather large current circuit would be impossible and necessary to devise for avoiding arc faults.
The other subject to be solved, caused by micro gravity is that how corona or partial discharges would behave especially in liquid insulation. As the case

in the arc phenomena, bubbles caused by partial discharges on the earth go up out of the surface of the liquid and can be exhausted outside. But in space this function does not work on the bubbles. They would stay floating around the electrode where the electric field was high enough to originate partial discharges or stay just on the electrode untill the bubble inflated enough big to explode. After explosion of the bubble, spread small bubbles behavior must be important. If they were attracted to the opposite electrode or conductor the line or ground flashover would be originated through floating small bubbles between the electrodes or conductors. There would be also the case that small bubbles due to weak electric field but near above the onset would accumulate with taking balance of bubble origination and absorption into the liquid. These phenomena are neccessary to be observed by using micro gravity experimental simulation. If difficult is the case to apply a circuit breaker, only the solid state circuit switching would be depended upon.

4. Transmission line for space use

In order to realize a higher voltage transmission line in space, withstand voltages of the space must be statistically determined through a lot of flashover voltages experimented in actual space or in a simulation chamber. The simulation is composed of superposed various single components to be taken into consideration that is, higher vacuum, plasma particles, outgas, evaporation, debris, strong ultraviolet and cosmic radiations, remaining gas of air, heat accumulation and its gradient. Except vacuum, all components will act negatively against electrical insulation strength. The effect of these complicated factors has a bad influence over insulation material body, surface condition and space gap withstand strength.
Without having information about which component influences to what degree and also about total influential degree on the space flashover voltage, the design of a space aerial transmission line is almost impossible to realize.
The important withstand voltage problem is existence in radical reduction of flashover voltages due to outgas and evaporation from material in addition to gases for various experiments and propelling power used in space. These gaseous densities produced are recognized as partial sparse gas layer built up on the surface of the material, to which the Paschen's Law can be applied and leading to unexpectedly low flashover or surface discharge occurrence. The old data of the Paschen's Law can be found in many text books on high voltage but new data are required to confirm for new or mixed gases which have much possibility to be used in space, as well as reviewing experimentally the fundamental gases under not only DC voltage but also other kinds of voltages.
Flashover data under nonuniform field is indispensable because most electrodes and material shapes are not made and not arranged to produce uniform electric field. After having surveyed newly the Paschen's Law, withstand surface voltage in terms of kV/cm would be able to be estimated. The results would make possible design an insulator string for setting conductor and a bushing for introducing power line into space ship or space factory.
Paschen's Law also should be applied to the design of flashover strength of the space gap itself as well as surface flashover in space especially in places that many influencial factors on flashover exist. The flashover voltage in this environment may be reduced further than sparse gases only.
A bushing is one of must important equipment in the electric system in space because it must be set at the boundary between space itself and sealed space

room and must have enough reliability on electrical strength and airtight and mechanical strength. Due to wide temperature change on the orbit night and day cycles, the selection of interior insulation material becomes important. In liquid insulation case, the phase change must be expected then to be difficult to adopt in space itself without any adiabatic surface material. Various kinds of insulations, such as gas, liquid, solid and compound insulations must be discussed in order to apply them into space utilization considering many space factors affecting insulation characteristics.

5. Research subjects in space high voltages to be done in Japan

The followings are main subjects to be clarified for considering and designing a space transmission lines.

Electrical Insulation in Space

(1) Fundamental Phenomena on High Voltages
- (a) Withstand voltage of space itself and Influence of triggering factors due to plasma, cosmic rays, UV rays, debris, outgas, evaporation, remained gases
- (b) Paschen's Law confirmation for various gases and under various space factors
- (c) Potential charging up
- (d) Surface discharge
- (e) High frequency interference level
- (f) Transient overvoltage level

(2) Transmission Systems between Generator and Load Sides
- (a) Possible highest voltage in space for bare and insulated conductors
- (b) Conductor supporting systems
- (c) Copper loss heat radiation method
- (d) Switching, trip, protection method

(3) Order of Merit for Insulation Method
- (a) Solid insulation
- (b) Liquid insulation
- (c) Gas insulation
- (d) Compound insulation

(4) Most Reasonable Power Sources in Space
- (a) AC low or high frequency
- (b) DC

6. References

(1) IEEE, IECEC Proceedings 1984, 1988, 1989
(2) Design Guide : Designing and Building High Voltage Power Supplies Vol I and II

3 Natural and Triggered Lightning

3.1 Lightning Observations in Finland and Theoretical Calculation of the Lightning Electromagnetic Pulse

T.J. TUOMI, T.H. SUMMANEN, and R.J. PIRJOLA

Finnish Meteorological Institute, Geophysics Department, SF-00101 Helsinki, Finland

This paper deals with lightning observations made in Finland by lightning flash counters for thirty years and by an LLP lightning location system since 1984. Estimation of flashes per unit area based on counter data cannot yet be performed since the operational range of the counters is unknown, being somewhere between 17 and 45 km. The counters seem to record more flashes than the LLP system does. A reason could be the low conductivity of the earth in Finland, which might deform the electromagnetic signal emitted by a lightning stroke too much during the propagation to pass the acceptance criteria of the LLP system. Signals incident on a LLP direction finder have been recorded, and theoretical calculations were started. The latter work also discussed in this paper has, however, so far only concerned a perfectly conducting earth thus giving no answer to the question of the effect of the low conductivity.

1. INTRODUCTION

Lightning flash counters have been used in Finland since 1959, and since 1961 we have systematic statistics [1]–[8]. The main purpose of the Finnish lightning observations is practical, which means that data are used for engineering studies of lightning effects on electric power systems, telecommunication networks etc., and for meteorology. Research work on lightning as a physical phenomenon has not been done much.

The first lightning flash counters in Finland were of the so-called Pierce-Golde type. Later a new flash counter known as SL 72 and suitable for Finnish weather conditions was developed, and since 1975 only SL 72 counters are used. A technical description of SL 72 is given in [5, pp. 40–42]. In recent years the number of counters in Finland has been about 50 to 60; in 1988 complete results were obtained from 47 counters. However, in 1989 the flash counter network was closed down except for eight counters in northern Finland (Lapland). This paper will deal with the main results obtained by the counters in Finland.

In 1984 the Finnish Meteorological Institute purchased an automatic lightning location system, so-called LLP (Lightning Location and Protection, Inc.), manufactured in the USA [5, pp. 46–47]. Its operation is based on direction finders (DF), the number of which was two in Finland in 1984–86. In autumn 1986 the third DF was installed, and the fourth in summer 1989. Now the LLP system covers well the southern half of Finland up to southern Lapland. In this paper Finnish LLP observations will be presented.

There is a clear difference between the Finnish flash counter and LLP results not yet completely understood. The question of the operational range of the flash counters, still somewhat uncertain [5, p. 12], is related to the disagreement between the two data sets. To make the matters clearer, a project was started some years ago to compare electromagnetic signals observed by an LLP DF with theoretically computed results. Until now, however, the model calculations performed and discussed in this paper have been rough due to the assumption of a perfectly conducting earth. Furthermore, the importance of the project has today diminished due to the improvements of the Finnish LLP system with four DFs.

Fig. 1: SL 72 lightning flash counters in Finland in 1988 [7].

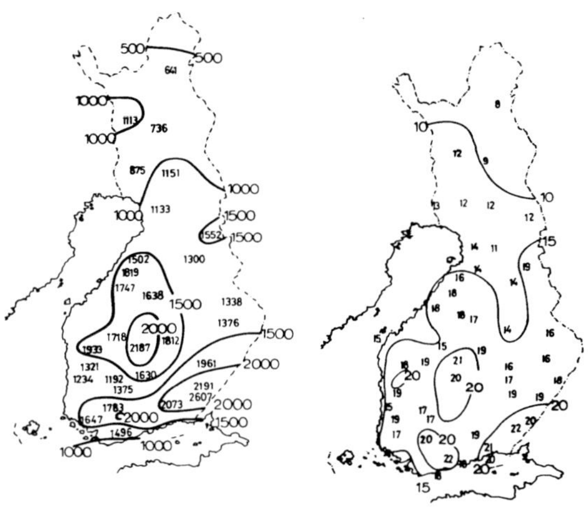

Fig. 2: Lightning flash counter statistics in Finland in 1961–1980 [2]. Left panel: yearly averages of counter observations. Right panel: yearly averages of counter lightning days (CLDs).

2. *LIGHTNING FLASH COUNTERS*

The map in Fig. 1 shows the 47 lightning flash counters in Finland which gave complete results in 1988. The stations are not located evenly, and they do not cover the whole area of the country, but a good general view of the regional distribution of lightning can certainly be obtained. The most essential insufficiency is in Lapland, but on the other hand, lightning is less frequent there.

Figure 2 presents statistics of lightning flash counter results in Finland in 1961–1980. The facts that the most active areas in this respect are southeastern and central Finland and that towards Lapland the number of flashes decreases are clearly demonstrated. A counter lightning day (CLD) (or a thunder day) included in the right-hand panel of Fig. 2 is defined according to an international agreement as a day when the counter in question has at least five recordings.

The period of thirty years of lightning flash activities in Finland is certainly long enough for statistical conclusions of the occurrence of lightning in the country. Estimation of lightning flashes per unit area requires a division of flash numbers, e.g. those shown in the left panel of Fig. 2, by the operational area of the counters. However, as mentioned in Section 1, there seems to exist a great uncertainty in the operational range of the Finnish counters. Traditionally it has been assumed to equal 17 km while later investigations indicate a value of even 45 km. Such a difference means a sevenfold uncertainty in the area. It is clear that the operational range is not a sharp boundary between observation and no observation, but stronger signals are recorded from larger distances, and a reasonable definition of the range corresponds to the area within which most of the flashes are and outside which most of the flashes are not observed. Anyway, the question of the operational range of the Finnish flash counters is very important to make the lightning statistics reliable and clear. It will be studied by furnishing flash counters with clocks and comparing counter data with lightning location system data (see Sections 3 and 4).

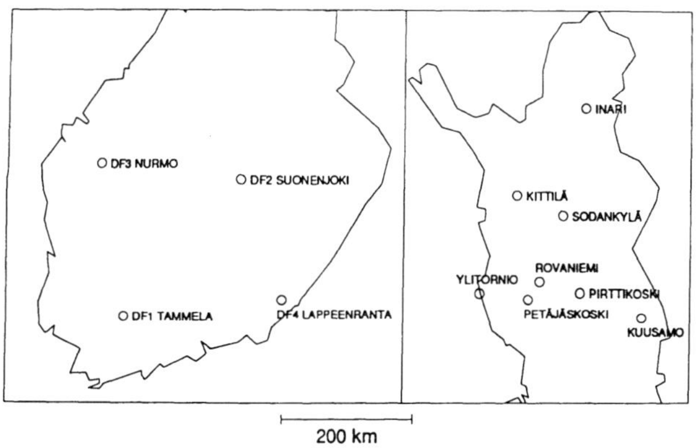

Fig. 3: Left panel: LLP lightning location system in Finland in 1989 [8]. Right panel: SL 72 lightning flash counters in Finland in 1989.

3. LIGHTNING LOCATION SYSTEM

The map in the left-hand panel of Fig. 3 presents the LLP lightning location system in Finland, and the right-hand panel shows the eight flash counters operating in 1989 to complete the coverage of the LLP system in Lapland. The Tammela (DF1) and Suonenjoki (DF2) direction finders were installed in 1984, and the Nurmo (DF3) and Lappeenranta (DF4) direction finders in 1986 and 1989, respectively. The central unit or the position analyzer (PA) collecting data from the DFs and computing flash locations and other parameters is situated at the weather service of the Finnish Meteorological Institute in Helsinki on the southern coast of the country.

The ranges of the DFs are about 400–500 km, and the three-DF system seemed to cover the southern part of Finland with an about 60–70 % detection efficiency [9] (see also

Section 4). Each DF has systematic bearing errors caused by nearby buildings, topographic features etc. We have studied them by analyzing a sufficient amount of flash data [6, pp. 32–35], [9]. Figure 4 presents the results, which are based on the three-DF configuration. Systematic errors are easy to correct, but the error bars in Fig. 4 indicate that there are significant random errors and probably a dependence of systematic errors on the distance. The average error in the final location of a flash may be represented by the average length of the shortest side of the error triangle formed by the bearings of the three DFs. The error is about 11 km [7, p. 31], but in one half of the cases the error remains below 8 km. Errors larger than 30 km, which considerably increase the average, seem to occur mostly in central Finland. It is also the area in which the electrical conductivity of the earth varies abruptly, possibly causing the large flash location errors (cf. Section 4). The above error figures apply to the real-time location data for which the locations are calculated as the intersection of the "best" pair of directions (i.e. of those intersecting closest to 90°). Application of optimization methods improves the location; for the present, they are used in the re-processing for all observations involving at least three DFs.

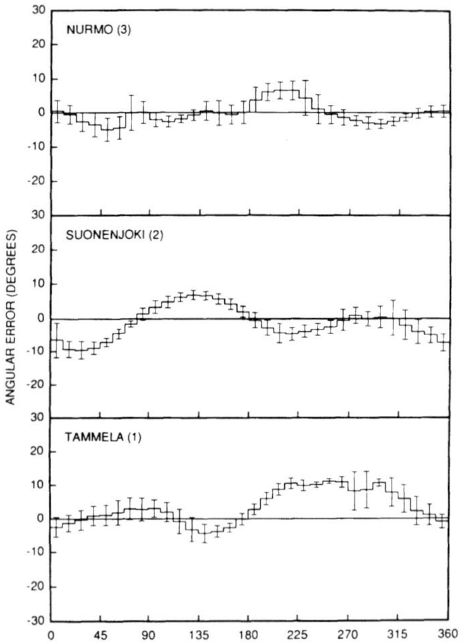

Fig. 4: Systematic bearing errors of the LLP direction finders at Tammela (DF1), Suonenjoki (DF2) and Nurmo (DF3) in Finland [9]. The angle goes to the east from the north direction.

Figure 5 shows the flashes (altogether 19602) per 100 km^2 located by the LLP system in 1989 which can be considered an average year as concerns lightning activity in Finland. The figure supports the above-mentioned conclusion that the most active areas are southeastern and central Finland.

The multiplicity, i.e. the number of return strokes in a flash, and the strength indicating the peak current in a flash are other parameters determined by the LLP system. In Finland, roughly 50 % of the ordinary negative flashes have only a single stroke, and for the less frequent positive flashes, which e.g. in 1989 accounted for about 10 % of all flashes, the percentage of single-stroke flashes is clearly higher. On the other hand, positive flashes are stronger on the average [5]–[8].

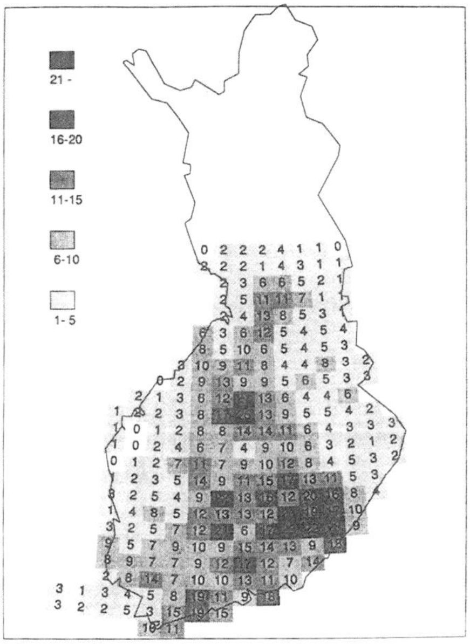

Fig. 5: Flashes per 100 km^2 located by the LLP system in Finland in 1989 [8].

4. COMPARISON BETWEEN THE RESULTS OF THE FLASH COUNTERS AND OF THE LOCATION SYSTEM

Comparison between flash-counter and location-system data for the relatively active summer of 1988 indicates the following empirical relationship to be roughly valid:

$$b = 100\, s \tag{1}$$

where b is the number of flashes recorded by a lightning flash counter and s is the number of flashes observed by the LLP system per 100 km^2 in the vicinity of the counter. This result is based on the data of 19 counters which lie in the best detection area of the location system and on periods when the quality of both the flash-counter and location-system data could be considered good. The coefficient 100 in equation (1) has a standard deviation of 15.

Expressed in a different way, equation (1) means that the operational area of the flash counters would be 10000 km^2 yielding an operational range of about 56 km. This value is in a complete disagreement with the previously assumed range of 17 km and also too large compared to 45 km (see Section 2). Thus, the counters clearly record more flashes than the LLP system does.

One reason for the difference between flash numbers observed by the counters and the LLP system can be that the counters also record a part of cloud-to-cloud and intracloud flashes, although both systems should be sensitive only to ground flashes. In Finland, about one third of all flashes are ground flashes. On the other hand, the counters do not record positive ground flashes at all.

Figure 6 shows the average conductivity of the earth at low and medium frequencies in Finland. Comparison of Fig. 6 to the results of [11] shows that the conductivity in Finland is typically clearly smaller than that in North America. Besides, the Finnish earth conductivity obviously has rapid spatial variations due to lakes. These facts indicate another possible reason for the difference between flash counter results and locations: The "peculiar" conductivity may deform the electromagnetic pulse emitted by a return stroke too much during the propagation to pass the acceptance criteria of the LLP DFs. Additionally, spatial variations of the conductivity might cause that the signal does not propagate straight.

ELECTRICAL CONDUCTIVITY OF THE GROUND IN FINLAND

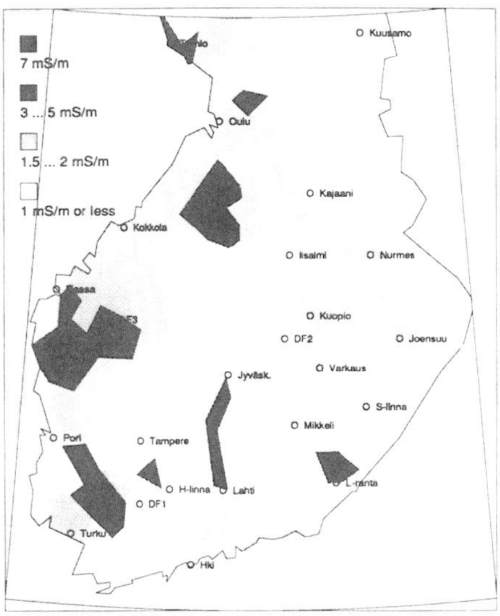

Conductivity of S and W coast water 10 ... 20 mS/m.

Fig. 6: Average earth conductivity at low and medium frequencies in central and southern Finland (modified from the map published by the Finnish Broadcasting Company (J. Laiho) in 1975, see also [10]).

We started a research project in 1987, in which our aim was to make theoretical studies of the lightning electromagnetic field and also directly measure the electromagnetic signal incident on a DF antenna. Figure 7 shows two examples of pulses recorded at Tammela (DF1). The upper curve was interpreted as a real lightning signal by the DF while the lower pulse was rejected. However, the regular form suggest that also the lower pulse was a genuine flash. Rough studies of the recorded pulses indicate that a rejection of a real lightning pulse is not at all exceptional, and so the LLP DF criteria might be relaxed in Finland. It has not yet been done, and it would certainly be a difficult job to ensure that man-made disturbances are still rejected.

5. THEORETICAL CALCULATION OF THE LIGHTNING ELECTROMAGNETIC FIELD

As indicated above, our practical reason for starting theoretical studies of the electromagnetic field caused by lightning was to investigate the low and rapidly varying conductivity of the earth in Finland. The work was, however, started from the beginning by assuming the earth to be a perfect conductor permitting the use of an image source. In fact, all our theoretical calculations performed until now only concern a perfect conductor, so the answer to the original practical question is still unknown. On the other hand, the importance of this kind of a study has decreased, because the use of four DFs is certainly improving the location system.

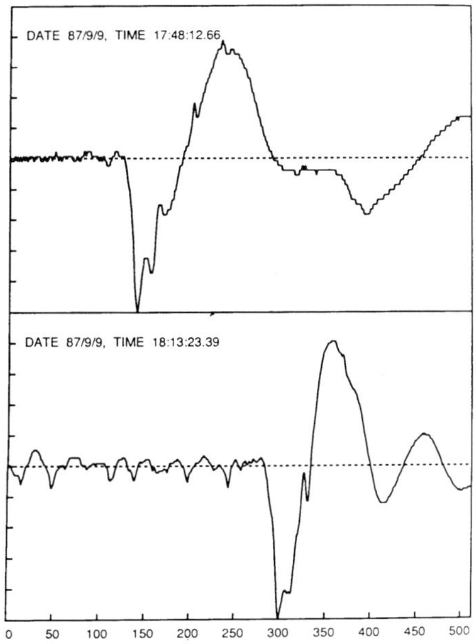

Fig. 7: Lightning electric field pulses recorded at Tammela (DF1) [6]. The vertical scales used are arbitrary. The upper pulse was connected with a located flash, i.e. it was accepted, while the lower pulse was not accepted by the DF.

Our theoretical calculations have also been based on the neglect of the curvature of the earth and on the use of a transmission line model (TLM) for the return stroke current in an infinitely thin lightning channel:

$$I(z,t) = I(t - z/v) \tag{2}$$

where t denotes the time, z is the spatial coordinate along the channel and v represents the constant propagation velocity of the current pulse, e.g. [12]. It should be noted that Maxwell's equations, which are, of course, the background of the calculations discussed now, implicitly take into account electric charges accumulating at points of divergence of the current. Thus e.g. charges appearing at the top of the lightning channel affect the fields. On the other hand, such an accumulation at the top is evidently a geophysical drawback, and the model could be developed by additional currents preventing the accumulation.

Using the assumptions described above, the lightning electromagnetic field and the effects of different model parameters were studied in detail both in the frequency and in the time domain in [13]. Special attention was paid to the differences between the exact solution and an approximate solution of the electric field where the former requires that the propagation velocity v of the current equals the speed of light $c = 3.0 \cdot 10^8$ m/s, and the latter is well acceptable at large distances from the lightning stroke. The current function mainly assumed was

$$I(t) = I_0 \,(\exp(-\alpha t) - \exp(-\beta t)) + I_1 \,(\exp(-\gamma t) - \exp(-\delta t)) \tag{3}$$

with $I_0 = 30$ kA, $I_1 = 2.5$ kA, $\alpha = 4.0 \cdot 10^4$ s^{-1}, $\beta = 8.0 \cdot 10^5$ s^{-1}, $\gamma = 1.0 \cdot 10^3$ s^{-1}

and $\delta = 2.0 \cdot 10^4\ s^{-1}$ [14]. For brevity, results presented in [13] are now referred to by discussing only two examples both of which involve this particular current pulse.

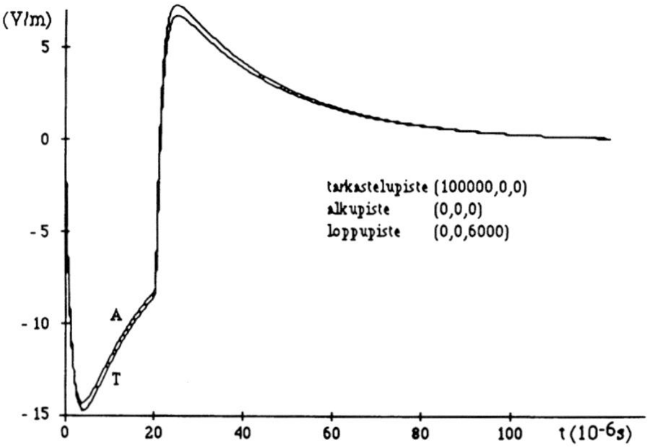

Fig.8: Vertical electric field as a function of time on the surface of a perfectly conducting earth [13]. The field is caused by a vertical lightning stroke starting from the earth at a distance of 100 km. The height of the lightning channel is 6 km, and the velocity of the upward-propagating current pulse equals the speed of light. The current pulse is expressed by equation (3) with the parameter values given in the text after the equation. The exact and an approximate solution are denoted by T and A, respectively.

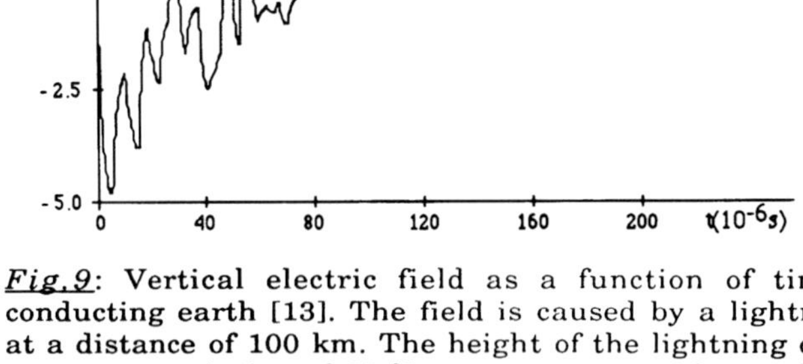

Fig.9: Vertical electric field as a function of time on the surface of a perfectly conducting earth [13]. The field is caused by a lightning stroke starting from the earth at a distance of 100 km. The height of the lightning channel is 6 km, and the channel is tortuous consisting of 28 linear segments with standard deviations of 150 m in the horizontal directions. The velocity of the current pulse propagating upwards along the channel equals one third of the speed of light. The current pulse is expressed by equation (3) with the parameter values given in the text after the equation.

Figure 8 shows the electric field on the earth's surface at a distance of 100 km from a vertical lightning stroke. The height of the lightning channel starting from the earth is assumed to be 6 km, and the (upward) velocity of the current pulse is $v = c$. Because the earth is perfectly conducting the electric field is, of course, vertical. The notations T and A refer to the exact and approximate solutions, respectively.

The situation in Fig. 9 is similar to that of Fig. 8 as concerns the distance between the stroke and the point of observation of the electric field and the height of the channel. However, in Fig. 9 the velocity of the current pulse along the lightning channel is $v = c/3$ permitting only an approximate solution. Also the lightning channel connected with Fig. 9 is tortuous since it is assumed to consist of 28 linear segments chosen using computer random numbers having a normal distribution. The mean value and the standard deviation connected with the displacements in the segments in the two horizontal directions are 0 and 150 m, respectively. In the vertical direction the corresponding numbers are 200 m and 50 m. It should be noted that due to the tortuousity the length of the lightning channel is greater than the height.

The smoothness of the curves in Fig. 8 compared to that in Fig. 9 is obviously due to the tortuousity of the lightning stroke in the latter figure. The fact that the behaviour of the electric field is in general more rapid in Fig. 8 than in Fig. 9 is caused by the higher propagation velocity of the pulse. Also the greater length of the channel in Fig. 9 has certainly some influence.

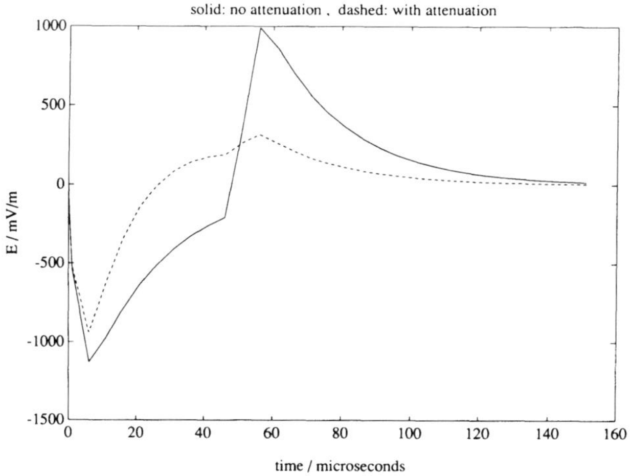

Fig. 10: Vertical electric field as a function of time on the surface of a perfectly conducting earth. The field is caused by a vertical lightning stroke starting from the earth at a distance of 100 km. The height of the lightning channel is 4 km, and the velocity of the upward-propagating current pulse is $8.0 \cdot 10^7$ m/s. The current pulse is expressed by equation (3) with $I_1 = 0$ and the values of the other parameters are given in the text after equation (4). The dashed line corresponds to equation (4) with the attenuation factor $\psi = 5.0 \cdot 10^{-4}$ m^{-1}, and the solid line refers to the case of no attenuation ($\psi = 0$).

Finally we will discuss a modified transmission line model for the return stroke current:

$$I(z,t) = I(t - z/v) \exp(-\psi z) \tag{4}$$

where a positive value of ψ implies an exponential attenuation of the current intensity as the pulse propagates along the channel. Exactly the same model has also been presented in [15]. Let us now use the pulse presented in [16] and obtained from equation (3) with $I_1 = 0$, $\alpha = 4.4 \cdot 10^4$ s^{-1}, $\beta = 4.6 \cdot 10^5$ s^{-1}. We further assume that $I_0 = 10$ kA, $v = 8.0 \cdot 10^7$ m/s and $\psi = 5.0 \cdot 10^{-4}$ m^{-1} corresponding to a decay length of 2 km. The dashed line of Fig. 10 then shows the electric field on the earth's surface at a distance of 100 km from a lightning stroke in a vertical channel starting from the earth and having a height of 4 km. For comparison, the solid line depicts the case of no attenuation, i.e. $\psi = 0$, but the other parameters have the same values. It is seen that the attenuation greatly affects the electric field.

6. CONCLUDING REMARKS

The time period of thirty years during which lightning flash counters have been operating in Finland is sufficient for statistical conclusions of the occurrence and regional distribution of lightning in the country. However, numbers of flashes per unit area cannot yet be given because there is a great discrepancy in the range of the counters, the possible values being between 17 and 45 km, which means a sevenfold uncertainty in the area. Solving the discrepancy will be one of our most important tasks in the future.

An LLP lightning location system was installed in Finland in 1984 containing first two direction finders. Now their number is four, and the system covers the southern half of the country. The system has showed its usefulness and is at present completely replacing the counter network. However, for continuity, it is necessary that the counter recordings and the LLP observations are comparable. Until now, there seems to be a clear difference between the numbers of flashes recorded by the two systems; the counters observe more flashes. A reason may be that the counters are also sensitive to cloud flashes, although both systems are planned to record ground flashes only.

Another possible explanation to the disagreement between the results of the two systems could be the low conductivity of the earth in Finland, which might deform the propagating lightning electromagnetic pulse too much to be accepted by the LLP system. To study it, we have recorded electromagnetic signals at an LLP direction finder, and preliminary observations seem to support the guess of too many rejected pulses. Theoretical calculations were also started, but until now they concern a perfectly conducting earth so that an answer to the main question has not been obtained. In the future the research should thus be extended to finitely-conducting earth models, and the effects of lateral conductivity variations would also be an important subject to be studied. The exact image theory developed recently [17] may be useful then.

7. ACKNOWLEDGEMENT

We wish to thank the Imatran Voima Oy power company for significant support in our studies on lightning.

8. REFERENCES

1. *Laitinen L (1976)* Lightning flash counter activity in Finland 1961...1975. *CIGRE-document* **33**-76 (TF 33-01.01) 16 IWD, Paris, France, August 1976, 10 pp.

2. *Laitinen L (1985)* Salamanlaskijatoiminta Suomessa vuosina 1961–1980 (= Lightning counters in Finland during the period 1961–1980). (Finnish with an English abstract), *Sähkö–Electricity in Finland* **58**, 4: 18–20

3. *Laitinen L. Leino K. Palva V (1971)* Lightning flash counter measurements in Finland 1961...1970. *Sähkö–Electricity in Finland* **44**, 9: 215–222

4. *Pirjola R (1978)* Lightning flash counter activity in Finland 1961...1975. *URSI XIX General Assembly*, Helsinki, Finland, Session E1, 1.8.1978, 9 pp.

5. *Tuomi TJ (1987)* Lightning observations in Finland 1984–1986. *Geophysical Publications*, Finnish Meteorological Institute, No. 4, 47 pp.

6. *Tuomi TJ (1988)* Lightning observations in Finland 1987. *Geophysical Public-*

ations, Finnish Meteorological Institute, No. 5, 37 pp.

7. *Tuomi TJ (1989)* Lightning observations in Finland, 1988. *Geophysical Publications*, Finnish Meteorological Institute, No. 8, 34 pp.

8. *Tuomi TJ (1989)* Lightning observations in Finland, 1989. *Geophysical Publications*, Finnish Meteorological Institute, No. 11, 39 pp.

9. *Tuomi TJ (1989)* Salamanpaikantimen paikannusvirheet (= Location errors of the lightning location system). (Finnish with an English abstract). In: Forsius J (ed): *Proceedings of XIV Geofysiikan päivät*, Helsinki, Finland, May 3–5, 1989, Geophysical Society of Finland: 67–71

10. *Koskenniemi O. Laiho J (1975)* Measurements of effective ground conductivity at low and medium frequencies in Finland. *E.B.U. Review* – Technical Part, Technical Centre of the European Broadcasting Union, 153, 4 pp.

11. *Fine H (1954)* An Effective Ground Conductivity Map for Continental United States. *Proceedings of the IRE* **42**: 1405–1408

12. *Uman MA (1977)* Calculation of the Electric and Magnetic Fields Produced by Close Lightning. In: Dolezalek H. Reiter R (eds): *Electrical Processes in Atmospheres*, Proceedings of the Fifth International Conference on Atmospheric Electricity held at Garmisch-Partenkirchen, West Germany, 2–7 September, 1974, Dr. Dietrich Steinkopff Verlag, Darmstadt: 597–606

13. *Summanen T (1988)* Salaman aiheuttaman sähkökentän teoreettinen mallintaminen (= Theoretical modelling of the electric field caused by lightning). (Finnish). *Thesis for the M.Sc. degree*, Department of Theoretical Physics, University of Helsinki, Finland, 78 pp.

14. *Le Vine DM. Gessel L (1984)* The influence of ground conductivity on the structure of RF radiation from return strokes. *Preprint Proceedings of the VIIth International Conference on Atmospheric Electricity*, Albany, New York, USA, June 3–8, 1984, American Meteorological Society: 401–406

15. *Nucci CA. Rachidi F (1989)* Experimental validation of a modification to the transmission line model for LEMP calculations. In: Dvorák T (ed): *Proceedings of the 8th International Zurich Symposium and Technical Exhibition on Electromagnetic Compatibility*, Zürich, Switzerland, 7–9 March, 1989, 76 M2: 389–394

16. *Bruce CER. Golde RH (1941)* The lightning discharge. *The Journal of The Institution of Electrical Engineers* **88**, Part II (Power Engineering), 6: 487–520

17. *Lindell IV. Alanen E (1984)* Exact Image Theory for the Sommerfeld Half-Space Problem, Part III: General Formulation. *IEEE Transactions on Antennas and Propagation* **AP-32**, 10: 1027–1032

3.2 A Possible Mechanism for Generation of Linear Lightning Flashes

A.I. GRIGORYEV and S.O. SHIRYAEVA
Faculty of Physics, Yaroslave University, Yaroslave, 150000 USSR

In spite of a long history of careful studies in linear lightning and the large number of publications dealing with it, it has not been clarified how the stepped leader is initiated, forms and propagates, as well as how lightning collects electric charge from individual drops in a thunderstorm cloud.

A flash has three stages [1]:(1) preliminary variations (~100 ms duration); (2) stepped leader (~10 ms); (3) return stroke (~0.2 ms). The present paper proposes a model for the preliminary stage of the lightnin flash based on the concept of charged drop instability in electric fields [2-4]. It is assumed that the preliminary lightning flash stage is associated with a plasma region (PR) forming in the cloud at the level of wet hail growth and intensive separation of electric charges, 4 to 5 km. PR appears during a corona discharge generated owing to charged drop instability and wet hail pellet instability in an electric field that is a superposition of a quasistationary intracloud field $\vec{E}_0$ and the Coulomb field of a flying large ($R \sim 1$ cm), strongly charged pellet of hail.

The idea of a lightning flash being initiated by a corona discharge from a set of closely spaced large drops or water-covered pellets of hail was suggested earlier by laboratory experiments in instability of thawing hail pellets in strong electric fields (see, for example, [5]), but that idea was mere hypothesizing. At present, new data are available relating to instability of drops and hail pellets in electric fields, making it possible to provide more detailed descriptions of the initial stage in the generation of a lightning flash.

According to radar and acoustic observations, the lightning flash is initiated at about 4-5 km height with a mean intracloud electric-field intensity of $E_0 \approx 3$ kV/cm. Suppose a water-covered hail pellet is flying through a layer of negatively charged drops and water-covered ice nuclei at the height mentioned above at a velocity of $V = 10$ m/s. Let us presume that the pellet is rather large ($R = 5$ mm) and is carrying an electric charge of the order of the critical one in the sense of Rayleigh stability. In the total electric field: quasistationary intracloud E_0 and impulsive time-dependent (with impulse duration of ~ 1 ms) field of the charged pellet E_*, large (with $r_0 > 30$) drops and water-covered ice cristals with the concentration $\sim 10^2 - 10^3\ cm^{-3}$ may become unstable [2,3] and emit extremely small ($r > 0.01 \cdot r_0$) heavily charged droplets also unstable in the sense of Rayleigh [5]. They will disintegrate during the time of the order of their oscillations period, of mode with $n=2 \sim 10^{-3}\sqrt{\rho \cdot r^3/\sigma} \sim 10^{-7}$ s, each emitting about 100 still smaller droplets with the intensity of electric fields of their own charges near the surface, large enough for autoelectronic emission ($10^8 \div 10^9$ V/cm) [5]. The electrons emitted in a process of this sort at pressures ≈ 400 Tor will initiate electron avalanches whose typical linear dimensions are controlled by the dimensions of the space in the vicinity of the pellet where $E/p \gtrsim 20$ V/cm per Tor (in our case we have $\approx 2.5 \cdot R$). An avalanche has existence times $\sim 10^{-7}$s, the same time interval obtaining between the emissions of two electrons from the drop. A sequence of avalanches travelling along the same path will produce a plasma frame in the avalanche trace. If the frames of neighbouring avalanches overlap, a plasma region (PR) forms in the vicinity of the hail pellet with typical linear dimensions of $L \sim (3-5)$ R. The time of formation of a PR, including all processes that lead to its initiation, is $\sim 10^{-4}$s (that is, is about an order less than the transit time of a hail pellet through a layer of drops with thickness R). Polarization of the PR in $\vec{E}_0$ will enhance the intracloud quasistationary field in the vicinity of PR apexes oriented along the field up to a value causing instability of large ($r_0 \gtrsim 30 \mu m$) water drops, formation of electron avalaches, and lengthening of the PR along $\vec{E}_0$ accompanied by further growth of the field near the apexes. The lengthening of the positively charged end of the PR cloudwards is accompanied by branching-out where E_0 is strongly inhomogeneous, typical of an intracloud field and associated with inhomogeneities in the spatial distribution of the cloud charge.

When the negatively charged end of the PR has come to the lower edge of the thunderstorm cloud, the first stage of the lightning flash is at an end (that was the preliminary stage) and the second will begin, the stepped leader. According to [1], the duration of the first stage is ~ 0.1 s, hence one should be able to explain the fact that plasma in a PR is stable during such large intervals of time.

In the model proposed here, electric conductivity in a PR

is maintained by the electron current from the positively charged end of the PR towards the negatively charged end. Electrons are supplied into the PR due to the above instability of negatively charged cloud drops and formation of electron avalanches in the strong electric field of the positively charged end of the PR. That means that the PR plasma is close to that of the positive column of a glow discharge in the physical characteristics. It should however be borne in mind that the above phenomenon takes place at pressures $\gtrsim$400-500 Tor at which the laboratory plasma of the positive column contracts due to ionization-overheating instability during typical times of $\sim 10^{-6}$ s. However, under the conditions of a PR growing in a thunderstorm cloud, there is an effective mechanism inhibiting the instability. The mechanism is due to evaporation of water drops that inevitably get into the PR and are pulverized inside it due to Rayleigh instability with a continuous re-charging up to the critical charge by the electron current frlowing within the PR. It should also be noted that the high humidity typical of a PR that is growing in a thunderstorm cloud maintains free electron concentration at $\sim 10^{13}$ cm^{-3} due to retarded recombination during the time $\sim$10 ms [6].

REFERENCES

1. W.Beasley, M.A.Uman, and P.L.Rustan. Electric fields preceding cloud-to-ground lightning flashes. J.Geophys.Res., 1982, v.87, No.C7, p.4882-4902.

2. A.I.Grigoryev and O.A.Senkevich. On the mechanism of instability developing in a fluid drop in an electric field. Izv. Acad.Sci.USSR, Mekhanika zhidkosti i gaza, 1985, No.6, p.10-15.

3. A.I.Grigoryev. On the explanation of the phenomenon of the "Kuril light", Meteorologiya i gidrologiya. 1988, No.5, p.67-75.

A.I.Grigoryev and S.O.Shiryaeva. Parameters of electrostatic fluid pulverization. Izv.Acad.Sci.USSR, Mekhanika zhidkosti i gaza, 1988, No.2, p.5-13.

5. V.A.Dyachuk and V.M.Muchnik. Corona discharge of water-covered hail pellets as the principal mechanism of lightning flash initiation. Doklady Acad.Sci.USSR, 1979, v.248, No.1, p.60-63.

6. S.F.Balandin, Yu.D.Kopytin, I.A.Tikhomirob et al. Long-lived aerosol plasma initiated by the radiation of an impulsive CO_2 laser. Zhurnal tekhnicheskoi fiziki, 1988, v.58, No.2, p.324-327.

3.3 The Characteristics of Winter Lightning Strokes Observed with Video Camera Systems

Y. GOTO[1] and K. NARITA[2]

Department of Electrical Engineering, Tohoku University, Sendai, 980 Japan[1] and Miyagi Polytechnic College, Miyagi, 987-22 Japan[2]

Abstract. Automatically driven video camera systems were developed and used for the observation of winter lightning. It was confirmed that the systems were very useful for the research of exceedingly accidental lightning phenomena. Until December 1988, 120 winter lightning flashes struck an isolated tower were taken pictures with the developed systems. The characteristics of winter lightning discharges to the tower were made clear.

1. INTRODUCTION

Many winter lightning occur over the northwest coast of Honshu Island facing the Sea of Japan. It is very important for designing protective measures in electric power facilities and equipments to make clear the characteristics of winter lightning, because many electric power plants and overhead transmission lines have been constructed in the area during the last two decades. We have been observing the winter lightning at Maki, Niigata Prefecture, Japan, using an isolated meteorological observation tower jointly with Tohoku Electric Power Company since 1976 [1],[2],[3],[4],[5]. To take the pictures of accidental lightning phenomena, two kind of video camera system which is automatically driven were constructed and used simultaneously at the two places since the winter of 1984. Later, the similar system was added at the third point.

Until December 1988, 120 times of winter lightning flashes struck the tower were taken pictures with those systems.

This report describes the developed video camera systems and the characteristics of winter lightning discharges with the systems.

2. OBSERVATION POINTS AND AUTOMATICALLY DRIVEN VIDEO CAMERA SYSTEMS

Winter lightning flashes struck the meteorological observation tower have been observed at Maki which is located about 30 km SW from Niigata City (37°46' NL., 138°49' LE.). The tower of about 150 m height is constructed on a ridge of 125 m above mean sea level and about 400 m apart from the coast of the Sea of Japan.

The shooting point A for the video camera system is about 250 m apart from the tower in west direction and at this point there is main lightning observation station in which other many kinds of lightning observing systems have been set up.

The point B is about 500 m apart from the tower in WSW direction and the point C is about 1.4 km apart from the tower in NNE direction. The geographical situation near the observation points around the tower is shown in Fig. 1.

An automatically driven video camera system consists of monochrome TV camera, video tape-recorder, video timer, lightning sensor and sequence controller with power source as shown in Fig. 2.

The system starts automatically when an electric field strength exceeds a certain level and runs for a preset period. The electric field is measured with a field mill or a needle corona electrode. In addition, date and time are simultaneously recorded on the images for easy correspondence to the records of lightning current waveform and variation of electrostatic field.

At the points A and B, Visicon TV cameras have been used. However they need long time until they attain normal sensitivity. Moreover, their image planes are easily burned by extremely strong light for instance the sunbeam, and the image tube has a long afterimage effect. So at the point C, CCD TV camera has been used. It has many advantages such as no image distortion, short lag, no image burning, no afterimage effect, no preheating and quick start (about 0.5 sec.) although it is very expensive.

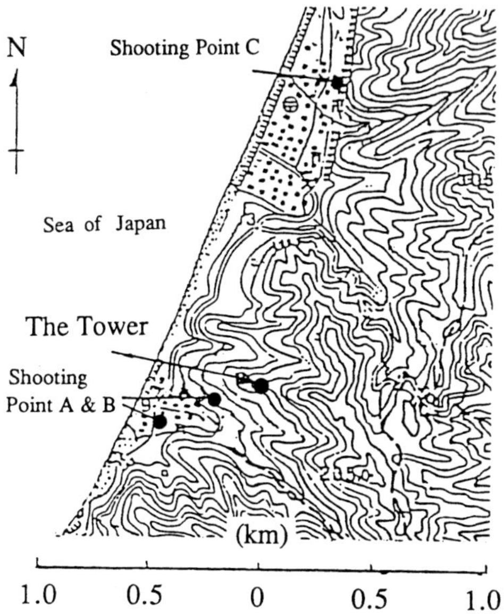

Fig. 1. Map around the observation tower and shooting points.

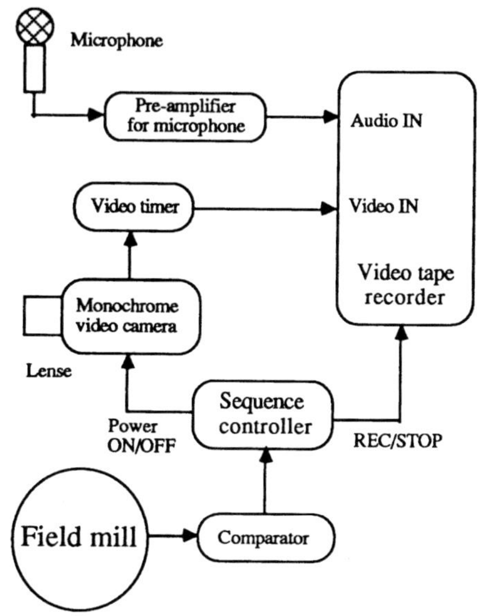

Fig. 2. Schematic block diagram of typical video camera system.

3. LIGHTNING PICTURES TAKEN WITH THE CAMERA SYSTEMS AND RESULTS

Figure 3 shows a series picture of the most branched flash taken with the video camera system at the point C. The luminosity of each channel has changed at the each field. But the main channel located center of each picture is very short and luminous from start to last field.
Figure 4 shows another series of the lightning flash which has two long horizontal channel.
Figure 5 shows the series picture of the flash which has the longest duration. About half of winter lightning flashes struck the tower had branched discharge channels. The histogram with the branched channel number of 120 flashes is shown in Fig. 6.
Branches of lightning flash always point in the direction of propagation of the lightning stroke independently of the polarity of the stroke when branching is observed [6]. Most of winter lightning discharges striking the tower were growing upward.
The duration of winter lightning discharge was estimated by counting the number of continued luminous flame fields took by CCD camera (1 field = 33.3 ms). Figure 7 shows the histogram of the flash duration as the number of luminous channel fields. They distributed between 1 and 18 fields and its median duration was 207±33 ms which was much longer than summer one. From this result, the continuing current is supposed to be flowed for very long time in the luminous channel.
Winter lightning images taken with the systems show the heights of thundercloud base were very low. The form of the channel sometimes could not be distinguished, because the top portion of the tower was covered by thunderclouds.
The heights of thundercloud base were estimated about from 200 m to 800 m above mean sea level. Even though the taken lightning picture shows that lightning channels progressed up to the top in the picture, it could not estimate its height. Because the channels bent to this camera side.
The pictures taken at the point C show the whole channels under the cloud base.On the other hand it has been taken only the channel in the vicinity of the tower at the points A and B, because of the short distance from the tower. The three dimensional shape of the whole channel could not be obtained except the channels near the tower. Nevertheless,the channel shape could be estimated with the pictures taken simultaneously from two or three directions. Figure 8 shows that the estimated channel shapes of the flash as shown in Fig. 3.
43 flashes and it's 71 stroke channels were observed from at least two directions and the incidence angle which is shown Fig.9, could be measured. Fig. 9 (a) shows the frequency distribution of the incidence angle of winter lightning stroke to the tower and Fig. 9 (b) shows the cumulative frequency distribution. The median angle is about 30 deg. and much larger than summer one [7].

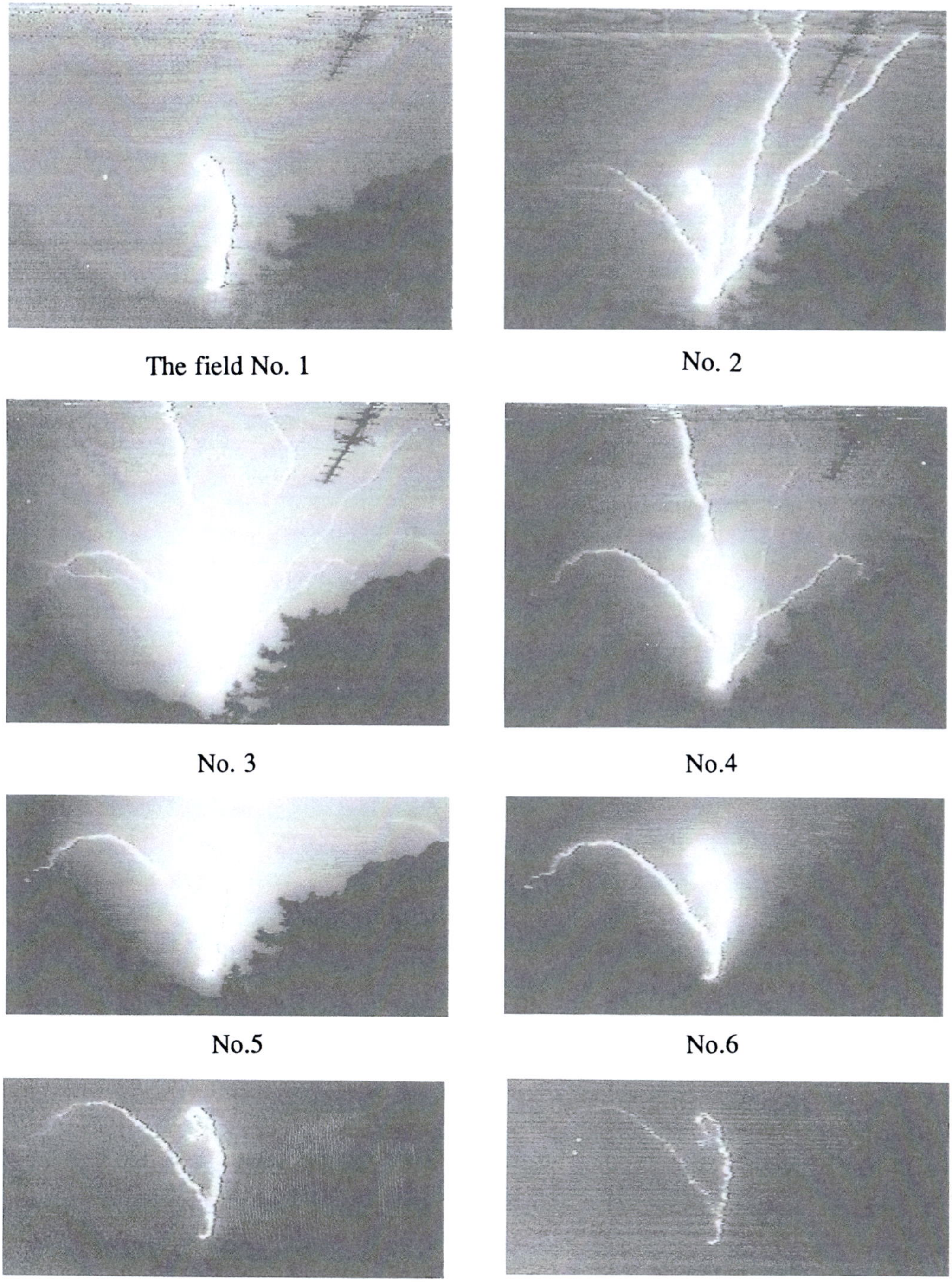

Fig. 3. A series video picture of winter lightning flash having the most brached channel taken at the point C. (19:30:07 JST, 30 Oct., 1986)

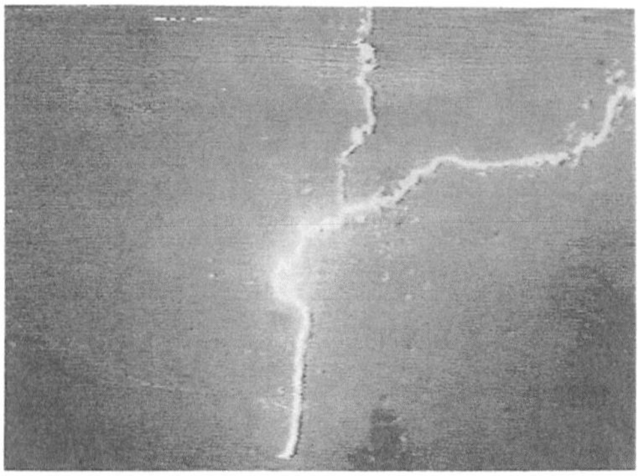

The field No. 1

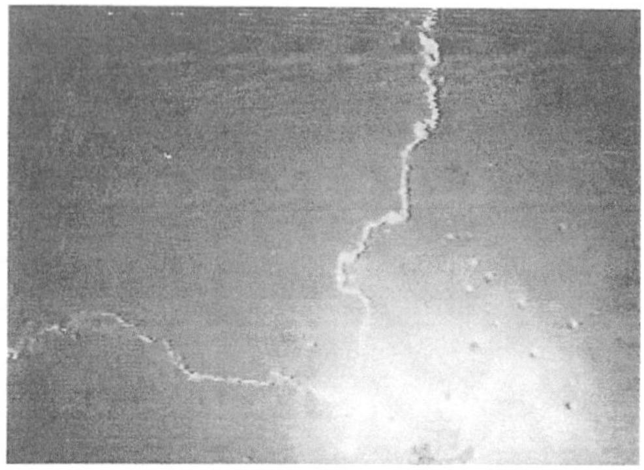

No. 2

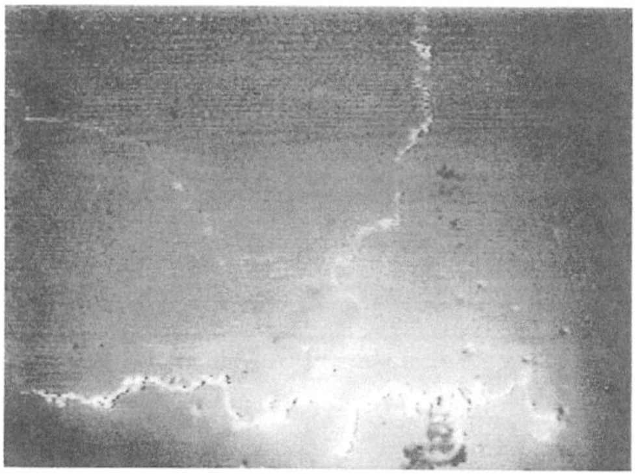

No. 3

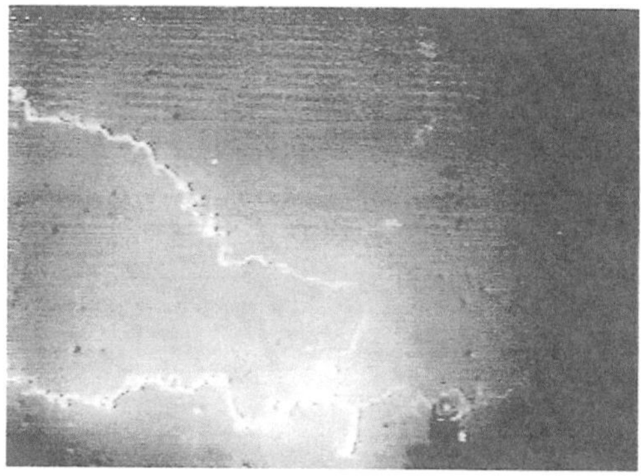

No.4

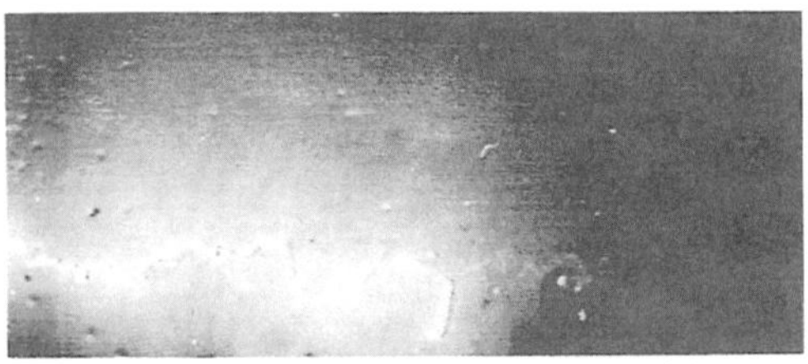

No.5

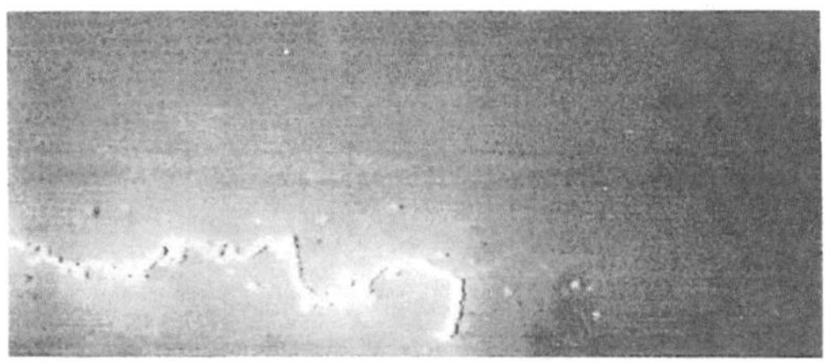

No.6

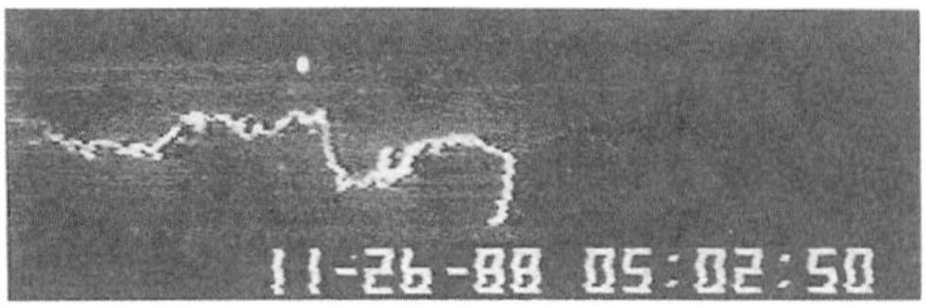

No. 7

No. 8

Fig. 4. A series video picture of winter lightning flash with two horizontal channels taken at the point C. (5:02:50 JST, 26 Nov., 1988)

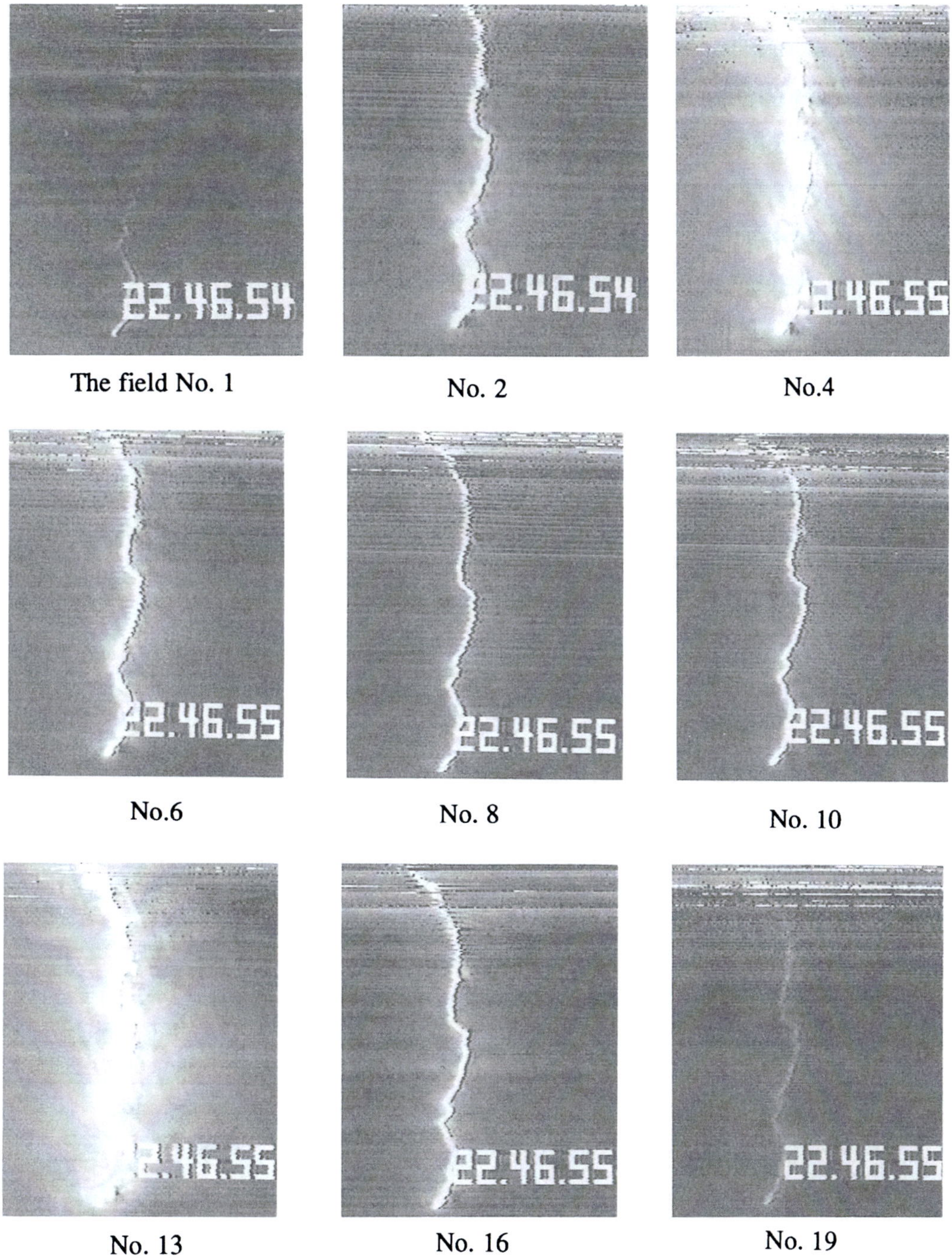

Fig. 5. A series video picture of winter lightning flash having the longest duration taken at the point C. (22:46:54 JST, 25 Dec., 1987)

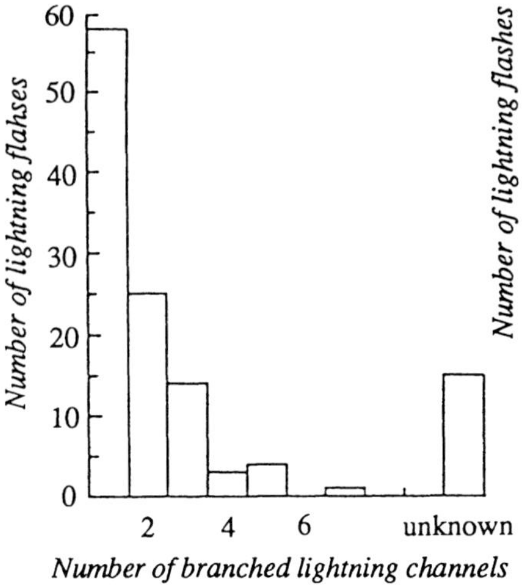

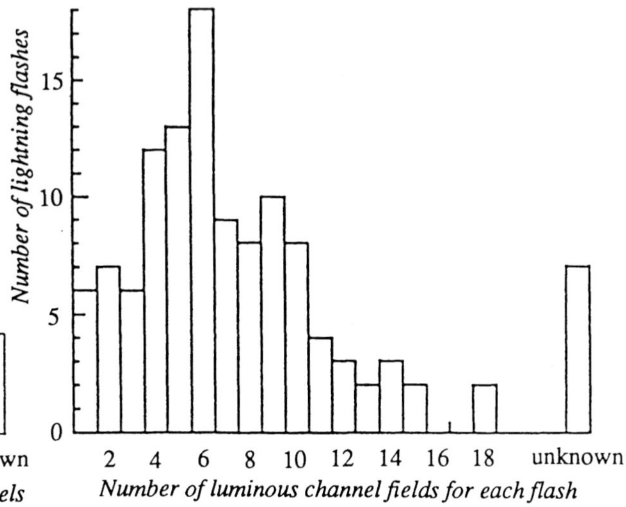

Fig. 6. Histogram of the branched numbers of 120 flashes.

Fig. 7. Histogram of the observed flash duration as the number of a series luminous channel field (1 field = 33.3 ms).

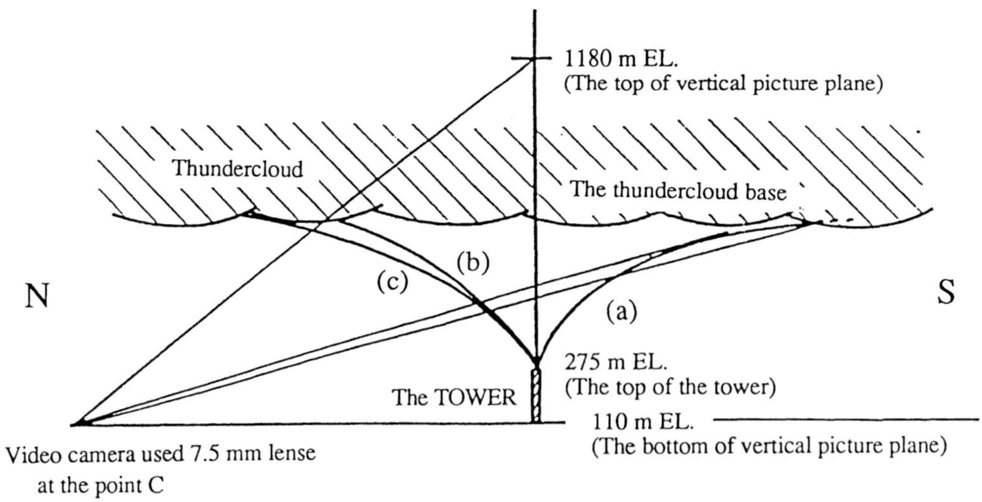

(a) The main stroke channel developping to the opposite side of the camera.
(b),(c) Other sub-stroke channel developing over this side of the camera.

Fig. 8. Side view of the video camera field with 7.5 mm lense taken at the the point C and estimated channels shown in Fig. 3.

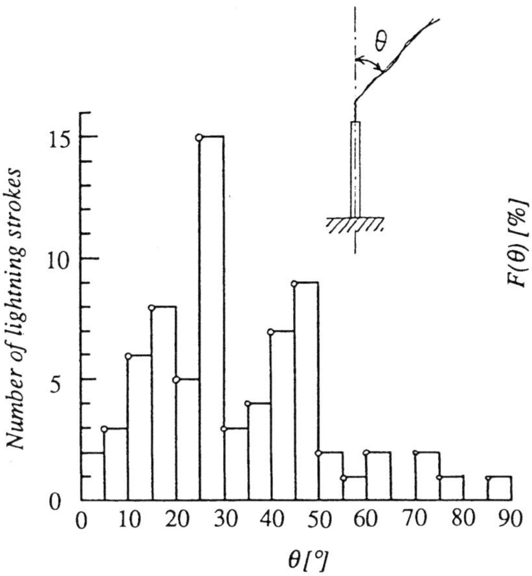

(a) Frequency distribution of incidence angles of a lightning stroke to the tower.

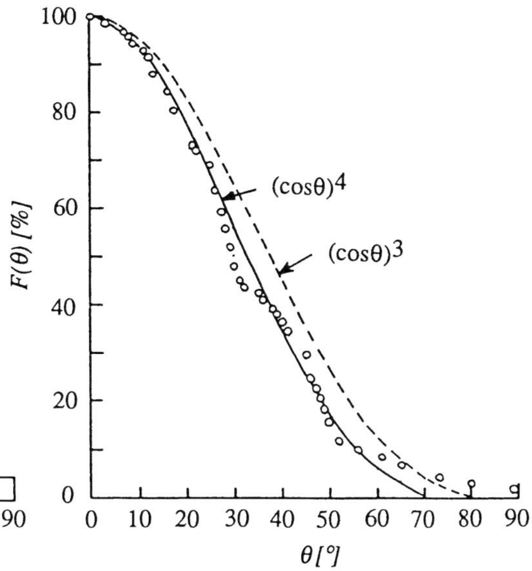

(b) Cumulative frequency distribution of incidence angles of a lightning stroke to the tower.

Fig. 9. The distribution of incidence angle.

The distribution of the incidence angle, $F(\theta)$, is approximated with the following function,

$$F(\theta) = (\cos\theta)^4$$

On the other hand, Mr. Miyake reported that the function $F(\theta)$ could be approximated with $(\cos\theta)^3$ [8].

4. CONCLUDING REMARKS

The automatically driven video camera systems were developed and used for the observation of winter lightning. It was confirmed that the systems were very useful for the research of exceedingly accidental lightning discharges.
From the research for winter lightning struck the tower with the systems following noteworthy items are point out:

(1) The height of winter thundercloud was very low. The cloud base was about 0.2 - 0.8 km high above mean sea level. Sometimes the tower top was hidden by thunderclouds.
(2) The most winter lightning flashes struck the tower were upward developing flashes which branched toward the base of thunder-clouds and the flashes show very long horizontal section frequently.
(3) The flash durations for channel luminosity were observed in the range between 1 and 18 fields with CCD TV camera at the point C which was 1.4 km apart from the tower and its median duration was 207±33 ms which was much longer than summer one. From this result, the continuing current is supposed to be flowed for very long time in the luminous channel.
(4) The incidence angle of winter lightning to the tower were widely distributed and its median value was about 30 deg., which was much larger than summer one.
(5) The cumulative frequency distribution function of the incidence angle, $F(\theta)$, could be approximated with $(\cos\theta)^4$.

Acknowledgments

A part of this work was supported by Grant-in-Aid for General Scientific Research No. 62550193 of the Ministry of Education, Science and Culture, Japan (1987, 1988).
The authors sincerely thanks Mr.H. Komuro and his associates of Tohoku Electric Power Co.,Inc. for their support and help in carrying out the field work.

References

1. Goto, Y., K. Narita and F. Naito (1983) : The digital recording system for lightning currents and results obtained for winter thunderstorms, Res. Lett. Atmos. Electr. Jpn., 3, 27-32.
2. Goto, Y., K. Narita and M. Naito (1984) : Spectral analysis for electrostatic field changes produced by winter thunderstorms, Res. Lett. Atmos. Electr. Jpn., 4, 21-28.
3. Goto, Y., K. Narita and R. Funayama (1986) : Tall tower triggered lightning during winter, 8th International Wroclaw Symposium on Electromagnetic Compatibility, part 1, 235-244.
4. Sawada, S., H. Komuro, Y. Goto and K. Narita 1988) : Current measurement of lightning superbolt at the Japan Sea coast, International Aerospace and Ground Conference on Lightning and Static Electricity, Oklahoma City, 4B. 1, 103-110.
5. Narita, K., Y. Goto, H. Komuro and S. Sawada (1989) : Bipolar lightning in winter at Maki Japan, J. Geophys. Res., 94, 13191-13195.
6. Berger, K. (1967) : Novel Observations of Lightning Discharge : Results of Research on Mount San Salvatore, J. Franklin Inst., 283,4 83-525.
7. Takeuti, T.(1987) : Phenomena of Lightning Discharge,Nagoya University Shuppankai, Nagoya,119-120. (in Japanese)
8. Miyake K. et al (1989) : The characteristics of winter lightning stroke to the high structures, C.R.I.E.P.I. comprehensive report, No. T10, 67-70. (in Japanese)

3.4 Anomalous Features of Electromagnetic Pulses Caused by Lightning Return Strokes During Winter Thunderstorms in Japan

K. YAMAMOTO[1], Y. AIKI[1], K. MATSUURA[1], Z-I. KAWASAKI[1], H. KINOSHITA[2], and I. ARIMA[3]
Osaka University, Osaka, Japan[1], Osaka Institute of Technology, Osaka, Japan[2], and Gifu University, Gifu, Japan[3]

ABSTRACT Electromagnetic pulses caused by lightning return strokes have been observed during the thunderstorms seasons both in summer and in winter to understand the physics of the lightning discharge. We notice that the signatures of electromagnetic pulses caused by negative lightning return strokes, which we have observed during summer, is not far from the typical signatures reported by other investigators. On the other hand we find that the signatures of electromagnetic pulses caused by the winter positive lightning return strokes are occasionally different from those of the summer negative return strokes. That means, the signatures detected by N-S and E-W magnetic loops are often different from each other. We believe this feature is due to the inclination of the channel of positive lightning return strokes. The video pictures which were recorded simultaneously with magnetic fields confirm our surmise.

1. INTRODUCTION

Multi phases accidents on the main 275-500 kV electric power transmission systems which are passing the Hokuriku Coast along the Sea of Japan often occur due to the lightning strikes during winter thunderstorm seasons [1]. The performance of the accidents is not understandable through the experiences. Since the 500 kV power transmission systems include the nuclear power plant, the understanding of the main factor of the accident and the protection method from the lightning hazard are needed urgently.

The investigation of the lightning physics has been mainly on the lightning during the summer. In addition to the above mentioned background Takeuchi [2] discovered the representative characteristics of the lightning during winter thunderstorms, and the investigation of the winter thunderstorms becomes the current topics among Japanese scientists.

The authors has begun the winter thunderstorm observations from 1982 at so-called Yamanaka Pass which is situated to the south-west region of the Hokuriku District to understand the lightning physics during the winter. For the purpose of the comparison between the summer and the winter, the observation of the summer thunderstorm has been carried out from 1983 in Osaka University campus. Until now we reported many observed data on lightning channels, electric fields, and magnetic fields [3]-[5].

In this paper the new representative features of the magnetic fields is presented.

2. OBSERVATION ITEMS AND METHODS

In this observation main items are taking video pictures of lightning channels and measurements of magnetic field changes, electric field changes and thunder. In this section the observation systems are described.

(a) Magnetic Field Change Measurement

We have measured magnetic field changes due to lightning discharges, mainly due to main strokes, in order to estimate a lightning stroke current. Though Norinder and Dahle [6] presented the apparatus for magnetic field changes, we followed Krider and Noggle to set up our own system [7][8]. That is, we have integrated the output of loop antennas constructed of coaxial cables. Figure 1 shows schematic diagram of measurement systems. The output of differential integrator has been amplified and has recorded by data recorder (SONY, FMR-1219, f:20Hz-1MHz). Our two loop antennas have been installed in the north-south direction and in the east-west direction. These antennas take forms of a square, each side of which is 50 cm, and these faces are vertical and the height from the ground level to the bottom side of the antennas is 1 m. Figure 2 shows relative responses of measuring system. We see from Fig.2(b) that the measuring system faithfully responds in onset time 1 μsec. The time constant of the integrator is 1.4 msec, and we see an attenuation of about 13 % in 200 μsec from Fig.2(a).

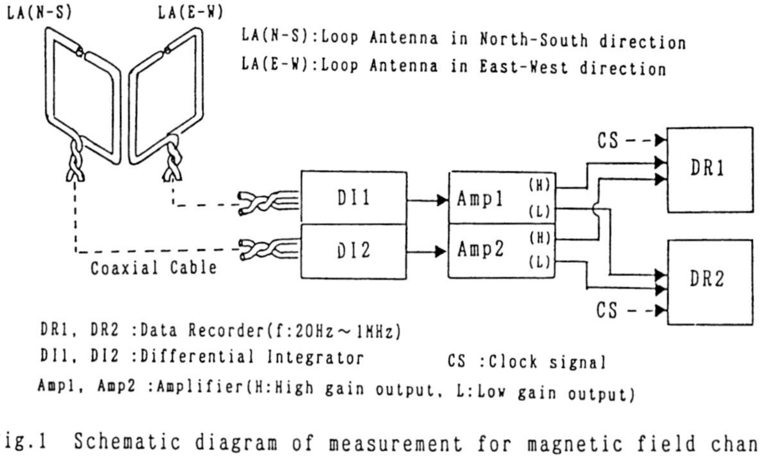

Fig.1 Schematic diagram of measurement for magnetic field change.

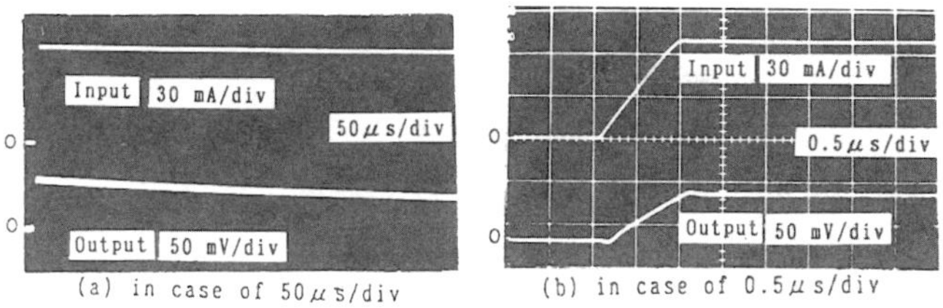

Input :Input current waveform of the other Loop Antenna along N-S Loop Antenna or E-W Loop Antenna

Output :Output of Amplifier

Fig.2. Relative Responses of measuring systems.

(b) Observation Of Lightning Channel

We have used two video cameras (MATSUSHITA WV-1550, WV-341) in order to understand the lightning stroke channels. We can control the direction of video camera to get the various views. It is noticed that each frame of video pictures corresponds to 1/30 sec.

Other items are followings. In electric field change measurement, quite slow changes due to movements of thunderclouds have been measured by a field mill, and electric field changes due to lightning discharges have been measured by two disk antennas (fast antenna, slow antenna). We have measured thunder by three microphones installed at each point of a horizontal equilateral triangle, each side of which is 3.5 m, and the height from the ground level is 1 m.

Since the magnitude of magnetic and electric fields changes and thunder vary widely, we have recorded each item in two channels with different sensitivity. And each recorder are synchronized.

3. OBSERVATION RESULTS

In Fig.3 we show typical magnetic field changes produced by lightning ground strokes (first strokes of negative lightning) in summer storms (at distances less than 10 km). The top trace in each figure was measured by the N-S loop antenna, and the bottom by the E-W loop antenna. Typical magnetic field changes have the rise times; 1-10 μsec, the periods to half of peak; 10-100 μsec, so the period to half of peak is longer than the rise time. These values are similar to that in Norinder and Dahle [6] measured at the same distances. The polarity of the trace by the N-S loop antenna and the polarity by the E-W loop antenna are different in some cases depended on the lightning striking point, however the shapes of the traces are same each other. And the periods to peak of the each traces can be said to be the same.

Figure 4 shows magnetic field changes produced by lightning ground strokes in winter. The traces in this figure except Fig.4(d),(f) have been measured in the lightning strikes to the tops of the steel

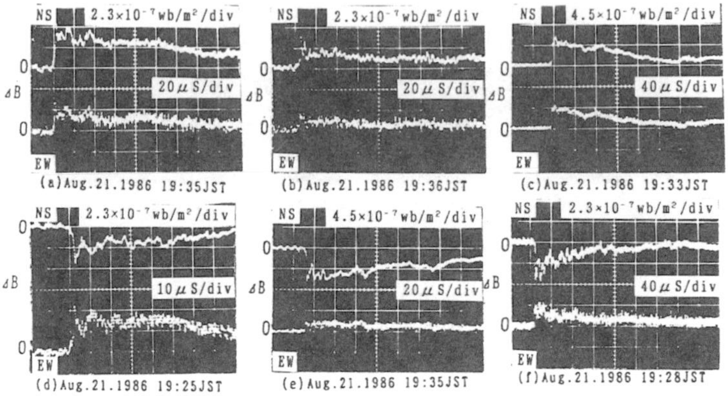

Fig.3. Magnetic field changes produced by lightning ground strokes in summer storms (at distances less than 10 km).

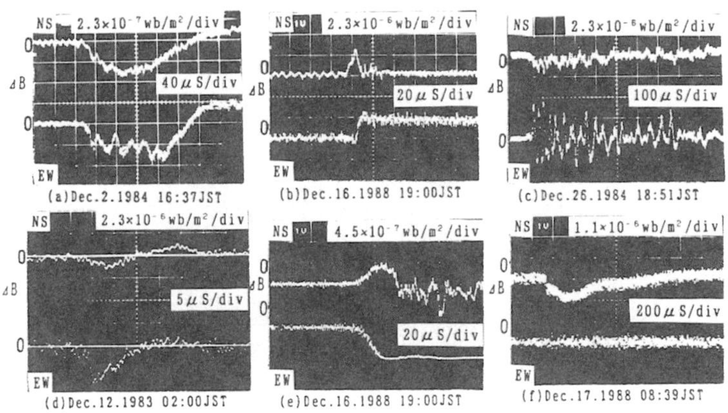

Fig.4. Magnetic field changes produced by lightning ground strokes in winter storms (at distances less than 3 km).

towers, the heights of which from the ground level are 50-70 m. From Fig.4 it is found that the magnetic field changes caused by winter lightning are complicated in comparison with that caused by summer lightning. In particular the magnetic field change by the N-S loop antenna is different from the change by the E-W loop antenna. For example, in the case of Fig.4(b) that was produced by the main stroke to the tower showed in Fig.8(c), the magnetic field change by the E-W loop antenna had a quite longer duration of wave tail than a duration of wave front (the duration of wave front; 8 μsec, the duration of wave tail; more than 100 μsec), but the change by the N-S loop antenna had very short tail (the duration of wave front; 10 μsec, the duration of wave tail; 8 μsec). And we see that a time lag of the each peak time in the traces measured by the N-S loop antenna and the E-W loop antenna was 8 μsec. These complicated magnetic field changes have been scarcely measured in the summer lightning.

We have measured the magnetic field change that had continuous pulse changes superimposed on the main change for about 4 msec (Fig.5(a)(i)) in the winter observation on 1988. There is the continuous pulse changes at Ps part in the figure. Figure 5(b) shows the electric field change measured by the fast antenna at that time. It is considered that this change was due to the positive lightning ground stroke from a sharp negative change (at R part in the figure) that is probably due to the main stroke, the lightning channel in the video picture and the lightning channel reconstructed from thunder [5]. Most of these pulses have time intervals of each pulses; 20-140 μsec, rise times; 2-20 μsec, durations of each pulses; 10-100 μsec. It is found that some of the pulses in the figure are similar to a magnetic field change due to lightning ground stroke in summer. We present an example in Fig.5(a)(ii) that was extended at the point of the arrow on Fig.5(a)(i). We can see these continuous pulses in magnetic field changes, which are considered being due to intracloud discharges.

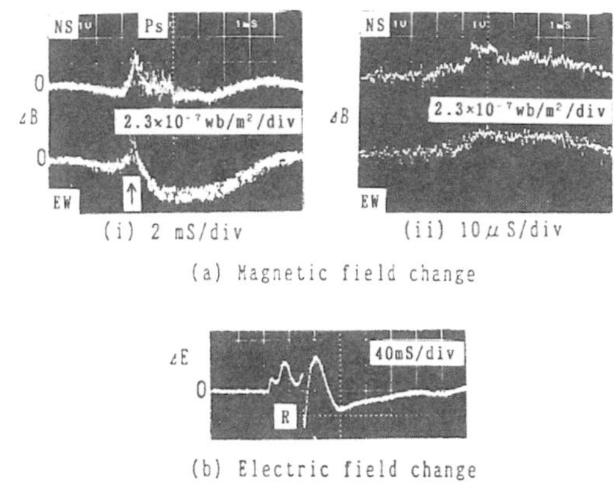

Fig.5. Magnetic field change with successive pulses and electric field change in winter lightning (at 20:20JST on Dec.16.1988).

4. DISCUSSIONS

The discrepancy of signatures between the N-S and E-W magnetic field components, which is mentioned in previous sections, is supposed to depend on mainly the complex configuration or the inclination of the return stroke channels. It is clear that if the lightning channel changes its progression direction at some finite height above the ground level, the incident angle of the radiated magnetic fields to the loop antennas varies extremely after the changing of progression direction. Though the ratio of the N-S and E-W magnetic field components basically depends on the azimuth, the azimuth itself varies according to the progression of lightning current in the above mentioned situation.

Fig.6. Video camera photographs of lightning ground strokes in summer storms.

To confirm this surmise, the statistics for the striking angles of lightning channels are obtained under processing of the video pictures which were recorded simultaneously with E and B fields. In Fig.6 typical lightning channels' pictures during summer are shown. In Fig.7 the distributions of two dimensional striking angles of lightning channels are shown, and these are obtained from video pictures. The definition of striking angles is show in Fig.7. From Fig.6 and Fig.7 it is noticed that the most probable striking angle is less than 20 degrees. Moreover it is noticed that the channel configuration is rather straight like. On the other hand during the winter we often observed horizontally progression channels with quite short vertical channels as shown in Fig.8. For example Fig.8(a) and (b) show the several horizontally progression branches at lower than 200 meters above the ground level and 70 degrees inclination of the channel. These are more complex configuration than those of during the summer.

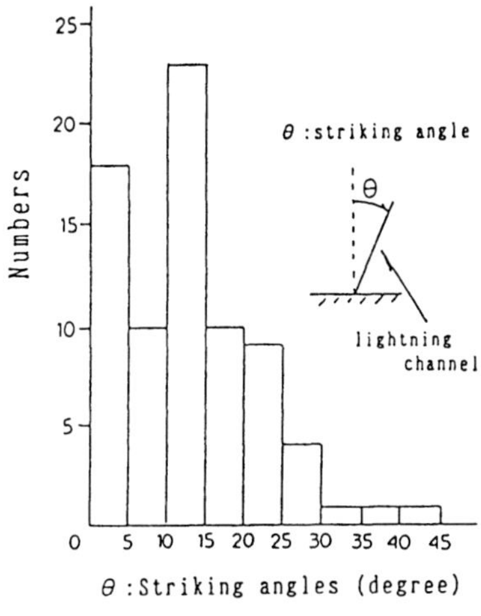

Fig.7 Frequency distribution of striking angles in summer lightnings

During the summer we observed only one lightning channel which inclined 70 degrees at 130 meters above the ground level. The polarity of this flash was confirmed to be positive by the electric field change as shown in Fig.9. In the same Figure magnetic fields (a), the video picture and its sketch (c) are also shown.

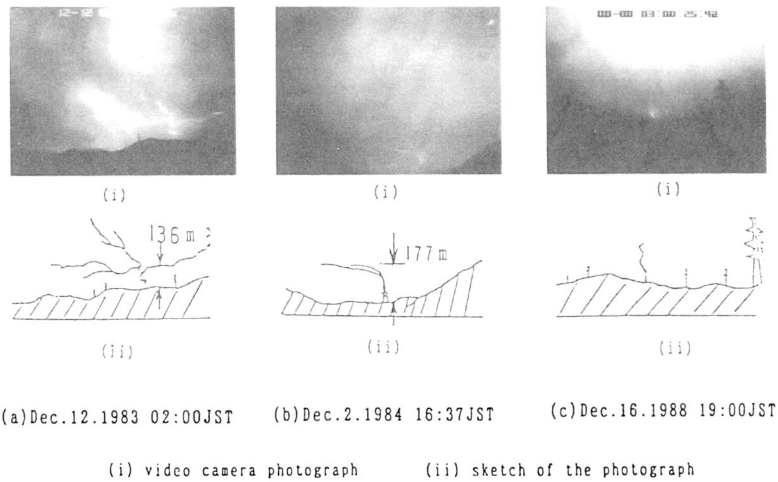

Fig.8. Video camera photographs and their sketches of strikes to tower tops in winter lightning.

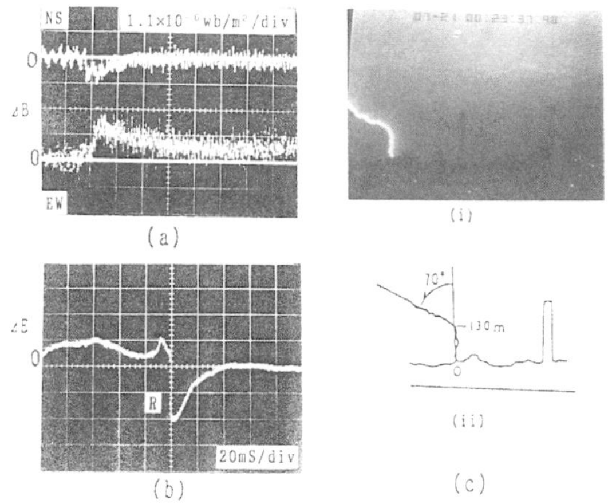

Fig.9. Magnetic field change (a), electric field change (b) and lightning stroke channel (c) of positive ground flash in summer storms (at a distance of 1.8 km, at 00:23JST on Jul.21.1984).

It is noticed that the rise times of N-S and E-W components of the magnetic fields are not the same. The difference of the rise times is about 20 μsec.

Under the above mentioned situation, low altitude inclination influences the waveforms and their amplitudes. That means the conventional estimation of the lightning current using E or B fields is not reliable. It concludes that the channel configuration should be considered.

5. CONCLUSIONS

The magnetic fields changes observed during the winter showed the more complex signatures than those of during the summer. One of the examples of this meaning is that the cross-correlation of the wave forms between N-S and E-W loop antennas was sometimes very low. We conclude that this feature depends on the lightning channel configurations. During the winter return stroke channels often changed their progression directions at the quite low height above the ground level, such as less than 200 meters. Moreover they had many branches. Though the estimation of the lightning currents is necessary for the the protection of the power transmission, the conventional current amplitude estimation by the observed fields is not accurate during the winter. As the conclusions for this purpose the influence of the configuration of the return stroke channels should be taken into account.

REFERENCES

1. The Committee for Investigating Characteristic of Lightning of IEE JAPAN (1988), The latest trend and problems on lightning research, Technical report of IEE JAPAN, (II)278, 1-78.

2. T.Takeuchi (1978), Research of the lightning during winter thunderstorms in the Hokuriku Coast, Journal of IEE JAPAN, Vol.98-12, 1156-1159.

3. H.Kinoshita et al.(1984), Lightning observation in winter and summer, Transaction of IEE JAPAN, HV-84-36, 51-59.

4. H.Kinoshita et al.(1985), Lightning observation in winter and summer (II), Transaction of IEE JAPAN, HV-85-15,55-64.

5. K.Yamamoto et al.(1989), Lightning observation in winter and summer (III), Transaction of IEE JAPAN, HV-89-51, 39-47.

6. Norinder and Dahle (1945), Measurements by frame aerials of current variations in lightning discharges, Ark.Mat.Ast.Fys. Bd32A, 1-70.

7. Krider and Noggle (1975), Broadband antenna systems for lightning magnetic fields, Journal of Applied Meteorology, Vol.14, 252-256.

8. M.Ishii et al.(1983), Transaction of IEE JAPAN, 986, 1246.

3.5 VHF/UHF Radiation from Positive and Negative Lightning

M. NAKANO, Z-I. KAWASAKI, T. TAKEUTI, T. SIRAI, and Y. KAWAMATA

Research Institute of Atmospherics, Nagoya University, Aichi, 442 Japan

ABSTRACTS

Measurements of VHF/UHF radiation from positive and negative lightning flashes were made for close lightning in summer and winter together with the electric field change measurements. Strong radiations are associated with return strokes of a ground flash in summer, but in winter more than 50% of return strokes are not followed by detectable VHF/UHF radiation. The radiation intensity from positive return strokes in winter is several times as small as that from negative ones in summer. The time delay of VHF/UHF bursts from the onset of the electric field change due to a return stroke is found to be related to the distance to the flash. In winter, the radiation from other processes than the return stroke in the negative ground flash is more intensive than that in the positive ground flash.

1. INTRODUCTION

VHF/UHF radiation from lightning discharges is strongly related to the fine structures of breakdown processes in the clouds. Measurements of VHF/UHF radiation are usefull for the study of micro discharge processes in clouds, because optical outputs from such processes are not detectable on the ground. The source location of VHF/UHF pulses generated in the clouds is also usefull for the prediction of the strike point on the ground. VHF/UHF techniques of the source location have been developed by several investigators ([1],[2], [3],[4],[5]). The characteristics of VHF/UHF radiation have been also given in many literatures ([6],[7],[8],[9],[10]). All investigations have been made for lightning only in summer thunderstorms. In winter season, many lightning flashes have been observed in the Hokuriku coast, Japan, and winter lightning shows some peculiar features ([11],[12],[13]). For example, the occurrence of a positive ground flash is very high in winter thunderstorms.

The objectives of the present study are to measure RF radiation from the positive ground flash in winter thunderstorms, and to compare the natures of RF radiation with those which has been given in the literatures for the negative ground flash in summer. The comparison of the radiation from the negative ground flash in winter with that in summer is also interesting.

2. MEASUREMENTS

Measuring system is shown in Figure 1. We selected the frequencies of 500 MHz in UHF band and 50 MHz in VHF. The bandwidth of the receiver is 300 kHz for both freqencies. The electric field change due

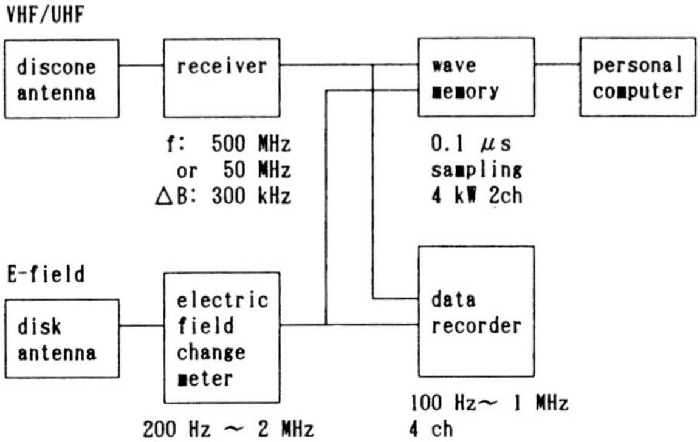

Fig. 1. Schematic diagram of measuring system.

to lightning was also measured simultaneously by an electric field meter referred to as a fast antenna. The data were stored on the two channels digital wave memory of 4kW for each channel, and transferred to the floppy disk. The sampling period of the wave memory was choosen to be 0.1μs in the present study. The data were also recorded with four channels data recorder which has the frequency response from 100 Hz to 200 kHz.

The measurements have been made in the summer and in the winter of 1988. Typical examples of the electric field change and UHF radiation in the return stroke phase are shown in Figure 2. Upper traces show the electric field changes and lower traces UHF radiation in arbitrary scale respectively. Figure 2 (a) shows the negative return stroke in summer, and shows the step pulses in E-field change due to stepped leaders before large rapid change corresponding to the return stroke. In this example, UHF radiations corresponding to leader step pulses were not detected, and strong radiation was detected several microseconds after the onset of the return stroke field change. Figure 2 (b) and (c) show negative and positive return strokes in winter lightning respectively. In these two examples, UHF radiations associated with the return stroke like Figure 2 (a) were not detected. Many return strokes of ground flashes in winter are not followed by detectable UHF radiation bursts.

3. DELAY OF VHF/UHF BURSTS FROM THE ONSET OF RERURN STROKE

The VHF/UHF radiation bursts are delayed from the onset of the rapid electric field change caused by a return stroke. Figure 3 shows the occurrence percentage of the delay time of VHF/UHF bursts in summer ground flashes. The mean values are 17μs in UHF range and 27μs in VHF respectively. The distances to the flashes are estimated within 25km for UHF data and within 40km for VHF data, which are deduced by the lightning location system. Figure 4 shows the relationship between the delay time and the distance to the flash. The delay times of the RF bursts are seemed to be correlated with the distance except a few extreme points both in UHF and in VHF ranges. The delay time is roughly proportional to the distance to the flash, and the regression lines are shown in Figure 4. These results are consistent with the idea in which radiation sources are distributed along the return stroke channel and the effect of finite conducting earth on the propagation is significant for the RF amplitude on the ground ([14], [15]).

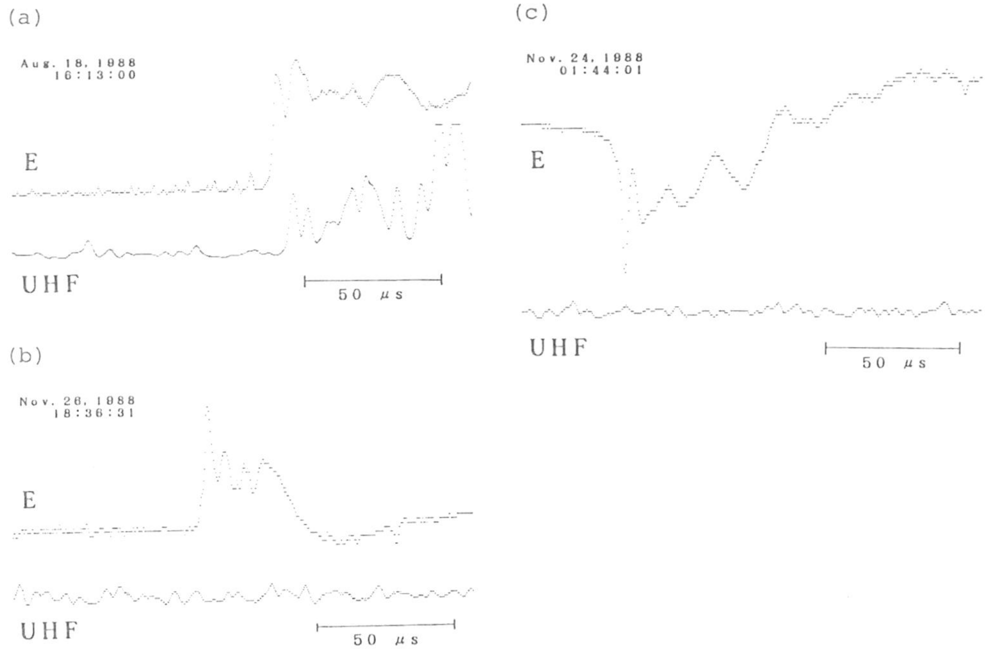

Fig. 2. Typical examples of the electric field changes (upper trace) and UHF radiation (lower trace) in the return stroke phase.
(a): negative return stroke in summer
(b): negative return stroke in winter
(c): positive return stroke in winter

4. PEAK AMPLITUDE OF VHF/UHF BURSTS

As shown in Figure 2, many lightning return strokes during winter thunderstorms are not followed by the VHF/UHF radiation bursts. The occurrences of return stroke-associated VHF/UHF bursts are summarized in Table 1. In summer thunderstorms, the occurrence frequency of return strokes which are not followed by RF bursts is only a few percent. On the other hand, in winter, the occurrence frequency exceeds 50 % both in negative and in positive return strokes. VHF/UHF radiations associated with the return stroke in summer ground flashes are very intensive, and the average intensity is 12 μV/m/kHz at 500 MHz in 25 km range and 140 μv/m/kHz at 50 MHz in 10 to 45 km range. These values are almost the same as those given in the literatures.

Next, the amplitude of UHF bursts associated with the return stroke in winter is compared with that in summer. In winter thunderstorms, the distances to flashes were not observed. Thus, in order to eliminate the effect of the distance on the amplitude, the ratio of the peak amplitude of UHF bursts (U) and the amplitude of electric field change (E) is calculated. The peak amplitude of the electric field change (E) due to a return stroke is approximately proportional to

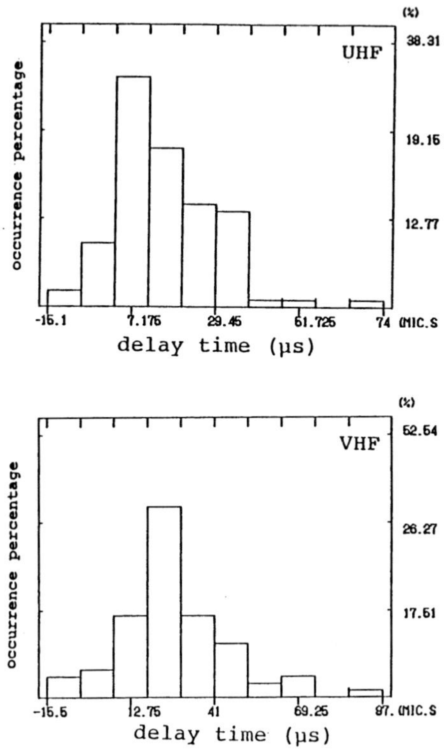

Fig. 3. Histgrams of the delay time of VHF/UHF bursts.

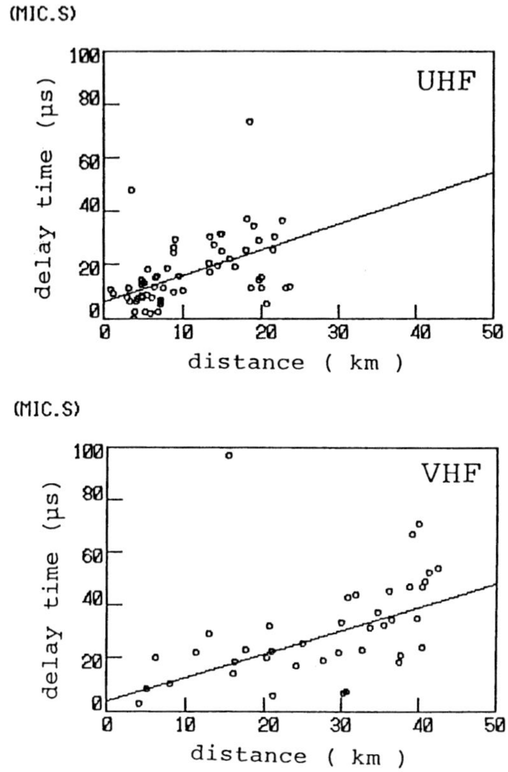

Fig. 4. The relationship between the delay time of VHF/UHF and the distance to the flash.

Table 1. The occurrence of return stroke-associated VHF/UHF radiation.

	Polarity	Return stroke-associated VHF/UHF radiation Yes	No
Summer	-	102	4
	+	5	0
Winter	-	18	26
	+	4	6

the peak current of a return stroke (I) and inversely proportional to the distance to the flash (D) for the distant flash, that is, $E \propto I/D$. Assuming that the peak amplitude of UHF burst (U) is proportional to some function of the peak current (F(I)) and inversely proportional to the distance (D), that is, $U \propto F(I)/D$, the ratio, U/E, depends only on the source current, that is, $U/E \propto F(I)/I$. The ratio is calculated for the data in winter and summer thunderstorms.

The results are given in Table 2. Because of the lack of good data for negative return stroke field change in winter, the comparison is made between negative return strokes in summer and positive return strokes in winter. Table 2 shows that UHF bursts associated with the positive return stroke in winter is several times as small as that associated with the negative return stroke in summer.

Table 2. The amplitude ratio of the return stroke field change to the UHF burst.

	Summer (negative)	Winter (positive)
U/E	3.5 ± 2.1	0.6 ± 0.1

5. DISCUSSION

In winter ground flashes, more than 50 % positive return strokes are not followed by detectable VHF/UHF radiations. The intensity of return stroke-associated UHF bursts in the positive ground flash in winter is several times smaller than that in the negative ground flash in summer, as given in Table 2. VHF/UHF radiations from other processes than the return stroke in winter thunderstorms are also analysed in several typical ground flashes. Examples are shown in Figure 5. Upper trace and lower trace in Figure 5 (a),(b) are UHF and the electric field change respectively. Large changes marked by R in the electric field changes correspond to return strokes. The radiations from preliminary breakdown, leader strokes and interstroke processes in negative ground flashes are more intensive than that in positive ones. The positive streamer in clouds is seemed to generate weak RF radiation because of continuous developments of streamers. Quantitative analysis is remained as further study. Although the statistical data of RF radiation associated with negative return strokes in winter were not obtained in the present study, the difference given in Table 2 may be due to the polarity of a ground flash.

The source locations of RF radiation pulses from a ground flash are not measured in the present study. In order to investigate the nature of RF radiation from a positive ground flash, the use of a location system for UHF radiation in clouds is very helpful. It is also helpful for understanding the nature of the big return stroke current and the large continuing current in a positive ground flash in winter which sometimes cause very serious damages in the electrical power transmission line and the telecommunications systems.

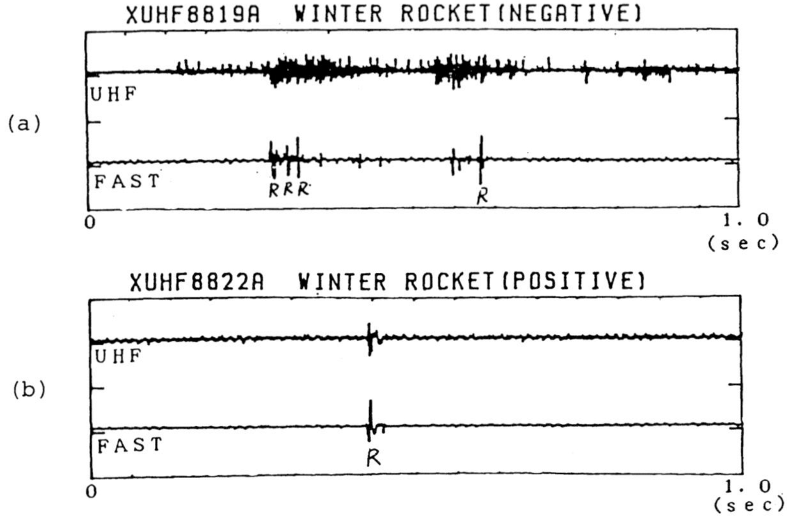

Fig. 5. Examples of UHF radiation (upper trace) and the electric field change (lower trace) during discharges.
(a): negative flash, (b): positive flash

Acknowledgement
The authors would like to thank M. Nagatani and H. Nakada of Nagoya University for their help in the instrumentaions and the observations, and K. Saikawa and M. Suzuki of Kansai Electric Power Co., Inc. for their support to the observations. The authors are grateful to K. Horii of Nagoya University for his valuable discussions.

REFERENCES

1. Proctor D.E. (1971) A hyperbolic system for obtaining VHF radio pictures of lightning, J. Geophys. Res., 76, 1478-1489.

2. Warwick J.W., Hayenga C.O., Bronsnahan J.W. (1979) Interferometric directions of lightning sources at 34 MHz, J. Geophys. Res., 84, C5, 2457-2468.

3. Rustan P.L., Uman M.A., Childers D.G. (1980) Lightning source locations from VHF radiation data for a flash at Kennedy Space Center, J. Geophys. Res., 85, C9, 4893-4903.

4. Hayenga C.O., Warwick J.W. (1981) Two-dimentional interferometric positions of VHF lightning sources, J. Geophys. Res., 86, C8, 7451-7462.

5. Richard P., Auffray G. (1985) VHF-UHF interferometric measurements, applications to lightning discharge mapping, Radio Science, 20, 171-192.

6. Brook M., Kitagwa N. (1964) Radiation from lightning discharges in the frequency range 400 to 1000 Mc/s, J. Geophys. Res., 69, 2431-2434.

7. Takagi M. (1969) VHF radiation from ground discharges, Proc. Res. Inst. Atmos. Nagoya Univ., 16, 163-168.

8. Hayenga C.O. (1984) characteristics of lightning VHF radiation near the time of return strokes, J. Geophys. Res., 89, D1, 1403-1410.

9. Richard P., Delannoy A., Labaune G., Laroche P. (1986) Results of spatial and temporal characterization of the VHF-UHF radiation of lightning, J. Geophys. Res., 91, D1, 1248-1260.

10. Proctor D.E., Uytenbogaardt R., Meredith B.M. (1988) VHF radio pictures of lightning flashes to ground, J. Geophys. Res., 93, D10, 12683-12727.

11. Takeuti T., Nakano M., Yamamoto Y. (1976) Remarkable characteristics of cloud-to-ground discharges observed in winter thunderstorms in Hokuriku area, Japan, J. Meteor. Soc. Japan, 54, 436-440.

12. Nakano M. (1979) The cloud discharge in winter thunderstorms of the Hokuriku coast, J. Meteor. Soc. Japan, 57, 444-451.

13. Brook M., Nakano M., Krehbiel P., Takeuti T. (1982) The electrical structure of the Hokuriku winter thunderstorms, J. Geophys. Res., 87, C2, 1207-1215.

14. Le Vine D.M., Gesell L. Kao M. (1986) Radiation from lightning return strokes over a finitely conducting earth, J. Geophys. Res., 91, D11, 11897-11908.

15. Cooray V. (1986) Temporal behaviour of lightning HF radiation at 3 MHz near the time of first return strokes, J. Atmos. Terr. Phys., 48, 73-78.

3.6 Artificially Triggered Lightning Experiments for Winter Thunderclouds

K. NAKAMURA[1] and K. HORII[2]

Nagoya Institute of Technology, Nagoya, 466 Japan[1] and Nagoya University, Nagoya, 460 Japan[2]

Abstract - The artificially triggered lightning experiments by rocket has been carried out in Japan in winter. Totally eighty four strikes have been obtained during ten seasons from 1977 to 1987 except 1984. In this paper the triggering methods are classified into four types and the results by those methods with the currents and the photographs in the triggered lightning experiments are discussed.

INTRODUCTION

The triggered lightning discharge is one of the artificial lightning discharge occurred along a thin steel wire carried by a small rocket from the ground. The bottom altitude of the winter thunderclouds is about 1000 m at most, then the triggering discharge happens at the height of 100 to 200 m above the ground.

In the early period of the experiment the steel wire was directly grounded, and then the triggered discharge was supposed to be similar to the natural one which begins with the upward leader from a tall tower. Due to the current measurements and the optical observations with a high speed framing camera and a streak camera, the differences were made clear between the triggered and the natural lightning.

When a series impedance is connected to the wire, an abnormal discharge frequently occurred, where the main discharge was not along the vaporized wire, but through the free air apart from the wire. In this discharge was observed the stepped leader propagated downward to the ground. Such a experiment with the abnormal discharge were applied to investigate the lightning protection effect with the ground wire of the transmission line. Ten lightning rods with 10 m height were distributed around the site and tried to get strikes on the rod. However, there has been no strike on the rod. There should be taken into consideration the both effects of the electrostatic shielding phenomena by the grounded steel wire and of the space charge emitted from the wire before the leader process. Then another new method was applied.

The new method was that the steel wire was insulated with a nylon line highly from the ground . The lightning occurred on the transmission line and tower. The downward stepped leader initiated from the bottom of the wire and propagated to the tower.

Seventy one lightning discharges have been obtained during eight seasons from 1977 to 1985 except 1984, carried out at the reclaimed plane land of Kahokugata, zero meter above sea level. Thirty nine lightning discharges have been obtained during the three seasons from 1986 to 1989 in Okushishiku mountain, where the altitude is about 1000m. There was equipped a test EHV transmission line.

This paper describes the summaries for eighty four triggered lightning obtained by the end of 1987. The contents are as follows, (1) triggering methods and the corresponding discharge aspects, (2) current and luminous phenomena, (3) lightning discharge test for some equipments.

TRIGGERING METHODS AND THE CORRESPONDING DISCHARGE ASPECTS

Triggering Methods

Four triggering methods have been employed as shown in Fig. 1. The No. 1 method was that the wire was grounded with almost zero impedance. In the No. 2 method a impedance circuit with a resistance and a reactance was inserted. The values of the resistance and the reactance were several kilo-ohm and the several milli-henry, respectively. In this case an abnormal lightning discharge often occurred. The main discharge took a different path apart from the vaporized channel of the steel wire. A short or a long gap were arranged between the bottom end of steel wire and the ground in the No. 3 method. Some test equipments were

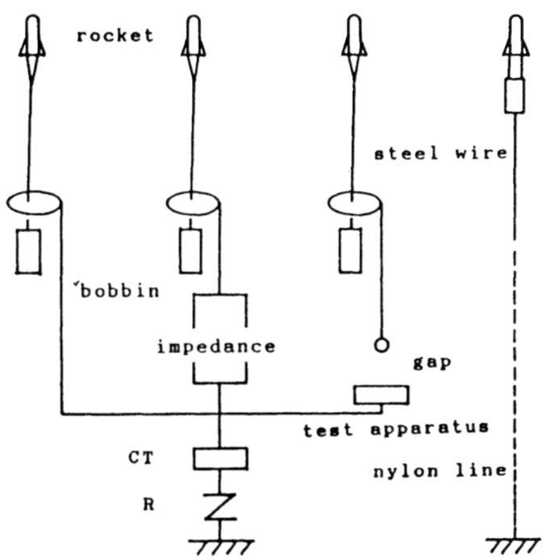

Fig. 1 Triggering methods

placed on the ground. In the No. 4 the wire was insulated from the ground with a nylon line of about 100 m distance. The lightning strikes occurred on the transmission line and tower.

Table 1 shows the results and the classifications of the eighty four triggered lightning. In the zero impedance method thirty six lightning were obtained. The success ratio was almost equal with the both polarities of cloud. One abnormal discharge was obtained only for the negative cloud. In the case of the impedance grounding eleven abnormal lightning occurred only with the negative clouds. The lightning of the No. 3 method were grouped in the category of the normal case. No. 4 were grouped in the abnormal case because with the similar downward stepped leader.

Table 1 Content of triggered lightning

kinds of lightning	polarity of cloud	number of success: zero impedance grounding	impedance grounding	gap	power line
normal	positive	17	2	7	-
	negative	18	15	4	-
abnormal	positive	0	0	0	4
	negative	1	11	0	5

Fig. 2 Triggered Lightning with zero impedance

Totally eleven striking test were tried for various equipments. The tests are described in the later.

Nine strikes have been obtained with a test transmission line. The five occurred at the tower and one separately occurred with a conductor, ground wire, rocket launcher and directly the ground, respectively.

Triggered Lightning with Zero Impedance Grounding

The resistance of the wire is 13.3 ohms/m (20°C), and the reactance is estimated to be 2-3 uH/m considering the image effect for the ground. The withstand voltage of the wire coated with a thin insulation material is about 15 kV DC. Contacting with a current collector ring, the wire was drawn out of a bobbin by rocket.

Figure two shows a still photograph of the triggered lightning of No. 80-21 for the negative charged cloud and simultaneous measurements of the predischarge and the main discharge current, the luminosity of infra-red and ultra-violet, and the streak photograph are shown in Fig. 3. The luminosity were measured by a photo-transistor and an ultra-violet sensor (UV-tron, Hamamatu Photonics Inc.), respectively. The figure 3-a shows the corona discharge current. The current increases with ascend of rocket and some pulses appear when the corona discharge current comes to around five milli-amperes. Four pulses of b, c and d are seen before triggering. Those pulses do not imply the net value because of saturation of the sensor. The three pulses of a, b and c correspond to those of a', b' and c' shown in Fig. 3-b, respectively. According to the figure those pulse are in the order of ten through hundred amperes. The charge quantity of the early pulse current is in the order of microcoulombs as usual. The pulses frequently appeared and gradually increased to one through five hundred millicoulombs before the triggering.

The main discharge current is shown in Fig. 3-c. The current continues about four hundred milliseconds. There are seen many pulses within. Such pulses were often seen in the negative discharge current.

As shown in the streak photograph of Fig. 3-f, the leader process continued more than 20 ms and the main discharge occurred 1 ms after the wire vaporization. The leader, however, was going on to the clouds. The velocity of this positive leader was in the order of 10^5 m/s. In case of the negative upward leader the velocity was in the order of 10^5 to 10^6 m/s.

As indicated in Fig. 3-d and 3-e, infra-red and ultra-violet lays were observed during the main discharge current. In another discharge were observed intermittently the ultra-violet lay during the leader process.

Triggered Lighting with Impedance Grounding

As shown in the table 1, the abnormal lightning often occurred on the negative cloud. The upward leader appeared to propagate to the cloud as well in the zero impedance grounding. After the wire vaporization the downward stepped leader began to propagate from the vaporized channel to the ground.

In order to trigger the lightning strike to a distribution line or a lightning rod, a test line with 500 m distance and 5 m height was equipped and ten lightning rods with 10 m height were distributed. Among the three abnormal lightning the one occurred on the line conductor and the two were on the ground, 9 m and 16 m distant apart from the nearest rods, respectively. These lightning occurred within the area of 60-degree protection angle.

Fig. 4 shows a typical example of the abnormal lightning struck on the ground. Fig. 4-b is a hole on the snow ground. Fig. 5 indicates a streak

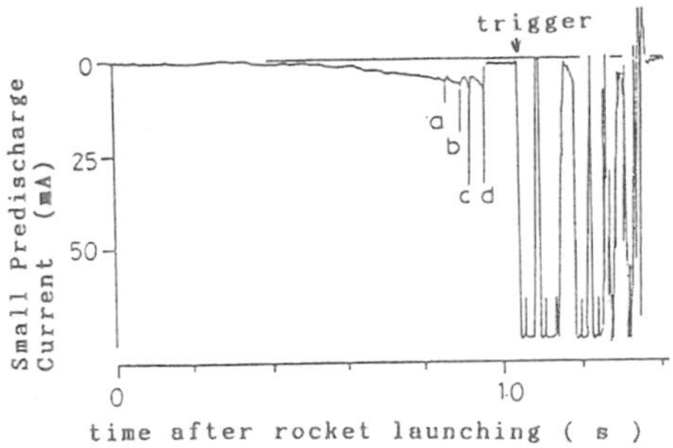

(a) small predischarge current

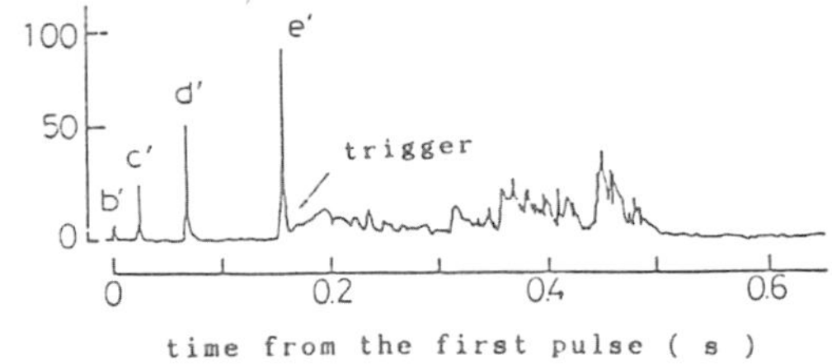

(b) pulse predischarge current

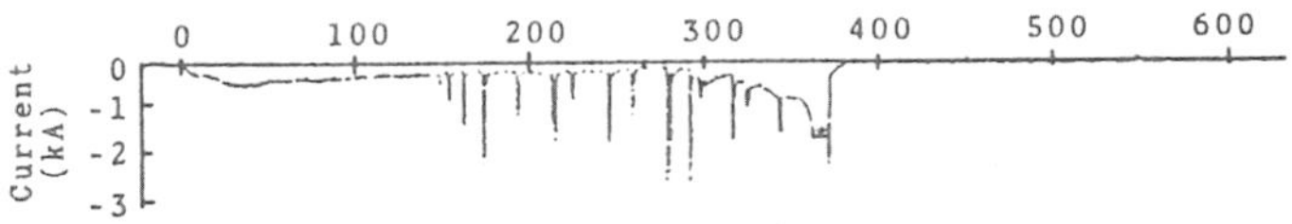

(c) main discharge current

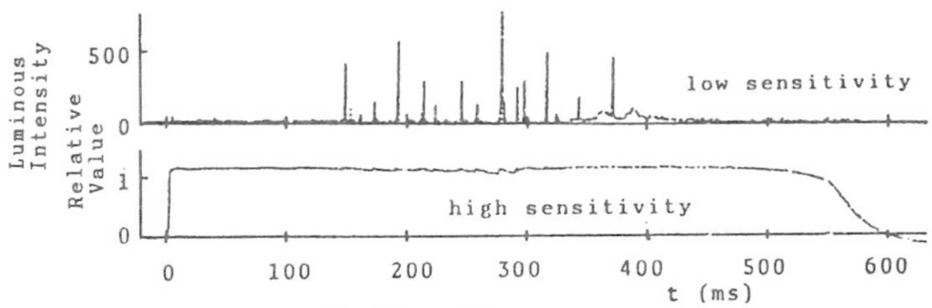

(d) infra-red lays

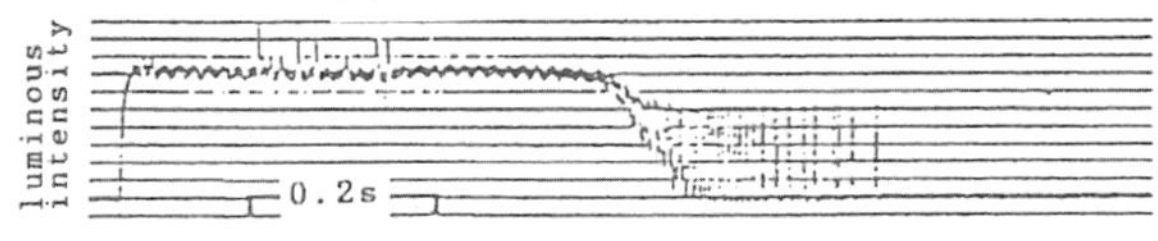

(e) ultra-violet lays

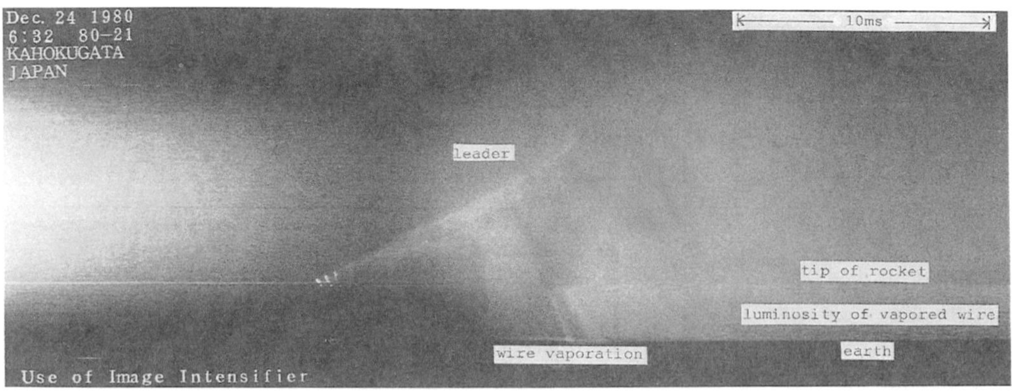

(f) streak photograph

Fig. 3 Simultaneous measurements of current, luminous change and streak photograph

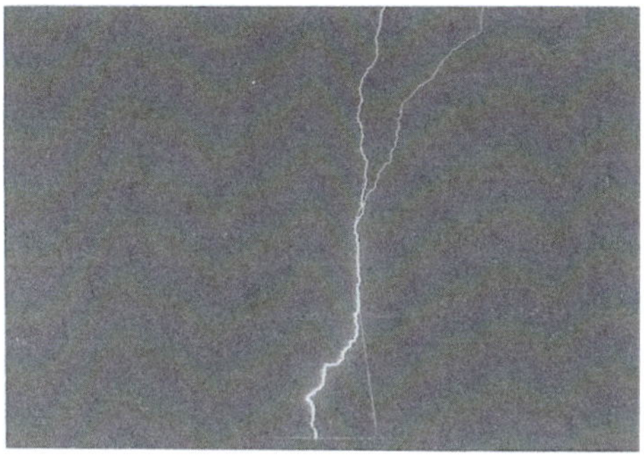

(a) abnormal lightning

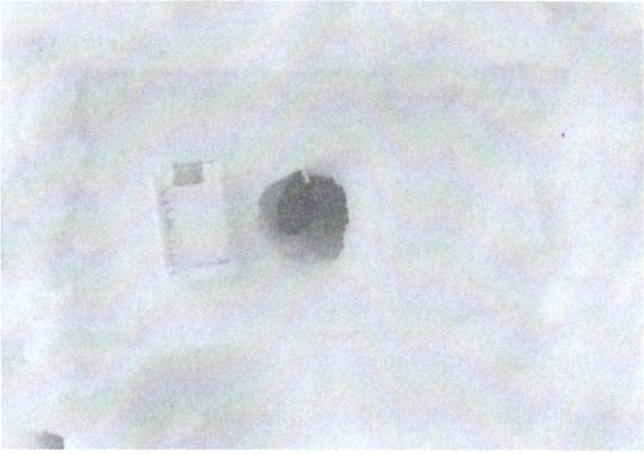

(b) a hole by lightning discharge

Fig. 4 Abnormal lightning

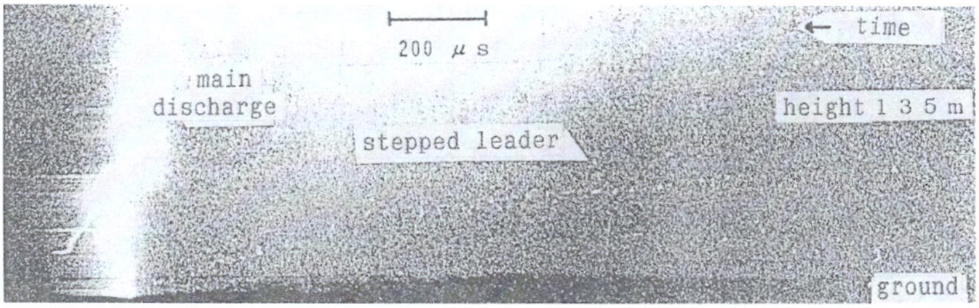

Fig. 5 Downward stepped leader in the abnormal lightning

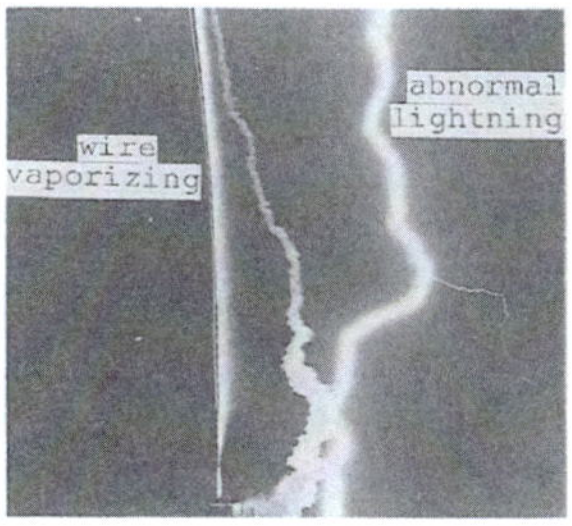

(a) still photograph

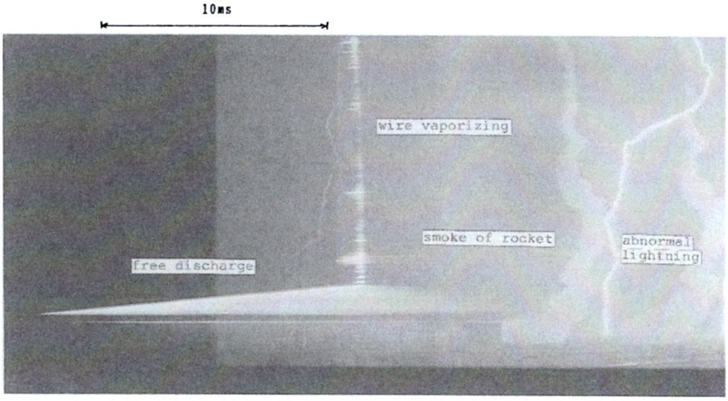

(b) streak photograph

Fig. 6 Discharge aspects between the wire end and the current collector

photograph just before the downward stepped leader started from the height of 135 m and propagated to the ground, and the number of steps counted forty three, the average step length was 3 m, and the velocity was in the order of 10^5 m/s. Fig. 6 indicate a discharge aspect between the wire bottom and the tip of current collector during the leader process. The distance of the discharge was about 3 m.

Triggered Lightning with Nylon Line

Fig. 7-a and 7-b show a triggered lightning struck on the top of the tower and the discharge current measured, respectively. The heights of the bottom of the wire and the top of the tower were 100 m and 60 m, respectively. The measured current peak was +18 kA and 53 C. Fig. 8 shows a steal photograph (Fig. 8-a) and a streak photograph (Fig. 8-b). In the streak photograph first the downward stepped leader begins from the bottom of the steel wire and propagated downward to the tower. After vaporization of the wire the upward leader went up to the cloud from the tower. By means of this type of the experiment it is available to discuss the striking distance.

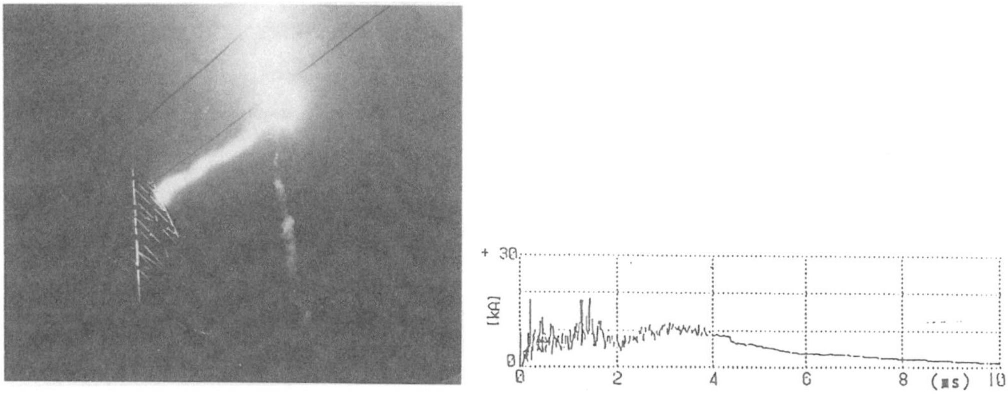

(a) still photograph (b) discharge current

Fig. 7 Triggered lightning struck to the top of tower

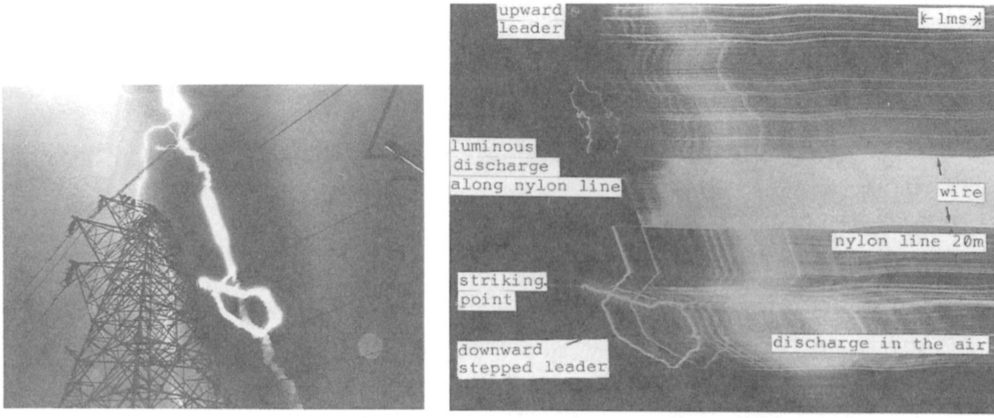

(a) still photograph (b) streak photograph

Fig. 8 Triggered lightning struck to the arm end of tower

Lightning Discharge Test for Equipment

The triggered lightning experiment is available to make discharge test to check the withstand characteristics of equipment. Lightning discharge was loaded on the test equipment through a gap. A 5 cm gap was take for a ground wire, 10 through 20 cm gaps for a surge arrester, and 6 through 8 m gaps for a automobile. Fig.9 shows an example for an automobile. The lightning struck the antenna tip of the car and the current flew through the body metal and the tire surface into the ground. A live rabbit has been put in the cabin. There has been found no damage with the rabbit.

Fig. 9 Lightning discharge test to an automobile

CUMULATIVE DISTRIBUTION OF MAIN DISCHARGE CURRENT AND LUMINOSITY DURATION

The cumulative distribution of the main discharge current with seventy three data is shown in Fig. 8. The distribution is divided into three groups of smaller, middle and large current. The average value is about 10 kA, which is comparatively one third smaller than that of the natural case. The distribution of luminous duration of the lightning flash with fifty three data is indicated in Fig. 9. These durations were obtained by the streak and the high speed framing photographs, respectively. The average time was about 200 ms. The distribution has two groups of short and long times. No correlation can be found between the current and the luminous duration.

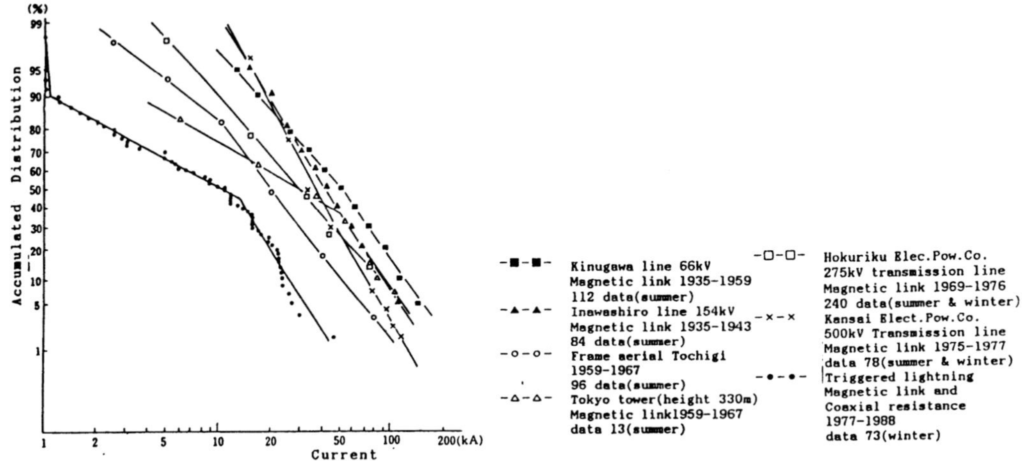

Fig. 10 Accumulated distribution of the triggered lightning currents

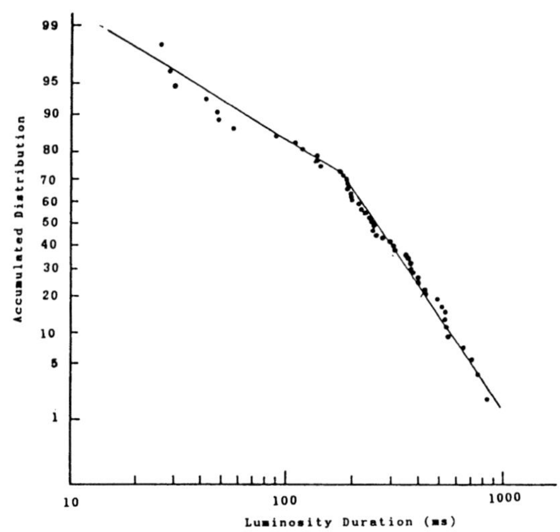

Fig. 11 Accumulated distribution of the luminosity durations

CONCLUSIONS

Totally eighty four triggered lightning have been obtained by the end of 1987. The main results of the lightning experiment obtained for winter thunderclouds with four triggering methods are as follows.

(1) In the zero impedance grounding thirty six lightning were obtained. The rate of success was almost equal for the both of positive and negative clouds. The predischarge process continued for one through two seconds before the main

lightning discharge.

(2) The abnormal lightning often occurred with the impedance grounding and for the negative clouds.

(3) In the experiment using a long nylon line seven lightning strikes occurred on the test transmission line and tower. This triggering method is quite available to get the strike on the transmission line and to discuss the striking distance.

(4) The lightning discharge test could be carried out for the various equipments by arranging a gap.

(5) The average values of the discharge current and of the luminosity duration were 10 kA and 200 ms, respectively.

ACKNOWLEDGMENTS

The authors would like to express their thanks to Prof. Y.Kito (Nagoya University), Prof. G.Ikeda, Dr. S.Sumi (Chubu University), Dr. M.Yoda (Aichi Inst. of Technology), Prof. S.Aiba (Kanazawa Inst. of Technology), Mr. H.Skurano (Ishikawa Tech. College), Mr. K.Wakamatsu (Numazu Tech. College) and Hokuriku Electric Power Company, Chubu Electric Power Company, and others in the triggered lightning experiments for their cooperations. This work was supported by the Grant in Aid for Scientific Research from the Ministry of Education of Japan.

REFERENCES

1. K.Horii (1986) Experiment of triggered lightning discharge by rocket. Korea-Japan Joint Sympo. on Elect. Material and Discharge 8-3

2. K.Horii. G.Ikeda (1985) A consideration on success conditions of triggered lightning. 18th ICLP 1-3

3. K.Horii. H.Sakurano (1985) Observation on final jump of the discharge in the experiment of artificially triggered lightning. IEEE trans. on Power Apparatus and Systems. PAS-104. 10. 2910-2917

4. Y.Kito. K.Horii. Y.Higashiyama. K.Nakamura (1985) Optical aspects of winter lightning discharge triggered by the rocket-wire technique in Hokuriku district of Japan. J. of Geophy. Res. 90. D6. 10647-10680

5. K.Nakamura. H.Sakurano. S.Aiba. K.Horii (1986) Predischarge Current and its effect on the electric field in the lightning triggered by a rocket. J. of Inst. Elect. Eng. 107-B. 8. 381-388

6. P.Hubert. P.Laroche (1984) Triggered Lightning in New Mexico. J. of Geophy. Res. 89. D2. 2511-2521

3.7 Boundary Layer Model and Calculation for Horizontal Thundercloud Electrification Preceding Natural and Rocket-Triggered Lightning

G.C. DIJKHUIS
Zeldenrust College and Convectron N.V., Terneuzen, Holland

ABSTRACT

An electro-hydrodynamic model is presented for horizontal thunderclouds in the atmospheric boundary layer combining turbulent shear flow with space charge from fair-weather electricity. Horizontal electrification is obtained from circulation of charged eddies carrying a spin current through wind-sheared air layers. Upstream deflection of downward field lines by eddy circulation in stratified space charge is found compatible with available fair-weather data. In weather conditions with reduced ohmic conductivity the spin current accumulates positive charge on leading cloud sections as seen in winter thunderstorms above the Sea of Japan.

The Poisson equation is solved on a square mesh for dipolar thundercloud electrifications approached by wire-trailing rockets. Equipotential plots show electric field reversal at ground level in bifurcation points of the zero potential line. Field strength plots identify dipolar clouds by a saddle point, and show regions of field reversal as electric cusps.

INTRODUCTION

Two recent articles by H. Kikuchi [1,2] present an electric model for lightning discharges triggered by wire trailing rockets fired at winter thunderclouds above the Sea of Japan. For these winter storms Takeuti et. al. [3] have established a wind-sheared horizontal cell structure at low altitude, with negative charge cloud regions trailing , and positive cloud regions leading in the prevailing wind direction. Such horizontal cloud electrification is in marked contrast with the more usual vertical cell structure and electrification in updraft-dominated thundercloud models [4,5].

Lightning strokes triggered from horizontal thunderclouds by a wire-trailing rocket usually show a bipolar current wave form. Kikuchi details one event beginning with a broad return stroke carrying positive charge to ground, followed by several narrower strokes with opposite polarity [6]. This bipolar flash has a much longer time interval between first and second stroke than successive strokes in a multiple, but monopolar flash [7]. Streak photographs and electric field records also revealed an intercloud (or intracloud) discharge just before the second ground stroke.

On sharp edges such as the tip of a wire-trailing rocket , electrostatic theory for vacuum fields gives singularities with infinite field strength, and dipolar clouds give points or lines with zero field intensity. Kikuchi terms the low-field regions around such zero-field points "electric cusps" by analogy with low-field regions in magnetized plasmas. With sufficient electric conductivity at ground level, thundercloud field lines meet induced surface charge at right angles. The electric cusp separates regions with opposite surface charge and vertical field direction . Kikuchi's 2D field plots of a near-horizontal thundercloud show one cusp at ground level. A similar plot for dipolar electrification in vertical direction gives two cusps at equal distance from the symmetry axis. 2D contour plots of electric potential show the cusps as bifurcation points of equipotential lines. A grounded wire behind a rocket bifurcates the surface line of zero potential, creating an artificial cusp region around the point of departure.

The fair-weather atmosphere contains free ions and electrons created by cosmic rays, radioactive decay and corona discharge from pointed objects at the surface [8]. The fair-weather space charge of air is positive, and decreases upward, but with uncertain magnitude and range, especially above sea. Free charges in air prevent vacuum field singularities near sharp edges by reducing field intensity through a Debye screening layer as in electrolytes and plasmas. Kikuchi estimates the Debye sphere around a rocket tip at ca. 1 m radius.

Ions attached to aerosol particles connect field lines with static or moving air masses.

Along the rocket tip and cusp boundary Kikuchi expects "electro-hydrodynamic wind" following from combined electro-hydrodynamic (EHD) equations of motion analogous to Alfven's MHD-equations in magnetized plasma. These EHD-equations should transpose MHD field line merging and reconnection associated with solar flares to corresponding EHD phenomena in terrestrial lightning.

WIND SHEAR

Kikuchi's wind-sheared thunderclouds form in turbulent air flow roughly half way up in the atmospheric boundary layer with a typical height of 1000 m. Standard boundary layer theory provides a universal velocity profile matching unsheared winds at high altitude with zero wind velocity at ground level [9]. Wind shear is strongest adjacent to the surface where friction forces from molecular viscosity suppress turbulence in a thin

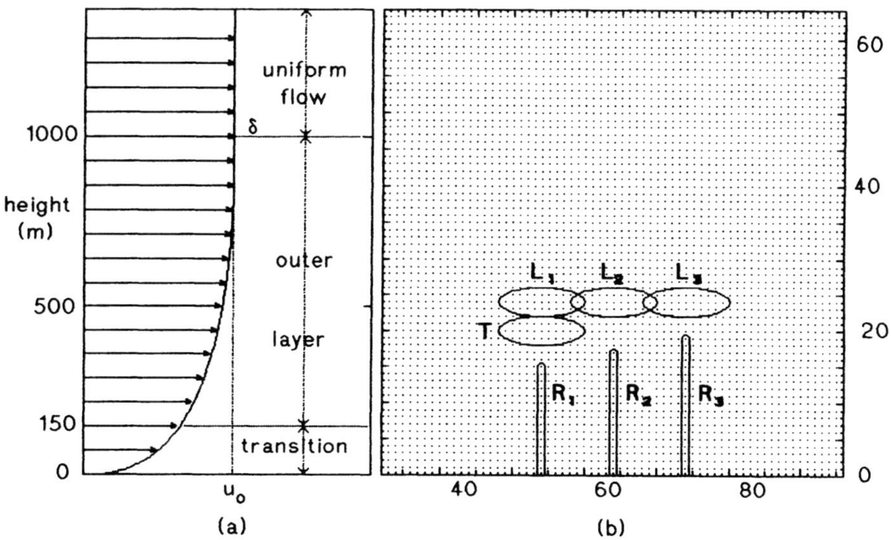

Figure 1a,b . Wind shear profile of atmospheric boundary layer (a) relative to horizontal thundercloud electrification (b) showing size and positions of negative trailing cloud T, positive leading clouds L_1, L_2, L_3 and wire-trailing rocket trajectories R_1,R_2, R_3; the dots in (b) are mesh points for electric potential and field calculations.

"viscous sublayer" with a linear velocity profile. At less than 1 meter above a smooth surface, Reynolds stresses from turbulent flow gradually replace molecular friction as dominant shearing force, and the boundary layer assumes a logarithmic velocity profile extending to ca. 150 m. The remaining "outer layer" has large eddies with mainly inertial forces leading to an approximately parabolic velocity profile measured by Hama [10]. With u_∞ for uniform wind velocity above height $y=\delta$, measurements of the velocity defect $u_\infty - u$ relative to friction velocity u_* confirm as universal boundary layer profile:

$$\frac{u_\infty - u}{u_*} = \begin{cases} 9.6*(1 - y/\delta)^2 & , \quad 0.15 < y/\delta < 1 \qquad (1a) \\ -2.44*\ln(y/\delta) + 2.5 & , \quad 0 < y/\delta < 0.15 \qquad (1b) \end{cases}$$

Logarithmic profile (1b) would make velocity u and its gradient singular as height y decreases to zero at the surface. Above a flat surface such as a wind tunnel or the sea, molecular viscosity suppresses turbulence and straightens the velocity profile inside a viscous sublayer with typical thickness 40 cm at geograhic latitude 35° for Japan. Larger surface roughness than above land increases surface shear stress above the molecular minimum, thereby bending the outer layer profile (1a) in Fig. 1 through friction velocity u_*.

A parabolic velocity profile like Eq'n (1a) arises in fully developed laminar flow between parallel plates from streamwise momentum balance $\mu d^2u/dy^2 = dp/dx$ with constant viscosity μ and pressure gradient dp/dx. A parabola also approximates Blasius' universal boundary layer profile derived from balancing viscous and inertial forces. For the turbulent boundary layer Prandtl replaces molecular viscosity by eddy viscosity $\ell_m^2 \, |du/dy|$, and takes mixing length $\ell_m = \varkappa y$ proportional with height y. Prandtl obtained the scale-free transition layer profile (1b) from height-independent turbulent friction $\varkappa y \, du/dy = u^*$. This inhomogeneous first-order differential equation has a homogeneous second-order equivalent $(d/dy)(y \, du/dy) = 0$. In the boundary-layer geometry of Fig. 1a, velocity-derivative $w_z = -du/dy$ is the only non-zero component of vorticity $\mathbf{w} = \nabla \times \mathbf{u}$. Thus Prandtl's boundary layer conditions admit $\nabla \times (\ell_m \mathbf{w}) = \mathbf{0}$ as equivalent vector equation for logarithmic velocity profile Eq'n 1b. A standard vector identity further gives $\ell_m \nabla \times \mathbf{w} = \mathbf{w} \times \nabla \ell_m$, showing that Prandtl's height-dependent mixing length $\ell_m = \varkappa y$ makes streamwise component $(\nabla \times \mathbf{w})_x = y^{-1} du/dy = \varkappa/y^2$ non-zero and positive in the transition layer. Stokes' vector theorem $\int \nabla \times \mathbf{w} \cdot d\mathbf{S} = \oint \mathbf{w} \cdot d\boldsymbol{\ell}$ transforms non-zero streamwise curl of vorticity into closed vortex lines in y-z-planes . Formation of such vortex loops from stretching and reconnecting vortex lines is the characteristic dynamic process of the transition layer [10].

BOUNDARY LAYER ELECTRIFICATION

Although thunderstorms are a local and transient phenomenon, their electrification maintains a global potential difference between earth and ionosphere against dissipating fair-weather currents. When diurnal variations in global thundercloud activity modify the ionospheric potential, all stations see simultaneous and comparable changes. For electrostatic calculations the ionosphere conveniently provides a second spherical equipotential surface concentric with ground level. Measurements reveal a downward fair-weather current density of $j_f \approx 3$ pA/m^2 , and a downward electric field with strength $E_0 \approx 120$ V/m at the surface, leading to $\sigma = j_f / E_0 \approx 2.5 * 10^{-14}\ \Omega^{-1} m^{-1}$ as ohmic conductivity for the lowest air layers [8]. Their positive space charge diminishes fair-weather field strength to 66 V/m at height 2 km, and 3.3 V/m at height 12 km. The available data conform to exponential height-dependence of vertical electric field $E_y = -E_0 * \exp(-y/\ell_e) \approx -120 * \exp(-y/3340)$ with electric scale length $\ell_e \approx 3340$ m , leading to electric field -89 V/m in the uppermost boundary layers of Fig. 1 at height 1 km. With vacuum permittivity ε_0 = 8.85 pF/m, charge density in the boundary layer follows from Gauss's electrostatic law through $\rho = \varepsilon_0 \nabla \cdot \mathbf{E} = \varepsilon_0 dE_y/dy = \rho_0 * \exp(-y/\ell_e)$, with $\rho_0 = \varepsilon_0 E_0 / \ell_e \approx 0.32$ pC/m^3 as charge density at ground level in fair weather. Most ions in air layers above land quickly attach to much heavier aerosol particles lowering charge mobility and electric conductivity. Cleaner air layers above sea seem to achieve a nearly equal conductivity with faster and longer-lived ions formed by cosmic radiation at roughly half of the land-based production rate [8].

Going downward in the atmospheric boundary layer, the vertical fair-weather current crosses turbulent air layers with increasing horizontal vorticity. Turbulent eddies with positive charge and intrinsic rotation rate magnetize the boundary layer parallel to its local vorticity. The magnetization amounts to $\mathbf{M} = \rho \ell_m^2 \boldsymbol{\omega}$ for mixing-length-size eddies combining space charge ρ with intrinsic angular velocity or spin field $\boldsymbol{\omega}$. The equivalent magnetization or spin current $\mathbf{J}_s = \nabla \times \mathbf{M} = \nabla \rho \ell_m^2 \boldsymbol{\omega}$ adds up with ohmic current $\mathbf{J}_\sigma = \sigma \mathbf{E}$ to a total current density $\mathbf{J} = \mathbf{J}_\sigma + \mathbf{J}_s$ given by:

$$\mathbf{J} = -\sigma \nabla \Phi + \nabla \times \rho \ell_m^2 \boldsymbol{\omega} \tag{2}$$

where Φ denotes the electrostatic potential of atmospheric electric field $\mathbf{E} = -\nabla \Phi$. The boundary layer of Fig. 1 has no other velocity components than velocity u in the positive x-direction. This makes z-component $w = (\nabla \times \mathbf{u})_z = -du/dy$ the only non-zero vorticity component , pointing in negative z-direction because the velocity profile of Fig. 1 has a positive slope du/dy. Accordingly, the aligned spin field $\boldsymbol{\omega}$ also has one non-zero component ω, with negative z-axis as direction. Charge density ρ and mixing length ℓ_m enter as positive scalar factors without effect on direction of magnetization $\rho \ell_m^2 \boldsymbol{\omega}$. The curl operator gives a streamwise $j_s = (\nabla \times \rho \ell_m^2 \boldsymbol{\omega})_x = d(\rho \ell_m^2 \omega)/dy$ as only non-zero component of the spin current $\mathbf{J}_s$. Thus for purely horizontal air flow the atmospheric boundary layer our charge transport equation (2) gives horizontal electrification as observed in Kikuchi's sea-borne thunderclouds. The positive charge observed on leading cloud sections requires spin current in positive x-direction. Our transport equation (2) couples sign of spin current to height-dependence of the three quantities making up magnetization $\rho \ell_m^2 \omega$.

Individual mixing-length size eddies spinning with uniform angular velocity ω have a circulation $\oint \mathbf{v} \cdot d\boldsymbol{\ell} = \pi \ell_m^2 \omega$. Ideal fluids conserve their circulation, and superfluids quantize it . With viscous stress weak relative to turbulent stress in the transition layer, eddy circulation $\pi \ell_m^2 \omega$ relaxes to an adiabatic invariant for eddies with spin ω exceeding collective rotation rate $\frac{1}{2}w$ from local vorticity w. Molecular viscosity ν locks laminar layers beneath sublayer height $y=\delta_0$ into a negative rotation rate $\omega_0 = \frac{1}{2}w_0 = -u_*^2/2\nu$ defining friction velocity u_* in boundary layer profile (1a,b). In the boundary layer geometry of Fig. 1, adiabatic invariance of eddy circulation and Prandtl's mixing length relation $\ell_{m,0} = \varkappa\delta_0$ at height δ_0 separate flow parameters $\varkappa$, δ_0, u_* and ν from charge density $\rho=\varepsilon_0 dE_y/dy$ in our spin current expression through $j_s = d(\ell_m^2 \omega\rho)/dy = \ell_{m,0}^2 \omega_0\, d\rho/dy = -(\varkappa^2\delta_0^2 u_*^2/2\nu)\,\varepsilon_0 d^2E_y/dy^2$. Horizontal electrification by spin current j_s drives an ohmic current $j_{\sigma,x} = \sigma E_x$ in opposite direction. Steady-state electrification demands zero total current $j_{\sigma,x}+j_s = \sigma E_x + j_s = 0$ giving a horizontal electric field strength $E_x = -j_s/\sigma$. Relative to its vertical fair-weather field $E_y = j_f/\sigma$, the boundary layer has electric field ratio $E_x/E_y = -j_s/j_f$, or:

$$\frac{E_x}{E_y} = \frac{\varkappa^2\delta_0^2 u_*^2}{2\nu}\,\frac{\varepsilon_0}{j_f}\,\frac{d^2E_y}{dy^2} \tag{3}$$

Here, von Karman constant $\varkappa$, sublayer height δ_0, friction velocity u_*, viscosity ν and electric conductivity σ combine into a positive scalar factor with magnitude depending on weather conditions and surface roughness. From downward current density $j_f < 0$, downward field $E_y < 0$, and convex E_y-profile $d^2E_y/dy^2 < 0$ in fair weather, Eq'n (3) gives an upstream horizontal field component $E_x < 0$ in the boundary layer geometry of Fig. 1. Thus our spin current mechanism generates horizontal electrification with field lines connecting positive charge on leading cloud regions with negative charge on trailing sections as observed in Kikuchi's sea-borne thunderclouds. Fair-weather current $j_f \approx 3$ pA, surface field $E_0 \approx -120$ V/m and electric scale height $\ell_e \approx 3340$ m give the electric parameter combination $(\varepsilon_0/j_f)\,d^2E_y/dy^2 = (\varepsilon_0 E_0/j_f \ell_e^2) * \exp(-y/\ell_e)$ in Eq'n (3) a maximum value $\varepsilon_0 E_0/j_f \ell_e^2 \approx 3.2 * 10^{-5}$ s/m^2 at surface height $y = 0$. Stronger winds increase boundary layer electrification (3) directly through friction velocity u_*, and indirectly through surface roughness rising above atmospheric sublayer height δ_o. Maximum free stream velocity $u_\infty \approx 100$ m/s in air, and sublayer height $\delta_o \approx$ 1 m limit the flow parameter combination in Eq'n (3) through $\varkappa^2\delta_o^2u_*^2/2\nu = \ell_m^2\,\omega_o < u_\infty\delta_o \approx 100$ m²/s above smooth terrain. Above sea, surface tension γ gives wave crests with height δ_o an inward surface stress γ/δ_o which we take as upper limit on skin friction from dynamic viscosity μ in adjacent air layers : $\mu u_*^2/\nu \le \gamma/\delta_o \approx 0.073$ Pa from $\gamma \approx 0.073$ N/m for water, and $\delta_o \approx 1$ m for sublayer width and maximum wave amplitude. Wind tunnel experiments give $\varkappa \approx 0.4$ as universal Von Karman constant. The US Standard Atmosphere has $\nu \approx 15$ mm²/s as kinematic viscosity, and $\mu \approx 18$ μPa*s as dynamic viscosity, their ratio $\mu/\nu \approx 1.2$ kg/m³ being mass density of air at sea level [8]. These values impose $\varkappa^2\delta_o\gamma/2\mu \approx 324$ m²/s as upper limit from surface tension of water on kinematic flow parameter combination $\varkappa^2\delta_o^2u_*^2/2\nu$ in Eq'n (3) above sea.

The fair-weather atmosphere with no wind has purely vertical electric field and current. While preserving the vertical direction of fair-weather currents at non-zero wind velocity, our boundary layer electrification mechanism bends downward electric fieldlines in upstream direction. Different directions for current density and electric field make boundary layer conductivity a tensor instead of a scalar quantity. By tensor relation $\mathbf{E} = \boldsymbol{\sigma}^{-1}\cdot\mathbf{j}$, horizontal field component E_x is proportional to vertical current density j_z through off-diagonal component $\sigma_{xz}^{-1} = E_x/j_z$ of tensor resistivity $\boldsymbol{\sigma}^{-1}$. Our estimates $\varepsilon_oE_o/j_f\ell_e^2 \approx 3.2 * 10^{-5}$ s/m² for the electric parameter combination, and $\varkappa^2\delta_o\gamma/2\mu \approx 324$ m²/s for the flow parameter combination in Eq'n (3) give field ratio E_x/E_y a maximum value $E_{x,o}/E_o \approx 0.01$ at sea level. The corresponding maximum angle $\tan^{-1}(E_{x,o}/E_o) \approx 0.6°$ for deviation of surface field from vertical direction makes our horizontal electrification mechanism insignificant for atmospheric boundary layers in fair weather.

Condensation of atmospheric water vapour has no great effect on turbulent flow in the boundary layer, but water droplets catching free electrons and ions reduce mobility of charges carrying ohmic currents through air. Steady-state electrification from spin current balanced by ohmic current at reduced atmospheric conductivity demands stronger horizontal fields in the boundary layer. Electric conductivity reduction by 5-6 orders of magnitude from water drops will raise our fair-weather maximum 1.2 V/m for horizontal field strength to thundercloud conditions with field threshold 500 kV/m for electric breakdown leading to a lightning discharge.

POISSON CALCULATION

Electric potential Φ of a static charge distribution ρ in vacuum with permittivity ε_0 satisfies Poisson's equation $\nabla^2\Phi = -\rho/\varepsilon_0$. As a linear second-order differential equation of elliptic type, Poisson's equation gives unique and stable solutions inside a closed surface with fixed potential (Dirichlet condition) or with fixed normal potential gradient. For our 2D electrification problem we discretize the Laplace operator $\nabla^2 = \partial^2/\partial x^2 + \partial^2/\partial y^2$ using standard finite-difference techniques from computational fluid dynamics [11]. With cartesian coordiates x,y the usual five-point computational "molecule" centered on mesh point (i,j) gives:

$$\Phi^{(k+1)}_{i,j} = \tfrac{1}{4}\{\Phi^{(k)}_{i+1,j} + \Phi^{(k+1)}_{i-1,j} + \Phi^{(k)}_{i,j+1} + \Phi^{(k+1)}_{i,j-1} - h^2\rho_{i,j}/\varepsilon_0\} \qquad (4)$$

where a distance h separates adjacent mesh points in both directions, and k=1,2,3... labels successive iterations of Liebman's numerical integration method [11]. Eq'n (4) updates potential $\Phi_{i,j}$ at mesh point (i,j) using two potentials $\Phi^{(k)}_{i+1,j}$ and $\Phi^{(k)}_{i,j+1}$ from the preceding iteration sweep, and two potentials $\Phi^{(k+1)}_{i-1,j}$ and $\Phi^{(k+1)}_{i,j-1}$ from the current iteration sweep. Each mesh point retains its initial charge density $\rho_{i,j}$ throughout the calculation. The calculation proceeds on a square domain with 256*256 mesh points. Boundary points at sea level and points inside the rocket trail start and stay at zero potential. Cloud electrification enters through mesh points with fixed charge density inside elliptic subregions shown in Figures 1, 2, 3. The iteration process stops when a new sweep changes potential less than 1 % at all mesh points.

Figures 2a,b,c show dipolar electrification geometry in thunderclouds with increasing wind shear. Contour plot 2d for unsheared cloud 1a shows two bifurcations B_1, B_2 in the zero-potential line at ground level corresponding to electric cusps C_1, C_2 in contour plot 1g for field strength. For sheared clouds 2b,c the domain of calculation retains one bifurcation B in potential plot 2e,f, and one electric cusp C in field strength plot 2 h,i.

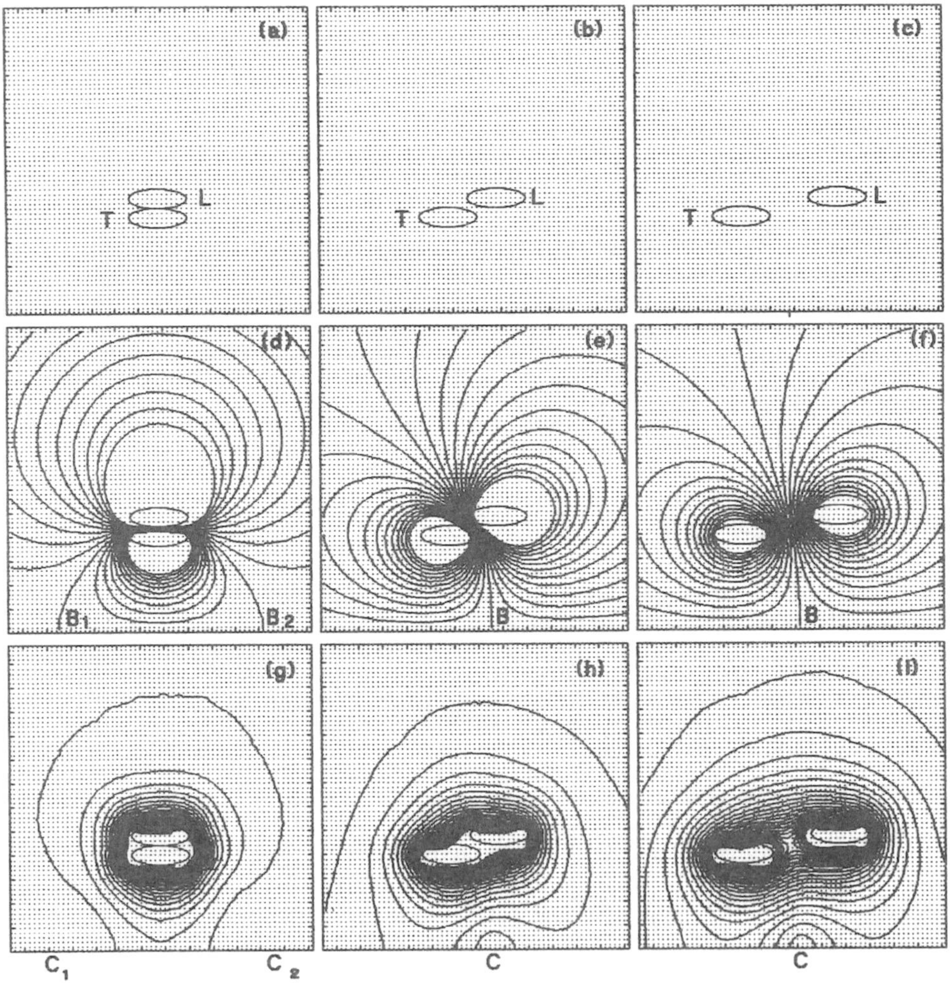

Fig. 2. Dipolar charge distributions on square meshes (a,b,c) show electrification geometry with leading cloud region L and trailing cloud region T carrying opposite charge; equipotential plots in (d,e,f) show bifurcation points B ; field strength plots in (g,h,i) show electric cusps C; (i) also shows a saddle point in between the charged regions.

Wire-trailing rockets for triggered lightning from electrification geometry 2c enter the Poisson calculation as grounded mesh points inside subdomains R in Figure 3a,b,c. Asymmetric rocket trajectories 3a,c move bifurcations B in potential plots 3d,f, and electric cusps C in field strength plots 3g,i upward from ground level. Rocket trajectory 3b fired from cusp position 2i keeps zero-potential bifurcation B in 3e, and electric cusp C in 3h at the point of departure.

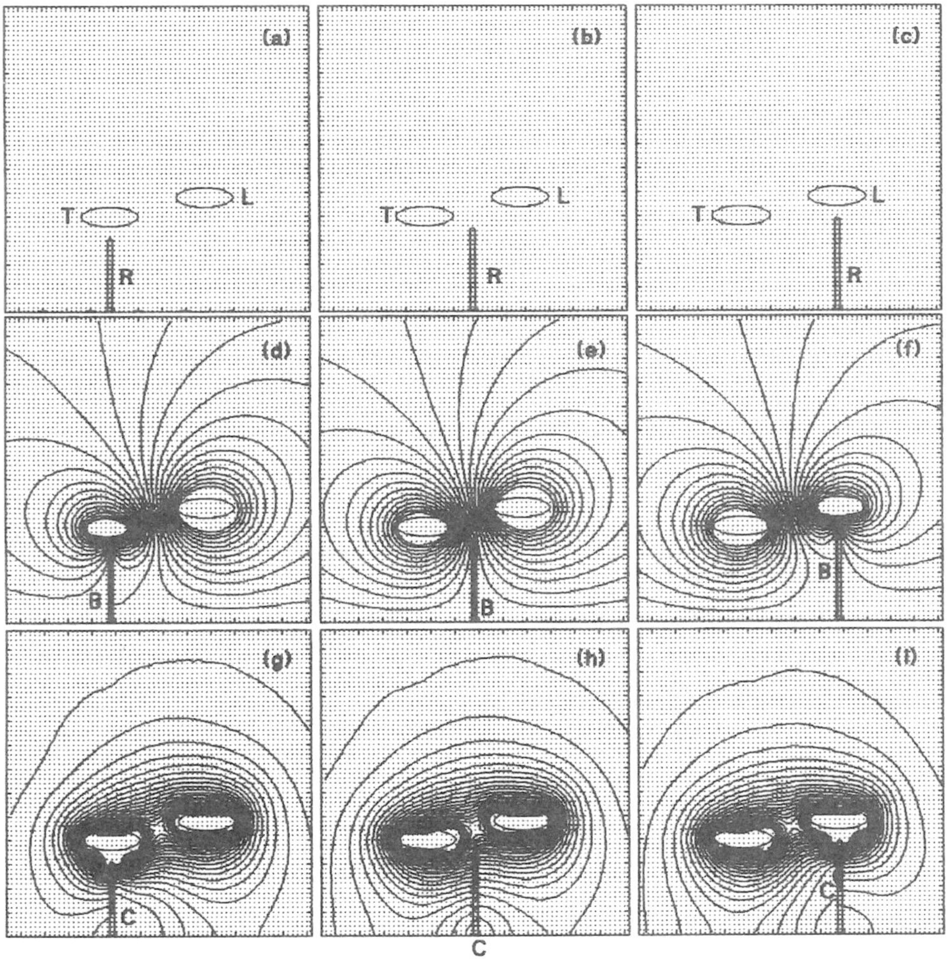

Fig. 3. Electrification geometry of wind-sheared cloud 2c approached by wire-trailing rocket trajectories R_1 , R_2, R_3 in (a,b,c) gives bifurcation points B in potential plots (d,e,f) and electric cusps C in field strength plots (g,h,i).

Our Poisson calculations were performed with a fast and flexible PC program developed by J. Pijpelink for ball lightning experiments with electric breakdown in nozzle flow [12].

REFERENCES

1. Kikuchi H (1986) Electric reconnection, critical velocity, and triggered lightning. In: Kikuchi H (ed) Proceedings of the second international workshop on the relation between laboratory and space plasmas. Springer-Verlag. New York Berlin Tokyo. p331-344
2. Kikuchi H (1985) A new model of triggered lightning. In: Proc. 6th Symp. Electromag. Comp. Zurich. p47-50
3. Takeuti T. Nakano M. Brook M. Raymond DJ. Krehbiel P (1978) The anomalous winter thunderstorms of the Hokoriku coast. J. Geophys. Res. 83: 2385-2394
4. Moore CB. Vonnegut B (1977) The thundercloud. In: Golde RH (ed) Physics of lightning. Academic Press. London New York San Francisco. p51-91
5. Browning KA (1982) Morphology and dynamics of thunderstorms. In: Kessler E (ed) Thunderstorms. NOAA, Dept. of Commerce, Washington DC, USA. p211
6. Kikuchi H (1986) Characteristics and theory of bipolar flashes observed from rocket-triggered lightning experiments. Janiszewski J. Moron W. Sega W (eds) Eighth international Wroclaw symposium on electromagnetic compatibility. Wroclaw, Poland p254-262
7. Berger K (1977) The earth flash. In: Golde RH (ed) Physics of lightning. Academic Press. London New York San Francisco. p119-190
8. Dolezalek H (1987) Atmospheric electricity. In: Weast RC (ed) Handbook of chemistry and physics (67th edn) CRC Press. Florida USA p. F149-151
9. Tennekes H. Lumley JL (1972) A first course in turbulence. The MIT Press. USA
10. Hinze JO (1979) Turbulence (2nd ed) McGraw-Hill Book Company. New York
11. Roache PJ (1972) Computational fluid dynamics. Hermosa Publishers. N.M. USA
12. Dijkhuis GC. Pijpelink J (1989) Performance of high-voltage test facility designed for investigation of ball lightning. In: Ohtsuki YH (ed) Science of ball lightning (firebal). World Scientific. Singapore. p. 325-337

4 *Atmospherics, Whistlers, and Emissions*

4.1 Sferics

H. VOLLAND
Radioastronomical Institute, University of Bonn, Bonn, Germany

ABSTRACT

A brief overview is presented about the excitation of sferics by lightning strokes and the propagation of these electromagnetic pulses within the atmospheric waveguide between ground and ionosphere.

INTRODUCTION

Sferics are electromagnetic pulses excited by lightning flashes, mainly by return strokes, and modified within the atmospheric waveguide between Earth and ionospheric D-layer, or within the magnetosphere where they propagate as whistlers. Sferics research is an exemplary case of multidisciplinary activity involving meteorology, atmospheric physics, ionospheric physics, magnetospheric physics, telecommunication and the theory of electromagnetic wave propagation.

Sferic signals were first discovered in 1895 by the Russian physicist Popov in St. Petersburg who connected a wire to a detector and heard noise of an impulsive character. Sferics have therefore been intimately related to radio science from the beginning.

If one observes the electromagnetic radiation of lightning strokes with broadband receivers, one notices impulsive signals at frequencies below about 100 kHz. At higher frequencies, however, it is no longer possible to distiguish between individual impulsive waveforms, and a broadband noisy spectrum appears (Fig. 1). Radio noise in this high-frequency band will affect the signal-to-noise ratio of any radio communication system. This is a subject in its own right including man-made noise and galactic cosmic noise [4, 15], but will not be treated here.

A short review of some aspects of sferics research will be given in this lecture, concentrating on the low-frequency band.

THE RETURN STROKE

A typical frequency spectrum of the electric field from a return stroke, observed at different distances from the source, is presented in Fig. 2. One notices a decay, almost inversely proportional to the frequency f, between about 10 kHz and 100 kHz. This decay becomes steeper at frequencies greater than 100 kHz and at distances above 200 km. The spectral maximum is near $f \cong 5$ kHz (curves A and B). This maximum indicates that the corresponding wavelength of $\lambda = c/f \cong 60$ km (c is the speed of light) of the dominant wave is several times larger than the typical length of the channel of a return stroke. The visible part of such channels has a typical length of about 5 km. Another part

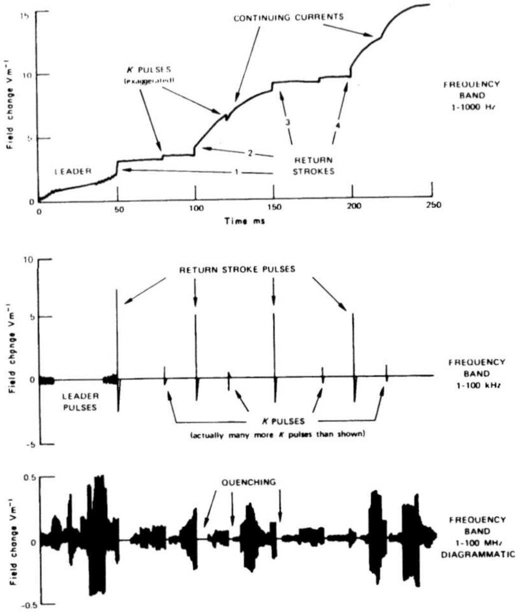

Fig. 1 Electric field changes due to flashes observed at about 50 km distance in three different bands [9].

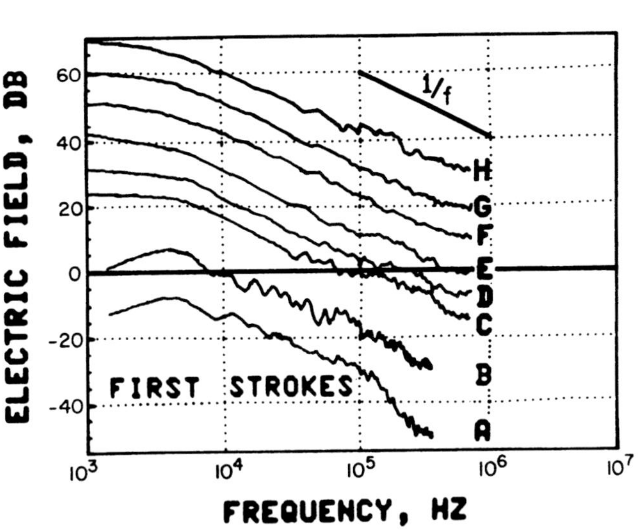

Fig. 2 Average spectra of electric fields of first return strokes at distances (A) 200 km; (B) 50 km; (C) 10 km; (D) 7 km; (E) 4km; (F) 2.5 km; (G) 1.5 km, and (H) 0.9 km [13].

of comparable magnitude may be hidden within the cloud and may have a significant horizontal branch.

If the wavelength of an electromagnetic wave is larger than the dimension of the propagation medium, the physics of the propagation must be derived from full wave theory. It does not make sense to consider a shock wave with a velocity of a fraction of the speed of light propagating up through a channel that has a length smaller than the wavelength of the dominant wave [16]. The ray theory concept breaks down in this case.

The lightning channel before discharge can be considered as a thin isolated wire of length l and diameter d, in which negative electric charge (electrons) has been stored. This electric charge is lowered to the ground when the wire reaches the electrically well-conducting Earth. In terms of electric circuit theory, one can adopt a simple waveguide model with a conductor C, where the charge is stored, a resistance R of the channel, and an inductance L simulating the electric properties of the channel (Fig. 3). The conductance G can be neclected in this case.

If the contact with the ground starts at time $t = 0$, the downward flow of the electrons at $t > 0$ corresponds to an upward directed electric current varying in time according to [19]

$$I = \bar{I}\,(e^{-\alpha t} - e^{-\beta t})\cos(Kz/l) \qquad (1a)$$

or

$$I = 2\bar{I}\,\sin\delta t\; e^{-\gamma t}\cos(Kz/l) \qquad (1b)$$

where $\bar{I}$ is the amplitude, $\alpha = \gamma + i\delta$, and $\beta = \gamma - i\delta$ with $\gamma = R/2L$, $\delta = (K^2/LC-\gamma^2)^{1/2}$, and $K = (2n-1)\pi/2$ with n a positive integer. The first expression (1a) - the famous Bruce-Golde formula [3] - is valid when δ becomes imaginary. An infinite number of individual modes of wavenumber n can be excited just as in the case of a piano string or any other resonant system. In practise, however, only the modes of lowest order ($n < 3$) are well defined because of the crude model approximations. The fundamental mode with $n = 1$ has a wavelength of $\lambda = 4l$ so that it does not "feel" the real tortuous channel configuration. The channel behaves like a quarter-wave antenna with maximum current at the bottom and zero current at the top. The higher order modes have wavelengths of $4l/3$, $4l/5$, etc. The storage of electric charge is a maximum at the top of the channel and zero at the bottom. Figure 4 shows the height configuration of current and charge for the two first modes. For mode two, charge is transported to the ground only in the lower part of the channel while the rest is internally redistributed. Mode one is therefore the most efficient one for lowing charge to the ground and heating the channel. It also has the longest lifetime and thus essentially determines the general waveform of the lightning return stroke. The higher order modes decay more rapidly and are not excited as strongly. They increasingly contribute to the high frequency radio noise with increasing mode number.

A more advanced model which considers the channel as a cylindrical wire of finite thickness, finite electric conductivity and finite height corresponds to a resonant electric waveguide that supports propagation of transverse-magnetic waves [17, 18]. It is possible in this case to relate the channel parameters height l and thickness d to the observed waveform parameters α and β. Given an observed sferic waveform, which is proportional to the time derivative of I in (1), one can then derive the channel parameters l and d. Of course, one

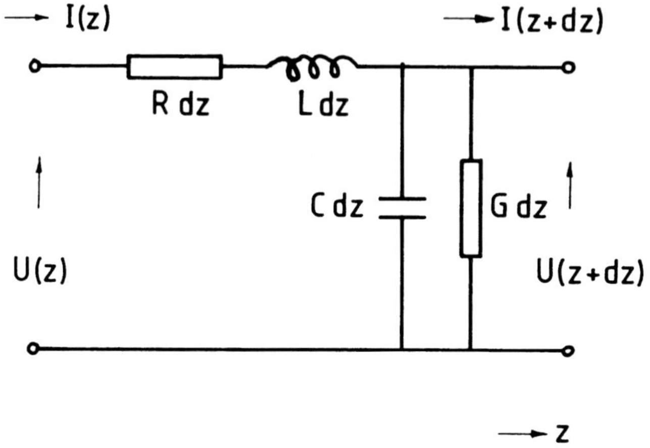

Fig. 3 Element of a transmission line with resistance per unit length R', inductance per unit length L', capacitance per unit length C', and conductance per unit length G'. U and I are voltage and current.

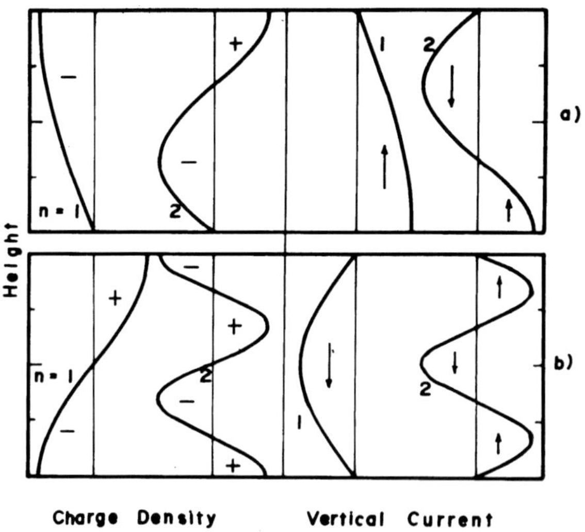

Fig. 4 Height distribution of electric charge q (left) and vertical electric current I (right) for the two modes of order n = 1 and 2. a) return stroke model; b) intracloud K stroke model.

should be cautious not to overinterpret such model, which gives only a crude simulation of the basic macroscopic process, leaving the details unexplained.

The theory shows that the parameters α and β can be real or complex depending on the electric properties of the channel. Equation (1a) yields an aperiodic waveform (type 1), eq. (1b) a damped oscillator (type 2), often referred to as a bipolar waveform in the literature [21]. Indeed, both waveforms have been observed, although the bipolar waveform seems to dominate. Typically, waveform (1a) exists for a resistance within the channel of R > 3 kΩ. Figure 5 shows examples of simulated waveforms, which can be remarkably similar to observed waveforms of return strokes and of intracloud K-strokes.

SFERIC WAVEFORM

In a first approximation, a lightning channel can be considered as a vertical electric dipole standing on a perfectly conducting ground. Its vertical radiation field is proportional to the time derivative of the electric current like that in eq. (1) (see Fig. 5):

$$E_z = - \frac{\mu l I}{2\pi\rho} (\alpha\, e^{-\alpha t} - \beta\, e^{-\beta t}) \cos(Kz/l) \qquad (2)$$

with μ the permeability of free space and ρ the distance between stroke and receiver.

Taking into account the finite conductivity of the ground, the theory of Sommerfeld [14] predicts an attenuation of the ground wave that increases with frequency and distance. A sferic waveform containing a broad spectrum of frequencies first loses its high frequency tail so that the spikes in Fig. 5 disappear with increasing distance and become rounded. This is seen in Fig. 6 where the vertical electric field of an intracloud K-stroke is calculated at a distance of ρ = 25 km according to eq. (2) (solid line: with γ = 0.1 μs^{-1}; δ = 0.2 μs^{-1}), and with a finite ground conductivity of σ = 0.001 S/m (dashed line). Further modification occurs if the finite channel length of 4.3 km and the height of the center of the channel at 5 km are taken into account (dotted line). We notice a finite rise time in agreement with the observations [21].

IONOSPHERIC TRANSMISSION FUNCTION

The ionosphere further modifies the waveform. Part of the electromagnetic energy of the sferic is reflected by the ionospheric D-layer at about 70 km (day) to 90 km (night) height. The atmospheric waveguide, of which the ground and the ionospheric D-layer are the boundaries, is dispersive. Wave theory must be applied to describe the propagation of the dominant very-low-frequency (VLF) signals and extremely-low frequencies (ELF) of a sferic, in particular at distances from the source larger than about 1000 km.

The vertical electric field of a pulse traveling in the waveguide is modified according to [5, 18, 22]

$$E_z(\rho,t) = \int \bar{E}_z(\rho,\omega)\, W(\rho,\omega)\, e^{-i\omega t}\, d\omega \qquad (3)$$

where $\bar{E}_z$ is the Fourier transform of the field over a perfectly conducting ground, and W is the transmission function of the dispersive atmospheric waveguide. W depends on distance and frequency

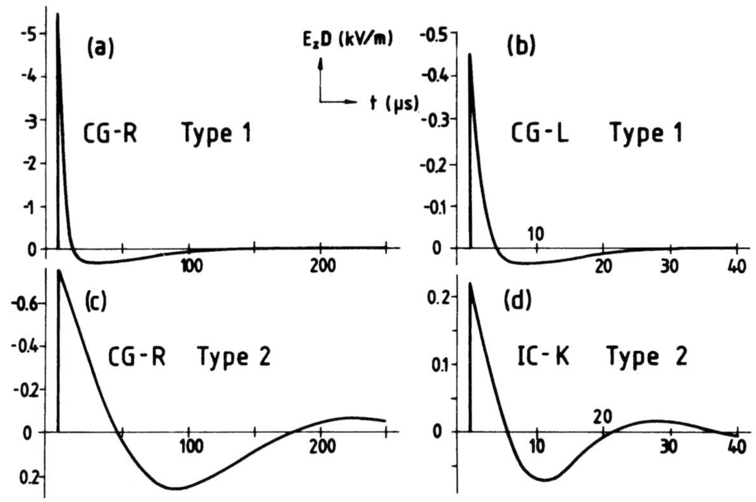

Fig: 5 Electric radiation fields of radiation from a vertical electric dipole. The electric field is normalized to a distance of 1 km. a) and b): return strokes of type 1 (aperiodic wave); c): return stroke of type 2 (damped oscillator); d): intracloud K stroke.

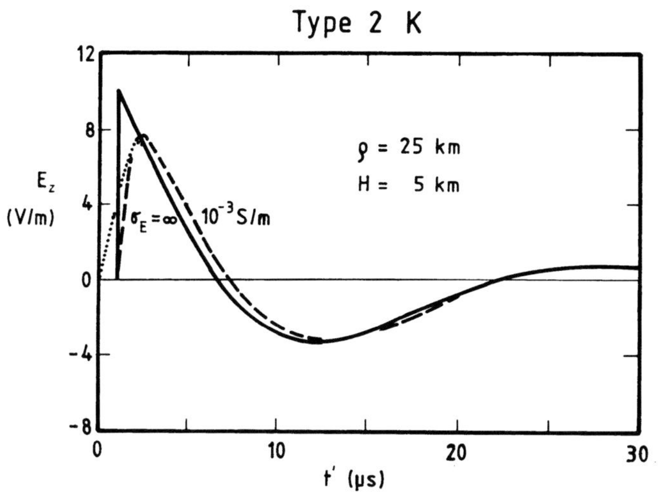

Fig. 6 Electric radiation field of a type 2 intracloud K stroke at a distance of 25 km. The center of the channel is located at 5 km height. Solid line: dipole approximation over a perfectly conducting earth. Dashed line: dipole over an imperfectly conducting earth. Dotted line: channel of finite length (4 km) over a perfectly conducting earth.

and, in addition, on the time of day and season, and on the geographic and geomagnetic conditions of the propagation path. In the case of waveform (2), one obtains

$$\bar{E}_z = -\frac{i\omega l \bar{I}\mu(\beta-\alpha)}{2\pi\rho(\alpha-i\omega)(\beta-i\omega)} e^{ik\rho} \qquad (4)$$

with $k = \omega/c$ the wavenumber, ρ the distance between sferic and receiver, and $\omega = 2\pi f$ the angular frequency of a particular wave.

The transmission function of the atmospheric waveguide can be approximated by applying the two concepts of ray theory and mode theory. Ray theory describes the propagation of a pulse as the sum of a directly arriving ground wave and sky waves reflected at the base of the ionosphere near 70 to 90 km height depending on time of day and season:

$$W = W_o + \sum W_n \qquad (5)$$

where W_o is the transmission function of the ground wave, and

$$W_n \cong 2\, R_i^{\,n} \sin^3\theta_n \; e^{ik(r_n-\rho)} \qquad (6)$$

is the transmission function of the n-th hop sky wave, θ_n is the angle of incidence, and r_n the phase path of the ray. In the VLF range ($3 < f < 30$ kHz) and at distances < 500 km, one may use the approximations

$$W_o \cong 1 \; ; \; R_i \cong -e^{-2\varepsilon \cos\theta} \qquad (7)$$

with $\varepsilon \cong 0.001\,\omega^{2/3}$ an empirical coefficient depending on frequency (ω in s^{-1}). The first hop arrives at a time $\Delta t = (r_1-\rho)/c$ later than the ground wave and has nearly the same pulse form. Pulses with pulse lengths < Δt can then be separated into a ground pulse and a first hop sky pulse at distances between about 200 and 500 km [7].

The multihop components become increasingly important at greater distances, so that the wave concept becomes more convenient in this case. The transmission function of the n-th wave mode is [4, 22]

$$W_n \cong \frac{1}{h}\sqrt{\frac{\vartheta}{\sin\vartheta}\,\frac{2i\pi\rho c}{\omega}}\; K_n\, e^{(-A_n+iB_n)\rho} \qquad (8)$$

with $\vartheta = \rho/a$ the polar distance between receiver and transmitter, h the virtual height of the waveguide, a the Earth's radius, K_n an amplitude factor, A_n an attenuation factor, and B_n a phase factor.

In the VLF range, the ground behaves to a first approximation like an electric wall ($R_e \cong 1$), and the ionosphere behaves like a lossy magnetic wall with R_i from (7). Full wave theory demands that a single wave mode obeys a certain phase relationship: a wave after being reflected at the surface and at the ionosphere must have a phase which is a multiple of 2π of the phase at its origin. The eigenvalue equation of the atmospheric waveguide has then the form

$$R_e R_i e^{2ikh\cos\theta_n} = e^{2in\pi} \qquad (9)$$

from which follows

$$\cos\theta_n = \frac{C_n}{1+i\varepsilon/kh} \qquad (10)$$

with $C_n = (2n-1)\pi/kh$.

In the ELF range, the ionosphere behaves like an electric wall ($R_i \cong 1$), so that $C_n = n\pi/kh$.

The phase factor in (8) is related to $\cos\theta_n$ by

$$B_n = k\left(\sqrt{1-\cos^2\theta_n} - 1\right) \cong -kC_n^2/2 \qquad (11)$$

EQUIVALENCE BETWEEN RAY THEORY AND MODE THEORY

The equivalence between ray theory and mode theory in the VLF range can be seen immediately by comparing the interference pattern of ground wave and first hop wave on the one hand, and first and second mode on the other hand. The interference minima of the two first modes according to (8), (10), and (11) are at

$$(B_1-B_2)\rho \cong \frac{\pi^2\rho}{h^2k} = (2m-1)\pi \quad (m = 1, 2, \ldots) \qquad (12)$$

Similarly, those of the two first rays are at (see eqs. (6) and (7))

$$k(r_1-\rho)-\pi \cong 2h^2k/\rho-\pi = (2m'-1)\pi \quad (m' = 1, 2, \ldots) \qquad (13)$$

The last interference minimum of ray theory (for $m' = 1$) is at $\rho \cong kh^2/\pi$. This is also the first interference minimum of mode theory ($m = 1$). At a frequency of $f = 15$ kHz and a virtual height of the atmospheric waveguide of $h = 70$ km (daytime conditions), one obtains $\rho_1 \cong 500$ km. The two modes interfere beyond this distance with interference minima equally spaced at 1500 km, 2500 km etc. Since the second mode is more strongly attenuated than the first mode, the first mode is dominant beyond about 1500 km distance. For $\rho < \rho_1$, only the ground wave and the first hop wave interfere. Because of the $\sin^3\theta$ dependence of the first hop mode in (6), this wave becomes of no significance at distances smaller than about 200 km where the ground wave dominates.

SCHUMANN RESONANCES

Mode theory becomes increasingly favorable with decreasing frequency. The extreme case occurs at the Schumann resonances. These resonances are the eigenfrequencies of the cavity between Earth and ionosphere. If one stretches this cavity to a rectangular box of length $2\pi a$ = 40000 km, one obtains resonance frequencies of transverse-magnetic waves which are given by

$$f \cong \frac{mc}{2\pi a} \cong 7.5\ m \quad \text{Hz} \quad (m = 1, 2, \ldots) \qquad (14)$$

Resonance peaks of electric and magnetic field intensities at 7.5, 15, 22.5, etc Hz have been observed [10]. They are generated by the lowest-frequency tail of lightning flashes, probably mainly from the "continuous currents" between adjacent strokes. The horizontal configuration of the vertical electric field and azimuthal magnetic field of the Schumann modes show relatively simple structures. Since three main sources of thunderstorm areas exist (Africa, South and Middle America, East Asia) with their daily variations peaking at local afternoon, the observation of these modes can serve as a measure of global lightning activity. From a measurement of mode splitting, which is due to the geomagnetic field, it is possible to determine the electric properties of the atmospheric waveguide and its boundaries [12].

PROPAGATION PROPERTIES OF THE ATMOSPHERIC WAVEGUIDE

Figure 7 shows the magnitude of the transmission functions $|W_n|$ of wave mode n = 0 (dashed lines) and n = 1 (solid lines) versus frequency for different distances. The zeroth mode is weakly attenuated at frequencies < 2kHz. The first mode has its attenuation minimum between about 5 and 20 kHz, depending on distance. Severe attenuation occurs in the transition range between zeroth and first mode near 2 kHz. The waveguide thus has two windows, one in the ELF band, another in the VLF band. This picture becomes somewhat modified due to the influence of the curvature of the Earth, the geomagnetic field and daily and seasonal variations.

Figure 8 shows the spectral field strength $|\bar{E}_z W|$ from (3) and (4) versus frequency at different distances for the return stroke in Fig. (2b). Since $W \cong 1$ for $f < 100$ kHz and $\rho < 200$ km, the upper curve in Fig. 8 is the magnitude of $\bar{E}_z$. The difference between the solid and the dashed line indicates the attenuation of the ground wave at frequencies > 100 kHz due to the finite electric conductivity of the ground. The bandpass effect shifts the maximum of the spectral amplitude from 4 kHz at distances < 200 km to 12 kHz at $\rho \cong 10000$ km.

From the evaluation of the phase factor B_n in (8), one finds that the group velocity increases with frequency at $f < 10$ kHz. This frequency dependence can be used to locate the sources of sferics at larger distances (> 1000 km) by measuring the difference in arrival time of the pulse train at two or more frequencies [11, 20]. Likewise, the frequency-distance dependence of the quantity $|\bar{E}_z W|$ (see Fig. 8) can be used for locating thunderstorm areas at greater distances.

WHISTLERS

Part of the electromagnetic spectrum of lightning discharges, mainly from the horizontal branches of their channels, can propagate along the geomagnetic field lines throughout the magnetosphere and can reach the conjugate point of the source. It may be reflected at the conjugate point and then travel several times to an fro along the same path. The upper limit of the frequency spectrum penetrating the magnetosphere is either the gyrofrequency of the electrons $f_{He} = 2.80 \times 10^{10}$ B (with B the geomagnetic field strength in T) or the plasmafrequency of the electrons $f_e = 9\sqrt{n}$ (with n the electron density in m^{-3}).

The group velocity of a signal propagating parallel to the geomagnetic field lines depends on frequency and on the properties of the magnetospheric plasma. It can be defined by a dispersion measure

$$D = \tau\sqrt{f} \cong 8.94 \times 10^{-14}\ L \langle\sqrt{n/B}\rangle \qquad (15)$$

with τ the group travel time at frequency f, L the length of the geomagnetic field line, and the brackets indicate an average along the propagation path. Typical numbers are $D \cong 50 - 100$, and $f \cong 3 - 10$

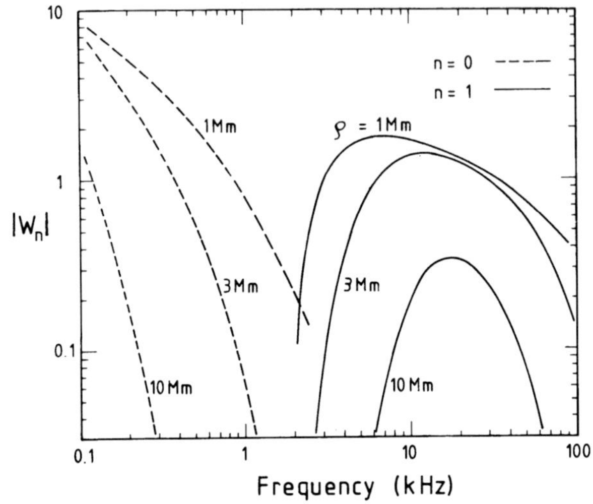

Fig. 7 Magnitude $|W_n|$ of transmission functions of zeroth mode and first mode vs frequency at distances 1000, 3000, and 10000 km during daytime conditions at midlatitudes.

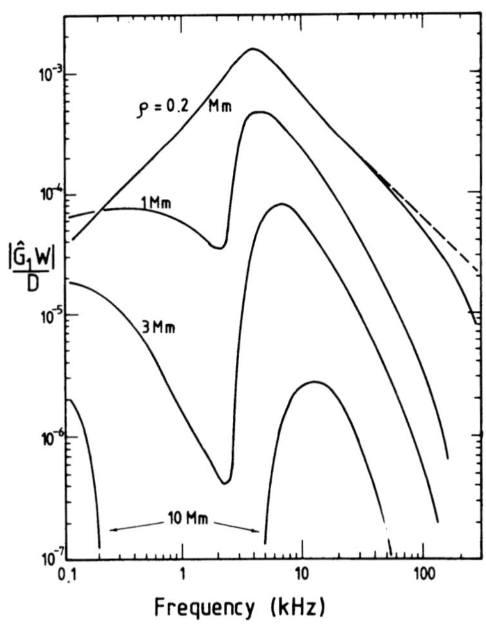

Fig. 8 Spectral field strength (magnitude) of type 2 return stroke in Fig. 5c vs frequency at different distances. The dashed curve gives the spectral amplitude at 200 km distance in the case of a perfectly conducting ground.

kHz. The higher frequencies arrive earlier than the lower frequencies, so that the received signal, containing a broad spectrum of waves, is heard as a whistling tone that changes in pitch. It is possible to derive the electron density of the magnetosphere from the observations of whistlers via eq. (15) [8].

The propagation conditions become more complicated if inhomogenities exist within the magnetospheric plasma. Several ions (besides hydrogen) may be involved in the dispersion of the signal, and wave-wave interaction may thus generate secondary whistler signals which can be observed by satellites. Finally, wave-particle interaction of a whistler signal with electrons, mainly within the equatorial area of the magnetosphere, may stimulate high-energetic electrons which can penetrate along the geomagnetic field lines down to ionospheric heights and give rise to various radiation phenomena [1].

CONCLUSION

We have discussed basic properties of the electromagnetic pulses excited by lightning strokes and propagating as sferics within the atmospheric waveguide between ground and the ionospheric D-layer, or as whistlers along the geomagnetic field lines within the magnetosphere. Sferics, as well as whistlers, have served as tools to derive the physical conditions of the propagation media. The observation of sferics is also a very convenient method to monitor global lightning activity. More information about sferics and related phenomena may be found in [2], [6], [16] amd [18].

LITERATURE

1. Goldberg RA. Barcus JR. Hale LC. Curtis SA (1986) Direct observation of magnetospheric electric precipitation stimulated by lightning. Journ. Atm. Terr. Phys. 48: 293

2. Golde RH (ed) (1977a) Lightning, vol. I and II. Academic Press, London.

3. Golde RH (1977b) Lightning and related phenomena. In Golde RH (ed) Lightning, vol. I. Academic Press, London. p.309

4. Hagn GM (1982) Man-made radio noise. In Volland H (ed) Handbook of Atmospherics, vol. I. CRC Press, Boca Raton. p.329

5. Harth W (1982) Theory of low frequency wave propagation. In Volland H (ed) Handbook of Atmospherics, vol. II. CRC Press, Boca Raton. p.133

6. Krider EP. Roble RG (eds) (1986) The Earth's Electric Environment. National Academic Press. Washington, DC

7. McDonald TB. Uman MA. Tiller JA. Beasly WH (1979) Lightning location and lower-ionospheric height determination from two-station magnetic measurements. Journ Geophys. Res. 84: 1727

8. Park CG (1982) Whistlers. In Volland H (ed) Handbook of Atmospherics, vol. II. CRC Press, Boca Raton. p.21

9. Pierce ET (1977) Atmospherics and radio noise. In Golde RH (ed) Lightning, vol. I. Academic Press, London. p.351

10. Polk C (1982) Schumann resonances. In Volland H (ed) Handbook of Atmospherics, vol. I. CRC Press, Boca Raton. p.111

11. Sao K. Jindoh H (1974) Real time location of atmospherics by single station techniques and preliminary results. Jour. Atm. Terr. Phys. 36: 261

12. Sentman DD (1989) Detection of elliptical polarization and mode splitting in discrete Schumann resonance excitations. Journ Atm. Terr. Phys. 51: 507

13. Serhan GI. Uman MA. Childers DG. Lin YT (1980) The RF spectra of first and subsequent lightning return strokes in the 1 - 200 km range. Radio Sci. 15: 1089

14. Sommerfeld A (1952) Lectures in Theoretical Physics, vol. VI. Academic Press, New York

15. Spauldin g AD (1982) Atmospheric noise and its effect on telecommunication system performance. In Volland H (ed) Handbook of Atmospherics, vol. I, CRC Press, Boca Raton. p.289

16. Uman MA (1987) The Lightning Discharge. Academic Press, New York

17. Volland H (1981) A waveguide model of lightning currents. Journ. Atm. Terr. Phys. 43: 191

18. Volland H (1982) Low frequency radio noise. In Volland H (ed) Handbook of Atmospherics, vol. I. CRC Press, Boca Raton. p.179

19. Volland H (1984) Atmospheric Electrodynamics. Springer. Heidelberg

20. Volland H. Schmolders M. Proelss GW. Schaefer J. (1987) VLF propagation parameters derived from sferics observations at high southern latitudes. Journ. Atm. Terr. Phys. 49: 33

21. Weidman CD. Krider EP (1979) The radiation field wave forms produced by intracloud lightning discharge processes. Journ. Geophys. Res. 84: 3159

22. Wait JR (1972) Electromagnetic Waves in Stratified Media. McMillan, New York

4.2 Polar VLF Emissions Observed by ISIS Satellites

T. ONDOH

Office for Space Science, Communications Research Laboratory, Tokyo, 184 Japan

Abstract

Polar VLF emissions received from ISIS-1 and -2 at Syowa station, Antarctica inculde polar cusp hiss,auroral hiss, VLF saucer and ELF hiss. High-latitude limit of ELF hiss approximately follows the auroral zone. A polar map of the occurrence rate for broad-band auroral hiss, obtained from ISIS VLF data, is compared to one for the occurrence locations of inverted-V electron precipitations, obtained from Atmospheric Explorer(AE)-D. The occurrence map for broad-band auroral hiss is qualitatively similar to that for the inverted-V electron precipitations ; concerning the low-latitude boundary and axial symmetry of the 10-22 hour geomagnetic local time(MLT) meridian. So, the polar occurrence map for broad-band auroral hiss is different from the auroral zone which has the axial symmetry of the noon-midnight meridian. The broad-band auroral hiss seems to be generated by inverted-V electrons precipitated from the boundary plasma sheet, but not by auroral electrons. The broad-band auroral hiss should be hereafter called by the name of the polar hiss. The frequency range of the broad-band auroral hiss is discussed in terms of whistler-mode Cherenkov radiation from inverted-V electrons.

1. Introduction

ELF hiss is usually observed in the plasmasphere[1,2], but it extends often to the auroral zone as observed by the ISIS satellites. Intense bursts of VLF emissions associated with auroral disturbances were often observed at high-latitude ground stations[3,4,5,6,7,8]. Helliwell[9] classified these VLF emissions as auroral hiss. Gurnett [10] first reported satellite observations of auroral hiss associated with precipitating electrons below 10 keV. Correlative analyses of satellite data have shown that auroral hiss is generated by intense fluxes of precipitating auroral electrons with energies ranging from 100 eV to about 40 keV[11,12]. Hughes et al.[13] made the polar occurrence map of 9.6 kHz auroral hiss using the Ariel-3 VLF magnetic field data. Ondoh[14,15] made the polar map of the occurrence rates for broad-band auroral hiss by omitting polar cusp hiss since the polar cusp hiss occurs irregularly in frequency, time and intensity.

This paper is concerned in high-latitude limit of the ELF hiss and a comparison of the polar occurrence maps for broad-band auroral hiss obtained from 347 ISIS VLF passes[14,15] and inverted-V electron precipitation made from 280 events[16,17].

The Cherenkov radiation generated from auroral electrons was discussed as the origin of auroral hiss[3,18,19,20]. The coherent Cherenkov radiation or the resonant Cherenkov instability was also discussed to explain auroral hiss[11,21,22,23]. Finally, the frequency range for broad-band auroral hiss is discussed in terms of whistler-mode Cherenkov radiation generated from the inverted-V electrons in the polar magnetosphere.

2. High Latitude Limit of ELF Hiss

Electric field data(50 Hz - 30 kHz) from low-altitude polar orbiters , ISIS-1(500 - 3500 km) and ISIS-2(1450 km) were received at Syowa station, Antarctica(geomag. lat. 69.7°S, long. 77.7°E) to study characteristics of polar ELF and VLF emissions for June,1976 to January, 1982. Narrow-band VLF intensity data were processed from magnetic tape outputs of ISIS electric field data by narrow-band dc amplifiers with a minimum reading circuit similar to a ground VLF hiss receiver. Charging and discharging time constants of the minimum reading circuits are 10 s and 10 ms respectively. The maximum range in intensity variation of narrow-band VLF data is about 30 dB in arbitrary scale, while the range of light and shade of f-t spectrum films is about 10 dB.

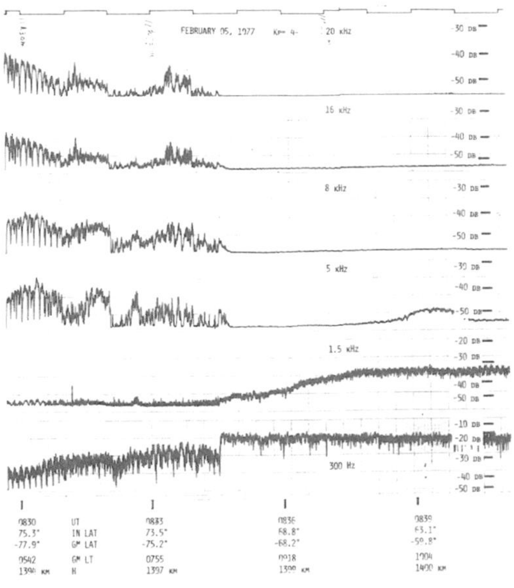

Fig. 1 Example of High-latitude limit of ELF hiss observed by ISIS-1 on February 5, 1977(K_p= 4-).

Figure 1 shows an example of ELF hiss whose high-latitude limit lies at invariant latitude 71° in the geomagnetic local morning. The ELF hiss suddenly became weak and broad-band VLF emissions appeared at latitudes higher than 71°. Thus, the ELF hiss often extends to the auroral zone , although the ELF hiss usually predominates in the plasmasphere[1,2]. Figure 2 shows an occurrence distribution of 76 high-latitude limits of the ELF hiss in geomagnetic invariant latitude vs. geomagnetic local time coordinate. Seventy six ELF hisses occurred in geomagnetic activity from K_p= 0 to K_p= 6, and most (90 %) of them occurred for K_p=1 to 4.

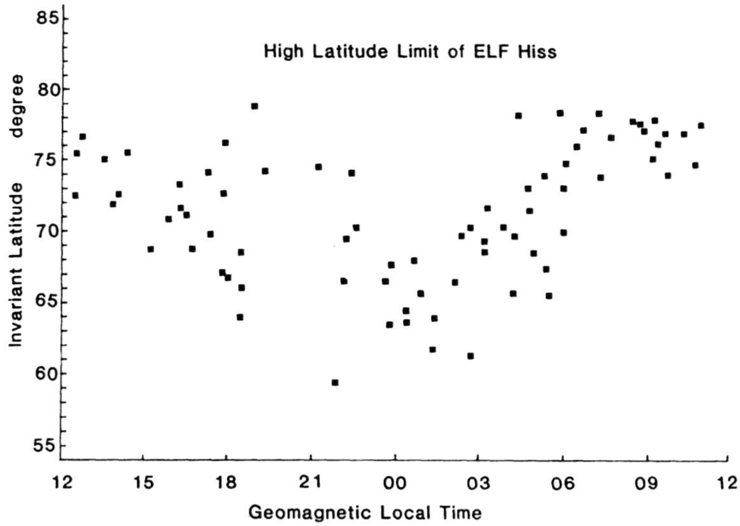

Fig. 2 High-latitude limits of ELF hisses observed by ISIS satellites

Figure 2 shows that the high-latitude limit of ELF hiss goes down to about 60° in invariant latitude around the midnight, and that it comes up to about 79° for the daytime(06 - 18 MLT). This implies that the high-latitude limit of ELF hiss follows approximately the auroral zone.

3. Broad-band Auroral Hiss Observed in the Polar Topside Ionosphere

Figure 3 shows time and spatial variations in the ISIS VLF electric field at 6 frequencies, 0.3, 1.5, 5, 8, 16 and 20 kHz received at Syowa station on October 31, 1978(K_p= 2+). Broad-band auroral hisses appearing in the 5, 8, 16 and 20 kHz bands can be seen during local geomagnetic morning, 04 to 07 MLT, between invariant latitudes of 73° and 75°. The broad-band auroral hiss is defined by simultaneous increases in intensity above 5 dB as compared to the quiet-time levels in the 5, 8, 16 and 20 kHz bands at high invariant latitudes above 60°, as shown by Fig. 3.

Figure 4 shows a polar map for the ratio of number of ISIS passes including broad-band auroral hiss to all the ISIS passes over a certain unit area(1° invariant latitude interval x 1 hour MLT), where 0.3,0.4, and 0.5 indicate the equicontours of the ratio[14,15]. In the geomagnetic local time interval 09-11 MLT, the occurrence rate of broad-band auroral hiss is the lowest of all other local times at invariant latitudes from 65° to 82°, because we omitted the polar cusp hiss in making the polar occurrence map for broad-band auroral hiss. The frequency-time spectrum of polar cusp hiss observed mainly in the 09-11 MLT interval is very irregular in frequency, time and intensity, while that of broad-band auroral hiss observed in other local times is steady and stable in frequency, time and intensity[24]. Figure 4 is the statistical result of 347 ISIS VLF passes observed for geomagnetic activities from K_p= 0 to 7, and the occurrence rate of ISIS VLF passes for K_p= 4 to 7 is 0.23. So, Fig. 4 represents an average occurrence map for broad-band auroral hiss during various geomagnetic activities ,but not during a specific one.

The high-latitude contour of occurrence rate 0.3 lies at an invariant latitude of about 82° for all local times.

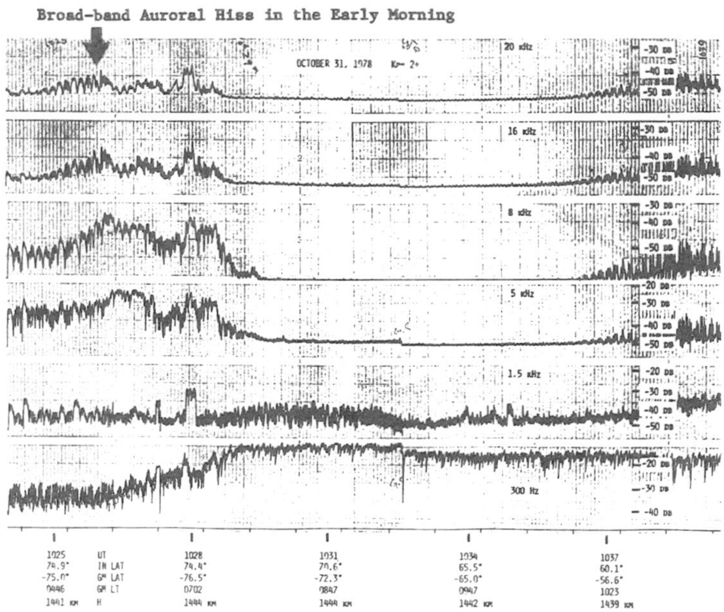

Fig. 3 Example of broad-band auroral hiss observed at invariant latitudes above 70° in local morning of October 31, 1978 (K_p= 2+).

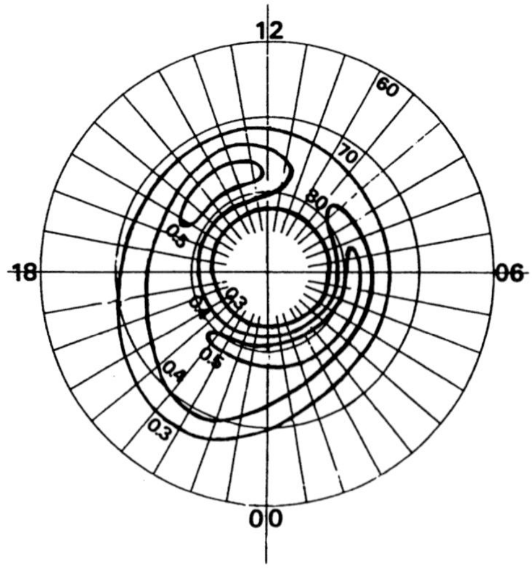

Fig. 4 Spatial map of occurrence rates for broad-band auroral hiss in coordinates of invariant latitude vs. geomagnetic local time. Broad-band hisses of 347 ISIS VLF passes were used.

The low-latitude contour of 0.3 lies at an invariant latitude of about 74° at 10 MLT and extends down to 67° at 22 MLT. Thus, the low-latitude contour for this rate is symmetric with respect to the geomagnetic meridian at 10-22 MLT. It should be noted that the DP-2 current system associated with the plasma convection in the magnetospheric equatorial plane also has parallel polar current flows at the 10-22 MLT meridian [25]. Consequently, the polar occurrence map for broad-band auroral hiss seems to be controlled by the magnetospheric convection through the inverted-V electrons precipitated from the boundary plasma sheet.

A polar occurrence frequency map compiled from Ariel 3 VLF data[13] for 9.6 kHz auroral hiss at magnetic field intensity above 10^{-12} gamma^2/Hz shows pronounced dawn-dusk asymmetry with a distinct occurrence minimum during local morning. A discrepancy between the Ariel and ISIS auroral hiss maps seems to be mainly due to differences between the 9.6 kHz magnetic field data and the multi-frequency electric field data, since the probability of detecting auroral hiss using ISIS multi-frequency data is much higher than when using Ariel fixed-frequency data. Also, number and period of ISIS VLF electric field data used for the polar map of broad-band auroral hiss are much greater than those of Ariel VLF magnetic field data used for the polar map of 9.6 kHz auroral hiss.

4. Comparison of Polar Occurrence Maps for Broad-band Auroral Hiss and Inverted-V Electron Precipitation

Particle precipitation in the high latitude is traditionally broken into three regions ; auroral oval, dayside cusp and polar cap. Electrons of a few keV precipitate from the central plasma sheet into the region of the diffused aurora. The precipitation usually varies smoothly and produces diffuse auroral patterns. The diffused auroral region in the high latitude ionosphere lies alongside of the low-latitude half-side of the auroral oval [26]. Correlative studies have clearly indicated that strong auroral hisses occur simultaneously with inverted-V events in the auroral oval and low-latitude side of the polar cap [11,12].

Figure 5 shows an example of this correlation for auroral hiss with the inverted-V electron precipitations first observed by Injun-5 in the polar topside ionosphere[11]. The observation was made at an invariant latitude of about 72° around 19 MLT. The observed electron energy may be above 40 keV. If one maps the region for the inverted-V events along the geomagnetic field lines on the magnetospheric equatorial plane, the occurrence region will correspond to the plasma sheet and magnetotail. Many inverted-V events in the dusk-to-midnight hemisphere have a maximum peak energy of above 10 keV. In the polar-cap region, however, the inverted-V's usually have much lower peak energies that range from about 300 eV to about 1 keV [16,17]. Precipitation near the high latitude boundary of the auroral oval, referred to as boundary plasma sheet precipitation, is characterized by discrete structures. Average energy of the boundary plasma sheet electrons is usually 200 - 300 eV during calm periods, but increases dramatically during the storm periods [27]. According to the AE-D observations of inverted-V events[16,17], the inverted-V electrons seem to mostly precipitate from the boundary plasma sheet, and they essentially occur all the time, independent of the geomagnetic activity, especially in the pre-midnight region. The occurrence of inverted-V events was independent of the altitude between 200 km and 800 km. The low latitude boundary for occurrence is a function of local time and gives the impression of following the auroral oval.

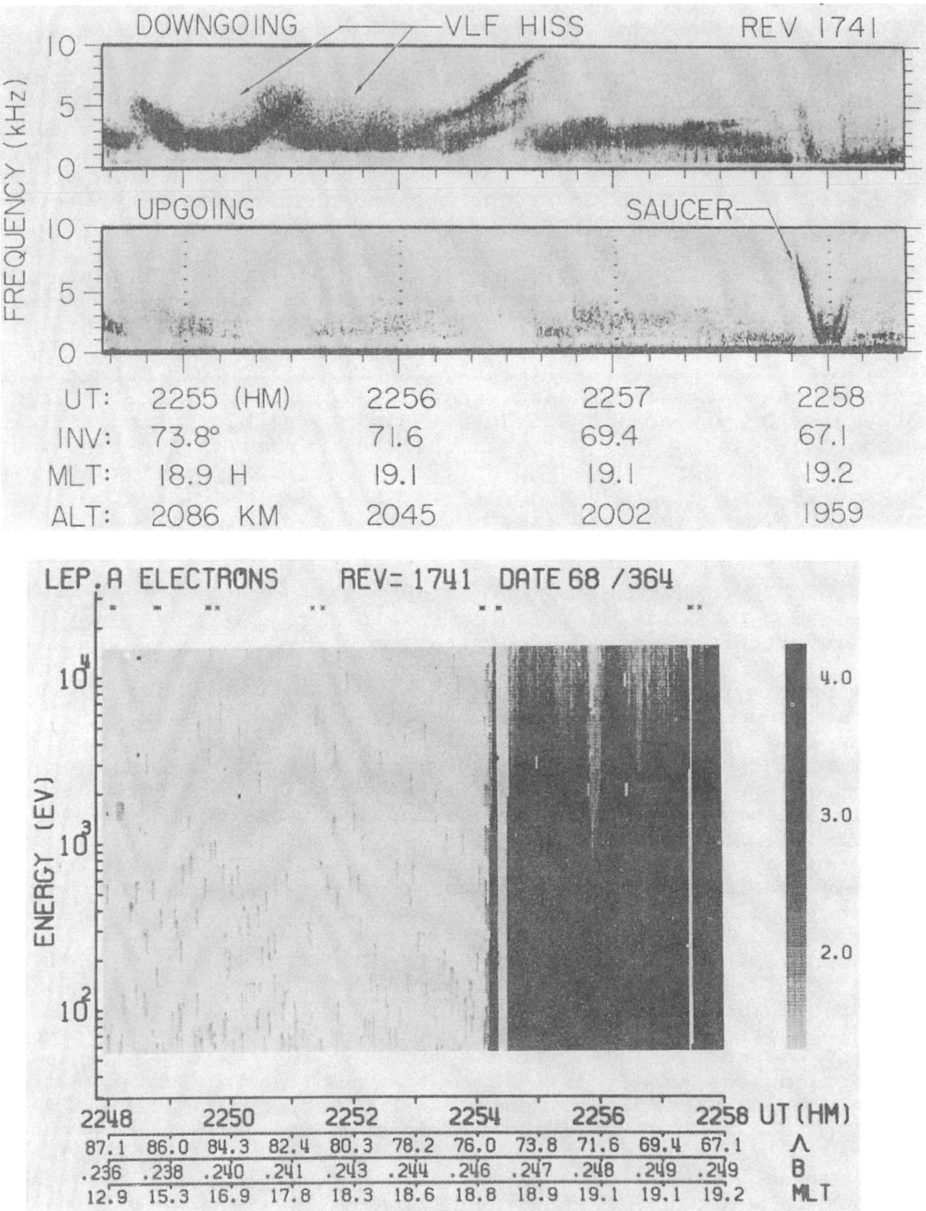

Fig. 5 Frequency-time spectrograms of electric(upper) and magnetic (middle) fields and energy-time spectrogram(lower) of the inverted-V electron precipitation events observed by Injun-5 (after Gurnett and Frank,1972). Downgoing auroral hiss is closely associated with inverted-V events.

Figure 6 shows a spatial occurrence map for 280 inverted-V events observed by AE-D with invariant latitude and geomagnetic local time coordinates[16]. The two thick lines, indicating low-latitude and high-latitude boundaries, have been drawn by the present author to compare this polar map with the one for broad-band auroral hiss observed during 347 ISIS VLF polar passes(Fig. 4).

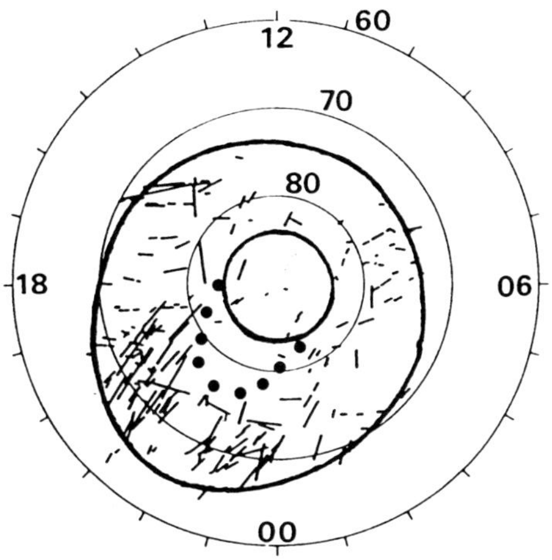

Fig. 6 Spatial occurrence map for 280 inverted-V events observed by AE-D with invariant latitude vs. geomagnetic local time coordinates (after Hoffman and Lin, 1981).

The high-latitude boundary of the inverted-V events may be better represented by the dotted curve elongated toward 22 MLT (Hoffman, 1988, Private Communication). This map shows the distribution of occurrence locations for the inverted-V events observed during various geomagnetic activities(K_p= 0 - 5), and the broad-band auroral hiss map shows the distribution of occurrence rates for broad-band auroral hisses during various geomagnetic activities. Both polar maps show axially symmetric axis at the 10-22 MLT meridian. They also show rare occurrence regions in the 09-11 MLT sector and the polar cap region. The low latitude contour of the 0.3 occurrence rate for broad-band auroral hiss is especially similar to the low-latitude boundary for inverted-V event occurrence. Both of these low-latitude boundaries lie at an invariant latitude of about 74° at 10 MLT and extend down to 66° at 22 MLT. Thus, the polar occurrence map for broad-band auroral hiss generally corresponds well to the one for the inverted-V events as seen during the simultaneous observation of strong auroral hiss and inverted-V electrons in the polar topside ionosphere[11], even though the former indicates occurrence rates and the latter occurrence locations.

Vasyliunas[27] projected the inner edge of the plasma sheet along the geomagnetic field lines on a polar map, shown by Fig. 7, for the auroral zones during geomagnetically quiet and disturbed conditions. Polar maps for these quiet and disturbed auroral zones show an approximately symmetrical axis for the noon-midnight meridian. The quiet (Q = 0) auroral zone lies from invariant latitudes 76° to 78° at the geomagnetic noon and from 70° to 73° around the midnight, respectively. Also, the disturbed(Q = 4) auroral zone lies from 74° to 76° at the noon meridian and from 65° to 72° around the midnight meridian, respectively. The broad-band auroral hiss region is wider than both of the auroral zones.

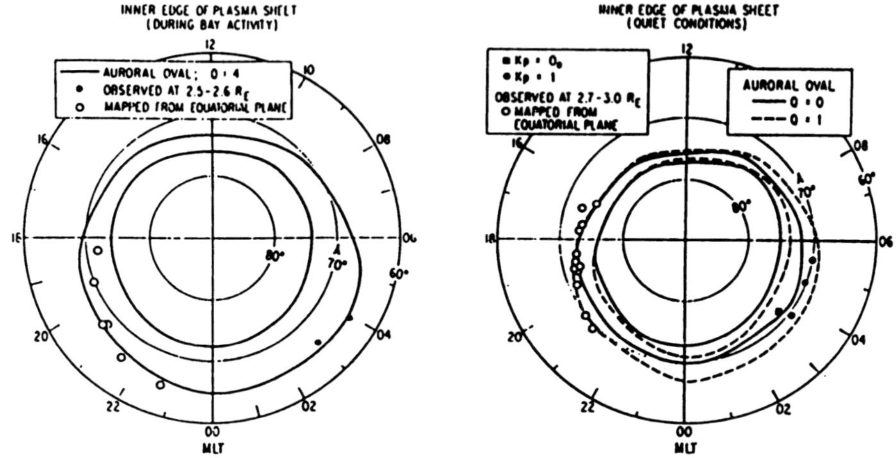

Fig. 7 Polar maps of auroral zones during geomagnetic disturbed (Q = 4,left) and quiet(Q = 0 & 1,right) periods. Ionospheric projections of the equatorial inner edges of the plasma sheet along geomagnetic field lines are also shown(after Vasyliunas, 1970).

Recently, Ondoh[28] has obtained the polar distributions of the occurrence rates for broad-band auroral hiss in geomagnetic quiet period (K_p= 0 - 1) from ISIS VLF narrow-band data by using the same criterion of broad-band auroral hiss as Fig. 4. These were statistically derived from 143 ISIS VLF passes received at Syowa station, Antarctica from January, 1980 to January 1983. The broad-band auroral hiss region with occurrence rates above 0.3 lies from invariant latitude 74° to 84° in geomagnetic local time interval 06-14 MLT, from 70° to 82° in 14-22 MLT and from 69° to 83° in 22-06 MLT, respectively. The low-latitude boundary of broad-band auroral hiss is higher in latitude on the dayside(06-14 MLT) than on the afternoon-night-morning side(14-22 MLT, 22-06 MLT), and it is approximately symmetric with respect to the 10-22 MLT meridian as well as Fig. 4. Thus, the latitudinal extent of broad-band auroral hiss region in geomagnetic quiet period(K_p= 0 - 1) is appreciably narrower than that of Fig. 4 including various geomagnetic activities(K_p= 0 - 7), but it is wider than the quiet-time auroral zone. Therefore, the polar map for broad-band auroral hiss differs noticeably from the auroral zone, though the former includes many ISIS polar passes during various geomagnetic activities and the latter is defined for a specific geomagnetic activity. In fact, the mean latitudinal width of inverted-V events is about 0.5° invariant latitude and an average longitudinal width appears to be much larger than the typical width, 0.1° of auroral arcs[16.17]. Of course, the auroral electron precipitation region partially overlaps the inverted-V electron precipitation one. The broad-band auroral hiss occurs above the ionosphere, while the aurora is observed at ground level or it occurs in the ionosphere. These imply that the broad-band auroral hiss is generated by the inverted-V electrons precipitated from the boundary plasma sheet above the ionospher, but not by the auroral electrons producing the auroras in the ionosphere. In other words, the broad frequency range of auroral hiss seems to correspond to the broad energy range of the inverted-V electrons. Thus, the broad-band auroral hiss should be hereafter, called by the name of the polar hiss.

5. Whistler-Mode Cherenkov Radiations Generated from Inverted-V Electrons in Polar Magnetosphere

Broad-band auroral hisses observed by ISIS satellites are downgoing whistler-mode waves in a frequency range of from 1 kHz to above 30 kHz, and they were continuously observed over a wide latitudinal range by the ISIS satellites[24]. The auroral hiss occurred in direct association with the inverted-V electrons[11]. Though inverted-V events occur in the general area of the auroral arcs, latitudinal width of the inverted-V electrons is much larger than the auroral arc width of less than 10 km as well as that of the broad-band auroral hiss. The latitudinal width of most inverted-V events are from 0.1° to 2° invariant latitude with a mean width of about 0.5°, and the average longitudinal width appears much larger than the typical auroral arc length of 0.1°[17]. Thus, it is no longer necessary to discuss the relationship between auroral hiss and the aurora.

Precipitating electrons are the free energy sources of certain plasma instabilities in a high-latitude magnetosphere. Since the auroral hiss is believed to be generated by the Landau resonance[21,22] in which the electron beam moves in the same direction as the wave, i.e., $\omega/k_{\parallel} = V$, the downward propagating auroral hiss should be generated by downward moving inverted-V electrons, where V is the parallel beam velocity to the field line and $k_{\parallel}$ is the parallel wave number to the field line. Then, the frequency range of the Cherenkov radiation generated by precipitating electrons with a broad energy range in the polar magnetosphere is discussed to explain the frequency range for downgoing broad-band auroral hiss. Whistler-mode refractive index, n is given by $n^2 = f_p^2/ff_H\cos\Theta$ for the conditions f_H, $f_p \gg f$ and $f_H\cos\Theta \gg f$, where f denotes whistler-mode wave frequency, f_H electron gyrofrequency, f_p electron plasma frequency and Θ wave normal angle to the geomagnetic field line[29].

For simplicity, we consider Cherenkov radiation from electrons with parallel velocity, V to the geomagnetic field lines. The frequency of the whistler-mode Cherenkov radiation, $f = \frac{f_p^2\cos\Theta}{f_H}\cdot\frac{V^2}{c^2}$ is derived from the whistler-mode refractive index and the Cherenkov resonant condition, $c/n = V\cos\Theta$, where c is light speed, n whistler-mode refractive index and Θ the wave normal angle to the field line. The frequency of this radiation is rewritten as $f = f_p^2\cos\Theta\cdot E/250f_H$ kHz, where $E = 250(V/c)^2$ keV is electron energy with parallel velocity, V. So, whistler-mode Cherenkov radiation has an upper limit frequency given by $f \leqq f_p^2\cdot E/250f_H$. This means that the higher frequency components of the radiation should be generated at the lower altitudes along the geomagnetic field line since the upper limit frequency of the radiation decreases with increasing altitude.

Next, we will calculate the upper limit frequency of the radiation as a function of electron energy, E at various altitudes along the two geomagnetic field lines for invariant latitudes 70° and 77°, using a dipole type electron gyrofrequency of $f_H = 896(R_e/R)^3(1 + 3\sin^2\Theta_m)^{1/2}$ kHz for a geomagnetic field of $H_o = 0.32$ gausses at the geomagnetic equator and the polar altitude profile for electron density (Fig. 8) obtained from the DE-1 plasma wave data and Alouette/ISIS data[30], where R is geocentric distance, R_e earth's radius and Θ_m geomagnetic latitude. The two geomagnetic field lines of 70° and 77° represent, respectively, the low latitude occurrence boundary and the active region for the broad-band auroral hiss as shown in Fig. 4. The thick curve

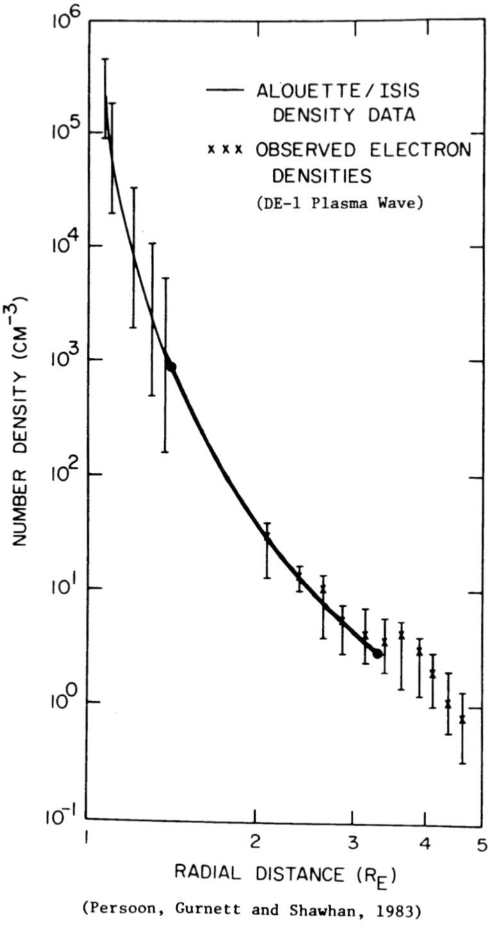

Fig. 8 Altitude profile of electron density in the polar magnetosphere obtained from Alouette/ISIS radio sounding data and DE-1 plasma wave data (after Persoon, Gurnett and Shawhan, 1983). The thick curve is the present author's interpolation.

between the Alouette/ISIS and the DE-1 data(crosses) in Fig. 8 was interpolated by the present author for this calculation and is the most realistic available electron-density profile in the polar magnetosphere. Uncertainties of electron densities in the nightside auroral zone are as high as ± 40 percents[31]. The upper limit frequency of the whistler-mode Cherenkov radiation was calculated at four geocentric distances; 1.32 R_e(altitude about 2000 km), 1.5 R_e(about 3200 km), 2.0 R_e (about 6400 km) and 3.0 R_e(about 13000 km) along the two geomagnetic field lines of invariant latitude 70° and 77°. Table 1 lists the key parameters for these calculations.

Table 1 Key parameters used to calculate the upper limit frequency of whistler-mode Cherenkov radiation in the polar magnetosphere.

R	Θ_m	f_H	min f_p	mean f_p	max f_p	N
On the field line at invariant latitude 70°						
1.32 R_e	66.9°	732.6 kHz	304 kHz	392 kHz	464 kHz	1900 ± 760/cm^3
1.5	65.2°	494.8	139	180	213	400 ± 160
2.0	61.1°	203.5	38	49	58	30 ± 12
3.0	53.7°	57.0	16	20	24	5 ± 2
on the field line at invariant latitude 77°						
1.32 R_e	75.0°	759.3 kHz	304 kHz	392 kHz	464 kHz	1900 ± 760/cm^3
1.5	74.0°	515.6	139	180	213	400 ± 160
2.0	71.5°	215.3	38	49	58	30 ± 12
3.0	67.1°	62.5	16	20	25	5 ± 2

Three kinds of the electron plasma frequency,f_p represents ones corresponding to the electron density uncertainties. Frequencies above 1 kHz for broad-band auroral hiss do not satisfy the whistler-mode condition $f_H, f_p >> f$ at 3.0 R_e on the two geomagnetic field lines.

Figures 9a and 9b show the relationship of the calculated upper limit frequency of the whistler-mode Cherenkov radiations to electron energy at the four geocentric distances on the two field lines of 70° and 77°, respectively. Vertical rods show error ranges of upper limit frequency of whistler-mode Cherenkov radiation corresponding to the electron density uncertainties of ± 40 percents. The error range of the upper limit frequency increases with increasing electron energy at the same geocentric distance, and also, it increases with decreasing geocentric distance at the same electron energy. The frequency range of the whistler-mode Cherenkov radiation at each geocentric distance is in a region lower than each oblique straight line. Same energy electrons can emit higher frequency components of whistler-mode Cherenkov radiation at lower altitudes on the same field line and, also the generation altitude region for lower frequency components is wider than that for higher frequency components. The upper limit frequencies on the 70° field line are approximately the same as those on the 77° one. The energy range of the inverted-V electrons associated with auroral hiss is below about 40 keV[16]. So, frequency ranges for whistler-mode Cherenkov radiation from electrons with energy below 40 keV are below 34 kHz at a geocentric distance of 1.32 R_e(altitude about 2000 km), below 12 kHz at 1.5 R_e(about 3200 km), below 2 kHz at 2.0 R_e (about 6400 km) and below 1 kHz at 3.0 R_e(about 13000 km) on both geomagnetic field lines. In other words, the lower frequency components below 2 kHz and 12 kHz of the whistler-mode Cherenkov radiation are respectively generated at altitudes below 6400 km and 3200 km in the polar magnetosphere. The higher frequency components below 34 kHz are generated at altitudes lower than 2000 km in the polar topside ionosphere. This result is considerably different from Shawhan's model [32] that V-shaped downgoing auroral hiss(1-30 kHz) is generated by

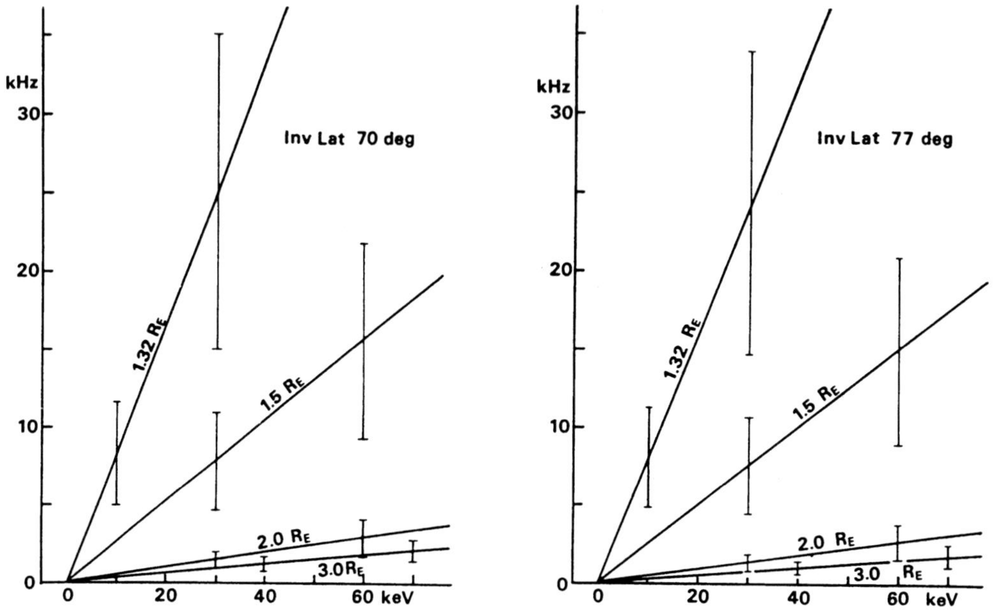

Figs. 9a & 9b Relationship of electron energy vs. upper limit frequency of whistler-mode Cherenkov radiations generated from electrons calculated by $f_p^2E/250f_H$ kHz at geocentric distances of 1.32,1.5,2.0 and 3.0 R_e on two geomagnetic field lines at 70° (9a) and 77°(9b). Vertical rods are error ranges.

inverted-V electrons(1-10 keV) at altitudes around 4000 km on auroral geomagnetic field lines. Altitude profiles for frequency spectra of downgoing auroral hiss should be observed to examine if the broadband auroral hiss has the altitude dependent upper limit frequency as expected by the whistler-mode Cherenkov radiation. However, such an observation is difficult since a polar orbiting satellite can only detect auroral hiss once for a specific geomagnetic field line.

Recently, auroral hisses with characteristic "funnel-shaped" f-t spectrum were observed by DE-1 over the auroral zone[33]. The funnel-shaped auroral hiss is very similar to the VLF saucer emission observed in the auroral zone by low-altitude polar orbiting ISIS satellites [23]. However, the funnel-shaped hiss appears in time scale less than 20 minutes(less than 10° in geomagnetic latitude) at DE-1 altitudes (15000 - 20000 km), and the VLF saucer appears in time scale less than 10 seconds(less than 1° in geomagnetic latitude) at ISIS altitudes (1000 - 3500 km). Both emissions have been explained by a propagation effect for whistler-mode waves with wave vectors near the resonance cone[23,33], though the time or spatial scale of the former is larger than that of the latter. At the ISIS altitudes, the VLF saucer normally occurs during a short period between curtain-shaped auroral hisses with irregular low-frequency end around 1 kHz, but at the DE-1 altitudes, the funnel-shaped auroral hiss occurs normally over the auroral zone. So, it is interesting if spatial and time scale difference of the two phenomena are simply due to a difference between the observed altitudes of VLF saucer and funnel-shaped VLF hiss.

Gurnett et al[32] compared ray path boundaries computed by the whistler-mode propagation model with observed f-t profile of a funnel-shaped auroral hiss event. The comparison shows that the funnel-shaped auroral hiss propagates upward from a localized source below the DE-1, and that the best fit radial distance of the low-altitude boundary of the upgoing auroral hiss source is about 1.7 - 1.8 R_e. Lin et al.[34] found several events in which upgoing electron beams were associated with funnel-shaped auroral hisses. Also, energetic ion beams above 1 keV were found inside the auroral plasma cavity in the region of lowest auroral electron densities by the DE-1[31]. However, studies of attenuation bands in VLF saucers show that source regions of VLF saucers observed in the polar topside ionosphere are located above or below the ISIS spacecrafts[35,36,37]. So, the funnel-shaped auroral hiss may be also produced by a downgoing source above the DE-1 spacecraft at altitudes below about 10000 km on polar field lines because the whistler-mode propagation condition can be satisdied only at frequencies below about 5 kHz in altitude range above 10000 km on the polar field lines.

Maggs[21] has suggested that a mono-energetic electron beam can excite coherent VLF hiss with propagation angles near the resonance cone, and that a coherent amplification mechanism can produce power fluxes of strong VLF hiss when f_H is less than f_p. This last condition can not be satisfied in the auroral plasma cavity of the polar magnetosphere. Furthermore, the broad frequency range of the downgoing auroral hiss should be interpreted as broad-band radiation generated from precipitating inverted-V electrons with a wide energy range at various altitudes in the polar magnetosphere. Thus, we have no appropriate theoretical model to explain the broad-band auroral hiss observed in the low-density polar magnetosphere.

6. Conclusion

High-latitude limits of ELF hisses follow approximately the auroral zone at all geomagnetic local times. A polar occurrence map for broad-band auroral hiss obtained from 347 ISIS VLF passes was compared with a polar occurrence map for 280 inverted-V electron precipitation events observed by AE-D. The polar map of the occurrence rates for broad-band auroral hiss is qualitatively similar to the one for the occurrence locations of inverted-V events; especially, for the low-latitude boundary and 10-22 hour MLT symmetric meridian. This relationship is the same as that in geomagnetic quiet periods(K_p= 0 - 1). So, the broad-band auroral hiss is closely related to the inverted-V electrons precipitated from the boundary plasma sheet. DE-1 observed that auroral VLF hiss appeared at altitudes above 10000 km on polar geomagnetic field lines[21,38].

The whistler-mode Cherenkov radiation generated by inverted-V electrons in polar magnetosphere was discussed as the source of downgoing broad-band auroral hiss. The upper limit frequencies of the whistler-mode Cherenkov radiation were calculated at geocentric distances of 1.32, 1.5, 2.0 and 3.0 R_e for two geomagnetic field lines at 70° and 77°, which respectively represent the low-latitude boundary and active occurrence latitude of broad-band auroral hiss. The results were approximately the same for the two field lines. The whistler-mode condition is not satisfied for all frequencies above 1 kHz for the broad-band auroral hiss at 3 R_e on the two field lines. The generation region for low frequency components of the downgoing whistler-mode Cherenkov radiation has a wider altitude than high frequency components along the same polar field lines. Frequencies above 1 kHz of downgoing broad-band auroral hiss seem to be explained by the whistler-mode

Cherenkov radiation generated from inverted-V electrons with energy below about 40 keV at altitudes below about 2 R_e in the polar magnetosphere.

The coherent generation mechanism for auroral hiss proposed by Maggs [21] is insufficient to explain the strong auroral hiss observed in the auroral plasma cavity above the polar ionosphere.

REFERENCES

1. Thorne R. M. Smith E. J. Burton R. Hlzer R. E. (1973) Plasmaspheric hiss. J Geophys Res 78 : 1581-1596
2. Ondoh T. Nakamura Y. Watanabe S. Murakami T. (1983) Plasmaspheric hiss observed in the topside ionosphere at mid- and low-latitudes. Planet Space Sci 31 : 411-422
3. Ellis G. R. (1957) Low-frequency radio emission from aurorae. J Atmos Terr Phys 10 : 302-306
4. Duncan R. A. Ellis G. R. (1959) Simultaneous occurrence of subvisual aurorae and radio noise bursts on 4.6 kc/s. Nature 183 : 1618-1619
5. Martin L. H. Helliwell R. A. Marks K. R. (1960) Association between aurorae and very low-frequency hiss observed at Byrd station, Antarctica. Nature 187 : 751-753
6. Jørgensen T. S. Ungstrup E. (1962) Direct observation of correlation between aurorae and hiss in greenland. Nature 194 : 462-463
7. Morozumi H. M. (1963) Semi-diurnal auroral peak and VLF emissions observed at the south pole. Trans American Geophys Union 44 : 798-806
8. Harang L. Larsen R. (1965) Radio wave emissions in the v.l.f. band observed near the auroral zone - I ; Occurrence of emissions during disturbances. J. Atmos Terr Phys 27 : 481-497
9. Helliwell R. A. (1965) Whistlers and Related Ionospheric Phenomena . Stanford Univ Press. p205
10. Gurnett D. A. (1966) A satellite study of VLF hiss. J. Geophys Res 71 : 5599-5615
11. Gurnett D. A. Frank L. A. (1972) VLF hiss and related plasma observations in the polar magnetosphere. J Geophys Res 77 : 172-190
12. Hoffman R. A. Laaspere T. (1972) Comparison of very-low-frequency auroral hiss with precipitating low-energy electrons by the use of simultaneous data from two Ogo 4 experiments. J. Geophys Res 77: 640-650
13. Hughes A. R. W. Kaiser T. R. Bullough K. (1971) The frequency of occurrence of VLF radio emissions at high latitudes. Space Res XI : 1323-1330 Berlin Akademie

14. Ondoh T. (1988a) Occurrence map of broad-band auroral hiss observed by ISIS satellites. Proc. NIPR Symp Upper Atmos Phys 1: 153-167

15. Ondoh T. (1988b) Polar occurrence map of broad-band auroral hiss observed by ISIS satellites. J. Radio Res Labs Japan 35 : 1-14

16. Hoffman R. A. Lin C. S. (1981) Study of inverted-V auroral precipitation events. In : Akasofu S.-I. Kan J. R.(ed) Physics of Auroral Arc Formation. Am Geophys Union Geophys Monogr 25: 80-90

17. Lin C. S. Hoffman R. A. (1982) Observation of inverted-V electron precipitation. Space Sci Rev 33: 415-457

18. Dowden R. L. (1960) Geomagnetic noise at 230 kc/s. Nature 187 : 677-678

19. Ondoh T. (1961) On the origin of VLF noise in the earth's exosphere. J Geomag Geoelectr 13: 77-83

20. Jørgensen T. S. (1968) Interpretation of auroral hiss measured on OGO 2 and at Byrd station in terms of incoherent Cerenkov radiation. J Geophys Res 73: 1055-1069

21. Maggs J. E. (1976) Coherent generation of VLF hiss. J Geophys Res 81 : 1707-1724

22. Maggs J. E. (1978) Electrostatic noise generated by the auroral electron beam. J Geophys Res 83 : 3173-3188

23. James H. G. (1976) VLF saucers. J Geophys Res 81: 501-514

24. Ondoh T. Nakamura Y. Murakami T. (1980) Characteristics of auroral VLF emissions observed by ISIS. J Radio Res Labs Japan 27: 141-156

25. Nishida A. (1968) Geomagnetic Dp-2 fluctuations and associated magnetospheric phenomena. J Geophys Res 73 : 1795-1803

26. Akasofu S.-I. (1981) Auroral arcs and auroral potential structure. In : Akasofu S.-I. Kan J. R. (ed) Physics of Auroral Arc Formation. Am Geophys Union Geophys Mongr 25: 1-14

27. Vasyliunas V. M. (1970) The polar Ionosphere and Magnetospheric Processes. Skovlii G. (ed) Gordon and Breach P 26

28. Ondoh T. (1990) Broad-band auroral hiss observed by ISIS in geomagnetic quiet conditions (to be published)

29. Storey L. R. O. (1953) An investigation of whistling atmospherics. Phil Trans Roy Soc London A 246: 113-141

30. Persoon L. A. Gurnett D. A. Shawhan S. D. (1983) Polar cap electron densities from DE-1 plasma wave observations. J Geophys Res 88 : 10123-10136

31. Persoon A. M. Gurnett D. A. Peterson W. K. Waite J. H. Jr. Burch J. L. Green J. L. (1988) Electron density depletions in the nightside auroral zone. J Geophys Res 93 : 1871-1895

32. Shawhan S. D. (1979) Magnetospheric plasma waves. In : Lanzerotti L. J. Kennel C. F. Parker E. H. (ed) Solar System Plasma Physics North-Holland pp 211-270

33. Gurnett D. A. Shawhan S. D. Shaw R. R. (1983) Auroral hiss, Z mode radiation, and auroral kilometric radiation in the polar magnetosphere : DE-1 observations. J Geophys Res 88: 329-340

34. Lin C. S. Burch J. L. Shawhan S. D. Gurnett D. A. (1984) Correlation of auroral hiss and upward electron beams near the polar cusp. J Geophys Res 89: 925-935

35. Watanabe S. Ondoh T. Nakamura Y. Murakami T. (1979) Attenuation band and electric field of VLF saucers. Antarctic Record 64 : 159-166

36. Horita R. E. James H. G. (1982) Source regions deduced from attenuation bands in VLF saucers. J Geophys Res 87: 9147-9153

37. Ondoh T. Watanabe S. Nakamura Y. (1984) VLF emissions observed by ISIS satellites during the IMS. Proc Conf Achievements of the IMS. ESA Sp-217 pp 525-528

38. Ondoh T. Nakamura Y. Watanabe S. Aikyo K. (1988c) Comparative study of VLF radio wave phenomena from DE-1 and ISIS satellites received at Kashima, Japan. J Radio Res Labs Japan 35: 27-42

4.3 Direction Finding of Magnetospheric VLF/ELF Emissions and Their Generation Mechanism

M. HAYAKAWA

Research Institute of Atmospherics, Nagoya University, Aichi, 442 Japan

Abstract The purpose of this report is to review our recent studies on the generation and propagation mechanisms of magnetospheric VLF/ELF emissions based on the direction finding measurements on board spacecrafts in order to demonstrate the potential use of the direction finding measurements in plasma wave studies. The plasma waves treated in this report are, (1) half-gyrofrequency VLF emissions, (2) ELF hiss in detached plasma regions of the magnetosphere, (3) plasmaspheric ELF hiss, and (4) VLF/ELF chorus. We then point out the technical problems to be investigated concerning the spacecraft direction finding measurement to be used for the future satellite experiments.

1. INTRODUCTION

Information concerning the wave normal directions of different magnetospheric VLF/ELF emissions is considered to be an invaluable tool in studying such wave properties and the mechanisms of their generation and propagation. Hence, the so-called "direction finding (DF)" to determine the wave normal directions of magnetospheric plasma waves is of potential use in the investigation of plasma waves. Several methods have been proposed for the DF and several papers have been published on the generation mechanism of different plasma waves by making full use of the DF measurements.

The purpose of this report is to review firstly our recent findings on the mechanisms of generation and propagation of different magnetospheric plasma waves as based on the DF measurements in order to demonstrate how important the usage of DF is in the study of plasma waves. Then,we point out the works associated with the DF to be utilized in the future satellite experiments.

2. DIRECTION FINDING RESULTS OF MAGNETOSPHERIC VLF/ELF EMISSIONS

Although there have been published papers dealing with the ground-based DF measurements, we will concentrate ourselves to the results by the spacecraft-based DF studies during the last decade. DF has been applied to different kinds of magnetospheric VLF/ELF emissions. In order to demonstrate how much the DF measurement has contributed to the better understanding of magnetospheric plasma waves, we have reviewed our recent findings based on the DF measurement for a few selected VLF/ELF emission types; (1) half-gyrofrequency VLF emissions, (2) ELF hiss in a detached plasma region of the magnetosphere, (3) plasmaspheric ELF hiss and (4) VLF/ELF chorus. Only the papers on which the present report is based, are cited in the references, and other papers appeared in the text are referenced therein.

2.1. VLF/ELF EMISSIONS AT FREQUENCY ABOVE ONE HALF THE ELECTRON GYROFREQUENCY (HALF-GYROFREQUENCY VLF EMISSIONS) [1,2,3,4]

This emission occurs at a frequency above one half the equatorial electron gyrofrequency, and this type of wave phenomena was first discovered by Coroniti et al.(1960). However,their wave characteristics (electromagnetic or electrostatic) and their generation mechanism have long been unsolved. Hayakawa et al. [1] are the first that have applied the DF measurement using the maximum likelihood and Means' methods to this wave phenomenon observed on GEOS-2 satellite in the vicinity of the geomagnetic equator. Figure 1 illustrates one example of the spectrogram in which the upper band emissions are the half-gyrofrequency VLF emissions, while the lower band indicates the chorus to be treated in Section 2.4. Figure 2 illustrates the corresponding DF result of the half-gyrofrequency VLF emissions which is reproduced from Hayakawa et al. [1]. Figure 1 shows that the emission frequency is mainly above one half the equatorial gyrofrequency, being indicative that the source region is located near the equator, and then Fig.2 indicates that the wave normal directions(θ) is very close to the oblique resonance cone (θ_{res}) angle. A combined consideration of the properties of wave frequency with the wave normal directions, has led us to consider that the half-gyrofrequency VLF emissions are quasi-electrostatic whistler-mode emissions. These experimental characteristics are checked with the theoretical generation mechanism proposed, and they are found to be very consistent with the theory of Harris-type electrostatic instability (Ohmi and Hayakawa [2]). Figure 3 is one example of their numerical calculations of the wave growth rate as functions of wave frequency and wave normal direction. The figure illustrates the variation of the most unstable region with increasing the density of hot plasma (n_h) when the the cold plasma density (nc) is fixed and f_H(electron gyrofrequency)= 2.5 kHz, which is consistent with our observational findings.

The corresponding off-equatorial DF for the same type of emissions has been performed on GEOS-1 satellite [3,4]. One example of the off-equatorial DF measurements by means of the wave distribution function method is illustrated in Fig.4, which indicates that the wave normals observed

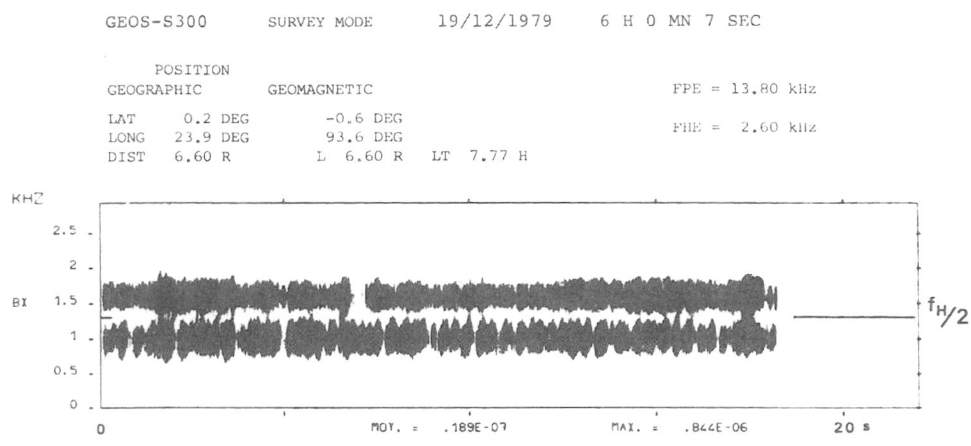

Fig.1 An example of the unstructured half-gyrofrequency VLF/ELF emissions (the upper band) observed near the geomagnetic equator. While,the lower band indicates the chorus emissions.

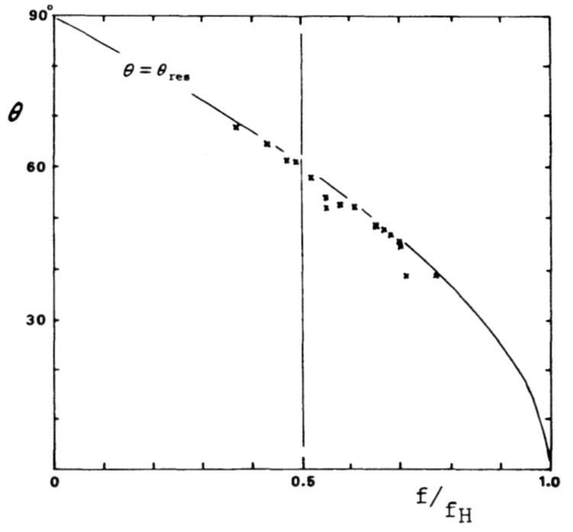

Fig.2 Direction finding result of the half-gyrofrequency VLF emissions shown in Fig.1. θ_{res} is the oblique resonance cone angle.

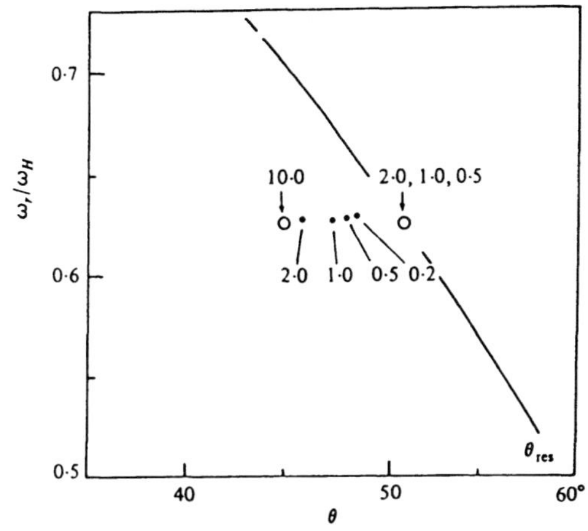

Fig.3 Variation of the most unstable region with increasing n_h when n_c is fixed (n_c=2.79 cm^{-3}). The values in the figure refers to n_h/n_c. •,$T_{\parallel}^{h}$=30 eV and o,10eV. $T_{\perp}^{h}/T_{\parallel}^{h}$=3.0 in both cases.

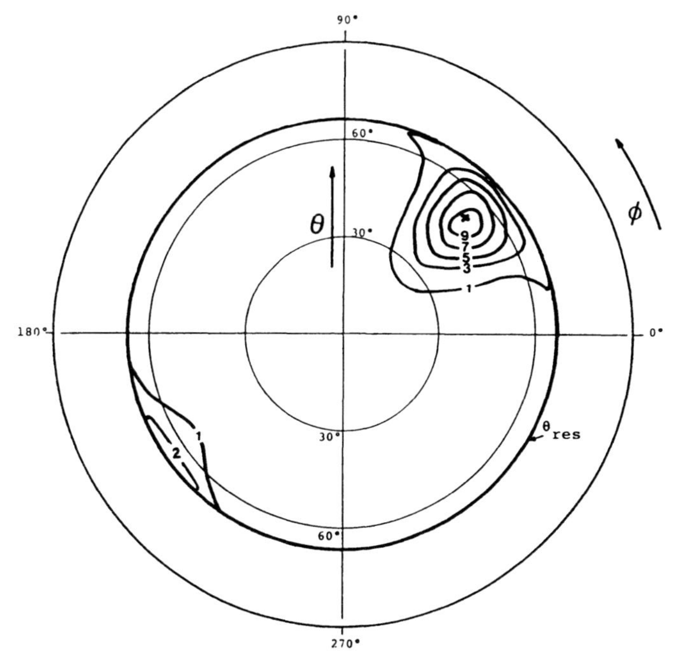

Fig.4 An example of wave distribution functions of the half-gyrofrequency VLF emissions observed far away from the geomagnetic equator.

are about 20° away from the θ_{res}. The rays whose wave normal directions are determined far away from the equator, are traced backward the equator where the emissions are possibly generated as indicated before. Then, it is found that the wave emissions have their frequency above one half the equatorial gyrofrequency with their wave normals very close to the oblique resonance cone (θ_{res}) there, and hence the equatorial generation mechanism we proposed is reconfirmed by these off-equatorial observations.

2.2. ELF HISS EMISSIONS IN DETACHED PLASMA REGION OF THE MAGNETOSPHERE (DP HISS) [5,6]

ELF hiss is known to be distinctly observed in detached plasma regions of the magnetosphere which are defined as regions outside the plasmapause where the cold plasma density is considerably enhanced compared with the ambient density (Chan,1974; Chan et al.,1974; Chan and Holzer, 1976; Cornilleau-Wehrlin et al.,1978; Hayakawa et al. [5]). And we refer hereafter to these emissions as "DP(detached plasma) hiss". Chan (1974) and Chan and Holzer(1976) have investigated the wave normals of DP hiss mainly in the off-equatorial region, and due to the off-equatorial observation they were unable to distinguish between the effects of generation and propagation. Hayakawa et al. [5] and Hayakawa [6] have determined the wave normal directions (wave distribution functions) of DP hiss in the equatorial plane of the magnetosphere which is considered to be the source region.

Figure 5 illustrates one example of DP hiss observed at the equator on GEOS-2 satellite at a particular L.T. when we have the enhanced plasma density, and Fig.6 indicates the wave distribution function in the form of wave energy contour map for DP hiss in Fig.5 at a frequency of 233 Hz. The previous morphological properties have been reconfirmed. The wave distribution functions of DP hiss consist of a single peak and the wave normal directions θ make small angles (less than 25°) with the Earth's magnetic field. The hypothesis of local generation at the equator seems to be supported by the experimental fact that the wave distri-

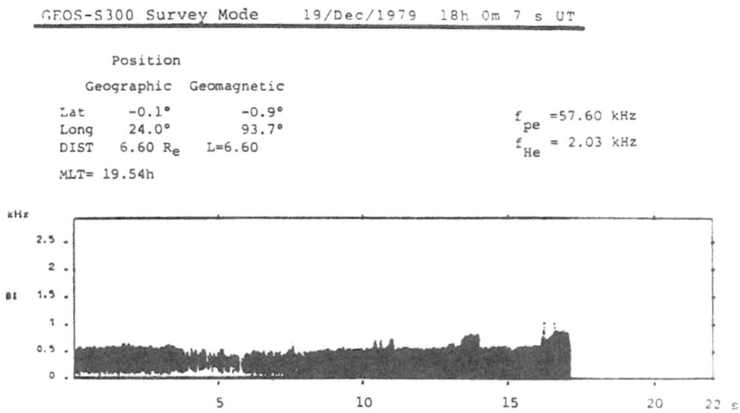

Fig.5 An example of the spectrogram of ELF hiss in a detached plasma region of the magnetosphere (DP hiss).

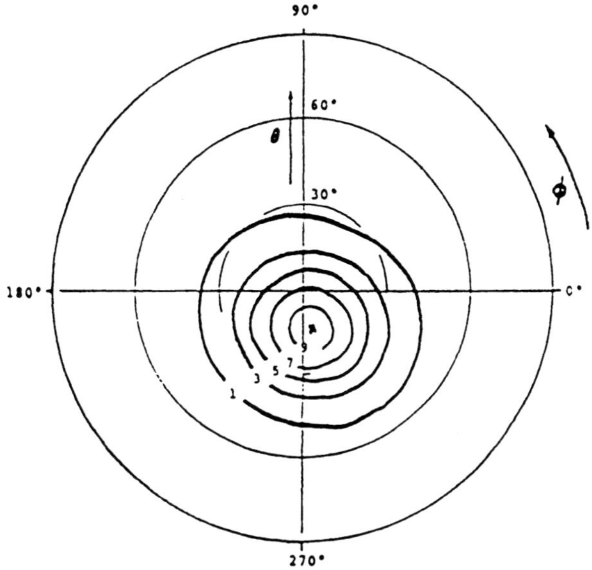

Fig.6 An example of wave distribution functions of DP hiss in Fig.5 at a frequency of 233 Hz.

bution functions are almost circular. Hence,we can conclude experimentally that the wave growth at the equator is strongest for the quasi-longitudinal propagation direction ($\theta \sim 0°$). The smallness of wave normal angles at the equator is favourable to both wave generation and ducted propagation down to the ionosphere. Chan(1974) has found that the wave normal directions of DP hiss at high geomagnetic latitudes (30°-50°) show a great scatter over an angular range from 2° to 40°, but they are mostly less than 35°, which are greater than the θ values at the equator, and furthermore he has found no latitudinal dependence of wave normal characteristics. So,it is likely that an additional scatter in wave normal directions at higher latitudes may be due to the scattering of wave normals by the density irregularities during the course of ducted propagation away from the equatorial source region. These DP hiss may be reasonably thought to remain trapped down to the ionosphere by the density enhancement of the detached plasma region.

Finally, many characteristics including the DF results at the equator and off the equator are successfully interpreted in terms of a quasi-linear electron Cyclotron instability with $\theta \sim 0°$, and the DF findings have substantiated the previous generation theory.

2.3. PLASMASPHERIC ELF HISS [7,8]

Plasmaspheric ELF hiss is a broadband and structureless whistler-mode noise persistent within the plasmasphere. Thorne et al. (1973) have made detailed analyses of the OGO 5 wave data, together with some DF results, which have led them to conclude that plasmaspheric hiss is generated with its wave normal aligned with the magnetic field, just inside the plasmapause by the Cyclotron instability due to medium energy (10-100 keV) electrons.

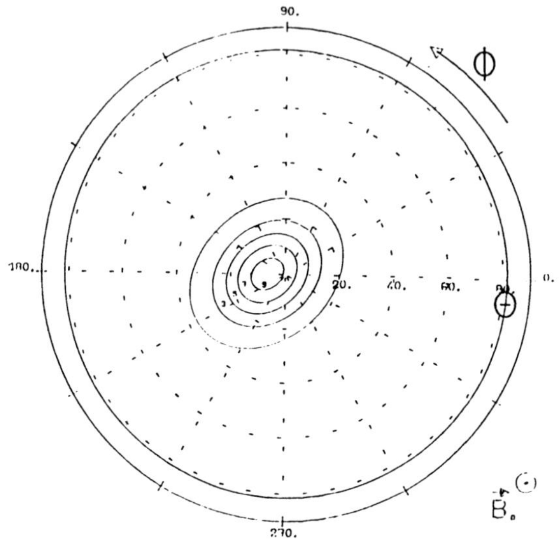

Fig.7 A wave distribution function of plasmaspheric ELF hiss on 5 August 1977. Frequency= 326 Hz.

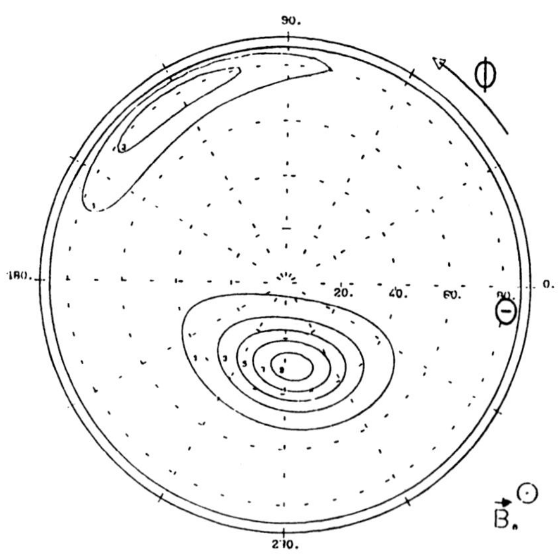

Fig.8 A wave distribution function of plasmaspheric ELF hiss on 28 September 1977. Frequency = 465 Hz.

Some of our GEOS-1 results on the wave normals of plasmaspheric ELF hiss observed near the equator are presented in Figs.7 and 8. Fig.7 refers to the event of 5 August 1977, and Fig.8, 28 September 1977. For the event of 5 August where the satellite is just at the inner edge of the equatorial plasmapause, the wave normal of ELF hiss at different frequencies as in Fig.7 is nearly aligned with the Earth's magnetic field. For other events including 28 September (Fig.8) for which the DF is made at the equatorial plane,but a little inside from the plasmapause by 0.3-0.5 Re(R_e: Earth's radius), two different groups of wave normal directions are found; one is a medium wave normal angle ranging from 20° to 50°, and the other is large wave normal angle θ=70°-80° which is around the Gendrin angle (θ_g) close to the oblique resonance angle (θ_{res}).

The DF results of our equatorial observations are used to test the previous theoretical model of Thorne et al.(1973). Their mechanism can be maintained by amplified waves being returned to the growth region after a reflection at the top of the ionosphere (ducted waves) (Kennel and Petschek,1966) or after a magnetospheric reflection (unducted waves). In the former process, too small an amount of wave energy is reinjected and in the latter only oblique waves generally considered as being only weakly amplified, are returned to the equator. Then, Thorne et al.(1979) have shown that the wave propagation including the plasmapause must be carefully considered in the growth rate computations, and they have proposed an importance of "cyclic waves", which have been amplified and followed by returning to the growth region after an internal reflection at the plasmapause. These cyclic waves will have the quasi-longitudinal wave normals at the equator so as to maintain the amplification. Just at the inner edge of the plasmapause (5 August), the wave normals are found to be nearly aligned with the magnetic field, which may provide a support to the existence of plasmapause-guided waves as suggested by Inan and Bell (1977) and Thorne et al.(1979), assuming an initial field-aligned wave normal at the equator just inside the plasmapause. Furthermore, Huang et al.(1983) and Church and Thorne(1983) have demonstrated that ray-path integrated gains are actually high along the plasmapause because the wave normals of the plasmapause-trapped waves are nearly along the magnetic field to allow further cyclotron amplification. However,on other two events of 4 and 28 September, many emissions are found to be highly turbulent hiss-type, and to consist sometimes of two plane waves. The larger θs range from 70° to 80°,slightly smaller than θ_g, while the medium θ values are widely distributed from $\sim$ 20° to $\sim$50°. The concept of cyclic ray paths might be reflected in the two wave peaks with smaller wave normals (15°-30°) for the events of 4 and 28 September on the basis of the comparison of our observed wave normal behaviours with the result from the three-dimensional ray-path calculations of the integrated gains by Church and Thorne (1983) for their model hot electron distributions. They have found that maximum amplification occurs for almost field-aligned waves in the outer plasmasphere and also they have found a general correlation between the zone of wave amplification and the class of waves that internally reflect from the plasmapause, following equatorial transits. For example, the net accumulated gain of the 500 Hz wave observed with $\theta \lesssim 30°$(which follow a cyclic trajectory) is much higher, by more than 40dB, than that with $\theta \sim 60°$(which exhibits no plasmapause encounter and suffers only the magnetospheric reflection) at two selected observation positions of L=4.5 and 3.0 in the case when the plasmapause is assumed to be 5.0. However, when we consider that the intensity of very oblique waves such as θ=60-80° is not so much decreased compared with that of longitudinal waves ($\theta \lesssim 30°$), very oblique waves cannot be explained by the mechanism of Church and Thorne(1983). Then,it is very important to compare the relative location of the observing point with respect to the plasmapause,with the L dependence of the class of ray trajectories, because there is a possible range of launch positions and frequencies for cyclic ray paths for which we can expect higher accumulated net gains for cyclotron instability. The ray-tracing study by Thorne et al. (1979) has yielded the range in frequency and in launch position for which we can expect multiple field-aligned equatorial transits, on the assumption of field-aligned initial wave normal at the equator. To be more quantitative based on their results, when the plasmapause is again assumed to be L=5.0, the cyclic trajectories at f=500Hz we are concerned with, are expected at the lauch L values from ~4.8 to ~4.5 (about 0.2-0.5 Re inside the plasmapause) and also from 4.2 to 3.6. For the two events of 4 and 28 September, the satellite location is 0.5 and 0.3 Re, respectively, inside from the observed plasmapause boundary, where cyclic orbits are anticipated to take place. Especially, in the case of 28 September,the plasmapause position is found to be exactly at L=5.0, which was adopted in the theoretical ray-tracing by Thorne et al.(1979). The present study has yielded that oblique waves are, on many occasions, found in the plasmaspheric ELF hiss, giving a further support to the recent finding by Lefeuvre et al.(1983). Church and Thorne's (1983) conclusion based on the cyclotron mechanism that field-aligned waves are most unstable, seems to be in contradiction with the detection of oblique waves in the region where cyclic orbits are expected to occur, which leads us to prefer to adopt the generation of oblique waves. However,ray paths are model dependent,it may be required to carry out the ray-tracings as done by Thorne et al.(1979) and Church and Thorne(1983) for the realistic magnetospheric profile observed simultaneously by the in-situ density measurement at the relevant time, before we come to the conclusion whether the generation of oblique waves is validated or not. Nevertheless,the support to oblique wave generation is given from the recent ground-based measurement at a low L (L=1.6) station (Hayakawa et al.1985).

Finally,it seems to the author that some of plasmaspheric ELF hiss emissions are generated by the electron cyclotron mechanism initially proposed by Thorne et al.(1973), but a considerable number of them are generated with large wave normal angles. This kind of wave generation with large wave normals for ELF hiss outside the plasmapause, has also been suggested by Lefeuvre and Helliwell(1985), but it is not certain whether their obtained result for the exohiss can be valid for plasmaspheric ELF hiss. The possible generation mechanism for oblique wave normals might be Cerenkov or Landau instability, and should be investigated in future. It is needless to say that further DF studies will be required for the better understanding of the generation mechanism of plasmaspheric hiss.

2.4. VLF/ELF CHORUS [1,9,10]

Chorus is an intense naturally occurring VLF emissions in the magnetosphere, which is characterized by its discrete spectra. The previous morphological studies have revealed that chorus is observed mainly at L values between the plasmapause and magnetopause and at all local times. The midnight sector has recently been considered to be an expecially interesting region for chorus because chorus is generated during magnetic substorms (Hayakawa et al.[1]). Then, chorus has been interpreted as being generated by substorm electrons with an anisotropic loss cone distribution and the experimental aspects are well explained by the theory of electron cyclotron instability (Hayakawa et al.[1], Goldstein and Tsurutani(1984)).

The equatorial DF result of chorus emissions is presented by Goldstein and Tsurutani(1984) and Hayakawa et al.[1], who have shown that the wave normals of the rising tone chorus having values of df/dt as normaly observed, take very small angles (5°-20°). These relatively small wave normal angles give further support to the loss cone instability as the generation mechanism as mentioned before. Hayakawa et al.[9,10] have recently made the off-equatorial DF for chorus and we will show those results with reference to their comparison with the equatorial ones.

The event analyzed is detected at a geomagnetic latitude of 17.4° and at an L value of 7.60 obviously outside the plasmapause. Figure 9 illustrates an ELF spectrogram of radio emissions up to 2.5kHz during this event, and it indicates a coexistence of different kinds of chorus structures with different df/dt's. In the figure we find two kinds of chorus; one is impulsive (or burstlike) and the other is the normal rising tone. Most of chorus emissions are found to exhibit a moderate df/dt of the order of ~ 1kHz/s, but there were not observed chorus emissions such as falling and constant frequency tones. The DF measurements have been performed for several intervals. Nineteen cases among successful 21 are found to represent single-peaked solutions in the wave distribution function. The (θ,ϕ) value of the peak of each wave distribution function was estimated and the result is summarized in Fig.10 in the form of θ vs wave frequency normalized by the local gyrofrequency. In the figure we have distinguished rising tones (indicated by Δ) from impulsive ones (x). The normalized frequency of the analyzed chorus emissions are found to lie in a range from 0.2 to 0.4. Figure 10 shows that there is no conspicuous relationship between the θ value and frequency in the case of rising tones (Δ) and they take larger θs in a range of 30°-50°; but a tendency for waves at higher frequencies to be travelling at a larger angle to the field for the impulsive chorus (x). The present result is directly compared with the corresponding off-equatorial DF measurement by Burton and Holzer (1974). They have found that θs for the dayside chorus are highly concentrated to a range less than 25°, but the distribution in θ becomes dispersed extending to $\theta \sim 60°$ when the geomagnetic latitude becomes above 25°. So, our Fig.10 is, generally speaking, consistent with their result. Burton and Holzer(1974) have ,however, only found normal rising tones, and the impulsive chorus is treated only in our paper.

Then, the relationship between the observed df/dt of chorus and the corresponding θ value is presented in Fig.11 for which θ is plotted as a function of df/dt. Two structures of chorus emissions are observed; i.e. rising tone and impulsive one. For the rising tones, df/dt is in a ran-

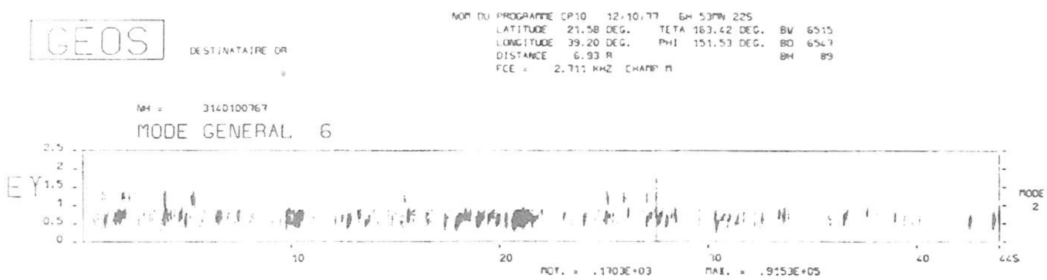

Fig.9 An example of ELF(0-2.5kHz) spectrograms of chorus emissions observed far away from the equator.

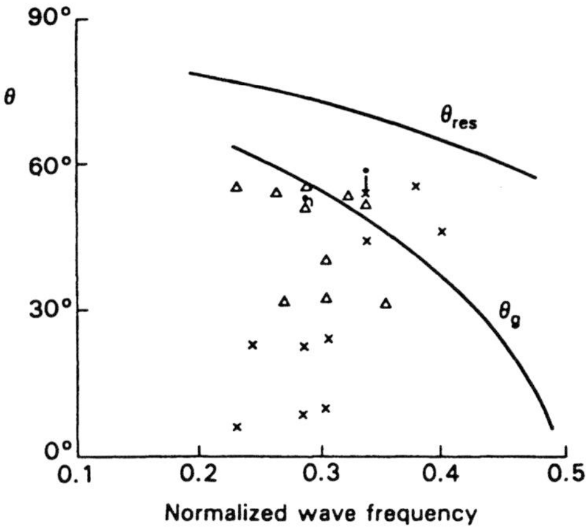

Fig.10 Relationship between θ and wave frequency normalized by the local gyrofrequency. x refers to impulsive chorus, while Δ refers to normal rising tone chorus.

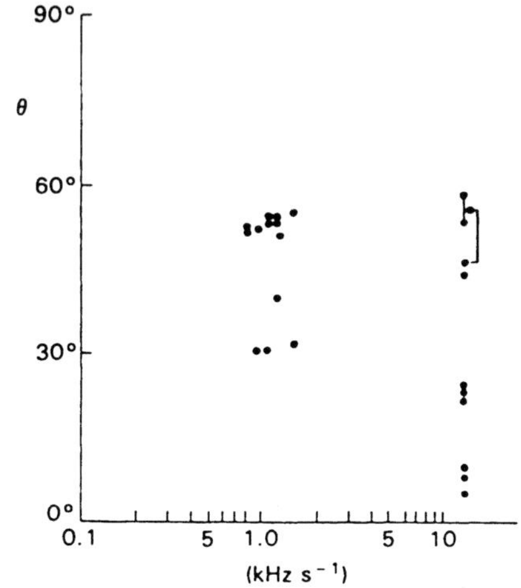

Fig.11 Relationship between θ and df/dt. Impulsive chorus is plotted just right of df/dt=10 kHz/s.

ge from 0.8 to 1.5 kHz/s and the θ is found to fall in a range from 30° to 55°. There is a large scatter in the θ of the impulsive chorus from 5° to ~60° but as mentioned before,there may be a frequency dependence of θ.

We discuss the behaviour of wave normal directions in the off-equatorial region as presented in this report, with respect to its comparison with the equatorial observations. First, we discuss the question of whether one wave or multiple waves are present in chorus. Goldstein and Tsurutani (1984) have compared the residuals of the one- and two-direction model fits to the observed spectral matrix and they have concluded that in most cases only a single plane wave is present. The present study based on the wave distribution function analyses has implied that about 90% of the events are composed of a single plane wave. This single wave observation is in agreement with Goldstein and Tsurutani (1984). In the studies of Burton and Holzer (1974) and Cornilleau-Wehrlin et al.(1976), they have adopted Means' method based on the hypothesis of the one-direction model, but their θ values seem to be reliable because most chorus events are due to the one-direction model even if they have erroneously determined wave normals of some chorus events if second waves are present. Within the source region near the equator, Goldstein and Tsurutani (1984) have found that the distribution in ϕ (azimuthal direction) appears to be isotropic, strongly implying that the satellite is located within the source region. In this report, two different types of chorus are observed: rising tones and impulsive ones. The df/dt of rising tones observed lies just around 1.0 kHz/s as in Fig.11, which is apparently typical at these L values and L.T.s from the work of Burtis and Helliwell (1976). The θ values of a chorus with df/dt of this order are found to be relatively small, less than 20° in the equatorial plane, which is likely to be an established fact. Compared with this result, Fig.11 seems to indicate that θs are considerably larger than those observed near the equator for rising tones, this being further supported by the previous results of off-equatorial DF by Burton and Holzer (1974). Although Burton and Holzer (1974) have concluded that the source region of chorus (falling and rising tones, but not impulsive) at daytime is at latitudes within 25° of the equatorial plane, it seems to us that the generation region of rising tone chorus observed at L.T.=10h in this report, is located at a latitude still lower than the satellite latitude of 17°. For the impulsive chorus, however, the θ value is greatly scattered in a wide range from 5°to 25° and from 45° to 60° in Fig.11, but the θ of the impulsive chorus near the equator makes larger angles with the magnetic field; because the generation mechanism of impulsive chorus is not well understood, we do not know at present which figure at the equatorial or at the off-equatorial region might reflect the wave generation distribution at the source region. One important point to mention here is related to the finding by Tsurutani and Smith (1977) of a high-latitude chorus in the dayside. They have indicated that a high-latitude chorus in the dayside appears to be generated locally in minimum B pockets, regions of local minimum magnetic fields found between 20° and 50° magnetic latitude which are caused by the compression of the dayside magnetosphere. The association of the chorus in this report with those high-latitude choruses would be interesting.

The DF has contributed very much to the better understanding of chorus, but the off-equatorial DF measurements,together with the equatorial ones, have stimulated the following problems to be investigated in future.

(1) For lower frequency (normalized frequency=0.1-0.3) rising tones with moderate df/dt s typical at the relevant L value as studied by Burtis and Helliwell (1976), their generation is due to the gyroresonance interaction between whistler-mode waves and counter-streaming electrons in the vicinity of the equator based on the following reasons. The first evidence is the equatorial DF studies. The other is the consideration of the interaction region by using the value of df/dt. Helliwell's(1967) theory suggests that the latitude of the interaction region is related to the slope,df/dt of an emission by the inhomogeneity of the medium. Using this theory in the relevant plasma parameters and taking df/dt=1 kHz/s, the latitude of the interaction region is found to be about 5°,so our Fig.11 as obtained at magnetic latitude of 17°,might indicate the presence of propagation effects. Ray-tracing computations will be greatly needed to confirm both the equatorial generation with $\theta \simeq 0°$ and the propagation effect from the equator to the spacecraft (Hattori et al. [11]).

(2) For rising tones at higher frequencies (normalized frequency=0.3-0.45), it is again plausible that the emissions are generated near the equator with small θs. However, Goldstein and Tsurutani (1984) have found a small concentration of θs at relatively large angles just around θ_g in the frequency range, 0.3-0.45. Of course, this concentration is not so conspicuous compared with the clear concentration of half-gyrofrequency VLF emissions at a specific angle of θ_{res} (Hayakawa et al.[1], Muto et al.[4]). There are a few theoretical studies on this point (Brinca(1972), Cuperman and Sternlieb (1974)) and Brinca(1972) has predicted maximum wave growth along the field for very low frequencies, but large off-axis growth with increasing frequency. Hence, further theoretical study on which kinds of magnetospheric conditions are required for oblique instability should be done in order to explain the possibility of this off-axis wave growth at higher frequencies below $f_H/2$. The θ results in the off-equatorial region (Fig.11) must be considered again with the aid of ray-tracing computations as mentioned in Point (1).

(3) More DF results for falling tone chorus events have to be accumulated in order to have a definite relationship between θ and df/dt, if any.
(4) Impulsive chorus at the equator is found to take large θ angles, but Fig.11 suggests that there is a large scatter in θ from nearly 0° to ~60°. The structures of normal rising, falling and nearly constant frequency tones seem to be explained in terms of the drifting oscillator model by Helliwell(1967). However, what is the generation mechanism of impulsive (burstlike) chorus emissions? Where are they generated and how are they propagated? Detailed study of this problem will be required.
(5) When dealing with the DF of chorus emissions in the off-equatorial regions in the dayside as done in this report and as further required in future, the association of the chorus emissions observed far from the equator with high-latitude chorus by Tsurutani and Smith(1977) will be investigated.

3. SUMMARY AND FUTURE WORKS

We have reviewed the generation mechanisms of four different kinds of magnetospheric VLF/ELF emissions as based on the DF measurements. The application of DF was extremely successful in elucidating the generation mechanism of half-gyrofrequency VLF emissions in Section 2.1. Next, the generation mechanism proposed for DP hiss was extensively elaborated by the equatorial DF in Section 2.2. The DF for rather old subjects of plasmaspheric ELF hiss and VLF/ELF chorus, has contributed much to the better understanding of each wave phenomenon, but it stimulated further interesting works as indicated in Sections 2.3 and 2.4. Hence, we made more extensive discussion for these two types of emissions than that of the former two types.

Several DF methods have been utilized in this review in Section 2, including Means' method, the maximum likelihood method and the wave distribution function (the maximum entropy method). The most general method of wave distribution function to estimate wave energy distribution is able to identify the simultaneous multiple propagation directions which are often encountered in satellite observations, but there are some problems even for this accepted method. In the following we will list several general technical problems concerning the DF system to be done in future.
(1) In the wave distribution function method, we use two parameters (i.e. prediction and stability parameters) for estimating how good the obtained result is, which helps us obtain the most optimum result. However, it is highly required to develop an "objective" way of yielding the optimum wave distribution function.
(2) In our review we have treated only the emissions in whistler-mode frequency range. However, when the wave frequency becomes either smaller than the proton gyrofrequency or above the electron plasma or gyrofrequency, it is possible that there exist two possible modes of propagation in certain frequency intervals. For this situation, we have to develop a DF system effective even for the presence of two possible modes of propagation. The study in this direction has been made by Hayakawa et al. [12,13], in which the frequency ramge below the proton gyrofrequency is dealt with. It seems hopeful and interesting to apply such a DF system to the low frequency radio emissions such as ELF waves observed on Aureol-3 satellite [12,13] and also to the micropulsations. Similar application to high frequency regime including the auroral kilometric radiations etc. would also be promising.

(3) Other direction of DF is the inclusion of hot plasma effect in wave distribution function method, because the previous methods are all based on the cold plasma theory. Ocarsson and Ronnmark[14] have developed the reconstruction of wave distribution functions in warm plasmas, which allows us to treat almost any linear wave mode and also to handle a large variety of plasma models including warm plasma. This new method seems to be very promising in the future satellite experiments.

REFERENCES

1. Hayakawa M. Yamanaka Y. Parrot M. Lefeuvre F (1984) The wave normals of magnetospheric chorus emissions observed onboard GEOS-2. J.Geophys. Res. 89: 2811-2821
2. Ohmi N. Hayakawa M (1986) On the generation mechanism of quasi-electrostatic half-gyrofrequency whistler-mode waves in the magnetospheric plasma. J. Plasma Phys. 35: 351-373
3. Muto H. Hayakawa M. Parrot M. Lefeuvre F (1987) Direction finding of half-gyrofrequency VLF emissions in the off-equatorial region of the magnetosphere and their generation and propagation. J. Geophys. Res. 92: 7538-7552
4. Hayakawa M (1989) Direction finding of half-gyrofrequency VLF emissions in the off-equatorial region of the magnetosphere. Proc.NIPR Symp. on Upper Atmosphere Phys. 1: 168-172
5. Hayakawa M. Ohmi N. Parrot M. Lefeuvre F (1986) Direction finding of ELF hiss emissions in a detached plasma region of the magnetosphere. J. Geophys. Res. 91: 135-141
6. Hayakawa M (1987) The generation mechanism of ELF hiss in detached plasma regions of the magnetosphere as based on the direction finding results. Mem. Natl Inst. Polar Res. Tokyo, Special Issue 47: 173-182
7. Hayakawa M. Parrot M. Lefeuvre F (1986) The wave normals of ELF hiss emissions observed onboard GEOS-1 at the equatorial and off-equatorial regions of the plasmasphere. J. Geophys.Res. 91: 7989-7999
8. Hayakawa M. Parrot M. Lefeuvre F (1987) The wave distribution functions of plasmaspheric ELF hiss: Geos 1 observation in equatorial region. Mem. Natl Inst. Polar Res. Tokyo, Special Issue 47: 157-172
9. Hayakawa M. Hattori K. Shimakura S. Parrot M. Lefeuvre F (1990) Direction finding of chorus emissions in the outer magnetosphere and their generation and propagation. Planet.Space Sci. 38: 135-143
10. Hayakawa M. Muto H. Shimakura S. Hattori K. Parrot M. Lefeuvre F (1989) The wave normal direction of chorus emissions in the outer magnetosphere. Proc. NIPR Symp. on Upper Atmosphere Phys. 2: 67-73
11. Hattori K. Ishikawa K. Hayakawa M (1990) Ray-tracing interpretation of wave normals of chorus emissions observed in the off-equatorial region of the outer magnetosphere. ibid, in press.
12. Hayakawa M. Lefeuvre F. Rauch JL (1985) The direction finding aboard Aureol 3 of ELF waves at frequency above and below the proton gyrofrequency. "Resultats du Project ARCAD 3 et des programmes recents en Physique de la magnetosphere et de l'ionosphere, Toulouse '84", Cepadues Edit. 499-507
13. Hayakawa M. Lefeuvre F. Rauch JL (1990) On the system of Aureol-3 satellite direction finding for ionospheric and magnetospheric ELF waves. Trans.Inst.Elect.Inform.Comm.Engrs.Japan, in press
14. Oscarsson T. Ronnmark K (1989) Reconstruction of wave distribution functions in warm plasma. J. Geophys. Res. 94: 2417-2428

4.4 Direction Finding of Very-Low-Latitude Whistlers and Their Propagation

M. HAYAKAWA[1], K. OHTA[2], and S. SHIMAKURA[3]

Research Institute of Atmospherics, Nagoya University, Aichi, 442 Japan[1], Department of Electronic Engineering, Chubu University, Aichi, 487 Japan[2], and Department of Electrical Engineering, Chiba University, Chiba, 260 Japan[3]

Abstract Spaced direction finding measurements of very-low-latitude whistlers were carried out in January 1988 at three stations in South China; Zhanjiang (geomag.lat. 10.1°), Guilin (14.1°) and Wuchang (19.4°). Two horizontal magnetic and one vertical electric field components were simultaneously recorded over a wide frequency range to enable comprehensive direction finding. The analyses of whistler data on several days have indicated that the ionospheric transmission latitude of equatorial latitude whistlers is restricted to the geomagnetic latitude range of 10-14°, and the presence of high occurrence of echo-train whistlers is a strong indication of their field-aligned propagation. The propagation mechanism of very-low-latitude whistlers is discussed.

1. INTRODUCTION

The propagation characteristics of whistlers at geomagnetic latitudes less than 20° are very poorly understood compared with those at latitudes greater than 20° [1]. Whistlers at lower latitudes are more susceptible to propagation conditions such as duct excitation, ionospheric transmission, etc. and therefore they are very useful for studying those problems which are still unsolved even for high- and middle-latitude whistlers.

Some important contributions to the study of these equatorial-latitude whistlers have recently been made by several workers [2,3,4,5,6], but their conclusions are sometimes contradictory to each other. In order to obtain further understanding of the propagation mechanism of very-low-latitude whistlers and of the associated plasma structure of the equatorial ionosphere, we have carried out spaced direction finding measurements in South China, covering the range of geomagnetic latitude from 10° to 20°. The purpose of this VLF campaign is described in detail in Hayakawa et al. [7], and very preliminary results have recently been published by Xu et al.[8]. A more detailed study than Xu et al.'s paper will be published in a separate paper by Hayakawa et al. [9], and so we will describe ,in this report, only its essential points.

2. CHARACTERISTICS OF VERY-LOW-LATITUDE WHISTLERS

The VLF campaign was carried out at three stations (Zhanjiang(abbreviated as ZJ)(geomag.lat.Λ=10.1°), Guilin (GL,14.1°), and Wuchang(WC,19.4°)) during the period of January 5-11, 1988. The three field components(two horizontal magnetic fields and one vertical electric) are measured in a wide frequency band from 2 to 6 kHz by means of crossed-loop aerials and a monopole antenna, and these enable us to carry out the comprehensive direction finding.

Two noticeable occurrence peaks on January 5 and 6 and two minor ones on January 9 and 11 have been analyzed extensively, and in the following we can summarize the important whistler characteristics obtained from these days.

(1) The geomagnetic latitude of the ionospheric exit region of very-low -latitude whistlers is localized to a range from $\Lambda=10°$ to $14°$. On the three days (January 5,9 and 11) the ionospheric exit region is just overhead at the station of ZJ ($\Lambda=10°$), and the goniometric triangulation on January 6 has yielded the ionospheric exit latitude, $\Lambda \sim 14°$. However, there are observed no whistlers whose exit latitudes are between $\Lambda=14°$ and $20°$.

(2) The extent of the ionospheric exit region of whistlers is found to be very stable and to have a radius of 40-50km. Also, no frequency dependence is recognized in the ionospheric exit region.

(3) When the whistlers have exited the ionosphere immediately overhead at the station ZJ, the wave polarization there is very close to right-handed circular. The same whistlers observed at WC have a polarization closer to linear.

(4) Propagation in the Earth-ionosphere waveguide after ionospheric transmission seems to exhibit a horizontal beaming around the magnetic meridian plane, and there is a tendency for the subionospheric propagation to be stronger toward higher latitudes than toward the equator.

(5) Echo-train whistlers (one example is shown in Fig.1) are found to occur at a surprisingly high rate such that more than 10% of the whistlers are accompanied by 3-hop components. Furthermore,the 2-hop(long) whistler components are often observed, as in Fig.1.

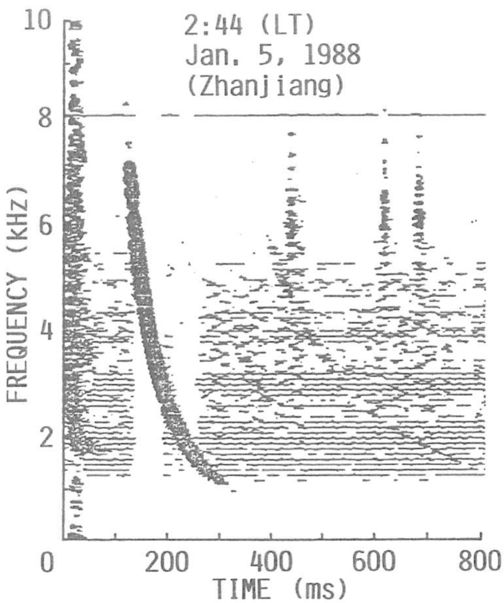

Fig.1 An example of an echo-train whistler observed at ZJ. The dispersion of the short whistler is 10.5 $sec^{1/2}$.

3. PROPAGATION MECHANISM OF VERY-LOW-LATITUDE WHISTLERS

We discuss the propagation mechanism of very low latitude whistlers,as based on the summary of the experimental findings in Section 2. We first compare the observational results with the features predicted from non-ducted propagation. The previous ray-tracing computations [3, 4,6] indicate that the geomagnetic latitude of the exit point of non-ducted whistlers is all the time just around $\Lambda=10°$,which seems to be consistent with the experimental fact (1). However, the extent of the ionospheric exit region of non-ducted whistlers which can transmit the ionosphere, is likely to vary greatly on different days because of the dynamically different density profiles and ,correspondingly, we would expect a wide range of variation in the distribution of the ionospheric exit region of whistlers. This is obviously in disagreement with the observational fact (2) indicating a high stability of the exit region. The next point difficult to interpret on the assumption of the non-ducted propagation is the horizontal beaming of the subionospheric propagation (experimental fact (4)), which is not allowed in non-ducted propagation because the propagation below the ionosphere must be focussed to the magnetic meridian plane. Then, the greatest difficulty for the non-ducted propagation is the surprisingly high occurrence of echo-trains given by the experimental fact (5). This fact would rather be a strong evidence in favour of field-aligned propagation. In support of our observational fact (5), Anderson et al.[10] have observed multiple hop whistlers on a rocket at a L value of $L\sim 1.06$,exactly at the same L value with that of ZJ. Then, the previous fact (4) might be reasonably explained by the field-aligned propagation, because the horizontal beam around the meridian plane would be formed as a consequence of the overlapping of the transmission cone and the trapping cone for the f field-aligned propagation.

As indicated just above,we have presented strong evidence that the larger amplitude whistlers, selected for analysis, have field-aligned propagation between the two hemispheres. However, Thomson's [5] results, where the VLF transmitter location, frequency, and power are well-defined (i.e. apparently not dependent on thunderstorms) may provide evidence indeed of non-ducted propagation in agreement with the ray-tracing studies. Non-ducted propagation provides the , at least, only explanation for a preferred geomagnetic latitude range of 10°-14° for the exit points. Furthermore, Faraday($L\sim 2.4$) observations of whistler-mode signals from NAA (24.0kHz) and NSS(21.4kHz) indicate that duct diameters are usually too low on low L-shells ($L < 2.5$) to support normal whistlers ($f < 10$kHz) propagation. Therefore, for the multi-hop whistlers,field-aligned guiding by an extended structures of many field-aligned filaments may be required.

It seems difficult, on theoretical grounds, to exclude non-ducted propagation (which almost certainly occurs at higher frequencies for VLF transmitters). Perhaps there is a two-stage process whereby initially non-ducted propagation (in this preferred latitude range,$\Lambda=10°-14°$) from a strong, localized thunderstorm leads to the creation of ionospheric and field-aligned irregularities and, eventually,self-sustaining ducted propagation in this latitude interval. Further details are descibed in Hayakawa et al.[9].

References

1. Hayakawa M. Tanaka Y (1978) On the propagation of low-latitude whistlers. Rev. Geophys. Space Phys. 16: 111-123

2. Ondoh T. Kohtaki M. Murakami T. Watanabe S. Nakamura Y (1979) Propagation characteristics of low-latitude whistlers. J. Geophys. Res. 84: 2097-2104
3. Andrews MK (1978) Non-ducted whistler-mode signals at low latitudes. J. Atmos. Terr. Phys. 40: 429-436
4. Liang BX. Bao ZT. Xu JS (1985) Propagation characteristics of nighttime whistlers in the region of equatorial anomaly. J. Atmos. Terr. Phys. 47: 999-1007
5. Thomson NR (1987) Experimental observations of very low latitude man-made whistler-mode signals. J. Atmos. Terr. Phys. 49: 309-319
6. Thomson NR (1987) Ray-tracing the paths of very low latitude whistler-mode signals. J. Atmos. Terr. Phys. 49: 321-338
7. Hayakawa M. Ohta K. Shimakura S. Xu JS. Bao ZT. Liang BX (1988) A proposal for multi-stationed direction finding measurements of low- and equatorial-latitude whistlers in China. Res. Lett. Atmos. Elect. 8: 31-35
8. Xu JS. Tian M. Tang CC. Hayakawa M. Ohta K. Shimakura S (1989) Direction finding of nighttime whistlers at very low latitudes in China: Preliminary results. Planet. Space Sci. 37: 1047-1052.
9. Hayakawa M. Ohta K. Shimakura S (1990) Spaced direction finding measurement of very-low-latitude whistlers and their propagation. Accepted for publication in J. Geophys. Res.
10. Anderson DN. Kintner PM. Kelley MC (1985) Inference of equatorial field-line-integrated electron density values using whistlers. J. Atmos. Terr. Phys. 47: 989-997

4.5 Frequency Dependence of Ionospheric Exit Points and Polarization of Daytime Whistlers at Low-Latitude

K. OHTA[1], M. HAYAKAWA[2], and S. SHIMAKURA[3]

Department of Electronic Engineering, Chubu University, Aichi, 487 Japan[1], Research Institute of Atmospherics, Nagoya University, Aichi, 442 Japan[2], and Department of Electrical and Electronic Engineering, Chiba University, Chiba, 260 Japan[3]

ABSTRACT

In order to study the propagation mechanism of whistlers, a new method of wide-band direction finding for whistlers, based on the simultaneous measurement of two horizontal magnetic field components and one vertical electric field component using the digital recorders and a fast Fourier transform analyser, has been adopted for daytime whistlers at low latitude Yamaoka (geomag. lat. 25°N). It is found that there is a negligibly small frequency dependence of the ionospheric exit point and polarization of daytime whistlers during each occurrence peak with duration of about 2 hours, which lends further support to their propagation in field-aligned duct in the magnetosphere for each occurrence peak.

1. INTRODUCTION

This study is concerned with the geomagnetic latitude range between 20° and 30° Hayakawa and Tanaka [1] have pointed out the importance of this low-latitude whistlers in the general whistler studies. Low-latitude whistlers suffer from extremely unfavorable propagation conditions associated with duct excitation, ionospheric transmission down to the ground, and so on; this, in turn, implies that the investigation of these low-latitude whistlers will surely provide useful information on the detailed points involved in those problems which are not well established even at higher latitudes. The most important evidence for this ducted propagation has been accumulated by our field analysis direction finding at a specific frequency around 5kHz, and on the basis of these direction finding results for many whistlers at this frequency, we have studied the characteristics of whistler ducts. However, since the information on the frequency dependence of the direction finding results has never been reported, we cannot know whether our previous findings based on the direction finding at a selected frequency are valid for all frequencies or not. So the complementary information on the frequency dependence of arrival direction and polarization of whistlers even in this latitude range is greatly needed, and the present paper reports on this frequency dependence with respect to its reference to the propagation mechanism.

2. WIDE-BAND DIRECTION FINDING FOR WHISTLERS

There are two different kinds of direction finding techniques to locate the region where VLF waves emerge from the lower ionosphere, enabling us to study the whistler ducts and magnetospheric plasma. One is the goniometer which is, principally, based on the amplitude ratio between the magnetic field components of linearly polarized incident wave on the crossed loop antennas [2]. A limitation of the goniometer is its susceptibility to polarization error for incident waves which are elliptically polarized. The second is the field-analysis method, which uses a vertical electric antenna in addition to the previous crossed loop antennas [3].
We have developed the so-called "field analysis direction finding" system based on the simultaneous measurement of three possible field components (two horizontal magnetic fields and a vertical electric field) [4], [5], [6].

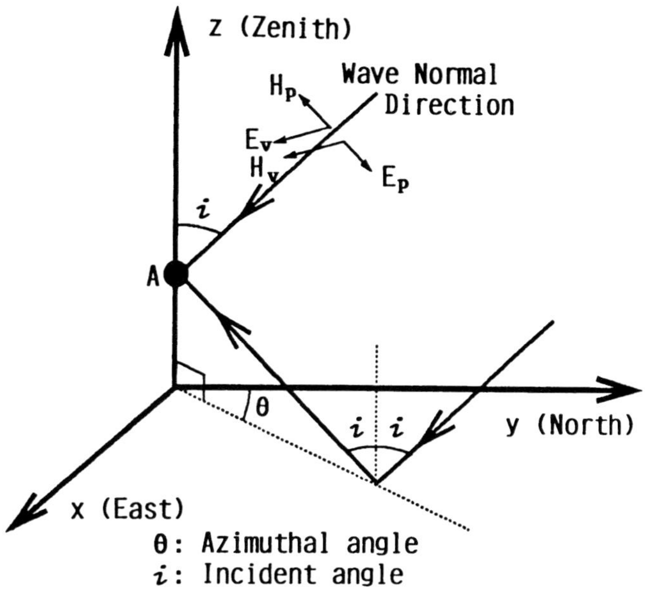

Fig. 1. The configuration of the problem. The downgoing wave is incident on the direction finder at A, with an incident angle i and with an azimuthal angle θ. The elliptically polarized wave is decomposed into two components; TE and TM modes, as shown in the figure.

Figure 1 illustrates the situation of the problem in which downgoing wave is incident on the direction finder at A, with an incident angle i and with an azimuthal angle θ. In the subsequent derivation of electromagnetic fields at the observing point A we have made the following reasonable assumptions. (1) The ground is flat and perfectly conducting, (2) the wave is, instantaneously, a monochronatic sinusoidal wave, and (3) the incident wave is generally an elliptically polarized plane wave. Now, the incident wave is decomposed into two modes; one has the magnetic field component perpendicular to the incident plane and the other has the electric field component perpendicular to the incident plane, as indicated in Fig. 1. These components are expressed by,

$$H_v = A_v \exp(j\omega t)$$
$$H_p = A_p \exp(j(\omega t - \alpha)) \qquad (1)$$

Where A_v and A_p are the amplitudes of respective components and α indicates the phase difference of the two components. Then, the polarization parameter, P in the wavefront is defined by,

$$P = H_p / H_v = u - jv \qquad (2)$$

The two horizontal magnetic fields picked up by crossed loop antennas, and one vertical electric field picked up by a vertical antenna are obtained. We put V_x, V_y, and V_z as the output voltages of the E-W and N-S loops and the vertical antenna.
We finally obtain the incident angle i and the azimuthal angle θ and the polarization parameter u and v.

$$\theta = \tan^{-1}(M_{y-z}\sin\phi_{y-z})/(M_{x-z}\sin\phi_{x-z})$$
$$i = \sin^{-1} 1/(M_{y-z}\cos\phi_{y-z}\cos\theta - M_{x-z}\cos\phi_{x-z}\sin\theta) \quad (3)$$

$$u = -(M_{x-z}\cos\phi_{x-z}\sin i + \sin\theta)/\cos i\cos\theta$$
$$v = M_{x-z}\sin_{x-z}\sin i/\cos i\cos\theta \quad (4)$$

Where M_{y-z} and M_{x-z} are the amplitude ratios of V_y and V_x relative to V_z and ϕ_{y-z} and ϕ_{x-z} are the corresponding phase differences.
However, these systems were operated at a specific frequency around 5kHz because of using analogue technique, and so the wide-band information on arrival direction and polarization is required for further elaboration of the propagation mechanism. The wide-band direction finding based on the four-parameter method was previously carried out by Cousins [3] by means of the analogue recording of three field components. So sophisticated adjustment is required in order to calibrate the errors in measuring the amplitude ratios and phase differences of these components.

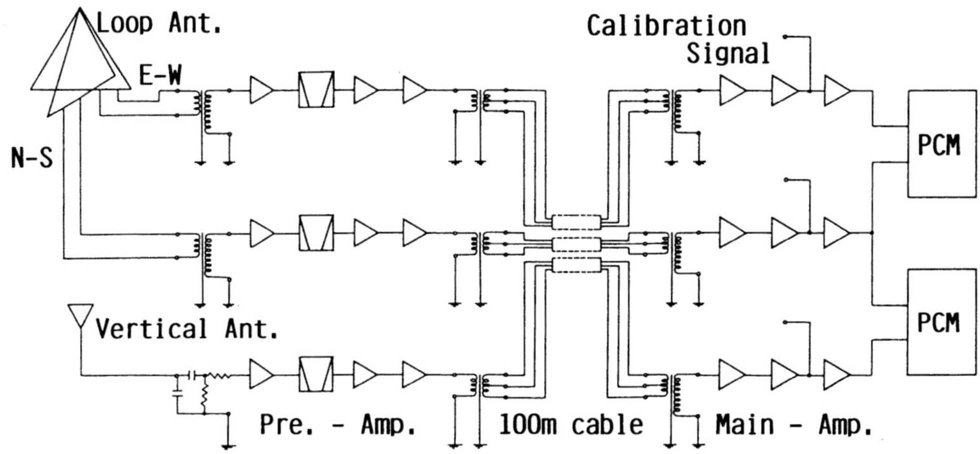

Fig. 2. The block diagram of the observing system.

These problems are completely and successfully solved here by a combined use of PCM (Pulse Code Modulation) recorders which have recently become more available and an FFT (Fast Fourier Transform) analyzer. Figure 2 shows the block diagram of the observing system by PCM recorder. The whistler signals induced by the crossed loop antennas and a vertical antenna are fed to the pre-amplifiers (Gain=60-80dB) then, to band pass filters (2-6kHz) in order to be free from the noise of VLF and power line. After that, the whistler signals are fed to the main amplifier (Gain=0-40dB) in our observatory with a cable (Length=100m). The calibration signals are fed to the main amplifier. PCM recorders are composed of two systems. One records the outputs of EW loop and NS loop signal, another records NS loop and vertical signal. After recording the whistler signal using PCM recorder, we can get the amplitude and phase differences by two channel FFT analyzer. The sampling time is 40μs. By the way of analysing 1024 data by FFT, we can get the amplitude ratio and phase differences up to a frequency of 10kHz with 25Hz resolution. The accuracy in measuring the amplitude ratio and phase difference between the two channels of less than 0.1dB and less than 1°, respectively, is easily achieved over a wide frequency range up to 10kHz. Hence the present system is considered to yield very accurate wide-band direction finding results by simple equipment that is easily handled.

3. FREQUENCY DEPENDENCE OF ARRIVAL DIRECTION AND POLARIZATION OF LOW-LATITUDE WHISTLERS

The observation was carried out during the interval of December 31, 1986, through January, 15, 1987, at Yamaoka Observatory (geomag. lat. 25°N) of Chubu University and was made continuously during daytime from 13:00 to 17:00 JST (Japanese Standard Time).

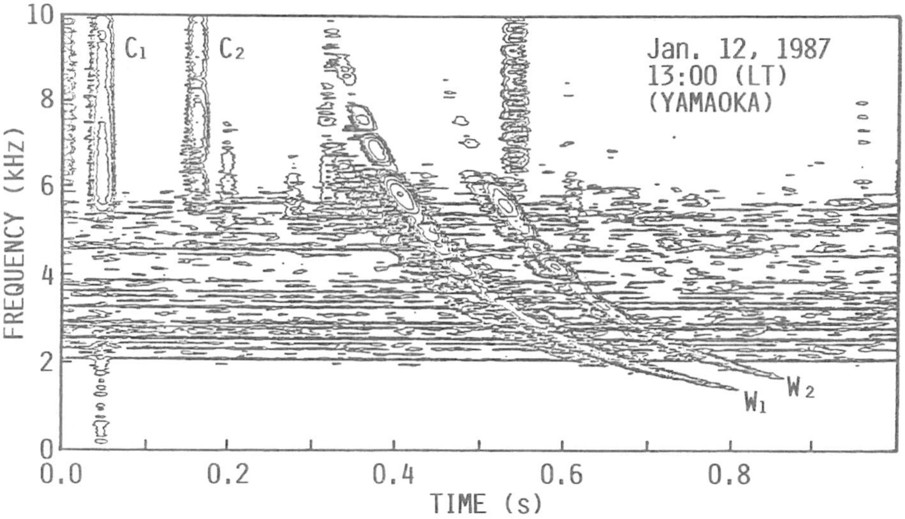

Fig. 3. An example of frequency spectra of daytime whistlers in the form of multiple flashes during the occurrence peak from 13:00 to 15:00 JST on January 12, 1987. The strong whistler, W_1 is chosen randomly and is subjected to further analysis.

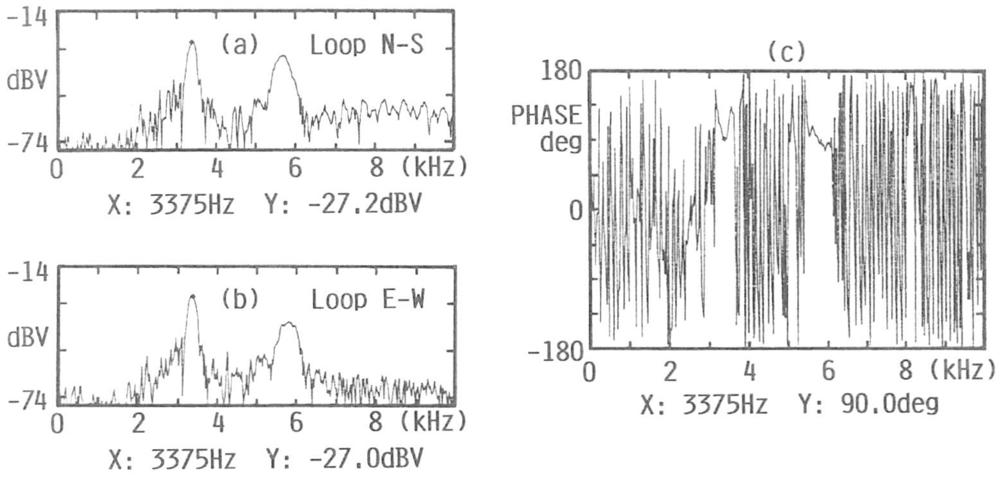

Fig. 4. An example of amplitude up to 10kHz of the whistlers W_1 and W_2 shown in Fig. 3. (a) shows the amplitude of the output of NS loop antenna and (b) shows that of EW loop antenna and (c) shows the phase difference between NS and EW crossed loop antennas.

As one of the most active days we have chosen January 12, 1987; on this day we received many whistlers during daytime, from 13:00 to 15:00 JST. The duration of this order is very typical for daytime whistlers in the latitude range from 20°N to 30°N [1], [7]. Figure 3 illustrates an example of whistlers (W_1 corresponding to the sferic C_1, and W_2 to C_2 observed at 13:00 JST, which indicates that they are in the form of multiple flashes with the same dispersion of $30sec^{1/2}$ during this occurrence peak. Figure 4 shows the analysed data up to 10kHz on amplitude ((a) shows the output of NS loop antenna, (b) shows the output of EW loop antenna) and phase difference, (c), between EW and NS loop antenna by FFT. The first peak of amplitude in Fig. 4 (a) and (b) shows the whistler W_1 at the crossing frequency point of 3375Hz and the second peak shows the whistler W_2 in Fig. 3. The phase difference between NS loop and EW loop antenna shown in Fig. 4 (c) is nearly 90° in the point frequencies of maximum intensity of W_1 and W_2. It means that the amplitude of NS and EW loop antenna are nearly same, and phase differences are nearly 90° even in the different frequency of different whistlers. These data suggest that whistlers W_1 and W_2 come down from the zenith of the observatory with a right-handed circular polarization.

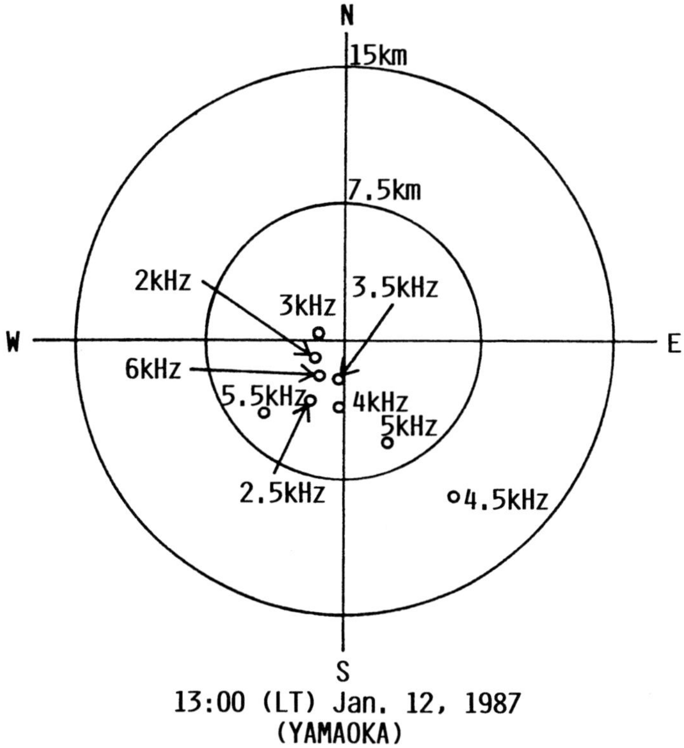

Fig. 5. Frequency dependence of ionospheric exit points of the whistler, W_1 indicated in Fig. 3, as based on the field analysis direction finding.

The frequency dependence of ionospheric exit points for a particular strong whistler, W_1, in Fig. 3 is summarized in Fig. 5 with frequency as a parameter where the ionospheric height is assumed to be 80km.
The accuracy inherent to the direction finding system of locating the exit points in Fig. 5 has been estimated to be less than 1° in the incident and azimuthal angles, corresponding to the error circle with diameter of a few kilometers.

It is found from Fig. 5 that the ionospheric exit points at different frequencies are highly concentrated within a radius of 15km, with most of them being located within a small area of less than 10km around the zenith of the station. Hence we can conclude that there is a negligibly small frequency dependence of the ionospheric exit points. Cousins has found many occasions for which there are a few specific arrival directions suggesting multiple ducts even for a single whistler at medium latitudes, which seems to be very different from the present characteristics observed at low latitudes.

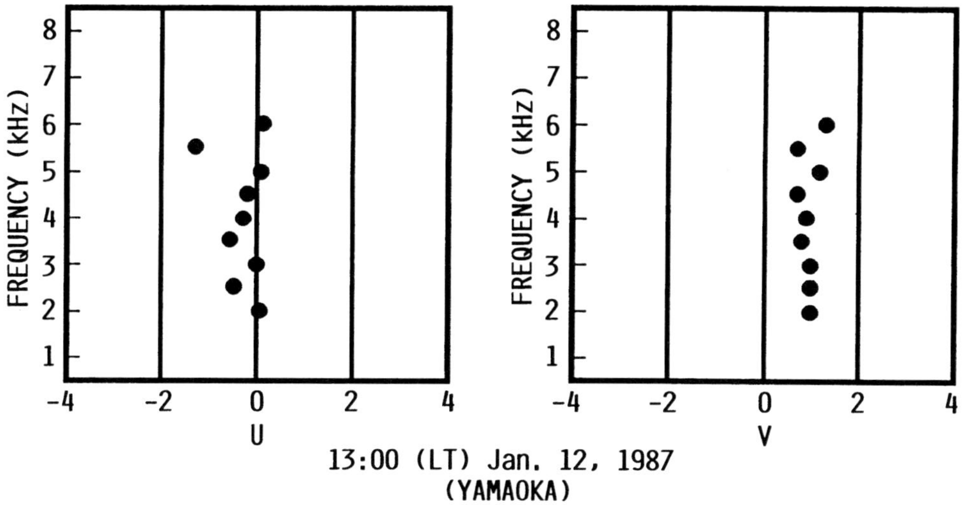

Fig. 6. Frequency dependence of the polarization parameter u and v for the same whistler as determined by the field analysis direction finding; $(u,v)=(0,1)$ indicates a right-handed circular polarization.

Figure 6 illustrates the corresponding frequency dependence of the polarization parameter for the same strong whistler. A right-handed circular polarization is designated by $(u,v)=(0,1)$. Figure 6 indicates that there is no definite tendency for u and v to vary with frequency and that the polarization is, generally speaking, approximately right-handed circular over a wide frequency range. We have presented the frequency dependence of the ionospheric exit region and polarization for the particular whistler, but this dependence is confirmed by examining about 10 intense whistlers during high activity.

4. CONCLUSION

Daytime whistlers in the geomagnetic latitude range of 20°N through 30°N are found to be attributed to ducted propagation on the basis of many experiments and theoretical analyses by Hayakawa and Tanaka [1]. Kimura studied the frequency dependence of ray paths in the inner plasmasphere in order to explain the drifting whistler cutoff phenomena and indicated that the unducted exit points in the opposite hemisphere varied over a range of about 2° of latitude for a frequency range of 1.0kHz to 6.0kHz for the magnetosphere-ionosphere model with horizontal gradients [8]. Later, Hasegawa and Hayakawa calculated nonducted ray paths for the inner plasmasphere including the realistic equatorial anomaly and found that the possible ionospheric transmission takes place at a latitude corresponding to our station of Yamaoka and that the spread in final latitude of possible ionospheric transmission is 1.0° to 2.5° (or 100km to 250km) for varying anomaly models [9].

However, the extent of distribution in Fig. 4 appears to be extremely small compared with the above theoretical ray-tracing prediction based on nonducted propagation, and we can consider that there is no frequency dependence of the ionospheric exit location. Hence we are led to rely on ducted propagation, and, further, the experimental result by Hayakawa and Ohtsu [10] is satisfactorily interpreted in terms of ducted propagation. The size of the distribution of the ionospheric exit region in Fig. 4 is estimated to be a radius of about 20km, which is consistent with the previous findings by Okada et al. [4], [5] Ohta et al. [6], and Hayakawa et al. [10]. This size is closely correlated with the duct dimension within the ionosphere.

Acknowledgmens The authors would like to express their sincere thanks to Prof. K. Bullough of Sheffield University for his useful discussions and suggestions and to Prof. H. Eguchi, Prof. O. Mikami, Mr. K. Baba of Chubu University and A. Kimpara, Emeritus Professor of Nagoya University for their support.

5. REFERENCE

1. Hayakawa, M., and Y. Tanaka [1978] On the propagation of low-latitude whistlers, Rev. Geophys. Space Phys., 16, 111.
2. Bullough, K. and J. L. Sagredo [1973] VLF goniometer observations at Halley Bay, Antarctica-I. The equipment and the measurement of signal bearing, Planet. Space Sci., 21, 899.
3. Cousins, M. D. [1972] Direction finding on whistlers and related VLF signals, Tech. Rep. 3432-2, Radiosci. Lab., Stanford Univ., Stanford, Calif.
4. Okada, T., A. Iwai, and M. Hayakawa [1977] The measurement of incident and azimuthal angles and the polarization of whistlers at low latitudes, Planet. Space Sci., 25, 233.
5. Okada, T., A. Iwai, and M. Hayakawa [1981] A new whistler direction finder, J. Atmos. Terr. Phys., 43, 679.
6. Ohta, K., M. Hayakawa and Y.Tanaka [1984] Ducted propagation of daytime whistlers at low latitudes as deduced from ground-based direction finding, J. Geophys. Res., 89, 7557.
7. Hayakawa, M., Y. Tanaka, and T. Okada [1983] Time scales of formation, life-time and decay of whistler ducts at low latitudes, Ann.Geophys., 1, 515.
8. Kimura, I. [1971] Drifting whistler cutoff phenomena -striation- observed by POGO satellites, Space Res., 11, 1331.
9. Hasegawa, M., and M. Hayakawa [1980] The influence of the equatorial anomaly on the ground reception of non-ducted whistlers at low latitudes, Planet. Space Sci., 28, 17.
10. Hayakawa, M., and J. Ohtsu [1973] Ducted propagation of low-latitude whistlers deduced from the simultaneous observations at multi-stations , J. Atmos. Terr. Phys., 35, 1685.

4.6 Waveform Analysis of Whistlers

H. HAGINO[1], T. OGAWA[2], and S. YANO[3]

Department of Electromechanical Engineering, Kagawa Technical College, Kagawa, 763 Japan[1], Department of Physics, Kochi University, Kochi, 780 Japan (Now at Science Laboratory International, Kochi, 780 Japan)[2], and Department of Electrical Engineering, Kochi National College of Technology, Kochi, 783 Japan[3]

ABSTRACT. Some of the whistlers displayed in the sonagram show that the signal amplitude changes with time and/or frequency. In order to explain this phenomenon the following theory is presented. It is concerned with the propagation of discrete whistler mode waves under the influence of energy flow direction differing from the direction of the wave normal and the earth's magnetic field.

The whistler waveforms were calculated in an integral form by means of the law of inverse Fourier-Laplace transform pair via the saddle point method for the delta function source. The calculated waveforms by using ionospheric parameters are compared with the observed ones on the ground in the frequency range upto 10 kHz - 1 kHz. It is shown as the result that the analytical result of whistler waveforms with polarized fading can explain the observed characteristics.

1. INTRODUCTION

The whistler waves are generally non-stational signals with amplitude and frequency modulations. Helliwell [1] collected in his textbook the ground whistler data in which some multiple-path and multi-flash whistlers indicated deep minima in amplitude at the same frequencies. Paymar [2] reported on the banded structure of whistler traces observed on OGO-4 satellite with the dispersion of around 90 $sec^{1/2}$. These banded whistlers were observed at around 40° geomagnetic latitude with the low latitude cutoff at ~30°, and were explained as they were caused by a mechanism of spectral reflection or transmission of the irregularities with an enhanced ionization in the ambient plasma. Ohta et al. [3] recently reported on patch type contour lines of the intensity in the Fourier spectra of whistler traces obtained using a FFT digital processor. They explained that this structure might be produced by the wave interference in a horizontally stratified irregularity in the F region ionosphere of downgoing whistler waves. Their investigation is based on the data in which the whistler traces measured within a short time interval (for example within a few seconds) indicated a similar banded structure and the minima in amplitude of whistler traces were roughly located at the same frequencies. Yano et al. [4] pointed out that the whistler wave fading was made by different rays penetrating into the ionosphere after several hops in the Earth-ionosphere waveguide in the source side hemisphere.

2. CALCULATION FORMULA

Vidmer et al. [5] proposed a calculation method of the whistler waves by using the saddle point method given by Budden [6], Tyras [7], and Felsen [8]. We follow their method with a variation in terms of the group refractive index of the medium.

The analytical conditions are as follows:

(1) The ionosphere is cold collisionless magneto-plasma in the F_2 layer. The ionospheric height is about 350 Km.

(2) The effect of the curvature of the Earth and the ionosphere is ignored. The Earth's magnetic field is static.

(3) The whistler waves propagate in the direction with a very small angle θ to the Earth's magnetic field B_0.

We introduce a case as shown in Fig. 1. It is shown that the direction of the phase velocity and the magnetic field line B_0 have an angle θ between them. This angle θ causes the angle ϕ between the phase velocity and the group velocity.

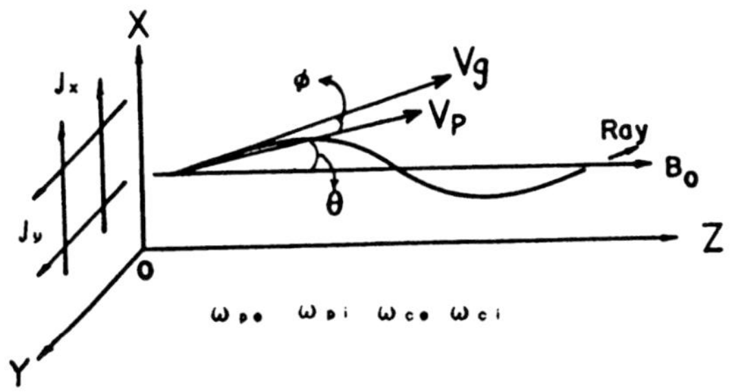

Fig. 1. Ray of a whistler originated from the sheet currents J_x and J_y and the definition of angles θ and ϕ. V_p and V_g are the phase and group velocities, respectively.

Following Stix [9], we use rotating coordinates to define the electric field and delta function source $J_{\mp}(t,z)$:

$$J_{\mp}(t,z)=\frac{1}{2}\left[J_x(t,z)\mp j\ J_y(t,z)\right] \tag{1}$$

$$J_x(t,z)=J_1\,\delta(t)\,\delta(z),\quad J_y(t,z)=J_2\,\delta(t)\,\delta(z) \tag{2}$$

$$E_{\mp}(t,z)=\frac{1}{2}\left[E_x(t,z)\mp j\ E_y(t,z)\right] \tag{3}$$

where the upper sign refers to right-hand polarization (RHP) and the lower sign to left-hand polarization (LHP). The current source density $J_{\mp}(t,z)$ consists of two delta function sources J_1 and J_2 which are dimensional constants; and $\delta(t)$ and $\delta(z)$ are the temporal and spatial delta function, respectively.

An instanteneous electric field of whistler waves, Real[$E_x(t,z)$], is generated from the temporal and spatial δ-function source shown in Fig. 1. In this case the next equation can be introduced.

$$\mathrm{Real}\,[E_x(t,z)] = \frac{1}{2}[\mathrm{Real}\,\{E_-(t,z)+E_+(t,z)\} + \mathrm{Real}\,\{E_-(t,z)+E_+(t,z)\}^*] \tag{4}$$

The group refractive index is substituted only in the second term in the right side, Real { }*. We use the Fourier-Laplace transform and then integrate over the k plane as described below, and then separate the refractive index in each mode of R and L from the R-L hybrid mode. This can be done becauce the propagation angle from the magnetic field line is very small. Then the electric field reduces to

$$E_{\mp}(t,z) = \frac{J_1 \mp jJ_2}{8\pi\varepsilon_0}\int_{-\infty-j\sigma}^{\infty-j\sigma}\frac{1}{2\pi j}\int_{-\infty}^{\infty}\frac{\omega^{-1}}{D_{\mp}^{e,i}(\omega,k)}\exp[j\{\omega(t-t_k)-kz\}]dkd\omega \tag{5}$$

The dispersion relation $D_{\mp}^{e,i}(\omega,k)$ for the cold collisionless magneto-plasma is given by

$$D_{\mp}^{e,i}(\omega,k) = \omega^{-2}\left[k^2c^2-\omega^2+\frac{\omega\omega_{pe}^2}{\omega\mp\omega_{ce}}+\frac{\omega\omega_{pi}^2}{\omega\pm\omega_{ci}}\right] \tag{6}$$

where ω_{ce} and ω_{ci} are the electron cyclotron frequency and oxygen ion cyclotron frequency, respectively. ω_{pe} and ω_{pi} are the electron plasma frequency and oxygen ion plasma frequency, respectively. C and ε_0 are the speed of light and the permittivity of vacuum in MKS unit.

By using the Fourier transformation we integrate Eq. (5) for $z\to\infty$. We apply the saddle point method. By substituting Eq. (5) into Eq. (4), we can calculate the electric field numerically.

$$\mathrm{Real}\,[E_x(t,z)] = \sum_{n=1}^{N} W_m(S_n)[\,\mathrm{Cos}\,\{b'(S_n)-m'a(S_n)-\pi/4\} + (1+\mathrm{Tan}^2\psi)^{1/2}\,\mathrm{Cos}\,\{ma(S_n)+\pi/4\}\,\mathrm{Cos}\,\{b(S_n)+\psi-\pi/2\}\,] \tag{7}$$

where

$$\left.\begin{array}{llll} S_n=\omega/\omega_{ce}, & \tau=\omega_{ce}t, & \zeta_z=z\omega_{pe}/C, & t_k=0 \\ X=\omega_{pi}/\omega_{pe}, & Y=\omega_{ci}/\omega_{ce}, & R=\omega_{ce}/\omega_{pe} & \\ \alpha=1+X^2, & \beta=1-Y, & Y=X^2 & \\ A_R(S_n)=(1-S_n)(S_n+Y), & M=\pi^{1/2}J_2/(8\pi\varepsilon_0 C) & & \end{array}\right\} \tag{8}$$

$$W_m(S_n)=2^{1/2}M\frac{R\omega_{ce}\ \{\alpha^{-3/4}A_R^{7/4}(S_n)\}\ \mathrm{Cos}\,\{m'a(S_n)\}}{\zeta_z^{1/2}\ \{-\beta S_n^2+(0.25\beta^2-3Y)S_n+\beta Y\}^{1/2}} \tag{9}$$

$$V_{GR}(S_n)=2RC\frac{A_R^{3/2}(S_n)\ \{R^2A_R(S_n)+\alpha\}^{1/2}}{2R^2A_R^2(S_n)+\alpha(\beta S_n+2Y)} \tag{10}$$

$$V_{PR}(S_n)=RC\frac{A_R^{1/2}(S_n)}{\{R^2A_R(S_n)+\alpha\}^{1/2}} \tag{11}$$

$$\frac{1}{V_{GR}{}^*(S_n)/C}=K(\phi)\ \{n(S_n)+S_n\frac{d}{dS_n}n(S_n)\}\ ,\ n(S_n)=\frac{1}{V_{PR}(S_n)/C} \tag{12}$$

$$K(\phi)=\{1+\mathrm{Tan}^2\phi\}^{-1/2}\ ,\quad \psi=\mathrm{Tan}^{-1}(J_1/J_2) \tag{13}$$

$$a(S_n)=RS_n\zeta_z\ \{K(\phi)-1\}\ C/V_{PR}(S_n) \tag{14}$$

$$b(S_n)=RS_n\zeta_zK(\phi)\ \{1-V_{GR}(S_n)/V_{PR}(S_n)\}\ C/V_{GR}(S_n) \tag{15}$$

$$b'(S_n)=b(S_n)/K(\phi) \tag{16}$$

$$m'=1-m \tag{17}$$

$V_{GR}(S_n)$ and $V_{PR}(S_n)$ are group velocity and phase velocity, respectively.

$\{V_{GR}\cdot(S_n)/C\}^{-1}$ is the group refractive index. $Y=X^2$ is the plasma's nutral condition. The saddle point frequencies S_n are given by the saddle point numbers of $n=1,2,$----. We use only one saddle point of number $n=1$. because only the low frequency is considered. The energy flow direction differs by a very small angle ϕ from the direction of wave normal. We refer the $E_x(t,z)$ compornent of the whistler mode with the RHP mode for $S_n \geqq 0$ and the LHP mode for $S_n \leqq 0$, we perform a similar analysis for the extraordinary whistler mode analysis the RHP mode for $S_n \geqq 0$ and the LHP mode for $S_n \leqq 0$ are considered. In the ordinary whistler mode analysis the RHP mode for $S_n \leqq 0$ and the LHP mode for $S_n \geqq 0$ are evanescent.

Thus we calculate both the first and second terms in the right side of Eq. (4). and combined both cotributions yield.

3. RESULTS OF CALCULATIONS.

Calculations were performed using Eq. (7). In the calculations the

Table 1. Ionospheric Parameters

Parameter	Value		Parameter	Value	
z	1608	Km	ζ_z	320×10^3	
n_e	1.12×10^{12}	m^{-3}	B_0	0.313	Gauss
f_{pe}	9.51	MHz	ω_{pe}	5.97×10^7	sec^{-1}
f_{pi}	55.4	KHz	ω_{pi}	3.48×10^5	sec^{-1}
f_{ce}	0.875	MHz	ω_{ce}	$1.75\pi\times10^6$	sec^{-1}
f_{ci}	29.7	Hz	ω_{ci}	1.87×10^2	sec^{-1}
R	0.092		V_A/C	5.83×10^{-4}	

typical parameter values were used as listed in Table 1, which are typical values of the day time ionspheric F_2 layer at a height of about 350Km. V_A is the Alfven velocity.

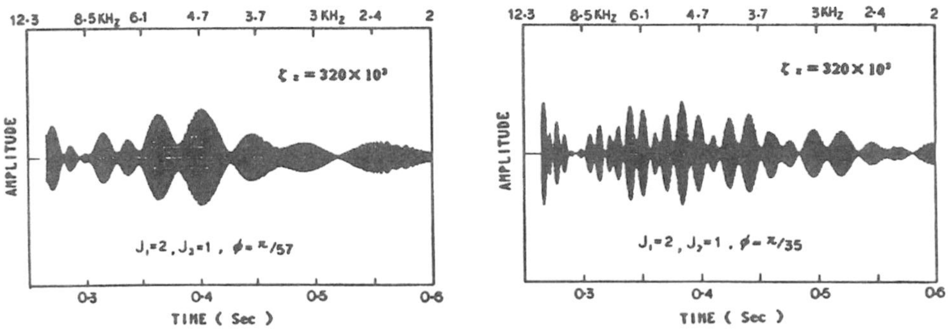

(a) (b)

Fig. 2. Whistler waveforms calculated with parameters in Table 1. m' was assumed to be 0.96.
(a) $\phi = \pi/57$, $J_1:J_2=2:1$.
(b) $\phi = \pi/35$, $J_1:J_2=2:1$.

Figures 2 (a) and (b) show the calculated fading patterns or modulation patterns of the whistler waves under the conditions of (1)~(3) with the parameters in Table 1. In Figs. 2 (a) the ratio $J_1:J_2$ of the source current was assumed 2:1, the angle $\phi = \pi/57 \fallingdotseq 3.2$ deg{$(\theta + \phi) \fallingdotseq 9.5$ deg}, and the field amplitude (Eq. (4)) is normalized by M. In Fig. 2 (b) the ratio $J_1:J_2$ of the source current is assumed 2:1, and the angle $\phi = \pi/35 \fallingdotseq 5.1$ deg {$(\theta + \phi) \fallingdotseq 15.4$ deg}.

4. COMPARISION WITH THE OBSERVED DATA

In Fig. 3 is shown an example of dispersion characteristics displayed by a sound spectrograph for a whistler observed at the Earth Observatory, Kochi University (33° 31' N, 133° 24' E) at 15:00 on 27 December, 1987.

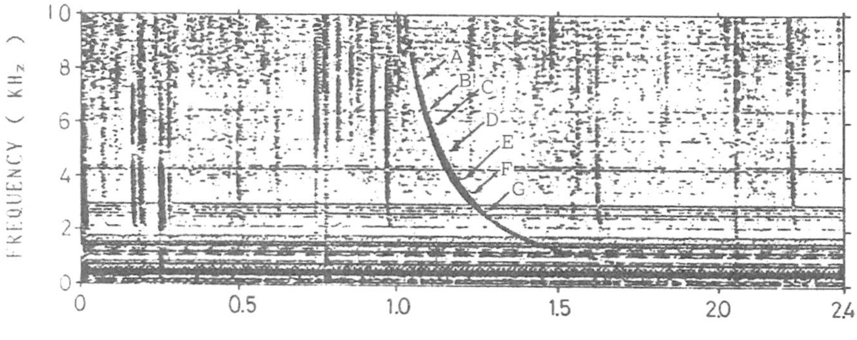

Fig.3. Example of sonagram of pure-tone whistler trace with modulation.

The signal amplitude of the whistler seems to change with time and frequency as seen at the points A ~G in Fig. 3. The whistler waveform

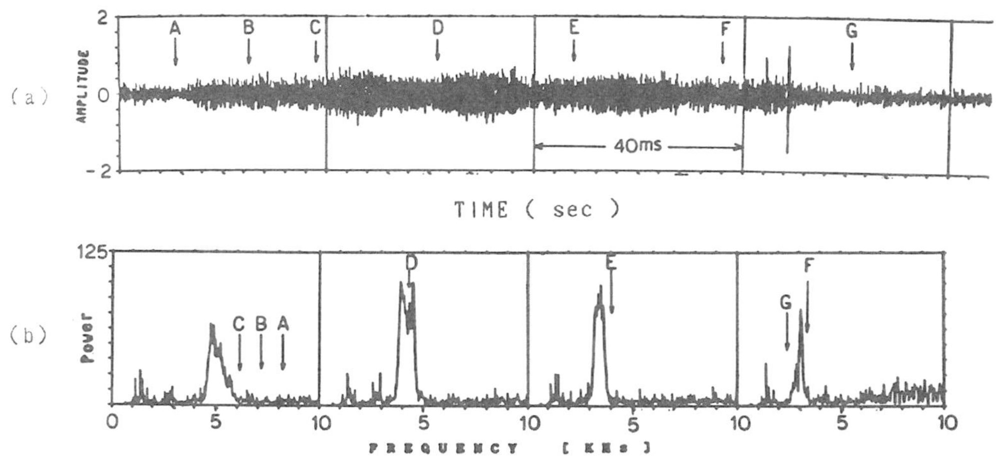

Fig. 4. Waveform and power spectra characteristics of the whistler wave shown in Fig. 3,

(a) Waveform measured by using the signal analizer with HPF set at 1.5kHz for the cutoff frequency and 40 ms of data length.
(b) Four individual power specras versus (a).

is displayed in Fig. 4 (a) corresponding to the whistler in Fig. 3. The waveform data (a) was obtained by using the signal analizer with a high pass filter (HPF) set at 1.5 kHz for the cutoff frequency and 40 ms of data length, The spectra data Fig. 4 (b) was corresponding to the whistler wave form in Fig. 4 (a) with same measured condition.

In Fig. 5 the result of calculated waveforms by using the present method with the parameters in Table 1. The signs A～G correspond to each other in Fig. 3 and 4.

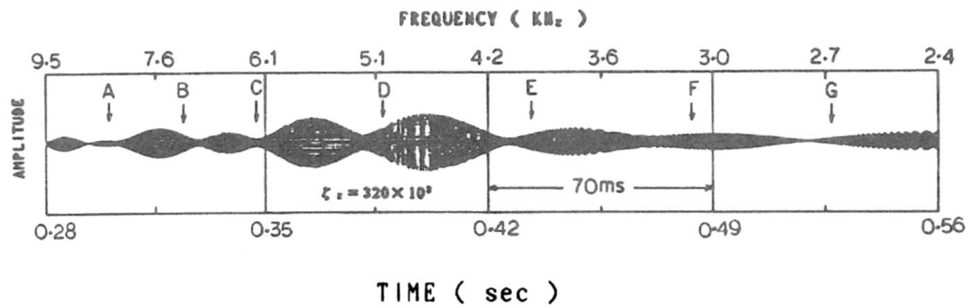

Fig. 5. Whistler waveform calculated with parameters in Table 1. $J_1:J_2$ was assumed to be 2:1 and $\phi = \pi/57$.

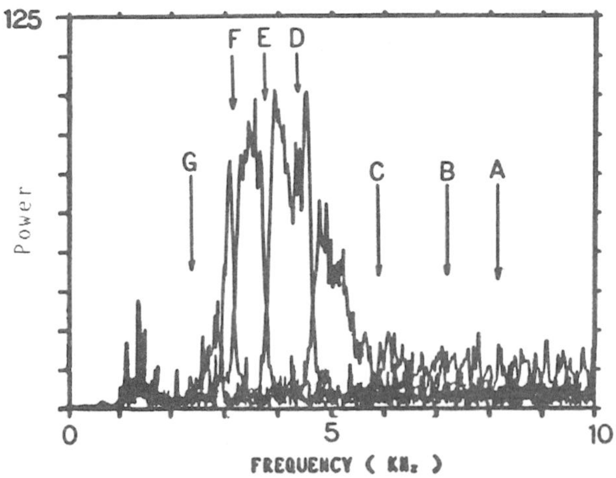

Fig. 6. Power spectra of the whistler shown in Fig. 3, showing beat type amplitude variation.

In Fig. 6 is given the power spectra of the whistler in Fig. 3. The spectra were overlaped by four individual spectra in Fig. 4 (b). The signs A～G correspond to each other in Figs. 3, Figs. 4 (a), (b) and Fig. 5.

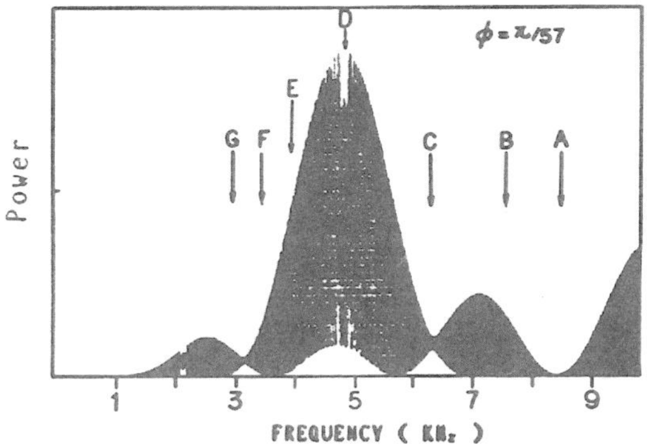

Fig. 7. Power spectra calculated by using Eq. (4) with parameters in Table 1. $J_1:J_2$ was assumed to be 2:1.

In Fig. 7 is given the calculated power spectra of the theoretical whistler in Fig. 5. The signs A~G correspond to each other in Fig. 3, Fig. 4, and Fig. 5. The variation of the spectral beat amplitude $|E_x(t,z)|^2$ with frequency was obtained by using Eq. (4).

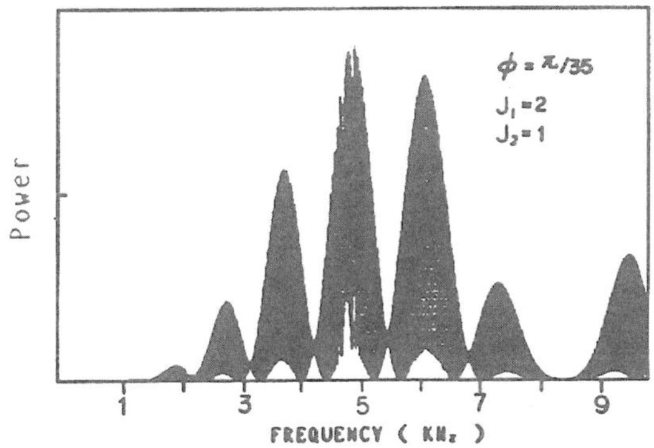

Fig. 8. Power spectra calculated by using Eq. (4) with parameters in Table 1. $J_1:J_2$ was assumed to be 2:1, and $\phi = \pi/35$.

In Fig. 8 is given the calculated power specra of the theoretical whistler in Fig. 2 (b). The three large beat type spectral amplitudes appeared.

4. DISCUSSIONS AND CONCLUSION

The calculation formulas of whistler waveforms are derived for the homogeneous medium using an arbitrary group refractive index, and approximated for the lower frequency whistler waves. It is pointed out that the modulation pattern may be made under the influence of energy flow direction differing from the direction of the wave normal and the static magnetic field. If there are whistler rays propagating with diferent angle of $\theta + \phi$, these rays have the different components of the group velocity from the component parallel to the direction of magnetic field. If these components differ by about few percent from the component of the magnetic field direction, then the modulation pattern of the whistler may be obtained. In this case we can calculate the modulation waveform patterns and the power spectral form with the frequency by using Eq. (7), and Eq. (4), respectively.

As results of comparing the calculated waveform with the observed one the following statements are noted.

1) The quick change of fading pattern may be produced due to the angle θ and ϕ.
2) The amplidude variations (the ratio of modulation) may occur by the inclination of relative to the Earth's magnetic field.
3) The occurrence of long and short fading patterns in the sonagraph may be given by the ratio of amplitude modulation.
4) The variation of the spectral amplitude with frequency may be produced by the ratio of amplitude modulation of waveform pattern and have high spectral peaks at $6kH_z \sim 3kH_z$.
5) The pass length of the whistler shown in Fig. 3 is about $Z=1608$ km in the ionosphere, The average all pass length (πz) to the conjugate point of Kochi may be estimated to be over 5053 km.

ACKNOWLEDGMENT

The authors would like to thank Dr. M. Hayakawa of Reseach Institute of Atmospherics, Nagoya University for his kind discussions and useful suggestions. We woud like to thank Prf. H. Volland of Radioastronomical Institute, University of Bonn for his useful suggestions and his papers. Whistler data analyzed in this paper were obtained with the Stanford University ELF/VLF Radiometer which was installed at the Earth Observatory, Kochi University. Thanks are due to Dr. A. C. Fraser Smith of Stanford University for his cooporation in the ELF/VLF project.

REFERENCES

[1] Helliwell RA. (1965) Whistlers and related ionospheric phenomena. Stanford University Press. California. pp. 83-179
[2] Paymar EM. (1972) Banded whistlers observed on OGO-4. Tec Rep. Stanford University. 3439-1
[3] Ohta K. Hayakawa M. Eguchi H. (1988) Ground-based observation of banded whistlers at low latitudes. Res Lett Atmos Electr. 8: 127-136
[4] Yano S. Ogawa T. Hagino H. (1989) Fading phenomena of whistler waves. Res Lett Atmos Electr. 9: 97-107
[5] Vidmer RJ. Crawford FW. Harker KJ. (1983) Delta function excitation of waves in the Earth's ionosphere, Radio Science. 18: 1337- 1354
[6] Budden KG. (1961) Radio waves in the ionosphere. Cambridge University Press. Cambridge. pp. 542
[7] Tyras G. (1969) Radiation and propagation of electromagnetic waves. Academic Press. New York. pp. 368

[8] Felsen LB. (1969) Transients in dispersiv media. I. Theory. IEEE Trans. Ant. Prop. AP-17: 191-199
[9] Stix TH. (1962) The theory of plasma waves. McGraw-Hill. New York. pp. 1-44

5 *Natural VLF/ELF Radio Noise*

5.1 Global Measurements of Low-Frequency Radio Noise

A.C. FRASER-SMITH, P.R. MCGILL, A. BERNARDI, R.A. HELLIWELL, and M.E. LADD
Space, Telecommunications and Radioscience Laboratory, Stanford University, Stanford, CA 94305, USA

ABSTRACT

We report illustrative results obtained by Stanford University's global survey of ELF/VLF radio noise (frequencies in the range 10 Hz – 32 kHz). Particular comparison is made between the noise measurements made at high (polar) latitudes with those at lower latitudes. Although most of the natural ELF/VLF noise observed everywhere in the world is lightning-generated, the high-latitude noise often contains additional components that are of magnetospheric origin. In the data we have examined, this noise consists predominantly of polar chorus, which is concentrated in the range 300 Hz–2 kHz. It produces a characteristic signature in the noise statistics. Less frequent occurrences of broad–band (auroral) hiss can occasionally mask most or all of the lightning-generated noise in the ELF/VLF range.

1. INTRODUCTION

As we have previously reported [1, 2], our Laboratory is presently conducting a global survey of extremely-low frequency (ELF) and very-low frequency (VLF) radio noise (specifically, the survey covers frequencies in the range 10 Hz – 32 kHz). Our three high latitude stations are Thule (TH; 76.5° N, 68.8° W) and Søndrestrømfjord (SS; 67.0° N, 50.1° W) in Greenland, and Arrival Heights (AH; 77.8° S, 193.3° W) in the Antarctic, and the magnitudes of the geomagnetic latitudes for these stations range from 77° (SS) up to 87° (TH), thus ensuring that their data include representative samples of ELF/VLF radio noise of magnetospheric origin (e.g., chorus and hiss), in addition to the lightning-generated noises (predominantly sferics) that typically dominate at our five lower latitude stations (Grafton, New Hampshire (43.6° N, 72.0° W); L'Aquila, Italy (43.4° N, 13.3° E); Stanford, California (37.4° N, 122.2° W); Kochi, Japan (33.3° N, 226.5° W); and Dunedin, New Zealand (45.8° S, 189.5° W)).

The radio noise statistics computed continuously at each of the stations consist of the average, root-mean-square (rms), maximum, and minimum amplitudes in 16 narrow frequency bands (5% bandwidth) distributed through the ELF and VLF ranges (Table 1). They are computed at the end of every minute from 600 amplitude measurements made at the rate of 10 per second on the envelope of the noise signal emerging from each narrow-band filter. Later processing of these data can, with little additional computation, give the V_d and F_a statistics. In addition, amplitude probability distributions (APD's) can also readily be derived from the sampled data. These various statistical quantities are widely used to characterize radio noise and they are described in several reports issued by the International Radio Consultative Committee, or CCIR [e.g., 3, 4].

Comparison of the noise statistics between the high and moderate-to-low latitude locations reveals many similarities and much stability of the statistics over time; but there are also some major differences. Many of the differences consist simply of expected changes in the average levels of the statistical quantities. However, some of the changes in the statistics are caused by differences in the nature of the noise, and in particular by the occurrence of magnetospheric noise. In the data we have examined, this noise consists most frequently of polar chorus [5], which consists of a band of hiss with rising tones (as originally defined in [5]), and in our measurements it is concentrated in the range 300 Hz – 2 kHz. It produces a characteristic signature in the noise statistics, which makes its presence relatively easy to identify. Less frequently, broad–band (auroral) hiss occurs, and on occasion it can be sufficiently strong to mask some or all of the lightning-generated noise in

TABLE 1. Center frequencies and bandwidths for the 16 channels of the ELF/VLF noise measurement systems.

Channel	Center Frequency	Bandwidth (5%)
1	10 Hz	0.5 Hz
2	30	1.5
3	80	4
4	135	6.75
5	275	13.75
6	380	19
7	500	25
8	750 Hz	37.5
9	1 kHz	50
10	1.5	75
11	2	100
12	3	150
13	4	200
14	8	400
15	10.2	510
16	32 kHz	1600 Hz

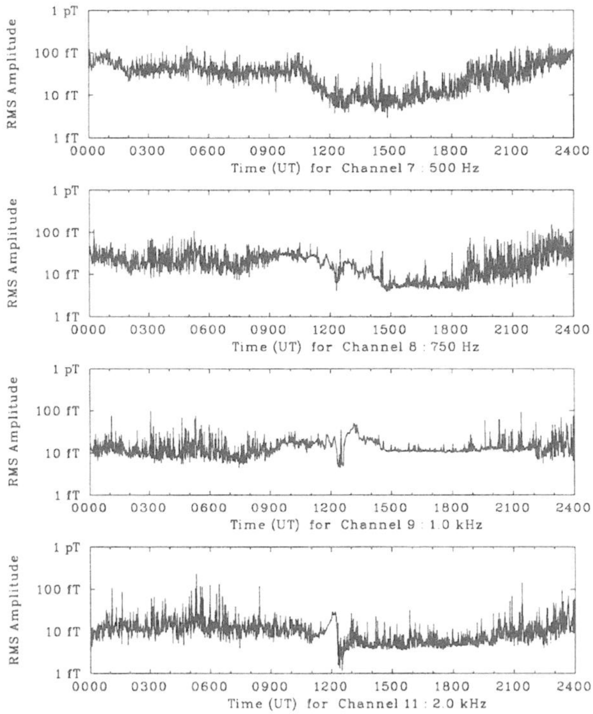

Fig. 1. Variation of the 500 Hz, 750 Hz, 1 kHz, and 2 kHz one-minute rms magnetic field amplitudes at Søndrestrømfjord, Greenland, during 13 November 1986 (UT). The applicable frequencies are shown under each panel. The amplitudes are given either in units of picoTesla (pT) or femtoTesla (fT; 1 pT = 10^3 fT = 10^{-12} T).

the ELF/VLF range. The noise statistics are less effective in distinguishing between this hiss and the lightning-generated noise, although the very strong auroral hiss events produce characteristic signatures in the statistics.

Even a partial presentation of the noise statistics being obtained by our survey would be outside the scope of this paper. We therefore concentrate on the presentation of illustrative results, with particular emphasis on the magnetospheric noise that is observed at high latitudes, sometimes quite commonly, and sometimes very strongly.

2. AMPLITUDE MEASUREMENTS

In Figure 1 we show the diurnal variation of the 500 Hz, 750 Hz, 1.0 kHz and 2.0 kHz one-minute rms amplitudes that were measured at Søndrestrømfjord on 13 November 1986. It was early winter at the measurement location; there were no local thunderstorms, and there were only a few hours of sunlight (local time at SS is 3 hours behind UT; thus 0300 UT corresponds to local noon). The data are typical in that they show considerable impulsiveness, or 'spikiness,' due to the transient and irregular nature of the sferics that are the predominant form of noise signal. Most of the time, at all frequencies covered by this study, plots of the daily variations of the one-minute average or rms amplitudes will resemble the data shown in the top panel of Figure 1, except for a general reduction in the impulsiveness for frequencies below $\sim$ 250 Hz.

In the three lower panels of Figure 1, it will be noticed that the impulsiveness of the data tends to go through a minimum in the interval 1000–2000 UT (a 10-hour interval very roughly centered on local noon), and for some smaller sub-intervals the character of the data change entirely. For example, during the interval 1100–1230 UT the impulsiveness of the 2.0 kHz amplitude data almost completely disappears and there is an abrupt change in the average level of the amplitudes. These changes are not typical of the normal sferic noise background for the chosen frequencies, nor are they typically observed at (1) frequencies above or below the range 300 Hz – 2.0 kHz, or (2) at middle and low latitudes. We have come to recognize them as signatures for the occurrence of polar chorus in the range 300 Hz – 2 kHz. The changes are even more clearly defined in plots of the 'voltage-deviation,' or V_d, statistic, which is a specific measure of the impulsiveness of noise. We will further discuss these occurrences of magnetospheric noise in a later section.

The one-minute average data illustrated in Figure 1 can be processed in many different ways to give additional information about the morphology of ELF/VLF noise, about its modes of propagation, and about its sources [e.g., 6, 7]. One important form of processing we use is to compute average or rms amplitudes over longer time intervals, usually one- or three-month intervals. Figure 2 illustrates one form of these longer averages, using data once again from Søndrestrømfjord. Taking all the SS one-minute average amplitudes for January 1987 (the middle of the northern winter), we have computed and plotted the average noise amplitude at each of our 16 measurement frequencies for each of the eight three-hour time divisions of a 24 hour UT day. The result is a set of eight spectral distributions which provide information about the diurnal variation of the ELF/VLF noise spectrum at Søndrestrømfjord in January 1987.

Taking a general view of the amplitude data in Figure 2, there is roughly an inverse relation with frequency that is typical of all the measurements we have made in our noise survey, and which also appears to be typical of a much broader frequency range including and extending on either side of the ELF/VLF range [1]. Looking at the data in more detail, we see considerable diurnal variation, with the largest amplitudes tending to occur around 0000–0300 UT and the smallest around 1200–1500 UT. The magnitude of the diurnal variation is frequency dependent: in the frequency range 3–8 kHz the largest average amplitude is nearly ten times greater than the smallest, whereas at 80 Hz there is little difference between the amplitudes. This variability may be solely a northern high latitude phenomenon, since it is not duplicated by the data from Arrival Heights for the same month (Figure 3) or for the month of June 1986, which is an equivalent winter month in the southern hemisphere (Figure 4).

The Arrival Heights ELF/VLF measurement system was the first of our noise survey systems to be set in operation and its data base is therefore the most extensive that is available to us. In

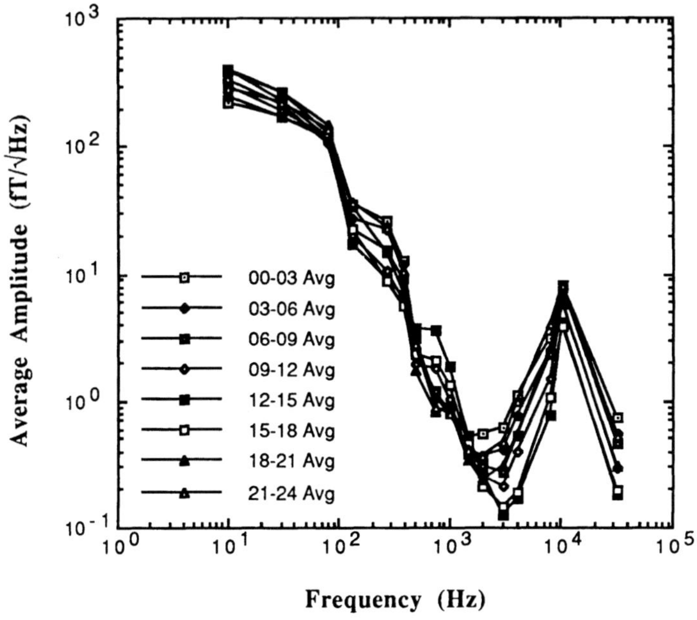

Fig. 2. Variation of the Søndrestrømfjord ELF/VLF noise amplitudes for the month of January 1987. Overall average amplitudes for each of the 16 narrow band frequencies are shown, and the data are broken down into eight 3-hour time blocks, starting with 0000–0300 UT.

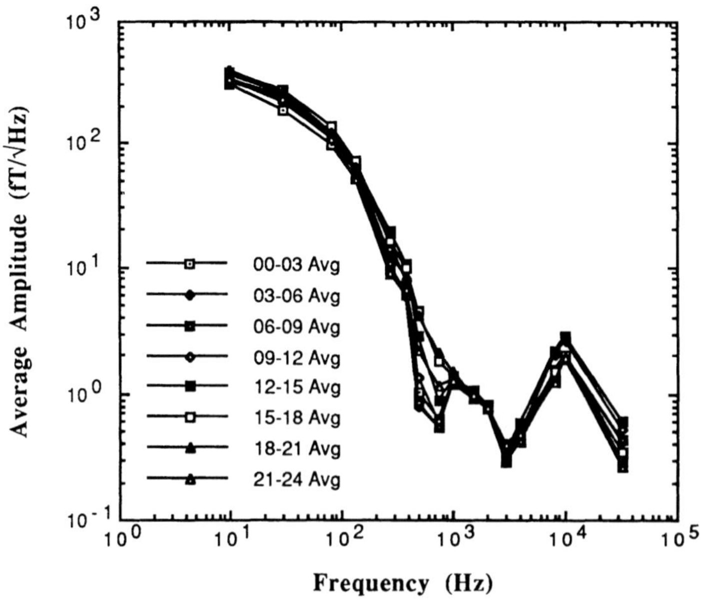

Fig. 3. Variation of the Arrival Heights ELF/VLF noise amplitudes for the month of January 1987. Overall average amplitudes for each of the 16 narrow band frequencies are shown, and the data are broken down into eight 3-hour time blocks, starting with 0000–0300 UT.

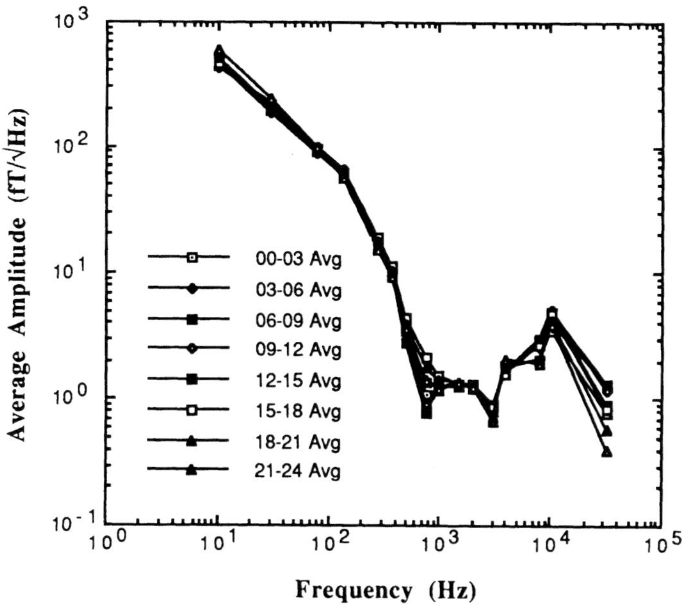

Fig. 4. Variation of the Arrival Heights ELF/VLF noise amplitudes for the month of June 1986. Overall average amplitudes for each of the 16 narrow band frequencies are shown, and the data are broken down into eight 3-hour time blocks, starting with 0000–0300 UT.

Figure 5 we illustrate the longer term variability of the ELF/VLF noise amplitudes by plotting their overall average values against frequency for each January and July in the two year interval starting January 1986. It can be seen that there is remarkably little difference between the amplitudes at frequencies below 1 kHz, but that there appear to be significant differences at the higher frequencies. However, even at the higher frequencies the year-to-year changes are not as marked as the figure suggests because the higher amplitudes are all measured during the southern winter. If we accept the evident seasonal variation, there is once again little difference in the amplitudes. This is particularly clear at 32 kHz, where the amplitudes for the three January months are nearly identical.

Finally, to illustrate some of the similarities and differences between the high and low latitude noise amplitudes, in Figure 6 we show the variation with frequency of the overall monthly average amplitudes of the June 1986 measurements at Thule, Søndrestrømfjord, and Arrival Heights, and the July 1987 measurements at Kochi. They are all summer season measurements except for those at Arrival Heights, which are taken during the southern winter. Despite the difference in season, there is close agreement between the Arrival Heights and Søndrestrømfjord monthly averages. The Thule averages also correspond reasonably well with those of the two other high latitude locations above 2 kHz, but at lower frequencies they are substantially higher. The Kochi noise amplitudes shown in the figure are roughly representative of the amplitudes that can be measured at middle and low latitudes during summer months. They are generally greater than the amplitudes measured simultaneously at high latitudes by some factor in the range 2–10, with the greatest differences occurring at the higher frequencies.

High ELF noise amplitudes have been a consistent feature of our measurements at Thule. Although Thule is particularly close to one of the geomagnetic poles, and therefore differs in that respect from our other measurement locations, there is no reason to expect higher ELF noise amplitudes near a geomagnetic pole and there is no previous record of such higher amplitudes being measured. At this time therefore we tentatively ascribe the high amplitudes to broad band ELF interference from the nearby air base.

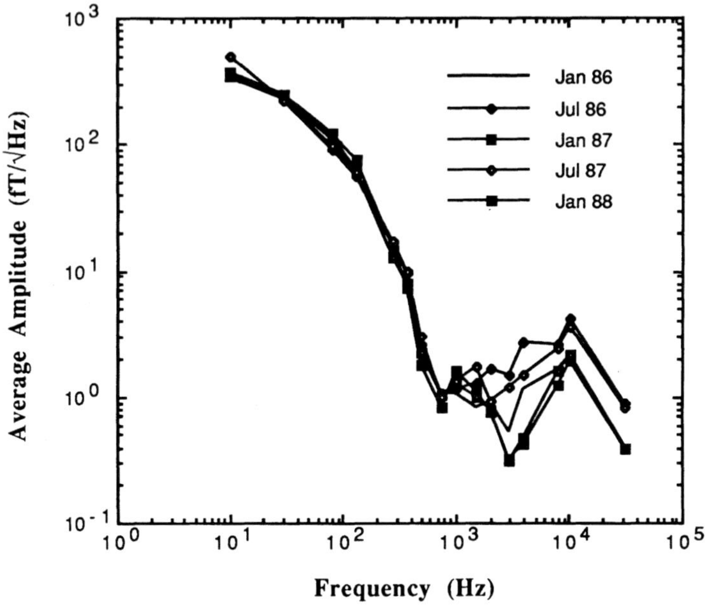

Fig. 5. Variation of the Arrival Heights overall average ELF/VLF noise amplitudes for each January and July in the two-year interval starting January 1986.

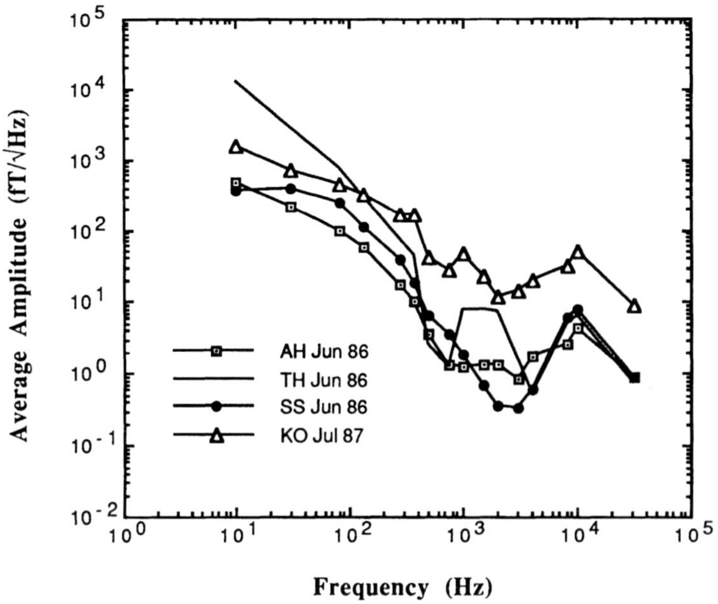

Fig. 6. Variation of the overall average ELF/VLF noise amplitudes measured at Thule, Søndreströmfjord, and Arrival Heights during June 1986, and at Kochi during July 1987.

3. V_d MEASUREMENTS

In Figure 7 we show a representative sample of our V_d measurements. The measurements were made at Søndrestrømfjord during June 1986 and overall monthly average values of the maximum, minimum, and average one-minute V_d's are shown for each of the 16 narrow band frequencies. A tendency for the values of V_d to increase gradually from a level near 1 at the lowest frequency (10 Hz) to a level near 10 at the highest frequency (32 kHz) is a feature of the measurements shown in the figure and it is typical of the measurements we have made on purely sferic noise.

The presence of magnetospheric noise generally results in reduced values of V_d in the frequency bands in which the (non-sferic) noise occurs. Since magnetospheric noise is not always observed, whereas sferics are always present (even though they may be masked by the other noise), the first evidence for magnetospheric noise in monthly plots of V_d such as the one shown here is observed in the plots of minimum V_d values. This can be clearly seen in Figure 7, where the pronounced dip in the minimum V_d values in the range 380 Hz to 2 kHz is the result of the occurrences of polar chorus during the month. As the magnetospheric noise increases in frequency of occurrence and intensity it begins to produce changes in the monthly average values of V_d as well as in the minimum values. Presumably, if there were further increases, the maximum values of V_d would begin to be affected as well, but we have not so far observed such sustained levels of non-sferic noise.

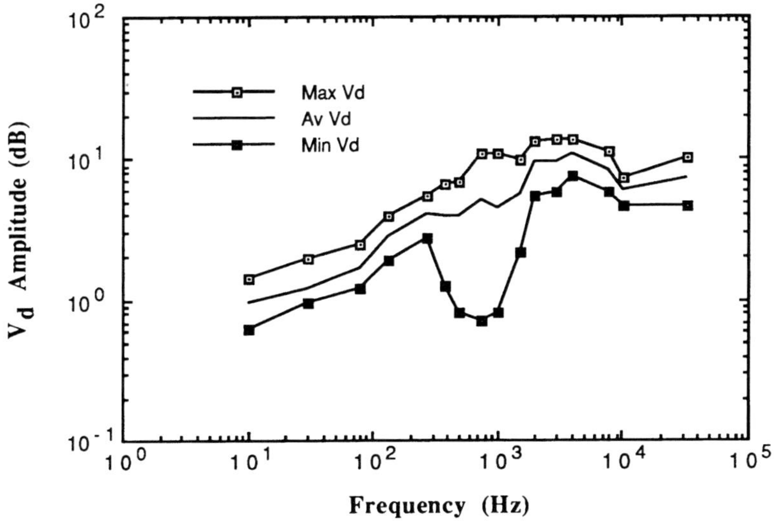

Fig. 7. Variation with frequency of V_d at Søndrestrømfjord for the month of June 1986.

4. EXAMPLES OF MAGNETOSPHERIC NOISE

We now show spectrograms of the two predominant forms of the non-sferic, or magnetospheric, noise that are observed at high latitudes. As we have mentioned, the most commonly observed form of magnetospheric noise at our high latitude stations is polar chorus, which is usually limited to the overall frequency range 300 Hz to 2 kHz. The other predominant form of magnetospheric noise is auroral hiss, which occurs over large portions of the ELF/VLF range, sometimes even extending up to frequencies of 200 kHz or more [8]. Examples of these two forms of ELF/VLF noise are shown in Figures 8 and 9.

The polar chorus shown in Figure 8 is part of an interesting, extended interval of activity at Søndrestrømfjord that started around 0900 UT on 13 November 1986 and which continued until after 1500 UT. It was not particularly strong and although its 'signature' in the plots of one-minute

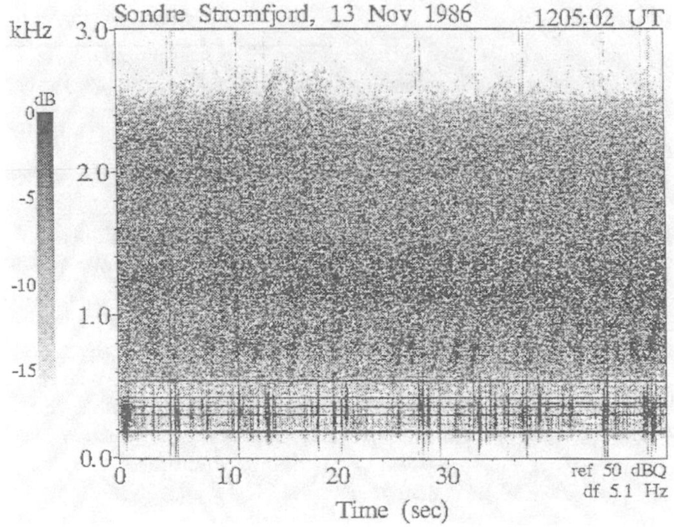

Fig. 8. A digital spectrogram of the ELF activity at Søndrestrømfjord during a roughly 50 second interval starting at 1205:02 UT on 13 November 1986. The strong horizontal lines toward the bottom of the spectrogram are harmonics of the local power supply frequency. The vertical lines are produced by sferics and the largely unstructured blackening from 500 Hz to 2.5 kHz is mostly hiss.

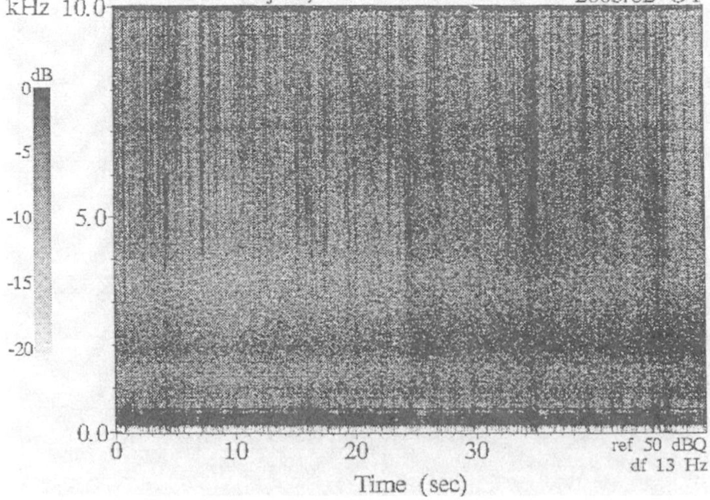

Fig. 9. A digital spectrogram of the ELF/VLF activity at Søndrestrømfjord during a roughly 50 second interval starting at 2005:02 UT on 16 November 1986. It is possible to see sferics, as well as some polar chorus around 400 Hz, but the entire display is dominated by auroral hiss, which produces the largely unstructured blackening extending over the entire 10 kHz frequency range of the display.

average amplitudes (Figure 1) is easily recognizable it is not nearly as marked as is often the case during these polar chorus events. The activity started as a series of quasi-periodic bursts of hiss, limited mostly to the frequency range 500 Hz – 1.3 kHz, and with a period of about 6 seconds between the bursts (which had a rising frequency characterisic). By 1200 UT the quasi-periodicity had essentially disappeared and the activity had increased both in intensity and in its frequency range, the upper frequency of which now approached 2.5 kHz. In addition to these characteristics, which can be seen in Figure 8, the activity developed a number of the rising elements typical of polar chorus. Since the activity only extended up to 2.0 kHz and above for about an hour, its 'signature' in the bottom panel of Figure 1 (for 2.0 kHz) is comparatively limited. The subsequent drop in the mean amplitude of the sferic activity is of great interest, since it suggests increased ionospheric absorption over a substantial region above Søndrestrømfjord.

The spectrogram shown in Figure 9 is quite extraordinary and it is shown here to make a point. The entire spectrogram, which covers a larger frequency range than Figure 8, is blackened by an occurrence of strong auroral hiss. Some sferics can be seen through the general blackening, as can a band of low-frequency polar chorus in the range 200–500 Hz. Other spectrograms of the activity show that the hiss extends up to around 16 kHz. Strong auroral hiss of this kind produces 'signatures' in our noise statistics that are similar to those of polar chorus, but with the exception that they extend to much higher frequencies. Since there are a variety of VLF navigation and communication transmissions above 10 kHz, auroral hiss has the potential to degrade these transmissions at high latitudes.

5. CONCLUSION

We have presented a number of quantitative examples of the ELF/VLF noise measurements made by our global array of ELF/VLF radio noise measurement systems and we have also presented some of the noise statistics that can be derived from the measurements. Most of the data displayed were obtained at high latitude locations. Our purpose in doing this was twofold: first, we wished to provide examples of data that are comparatively lacking, which is certainly the case for ELF/VLF radio noise data at high latitudes, and, second, we wished to emphasize the importance of magnetospheric noise, which has the potential to create difficult conditions for the reception of ELF/VLF transmissions at high latitudes.

ACKNOWLEDGEMENTS

This research was sponsored by the Office of Naval Research through Contract No. N00014-81-K-0382 and Grant No. N00014-90-J-1080. Support for the measurements at Thule, Greenland, was provided by Rome Air Development Center through Contract No. F19628-84-K-0043. Logistic support for the measurements at Søndrestrømfjord, Greenland, and Arrival Heights, Antarctica, was provided by the National Science Foundation (NSF) through NSF cooperative agreement ATM 88-22560 and NSF Grant DPP-8720167, respectively. We thank Professors T. Ogawa and M. Kusunose of Kochi University for their cooperation in the measurements in Japan; Mr. J. P. Turtle of Rome Air Development Center for his cooperation in the Thule measurements; and Dr. J. D. Kelly of SRI International for facilitating the measurements at Søndrestrømfjord.

REFERENCES

1. Fraser-Smith AC. Helliwell RA. (1985) The Stanford University ELF/VLF radiometer project: measurement of the global distribution of ELF/VLF electromagnetic noise. In : Proc 1985 IEEE Internat Symp on Electromag Compatability, IEEE Catalog No. 85CH2116-2 : 305–310.

2. Fraser-Smith AC. Helliwell RA. Fortnam BR. McGill PR. Teague CC. (1988) A New Global Survey of ELF/VLF Radio Noise. Conf. on Effects of Electromagnetic Noise and Interference on Performance of Military Radio Communication Systems, Lisbon, Portugal, 26–30 October, 1987. Published in AGARD Conference Proceedings No 420 : 4A-1–4A-7.

3. C.C.I.R. (1964) World distribution and characteristics of atmospheric radio noise. Report 322. International Radio Consultative Committee. International Telecommunication Union. Geneva.

4. C.C.I.R. (1988) Characteristics and applications of atmospheric radio noise data. Report 322–3. International Radio Consultative Committee. International Telecommunication Union. Geneva.

5. Ungstrup E. Jackerott IM. (1963) Observations of chorus below 1500 cycles per second at Godhavn, Greenland, from July 1957 to December 1961. In : J Geophys Res 68 : 2141–2146.

6. Fraser-Smith AC. McGill PR. Helliwell RA. Houery S. (1989) Radio noise measurments in the long-wave band at Thule, Greenland. Report RADC-TR-89-94. Rome Air Development Center. New York.

7. Fraser-Smith AC. Ogawa T. McGill PR. Bernardi A. Ladd ME. Helliwell RA. (1989) Measurements of ELF/VLF radio noise in Japan. URSI Symp on Environmental and Space Electromagnetics. Tokyo. Japan. (This publication).

8. Helliwell, RA. (1965) Whistlers and related ionospheric phenomena. Stanford University Press. Stanford. California.

5.2 Measurements of ELF/VLF Radio Noise in Japan

A.C. FRASER-SMITH[1], T. OGAWA[2], P.R. MCGILL[1], A. BERNARDI[1], M.E. LADD[1], and R.A. HELLIWELL[1]

Space, Telecommunications and Radioscience Laboratory, Stanford University, Stanford, CA 94305, USA[1] and Department of Physics, Kochi University, Kochi, 780 Japan (Now at Science Laboratory International, Kochi 780, Japan)[2]

ABSTRACT

This paper reports measurements of radio noise by a Stanford University ELF/VLF radiometer located near Kochi, Japan. The measurements provide information on the average and root-mean-square (rms) amplitudes in 16 narrow frequency bands (5% bandwidth) distributed through the range 10 Hz to 32 kHz. Later processing of the data gives the V_d and F_a statistics and amplitude probability distributions (APDs). Because Kochi is a low latitude station, the noise statistics tend to be more homogeneous and typical of lightning-generated ELF/VLF noise than those of higher latitude stations, where magnetospherically-generated noise may be present. On the other hand, the closer proximity to thunderstorms produces features that are not observed at higher latitudes.

1. INTRODUCTION

In September 1985 a Stanford University ELF/VLF radiometer (10 Hz to 32 kHz) was installed and set in operation at a location near Kochi, Japan (33.3° N, 226.5° W). The project was undertaken as a cooperative effort between Stanford University's Space, Telecommunications and Radioscience (STAR) Laboratory and the Department of Physics at Kochi University, and it has now provided much new information on the statistical nature of ELF/VLF radio noise in Japan. The radiometer has been described in detail elsewhere [1, 2, 3], so only summary details of its operating characteristics will be given here.

The ELF/VLF radiometer at Kochi is one of eight identical noise measurement systems that have been set up around the world to determine the global characteristics of ELF/VLF radio noise. Each instrument operates under the control of a small computer, which also computes a variety of noise statistics from the measurements. Specifically, the noise statistics computed continuously at each of the stations are the root-mean-square (rms), average, maximum, and minimum amplitudes of the noise in 16 narrow frequency bands (5% bandwidth) distributed through the range 10 Hz to 32 kHz. These noise statistics are computed at the end of every minute from a nominal 600 amplitude measurements made at the rate of 10 per second on the envelope of the noise signal emerging from each narrow-band filter. Later processing of these data can, with little additional computation, give the V_d ("voltage deviation") and F_a ("antenna noise factor") statistics. Similarly, the sampled data can be used to derive amplitude probability distributions (APD's). All these various statistical quantities are widely used to characterize radio noise and they will be found described in several definitive reports by the International Radio Consultative Committee, or CCIR [e.g., 4, 5]

To illustrate the basic form of the noise data, in Figure 1 we show simultaneous plots of the one-minute averages for 30 Hz, 500 Hz, 2 kHz, and 8 kHz, for the one week interval 6–12 July 1988. In these illustrative data the predominant feature is a well-defined diurnal variation that persists throughout the interval, although with some variation from day to day. Additional interesting features are a decrease in the impulsiveness of the changes in the 30 Hz amplitudes as the week progresses, and an increase in the 2 kHz and 8 kHz averages toward the end of the week.

A full presentation of all the ELF/VLF noise statistics for Kochi is beyond the scope of this paper. However, the statistics do not vary greatly once they have been averaged over times of the order of a month or more, and, as a result, the examples of the statistics presented here can be considered to be reasonably typical for both Kochi and for Japan as a whole.

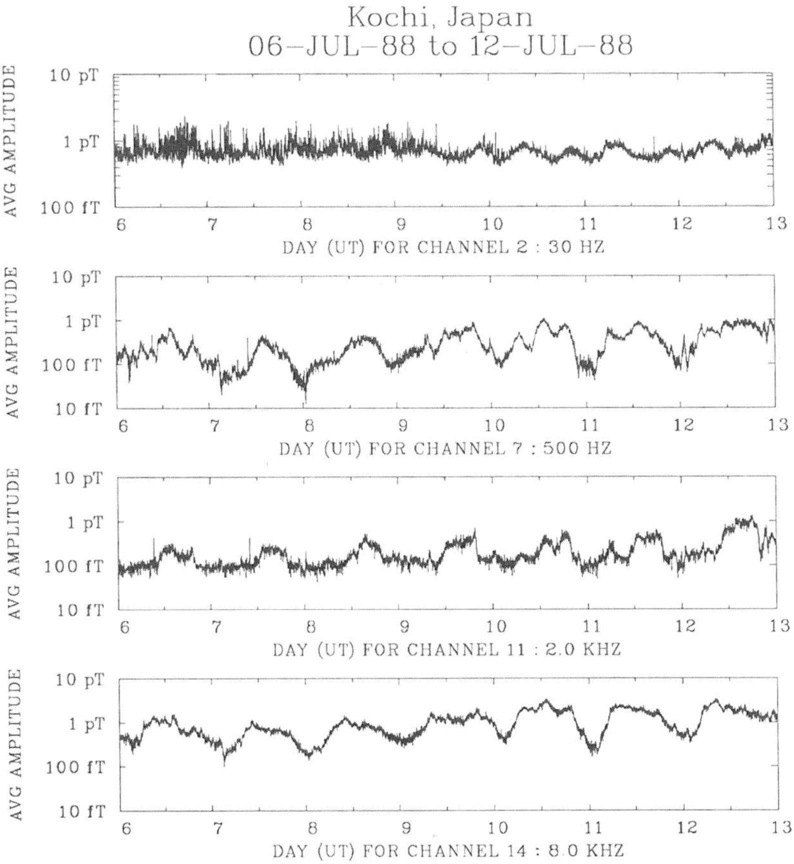

Fig. 1. Variation over the one week interval 6–12 July 1988 of the one-minute average ELF/VLF noise amplitudes measured at Kochi, Japan, for the four representative frequencies 30 Hz, 500 Hz, 2 kHz, and 8 kHz. Note the different amplitude scales.

2. AVERAGE AMPLITUDES

The data plotted in Figure 1 are typical of the basic one-minute averages computed by the Kochi radiometer and subsequently stored on magnetic tape. These one-minute averages can be further processed to provide average amplitudes over longer intervals of time. For example, in Figure 2 we show the overall average amplitudes for each of our 16 frequency channels for the entire months of January and July, 1988. We divide the data up into four time blocks (0000–0300 UT; 0600–0900 UT; 1200–1500 UT; 1800–2100 UT) in order to illustrate the variation of the average amplitudes on a diurnal basis. It will be seen that there is a general tendency for the amplitudes to decline with increasing frequency, in agreement with earlier measurements [e.g., 1, 6, 7], and that there is a relative minimum around 2–4 kHz, which is also expected [e.g., 1, 6, 7] and which is due to the earth-ionosphere waveguide cutoff in that frequency range. For both months, there is hardly any diurnal variation in the amplitudes at the lower frequencies, but in the range 750 Hz – 8 kHz the amplitudes for 1200–1500 UT and 1800–2100 UT, which are nighttime amplitudes in Japan, are noticeably higher than those for the day. It will of course also immediately be noticed that the average noise amplitudes for July all tend to be higher than those for January, with the biggest increases occurring at the higher frequencies. This is undoubtedly due to the greater thunderstorm activity in the vicinity of Japan during the summer rainy season, or "tsu-yu".

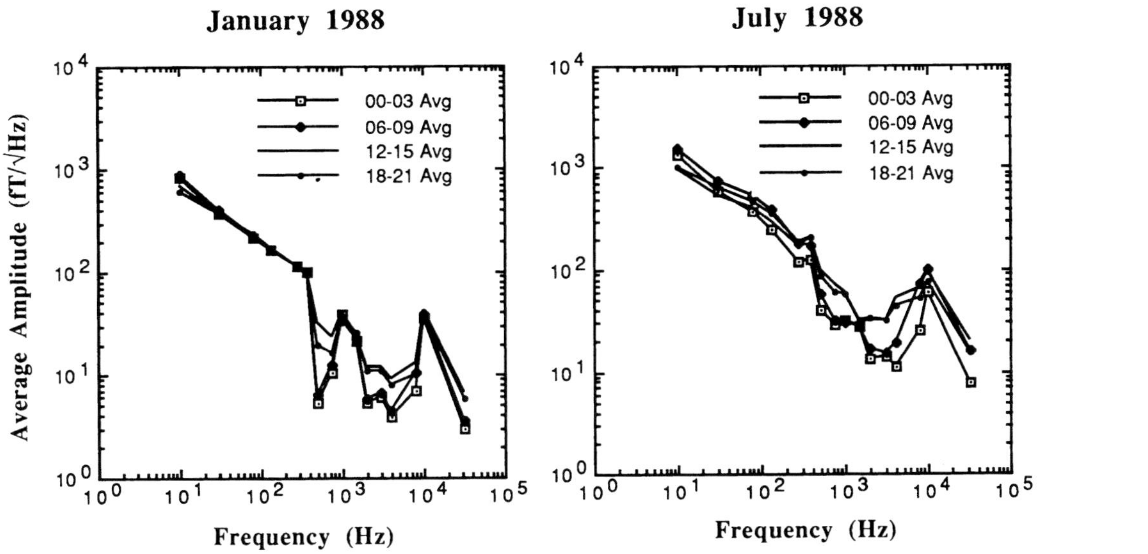

Fig. 2. Variation of the Kochi, Japan, ELF/VLF noise amplitudes for the months of January and July, 1988. Overall average amplitudes for each of the 16 narrow band frequencies are shown, and the data are broken down into four 3-hour time blocks, starting with 0000-0300 UT.

To show the seasonal difference in the ELF/VLF noise amplitudes in additional detail, and to give an indication of their possible variation from year to year, in Figure 3 we plot the overall average amplitudes (without division into time blocks) for July 1987 and for January and July 1988. The three plots closely resemble each other, but the amplitudes are closer in general for the two July plots and for these two latter plots the amplitudes are nearly identical for frequencies less than 500 Hz.

Finally, to illustrate the generally higher noise levels at Kochi as compared with those at higher latitude locations, in Figure 4 we compare the overall average amplitudes at Kochi for January and July, 1988, with the equivalent overall average amplitudes at Sondrestromfjord, Greenland (67.0° N, 51.0° W), for January 1987 (a winter month) and at Arrival Heights, Antarctica (77.8° S, 193.3° W), for January 1988 (a summer month in the Antarctic). Despite the differences in the seasons and locations, the two high latitude plots are very similar, but the noise amplitudes are 1–2 orders of magnitude less than those for Kochi.

3. V_d MEASUREMENTS

Although V_d is a well-known statistical measure of radio noise, few V_d data are available at low-frequencies. Because our radiometers compute both rms and average amplitudes, it is comparatively easy to derive values of V_d, which is defined to be the ratio in dB of the rms to average amplitude. The significance of this statistical measure at low frequencies is easily understood once it is remembered that most of the ELF/VLF noise observed at middle and low latitudes is generated by lighting and that it consists of many transient impulses, or 'sferics'. The amplitude distribution of the noise has a long tail on its high amplitude side corresponding to the occurrence of strong, nearby lightning (and possibly also to the occurrence of more distant 'superbolts' [8]). For these kinds of distributions, the rms amplitudes of the noise tend to be larger than the average amplitudes, and the ratio of the rms to average amplitudes, which is a measure of the distribution's skewness, or of the 'spikiness' of the noise, takes on values greater than unity.

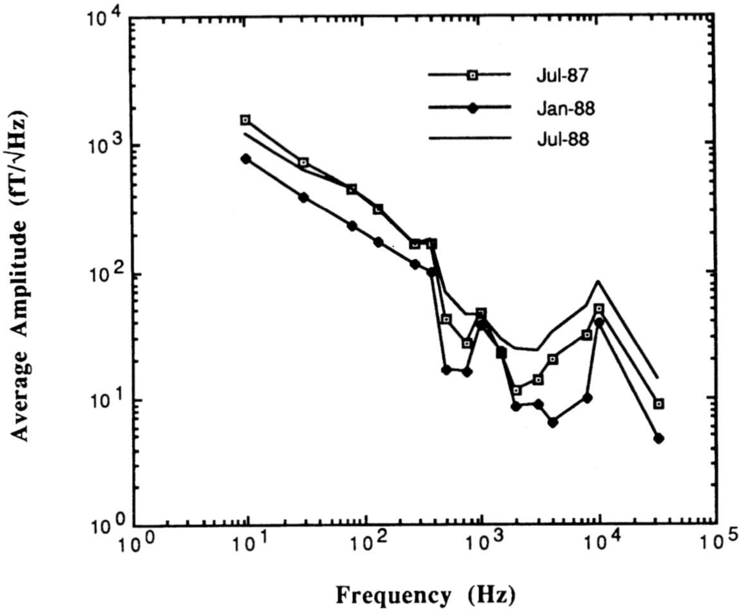

Fig. 3. Variation of the Kochi, Japan, ELF/VLF noise amplitudes for the months of July 1987 and January and July, 1988. Overall average amplitudes for each of the 16 narrow band frequencies are shown, without division into time blocks.

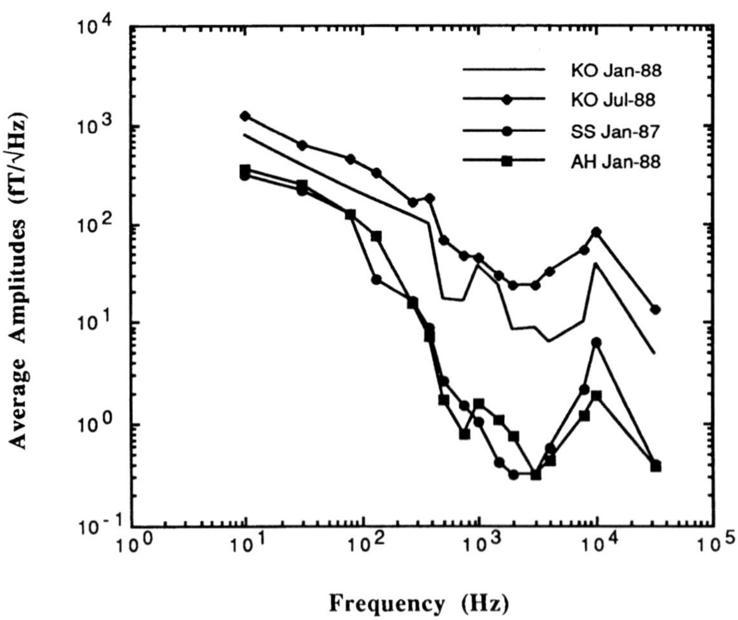

Fig. 4. Variation of the overall average ELF/VLF noise amplitudes at Kochi, Japan (KO), for the months of January and July, 1988, as compared with the equivalent overall average amplitudes at Sondrestromfjord, Greenland (SS), for January 1987 and at Arrival Heights, Antarctica (AH), for January 1988.

In Figure 5 we show three different plots of Kochi V_d measurements. Specifically, we plot the frequency variation of the average values of V_d measured at Kochi for each of the months July 1987 and January and July, 1988. It will be seen that there is an overall tendency for V_d to increase from low values of roughly 0.01–0.1 dB at the lowest frequencies (little spikiness) to values in the range 2–6 dB at the highest frequencies (considerable spikiness). However, for July 1987 and January 1988 there is also quite a remarkable minimum in the values of V_d for frequencies in the narrow range 750 Hz to 2 kHz, which is only marginally present in the data for July 1988.

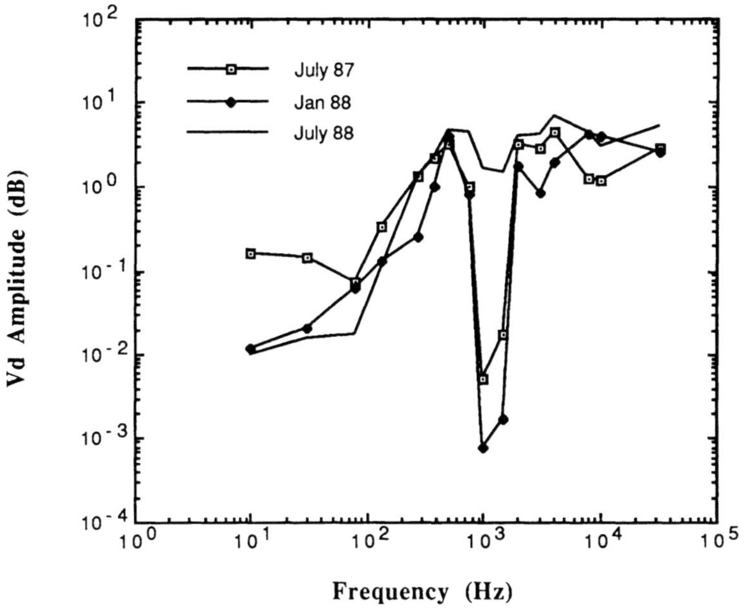

Fig. 5. Variation of the average ELF/VLF V_d amplitudes at Kochi, Japan, for the months of July 1987 and January and July, 1988.

4. AMPLITUDE PROBABILITY DISTRIBUTIONS

These distributions are widely used to predict the performance of radio systems operating in the presence of noise. However, there is considerable variety in the manner in which the APD's are displayed. We have chosen the form used by Watt [9] for our displays and thus we plot the percentage of times a particular amplitude measurement exceeds a certain value along the y-axis and the reference amplitude values along the x-axis. Figures 6 and 7 show two examples for Kochi. The first of these figures, Figure 6, shows APD's for the frequency 500 Hz for the months of January and July, 1988 (a winter month and a summer month), and the second figure, Figure 7, shows the corresponding APD's for the frequency 32 kHz. All the applicable amplitude measurements for each month are included in the plots; as a result, there are roughly 2.64×10^6 measurements per plot. It will immediately be noticed that the APD plots for July cover much higher amplitudes than do those for January, with the largest amplitudes (which occur rarely) being over a hundred times larger in July than in January. This tendency for ELF/VLF radio noise to have higher amplitudes in summer months is in agreement with the results presented above for the average amplitudes of the noise.

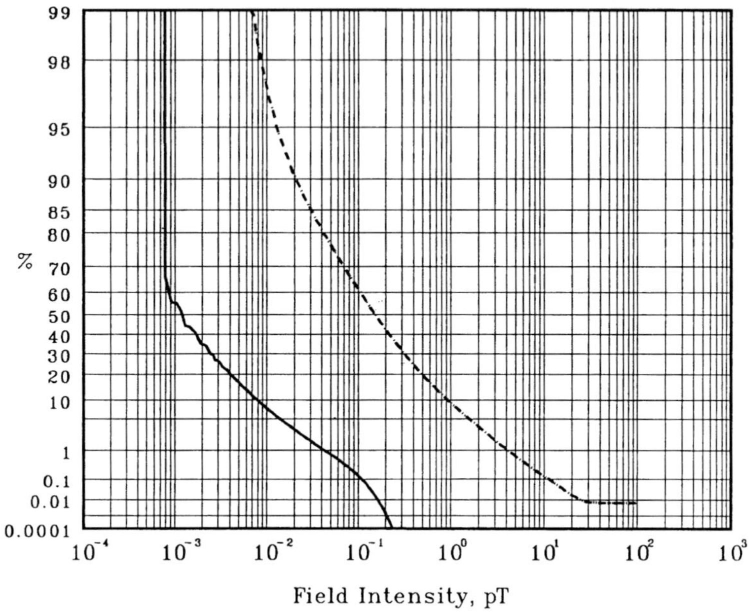

Fig. 6. Comparison of the amplitude probability distributions of the measurements made at Kochi at 500 Hz during January (winter; solid line) and July (summer; dashed line) of 1988. Over 2.6 million measurements are included in each curve. The percentage plotted is the percentage of these measurements that exceed the corresponding field intensity plotted on the horizontal axis.

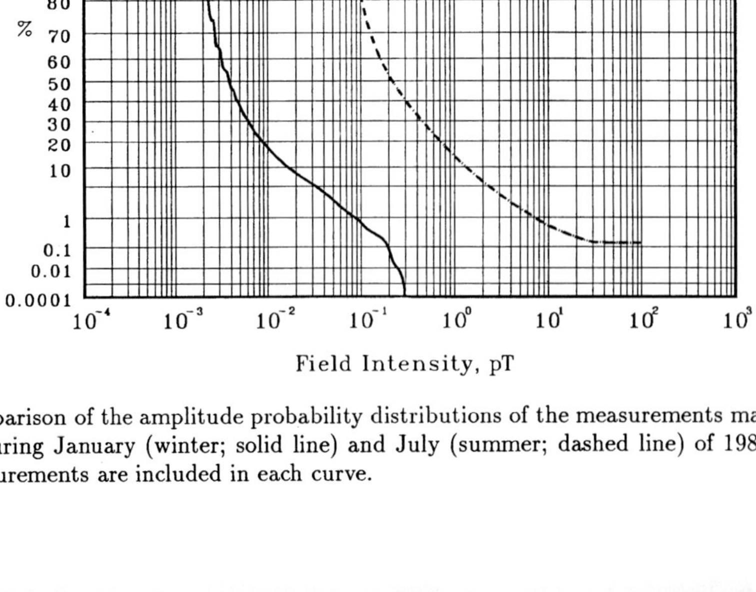

Fig. 7. Comparison of the amplitude probability distributions of the measurements made at Kochi at 32 kHz during January (winter; solid line) and July (summer; dashed line) of 1988. Over 2.6 million measurements are included in each curve.

5. DISCUSSION

Our measurements of ELF/VLF radio noise at Kochi have shown that the noise amplitudes tend to decline with increasing frequency, in agreement with earlier measurements at other locations. The decline is roughly proportional to 1/frequency. The voltage deviation, or V_d, statistic, on the other hand, tends to increase with frequency, thus indicating increasing 'spikiness' of the noise with increasing frequency. As an exception to this general trend, our measurements show a remarkable drop in V_d around 1 kHz.

The noise amplitudes at Kochi are roughly 3–10 times larger during the summer than during the winter, and, when taken as a whole, they are at all times substantially larger than those measured at high latitude locations. We attribute this difference to the close proximity of Japan to the thunderstorm regions that are the primary sources of the noise.

ACKNOWLEDGEMENT

We thank Professor Masahiko Kusunose of the Department of Physics at Kochi University for his cooperation, and Mr. Kevin Smith for his help with the radiometer installation near Kochi. The research at Stanford University was sponsored by the Office of Naval Research through Contract No. N00014-81-K-0382 and Grant No. N00014-90-J-1080. Logistic support for the Sondrestromfjord and Arrival Heights radiometers is provided by the National Science Foundation (NSF) through NSF cooperative agreement ATM 88-22560 and NSF Grant DPP-8720167.

REFERENCES

1. Fraser-Smith AC. Helliwell RA (1985) The Stanford University ELF/VLF radiometer project: measurement of the global distribution of ELF/VLF electromagnetic noise. In : Proc. 1985 IEEE Internat. Symp. on Electromag. Compatability, IEEE Catalog No. 85CH2116-2 : 305–310.
2. Fraser-Smith AC. Helliwell RA. Fortnam BR. McGill PR. Teague CC. (1988) A New Global Survey of ELF/VLF Radio Noise. Conf. on Effects of Electromagnetic Noise and Interference on Performance of Military Radio Communication Systems, Lisbon, Portugal, 26–30 October, 1987. Published in AGARD Conference Proceedings No. 420 : 4A-1–4A-7.
3. Fraser-Smith AC. McGill PR. Bernardi A. Helliwell RA. Ladd ME. (1989) Global Measurements of Low-Frequency Radio Noise. URSI Symposium on Environmental and Space Electromagnetics. Tokyo. Japan. (This publication).
4. C.C.I.R. (1964) World distribution and characteristics of atmospheric radio noise. Report 322. International Radio Consultative Committee. International Telecommunication Union. Geneva.
5. C.C.I.R. (1988) Characteristics and applications of atmospheric radio noise data. Report 322–3. International Radio Consultative Committee. International Telecommunication Union. Geneva.
6. Maxwell EL. Stone DL. (1963) Natural noise fields from 1 cps to 100 kc. IEEE Trans. Antennas Prop., AP-11 : 339–343.
7. Maxwell EL. (1966) Atmospheric noise from 20 Hz to 30 kHz. In : Sub-Surface Communications, AGARD Conf. Proceedings No. 20 : 557–593.
8. Turman BN. (1977) Detection of lightning superbolts. J. Geophys. Res., 82 : 2566–2568.
9. Watt AD. (1967) VLF Radio Engineering. Pergamon. London : 509–513.

5.3 Characteristics of LF Whistler-Mode Waves Observed in Conjugate Area at Low Latitudes

Y. TANAKA and M. NISHINO

Solar-Terrestrial Environment Laboratory, Nagoya University, Aichi, 442 Japan

ABSTRACT

Based on measurements of whistler-mode waves of Decca navigation signals in the magnetic conjugate area of the transmitters around Birdsville(L=1.55),Australia, characteristics of LF whistler-mode waves are indicated. The shift of transmitter frequency and the signal intensity increase are discussed mainly in relation to geomag-netic activities. Storm-related energetic electrons and magnet-ospheric electric fields penetrating into the low-latitude magneto-sphere may generate the intensity increase and the frequency shift. Also, the ducted propagation of LF whistler-mode signals is clarified.

1. INTRODUCTION

Whistler-mode signals from NLK(18.6 kHz) transmitter, Seattle were measured during eight years in Wellington, New Zealand(L=2.3). The signal amplitude increased on 2-3 days after the magnetic storms. The frequency of the whistler-mode signal shifted from the transmitter frequency, and the Doppler shifts due to the change of the phase path of the signal significantly varied during the nighttime depending on the magnetic activity. The average feature indicated a dusk positive shift and a dawn negative shift during quiet times[18]. The Doppler shifts were caused primarily, at least during quiet times, by drift of the magnetic field-alighned duct guiding whistler energies rather than by changing electron density along the path[33]. Thomson[35] found westward electric fields of < 0.4 mVm^{-1} inside of the plasma-sphere from the measurments of phase shifts and group delay times of the whistler-mode signals during quiet times. Andrews et al.[2] presented equatorial radial plasma velocities of tens of meters per second(azimuthal electric fields of one or two tenth of mVm^{-1}),and protonospheric coupling fluxes of a few times 10^{12} elm^{-2} $s^{-1}$from the observation of four nights. Moreover, Andrews[3] has revealed that coupling fluxes of 1-2x$10^{12}elm^{-2}s^{-1}$ contribute less than 20% to the measured Doppler shifts, most of which are therefore produced by cross-L drifts, from the 22 nights data during the months of November to February. The measurements of Doppler shifts of the whistler-mode signals from NAA(17.8 kHz) Culter, Maine were also carried out at Siple station(L=4), Antarctica[1]. The Doppler shifts and group travel times of two whistler-mode signals from NAA(24 kHz) and NSS(21.4 kHz) transmitters have been also measured in the magnetic conjugate region, Faraday station(L=2.3) Antarctica[27].

The L value of Moshiri observatory, Japan is 1.57, being situated in the outer edge of the inner zone radiation belt where higher energetic particles of 100 keV or less are trapped. Wave-particle interactions there are not well understood. However, the long-term VLF/ELF measure-ments at the low-latitude three stations (Moshiri,L=1.57;Sakushima,1.27;Kagoshima,1.20) ,Japan have shown that there are various kind of emissions generated at different L shells, including the VLF/ELF emissions in the vicinity of the plasmapause[28,8,10] and the VLF emissions due to interactions with relativistic electrons in the inner zone radiation belt[24]. In order to clarify experimentally the propagation of LF whistler-mode waves in the low-latitude magnetosphere and to elucidate the interaction of the LF waves with energetic electrons, the conjugate measurements of whistler-mode Decca signals were carried out in 1984[14,22] and also in 1986[32]. Previous to these measurements, conjugate measurements at low latitudes had never been attempted.In this paper, characteristics of LF whistler-mode waves at low latitudes are reviewed, based on results from the observations in conjugate area around Birdsville, Australia in 1984 and 1986[22,32,33].

2. MEASUREMENTS OF LF WHISTLER-MODE WAVES

Measurements of the whistler-mode signals from Decca transmitter (85.725 kHz,cw,1.2 kW) at Biei(L=1.54)(geographic coordinates,142.45° E,43.60° N),Japan were carried out during July-September,1984 at Birdsville(139.33° E,25.83° S),Australia,almost the magnetic conjugate point of the transmitter. Details of the measurement system of the Decca whistler-mode signals were represented by Iwai et al.[14]. So, the outline of the system is described. Signals are received by the crossed loop antennas with the planes directed to the geomagnetic NS and EW directions("NS" and "EW" antennas). To reduce interferences from impulsive atmospherics and Decca signal from Dampier(85.635 kHz) Western Australia, a narrow band reception($\Delta f=\pm 10$ Hz) centred at the carrier frequency(85.725 kHz) by means of a crystal filter was employed. The NS and EW outputs from the preamplifiers installed just below the antennas are led to the crystal filters($\Delta f=\pm 10$ Hz). The NS and EW narrow-banded signals centred at 85.725 kHz(fc) are fed to the frequency converters. The frequency converted outputs(5.725 kHz±10 Hz) are introduced to a VTR recorder connecting to a PCM processor, in order to record the waveforms of the NS and EW signals during 1 minute every 10 minutes from 7 h to 23 h UT(UT = JST(135° EMT) - 9 h). The outputs(5.725 kHz ± 10 Hz) are also converted down to 715.625 ± 10 Hz. Then, the output from the frequency converter for each of the NS and EW components is led to the digital filter bank consisting of 9 channel digital filters($\Delta f=\pm 0.2$ Hz), through which the signal is divided into 9 frequency components, corresponding to antenna input signals at 85.725 kHz, ±0.5,±1.0,±1.5 and ±2.0 Hz. The divided outputs of the total 18 channels are recorded continuously(14,22).

Measurements of LF signals transmitted from three Decca stations (Biei,85.725 kHz; Akkeshi,114.300 kHz; Wakkanai,128.588 kHz),Japan were carried out during July-September, 1986 in magnetic conjugate area around Birdsville(L=1.55),Australia. Figure 1 illustrates the magnetic conjugate points and L values of the Decca stations(Biei(BE), Akkeshi(AK) and Wakkanai(WK)),and the observing points(BE, Birdsville (BV), Clifton Hill(CH) and Mungerannie(MG)). The observing and conjugate points are found to be located approximately along the magnetic meridian plane except for the conjugate point of Akkeshi. At the main observing point(BV), magnetic fields at 85.725 kHz were received by the crossed loop antennas. Also, vertical electric field at 85.725 kHz was measured by using a vertical electric antenna(10 m high) with a radial counterpoise(15 m long). The NS,EW and vertical narrow-banded signals(85.725kHz±10Hz) were frequency-converted, and the frequency-converted components(5.725 kHz ± 10Hz) were continuously recorded in a VTR recorder[14,22]. The similar measurement systems were used for receiving magnetic fields at 114.300 and 128.588 kHz at the other observing points :Clifton Hill, Mungerannie and Biei's conjugate point,where the narrow-banded NS and EW components of magnetic fields at 85.725 and 128.588 kHz were recorded for 1 minute every 10 minutes. Simultaneous measurements at the two observing points at maximum and also Birdsville were, appropriately, made.

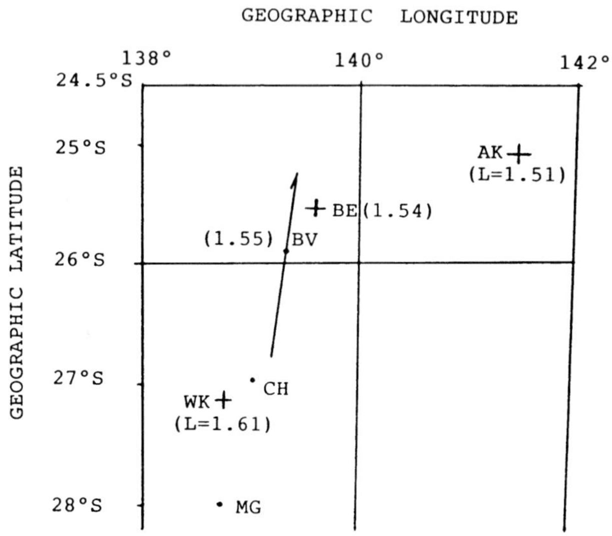

Fig. 1. Indicated are the magnetic conjugate points of Decca stations (Biei(BE),Akkeshi(AK) and Wakkanai(WK)),Japan, and the observing points(BE, Birdsville(BV), Chifton Hill(CH) and Mungerannie(MG)). The conjugate points and L values are calculated by means of the IGRF(1980) model. The arrow indicates the direction of the magnetic field(H component), and the declination is 7° in the area.

3. CHARACTERISTICS OF LF WHISTLER-MODE WAVES

Results from measurements of the Decca signals in the magnetic conjugate area in 1984 and 1986 are summarized as follows.
(1) The Decca signals were never observed during the daytime, being below the noise level. As the earth-ionosphere waveguide loss begins to decrease after sunset, the waveguide-mode signals usually appear and are observed during the nighttime. However, the EW components are considerably weak, due to the nearly linear wave polarization. Hence, the whistler-mode signals which revealed elliptical wave polarization could be explicitly identified on the EW channel.
(2) The whistler-mode signals were detected at 85.725 kHz at Birdsville, and they usually appeared at sunset and sunrise, and occasionally at night.
(3) They usually revealed the frequency shifts : a dusk positive shift(< 0.5Hz) and a dawn negative shift(> -0.5Hz), being independent of magnetic activity.
(4) Associated with magnetic storms,the EW intensity of the whistler-mode signal remarkably increases in the nighttime compared with the intensity at quiet times.

4. DOPPLER SHIFTS OF LF WHISTLER-MODE SIGNALS

Figure 2 shows an example of the Doppler shifts at moderately disturbed times(ΣKp=25- on August 25 and ΣKp=17 on August 26). The upper pannel indicates the temporal variation of the EW component at fc. The middle pannel indicates the Doppler shifts where strong whistler-mode signals with the positive shifts appeared during 9-10 hUT around sunset and during 15-17 h UT at the post midnight, and the negative shifts during 19-21 h UT around sunrise. Duration of each sub-event of

Doppler shifts is 2-3 hours. The leakage of the waveguide mode is seen at fc intermitently in the pre-midnight and the early morning. A significant change from the positive to the negative shifts is found around 19 h UT in the early morning. It is noted that the positive changes(∿ 0.6 Hz) of frequency shifts around the midnight (15 h UT) well correspond to the ones of the magnetic H and Z components.

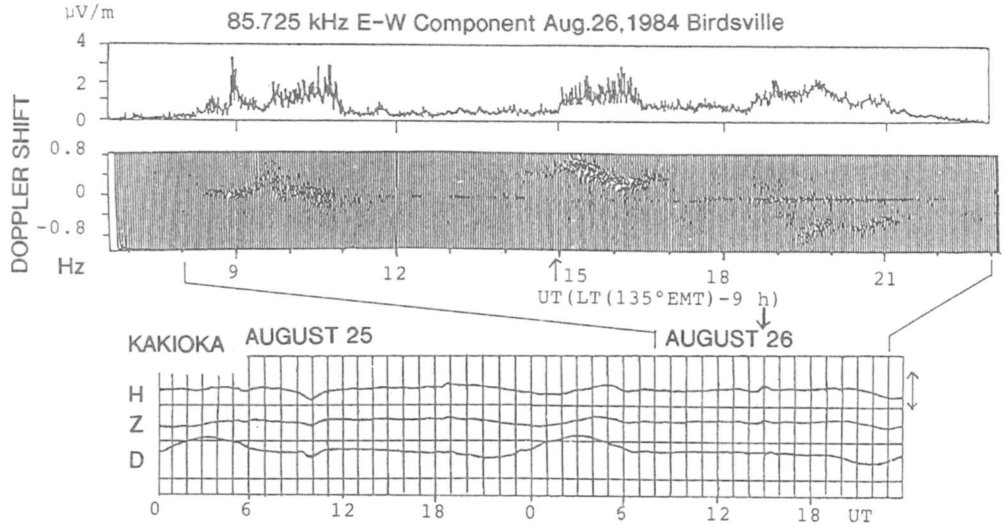

Fig. 2. An example of the Doppler shifts at mederately disturbed times. The upper panel indicates the temporal variation of the EW-component at 85.725 kHz. The middle panel indicates a computer-aided spectrogram scanned within ± 0.8Hz. Three component(H,Z,D) magnetograms at Kakioka (L=1.25) are shown in the lower panel. Upward and downward arrows indicate the positive change of the frequency and the corresponding magnetic variation, respectiely.(Reprinted with permission from [22]).

The phase path of a whistler-mode signal in wave lengths is given by

$$\phi = \frac{1}{\lambda_0}\int_{\text{path}} \mu\, ds = \frac{1}{\lambda_0}\int_{\text{path}} \frac{f_p}{\sqrt{f(f_H\cos\theta - f)}}\, ds$$
$$\approx 2\sqrt{f}\, D_0 \quad (f_H \gg f)$$

where ds=element of path, μ =phase refractive index, λ_0=free space wave length at frequency f, fp=plasma frequency, f_H=gyrofrequency, Θ =angle between wave normal and the magnetic field, and Do=dispersion of a whistler(sec $^{1/2}$). The Doppler shift Δf being the negative rate of change of the phase path is given by

$$\Delta f = -\dot{\phi} = -\frac{\partial\phi}{\partial B}\dot{B} - \frac{\partial\phi}{\partial N_T}\dot{N}_T - \frac{\partial\phi}{\partial R_e}\dot{R}_e$$

where B=magnetic flux density, N_T=flux tube total electron content between the northern and southern feet of the flux tube and Re=equatorial crossing distance of the field line[2]. Therefore, three possible causes of the frequency shift which could be responsible for the change in phase path are

(1) a change in the magnetic field strength along the path,
(2) a change in the electron density while the path remains fixed in

space,
(3) the duct drifting across a steady magnetic field due to the convection electric field.

The first possibility could be expressed by

$$\Delta f = -\frac{\partial \phi}{\partial B}\dot{B} \approx \frac{1}{2}\frac{\phi}{\langle B \rangle}\langle \dot{B} \rangle$$

where < > means a representative value. Impulsive-like magnetic changes observed in association with the Doppler shifts were less than about 50 nT. If appropriate values of B and $\dot{B}$ are taken as 10^4 nT at the apex of the field line and a maximum rate of the change as 50 nT/5 minutes, respectively, and the phase path is also estimated 2.93×10^4 for 85.725 kHz, referring to the average dispersion value of 50 $sec^{1/2}$ for whistlers at Moshiri(L=1.56) in July-August[9], the Δf is estimated as 0.24 Hz. So, comparatively abrupt frequency shifts with good correlation to impulse-like magnetic changes may be understandable. For gradual changes usually observed, the Doppler shifts due to the change in $\dot{B}$ may be negligibly small. Moreover, the change of 50 nT is about 0.5 % in the B, so that the magnetic field appears to remain essentially fixed in magnitude and position.

Thomson[34] indicated that the remaining possibilities (2) and (3) could be distinguished by considering the ratio of the change in phase travel time to the change in group travel time. However, the measurement of the Doppler shifts of the nearly continuous Decca transmitter signal cannot enable us to distinguish the causes (2) and (3). So, we estimate independently both the contributions of the causes (2) and (3) on the Doppler shift. The Doppler shifts due to the change in electron density along the fixed path may be given by

$$\Delta f = -\frac{\partial \phi}{\partial N_T}\dot{N}_T \approx -\frac{1}{2}\frac{\phi}{N_T}\dot{N}_T \qquad [2].$$

According to the above equation, the Doppler shifts can be estimated. From a quantitative estimation[22], it seems that the measured Doppler shifts around sunset and sunrise cannot be ,usually, satisfied merely by the upward and downward electron fluxes. Andrews[3] interpreted that the coupling fluxes contribute less than 20 % to the measured Doppler shifts, most of which is therefore produced by cross-L drifts at L=2.3.

The possibility (3) is given by

$$\Delta f = -\frac{\partial \phi}{\partial R_c}\dot{R}_c = -\frac{\partial \phi}{\partial R_c}V_c = -\frac{\partial \phi}{\partial R_c}\left(\frac{E}{B}\right),$$

$$\frac{\partial \phi}{\partial R_c} = \frac{\partial \phi}{\partial \theta_0}\times\frac{\partial \phi_0}{\partial R_c} = \frac{\phi}{S_M}\frac{\partial S_M}{\partial \theta_0}\frac{\partial \phi_0}{\partial R_c},$$

where Ve =duct drifting velocity across the steady magnetic field due to ExB drift, E=azimuthal electric field intensity, B=magnetic flux density at the equatorial plane, S_M=half length of the field line, and Θ_0=geomagnetic latitude on the ground. The approximate rèlation between Doppler shift and westward electric field at L=1.54 can be given by

$$f(Hz)=0.77\ E\ (mV\ m^{-1}).$$

The measured electric field producing the duct drift at L=2.3 at quiet times is, on the average, about 0.25 mVm^{-1} around sunset[3]. If that field is generated by a dynamo process in the ionosphere[17], and its amplitude is constant with latitude, the equatorial electric field would vary as $L^{-3/2}$[19]. The electric field would be 0.46 mVm^{-1} at

L=1.54, producing the frequency shift of about 0.35 Hz. Therefore the contribution of the electric field on the frequency shift around sunset seems to be rather larger than that of the electron flux at L=1.54, too. While, the comparatively sharp negative shifts around sunrise at L=2.3 is mainly due to the eastward electric field[3]. In additiol to the eastward electric field, a larger upward flux around sunrise as compared with a downward flux around sunset and in the nighttime[25] might be manifested on a trend of the frequency shift around sunrise, being a larger shift than those around sunset and in the nighttime.

Gonzalez et al.[7] observed large plasmaspheric electric field of 1.5-3 mVm^{-1} in the low-latitude plasmasphere(L=2) from the S3-3 satellite during strong geomagnetic activity(Kp>5), although the measured electric field was only radial component in the magnetic meridian. Those measurements imply the penetration of convective electric fields into the low-latitude plasmasphere at L=2. Andrews et al.[2] obtained the magnetospheric electric field from the VLF whistler-mode signals at L=2.3 on disturbed times(ΣKp=38) ; the dominant westward electric field of 0.2 mVm^{-1} in the pre-midnight, the fluctuating eastward electric field of about 0.2 mVm^{-1} in amplitude after the midnight(0-3h MLT) and the dominant westward electric field of 0.3 mVm^{-1} during 3-4h MLT. Carpenter et al.[4] observed the magnetospheric electric field at L=2.7 from the whistler observations in the recovery phase of the substorm; the dominant westward electric field of <0.3 mVm^{-1} in the post-midnight(1-3h M.L.T), the fluctuating eastward electric field of <0.4 mVm^{-1} in the early morning of 3h30-5h30m MLT. At moderately disturbed times, the frequency of the whistler-mode signals somewhat intensified during the nighttime at L=1.54 shifts to the positive, in general, and it varies with magnetic fluctuation. At severely disturbed times, the frequency of the whistler-mode signals deviates smoothly from a positive to the zero. These facts may suggest the penetration of westward convective electric fields during the nighttime into the low-latitude magnetosphere such as at L=1.54. The frequency shift during the nighttime on severely disturbed days appeared slightly smaller than that on moderately disturbed days, which may be due to the depression of flux tube total electron content in the low-latitude magnetosphere associated with magnetic disturbances[21].

5. PROPAGATION CHARACTERISTICS OF LF WHISTLER-MODE WAVES

Results from measurements of the Decca signals at different frequencies in conjugate area around Birdsville are summarized as follows.
(1) The whistler-mode signals were detected at 85.725 kHz at Birdsville, and they usually appeared at sunset and sunrise, and occasionally at night.
(2) The whistler-mode signals appeared at 85.725 kHz simultaneously at multiple observing points, but their intensities were different at different points except for the nearly same intensities at Birdsville and Biei's conjugate point apart ∿43 km northwards from Birdsville.
(3) The whistler-mode signals were not detected at 114.300 kHz at Birdsville.
(4) The whistler-mode signals were not explicitly identified at 128.588 kHz around the conjugate point of the transmitter.

Figure 3 shows an example of the whistler-mode signals observed at 85.725 kHz at Birdsville,Clifton Hill and Mungerannie. Figure 4 shows an example from measurements at three different frequencies at Birdsville, where the whistler-mode signals appear only at 85.725 kHz and they are intensified in association with magnetic disturbances.

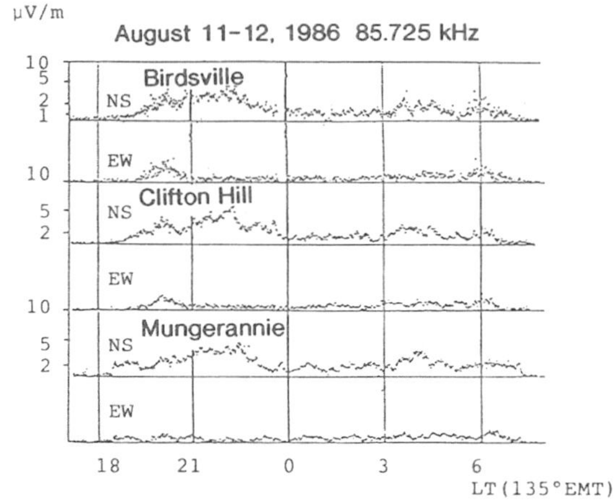

Fig. 3. An example of the whistler-mode Decca signals at 85.725 kHz observed at quiet times simultaneously at Birdsville, Clifton Hill and Mungerannie, where the signal intensities decrease slightly with increasing latitudes. (Reprinted with permission from [32]).

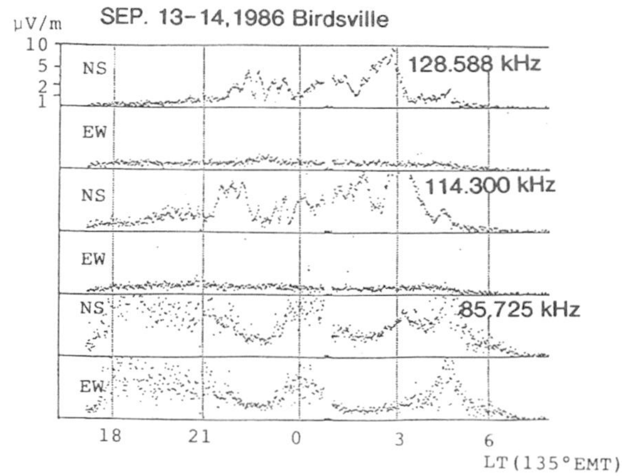

Fig. 4. An example of the whistler-mode Decca signals at the three frequencies observed at magnetically disturbed times at Birdsville, where the signals appeared only at 85.725 kHz almost continuously during the nighttime, and they were considerably intensified. (Reprined with permission from [32]).

5-1. DIRECTION FINDINGS

Waveforms of the NS,EW and vertical components(85.725kH ± 10Hz) were frequency-converted, and the frequency-converted components(5.725 kHz ± 10 Hz) were continuously recorded in a VTR. The recorded components are digitized and stored in a computer. Then, the amplitude ratios and phase differences among the three components are estimated, resulting in determining the arrival direction of the relevant signal[23]. As waveguide-mode components contaminate, in most cases, the waveforms of the whistler-mode components, direction findings(DFs) are available in limited cases when the relevant signals sufficiently dominate the waveguide-mode components and also the right-handed nearly circular polarization is confirmed. Obtained DF results indicate that the signals are propagated almost along the field lines down to Birdsville(32).

5-2. PENETRATION THROUGH THE IONOSPHERE

In order to discuss the occurrence of whistler-mode Decca signals, the wave attenuation in the lower ionosphere is firstly estimated by means of numerical calculations of the total transmission loss(reflection plus absorption) in the lower ionospheric models. The total transmission loss in the lower ionosphere below 120 km was calculated by means of a full wave computer programme[26,32]. The calculation result indicates that the loss is minimized for the exit angle parallel to the magnetic field and increases remarkably as the exit angle departs from the field direction. The rate of the loss increase is more enhanced with increasing frequencies. Estimation based on the calculation results with the assumption of ducted propagation can conclude that , for detecting the whistler-mode Decca signals at 85.725 kHz, the whistler ducts existing around the field line through the transmitter and/or at somewhat lower latitudes than the field line are essential, and also the exit angles almost parallel to the magnetic fields and/or rather inclined towards higher latitudes are needed. Therefore, the detectable region is localized, centred at the conjugate point of the transmitter, and such the locality increases with increasing frequencies. As is expected from such the estimation, the whistler-mode signals at 85.725 kHz appeared simultaneously at the multiple observing points, but their intensities were different at different points except for the nearly same intensities at BV and BE. The amplitudes of the whistler-mode signals simultaneously observed at the three observing points shown in Fig.3 can be reasonably explained in terms of the estimated transmission losses of signals penetrating the ionosphere down to the observing points with the exit angles deduced from DF results[32]. The whistler-mode signals at 128.588 kHz were measured at multiple points(BE,BV,CH & MG), but the signals were not explicitly identified even at CH near to the transmitter's conjugate point(WK), and also at MG at a somewhat higher latitude than WK. This may imply that non-identification is due to not the localized exit region from the ionosphere but the propagation mechanism of the relevant signals in the magnetosphere. As the conjugate point of the 114.300 kHz transmitter(AK) departs $\sim$ 230 km in the north-east direction from BV, the signals seem to be heavily attenuated below the detection level at Birdsville while penetrating the ionosphere with wave normals at large angles with respect to the magnetic field.

5-3. FREQUENCY DEPENDENCE OF WHISTLER-MODE PROPAGATION

Considering merely the transmission loss in the ionosphere, the whistler ducts existing around the field line intersecting the transmitter and/or at somewhat lower latitudes than the field line are essentially required for the reception of the relevant signals on the ground, as mentioned in the previous section . Hence, the conjugate propagation is rather reasonably assumed, and then the transmitter frequency(fc) is normalized by the electron cyclotron frequency at the magnetic equator(fHeq) by means of the IGRF(1980) model. The normalized frequencies(Λ=fc/fHeq) are 0.555, 0.404 and 0.317 for the three transmitter frequencies(128.588 kHz(WK), 114.300 kHz(AK) and 85.725 kHz(BE)), respectively.

On the other hand, for frequencies above fH/2 , a density trough is required for ducting[11]. However, for the trough ducting, the wave frequency must be above 0.5 MHz or higher in order to be ducted all the way down to the ionosphere. Thus, a crest irregularity and also a trough one can not trap the signals at 128.588 kHz all over the path. For frequencies below fH/2 , a crest irregularity is capable of ducting waves that have wave-normal angles less than a certain critical angle(Θ_0) given by $\cos(\Theta_0) = 2\Lambda$. For the signals at 114.300 kHz, the normalized frequency(Λ) at the magnetic equator is 0.404, and so only the signals with wave-normal angles less than 36.1° are trapped by an enhancement and propagated beyond the equator towards the southern hemisphere. If the duct base exists in the F-region of the ionosphere[30], the initial wave-normal angle at the duct entrance may be given by the angle between the vertical direction and the magnetic field,the value(34.5°) of which is less than the above-mentioned critical value(36.1°). Hence, a crest irregularity might be capable of ducting the signals all the way. For the signals at 85.725 kHz, the normalized frequency is 0.317 and the critical angle is 50.7° . And the initial wave-normal angle in the F-region is 34.0° ,so that a crest ducting is much feasible all the way.

Whistlers at low latitudes tend to be propagated to higher latitudes after exiting from ducts since only wave energies in an overlapping region of the narrow transmission cone around the vertical direction and the trapping cone around the magnetic field at a large angle from the vertical penetrate the ionosphere[9]. Referring to such the exiting of whistlers at low latitudes, the DF results indicating the exit angles almost parallel to the magnetic field may support the ducted propagation, too.

The ray path of the prolongitudinal mode is nearly symmetrical with respect to the magnetic equator, and the PL mode escapes from the lower ionosphere, due to negative latitudinal gradients of electron density. However, the PL mode never encounters a restriction of propagation dependent on wave frequency such as the ducted mode undergoes, and moreover the exit angles from the ionosphere are nearly uniformly distributed around the vertical direction[29].

6. WAVE-PARTICLE INTERACTIONS

6-1. STORM-ASSOCIATED INTENSITY INCREASE

At magnetically severely disturbed times, the whistler-mode Decca signals at 85.725 kHz appeared almost continuously during the nighttime and were considerably intensified, as shown in Fig. 4. The intensity increase was estimated with respect to the average threshold level for the detection of whistler-mode signals on the EW channel, identified at night at quiet times. It is found from the statistical investigation that the whistler-mode signals are intensified by more than 20 dB during the nighttime in association with severe magnetic disturbances.However,the occurrence of LF whistler-mode waves do not correlate significantly with current magnetic activity. A time-lag correlation between the whistler-mode occurrence(total duration per day) and the magnetic activity(ΣKp) up to 3 days prior to the whistler-mode occurrence for 57 data pair of the 1984 campaign indicates that the occurrence of whistler-mode signals has a maximum correlation with the activity at one-day lag. And the correlation is relatively increased(0.7 at maximum) with magnetic activity(22). While, the time-lag correlation for 1659 data pairs between the amplitude of the NLK(18.6 kHz) whistler-mode signal and the magnetic activity(Kp) at L=2.3 showed a gradual peak($\sim$0.2) by two days lag[18].

6-2. RESONANT PARTICLE ENERGIES

The increase of signal intensity must be caused by wave growth and/or improvement of propagation conditions, and the wave growth is firstly discussed. The wave growth mechanism is interpreted in terms of a model based on cyclotron resonance interaction, and the interaction region is located in a limited region centred at the magnetic equator, due to the less imhomogeneity of the medium[12]. So, energies of resonant electrons are estimated under the following conditions.
(1) Cyclotron resonance of whistler-mode waves with counter-streaming electrons at the equator.
(2) The cyclotron(gyro-) frequency model of electron density, such as $N(cm^{-3})$ = 10 fH(kHz) [11], where the gyrofrequency at the equator(fHeq) is given by $fHeq=856(1/L^3)(kHz)$.
Estimated are the energies of resonant electrons as functions of L value and the normalized frequency(fc/fHeq)[32]. For 85.725 kHz signals from Biei(L=1.54), the energies of resonant electrons are estimated 50 - 100 keV. On the other hand, measurements of VLF whistler-mode signals were carried out at the conjugate point(L=1.93) of the transmitter in Eastern USSR, in South Australia simultaneously during the LF observation in 1984 and also 1986 at Birdsville[31]. However, the intensity increase of the VLF whistler-mode signals was not identified. For the signals at 14.881 kHz, the energy of resonant electrons becomes ~ 600 keV. High energy particle measurements more than 0.19 MeV(electrons) and 0.58 MeV(protons) by EXOS-C satellite have demonstrated the enhanced particle precipitation less than 1 MeV centred near L=1.71, but no positive correlation is found with magnetic activity(Kp)[20]. These suggest that neither improvement of ducted propagation such as enhanced formation of ducts nor increase of high energy electrons of more than 500 keV occurred below L $\sim$ 2 in association with magnetic disturbances. Furthermore, no positive correlation of occurrence of natural whistlers observed at L=1.55 and 1.93 was identified with magnetic activity during the observation in 1984 and 1986, which also suggests no effect of magnetic disturbances on the ducted propagation. Hence, it is deduced that the intensity increase of the whistler-mode Decca signals at magnetically severely disturbed times is attributed to the wave growth caused by cyclotron resonance interactions of ring current electrons(50-100 keV) injected into the low-latitude magnetosphere below L $\sim$ 2.

6-3. STORM-RELATED ENERGETIC ELECTRONS

Magnetospheric substorm represents the intense, rapid conversion of stored magnetotail energy into hot plasma and energetic electrons(30-300 keV) in the outer magnetosphere[5]. Trapped energetic electrons (>68 keV) are abundant usually in the L shell region above 2[13]. At the severe magnetic storm, trapped low energy electrons(30-300 keV) invaded deep into the low-latitude magnetosphere such as L=1.3[16].

Figure 5 shows the time variations in Dst and the hourly value of foF2 normalized by its monthly median at Wakkanai(L=1.61),Japan for the largest magnetic storm during the LF campaign in 1984 and 1986, where thick and thin horizontal bars indicate the occurrence of whistler-mode signals with large and comparatively small intensities, respectively.

Investgated are temporal variations in electron count rates with energies of > 30, > 100, > 300 keV sampled by detectors at the sensor angles of 90° and 0° with respect to the local zenith at the satellite (polar orbiting satellite in nearly circular orbit at 815.5 km altitude) on orbits of 100°-180° in longitude centred at the meridian

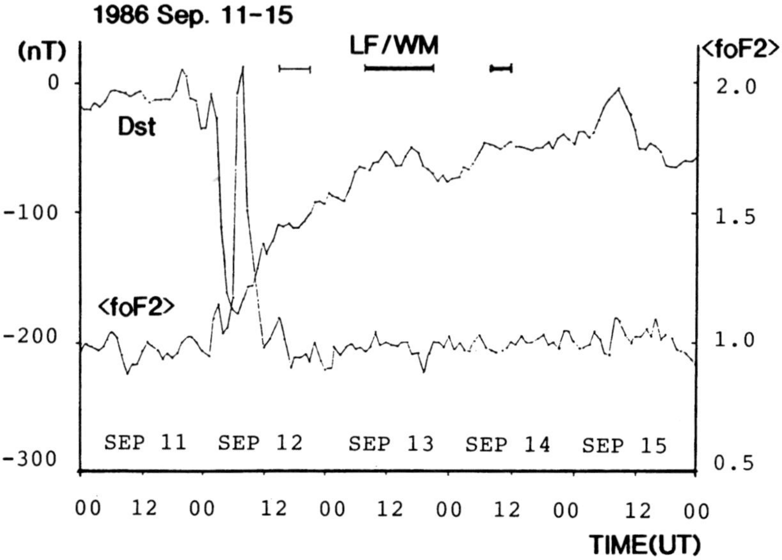

Fig. 5. Temporal variations in hourly values of Dst indices and foF2 normalized by monthly median at Wakkanai(L=1.61),Japan for the largest magnetic storm during the LF campaign. (Reprinted with permission from[33]).

plane including the observing point of LF signals. The following temporal variations between Dst and electron flux of > 30 keV detected by NOAA-6 in the low-latitude magnetosphere as far as L < 1.5 are identified for three magnetic storms with the maximum values of Dst indices exceeding 100 nT during the LF campaign. At magnetically quiet times, even electrons with energy of > 30 keV at the detection angle of 90° are much fewer in the low latitude below L- 2. The count rates in the maximum depression phase of the Dst indicate that electrons with energies more than 30, 100 and 300 keV remarkably increase all over the plasmasphere. However, the intensity of LF whistler-mode signals subsequently occurred at that night was rather small,as shown in Fig.5. The electron count rates on one day after the Dst maximum indicate that electron fluxes with energy of > 30 keV at the detection angle of 90° are abundant enough even in the low latitude below L 1.5. Whereas, electron fluxes with energy of > 30 keV at the detection angle of 0° are comparatively smaller. The enhanced LF whistler-mode signals were observed at night on that day at Birdsville. Even on the second day after the Dst maximum the enhanced LF whistler-mode signals were observed at night and the enhancement of electrons with energy of > 30 keV at the detection angle of 90° was also identified by the satellite as far as the low latitude below L $\sim$ 1.5. In another case when Dst maximum is 84 nT the similar trend is recognized.

Considering that Dst is a measure of energetic particle injection from the magnetotail at the time of magnetospheric substorm and is mainly attributable to the drift of quasi-trapped protons and electrons more than 100 keV, the above results necessarily require comparatively rapid (1 hour order or less) scattering of electrons drifting in the vicinity of the plasmapause, deep into the inner magnetosphere. Accodingly, energetic electron measurements by the satellite detected the enhancement of trapped electrons of > 30 keV in the low L shell region below 1.5 in association with medium world-wide magnetic storms, and simultaneously ground-based conjugate measurements identified the intensity increase in the LF whistler-mode signals as well as the frequency shift of the signals. The detection of abundant energetic electrons at the angle of 0° at the Dst maximum phase infers a rather isotropic distribution of pitch angles of energetic electrons. Electrons with small pitch angles less than the loss cone angles can precipitate down to the ionospheric le.el, resulting in enhanced ionization of the ionosphere. Comparatively small intensity increase of LF whistler-mode signals occurring at the Dst maximum phase and/or subsequently may be due to such a rather isotropic pitch angle distribution. Within one or two days after the Dst maximum, electron fluxes more than 30 keV at the detection angle of 90° are abundant enough except for the lowest L shell rewgion, whereas, the electron fluxes at the angle of 0° are much decreased, as compared with the fluxes at the Dst maximum phase. As a result, the LF whistler-mode signals can be enhanced, due to the wave growth caused by cyclotron resonant interactions with trapped energetic electrons.

6-4.WAVE-PARTICLE INTERACTIONS

According to Kennel and Petschek[15],the linear growth rate(γ) of whistler waves due to cyclotron resonance interaction is given by

$$\gamma = \pi\Omega_e\left(1-\frac{\omega}{\Omega_e}\right)^2\eta(V_R)\left\{A(V_R)-\frac{1}{\Omega_e/\omega-1}\right\},$$

where ω and Ω_e indicate the wave frequency and the equatorial electron cyclotron frequency. $\eta(V_R)$ means a quantity in proportion to the fractional concentration of the resonant electrons with respect to the total electrons. $A(V_R)$ is a measure of pitch angle anisotropy. For simplicity for accounting the anisotropy, a bi-Maxwellian distribution function with $T_\perp > T_\parallel$ for energetic electrons is chosen[6],

$$f(v) = \frac{n_1}{\pi^{3/2}U_\perp^2 U_\parallel}\exp\left(-\frac{v_\perp^2}{U_\perp^2}-\frac{v_\parallel^2}{U_\parallel^2}\right),$$

where n_1 is the hot electron density, $v_\parallel$ and $v_\perp$ the parallel and perpendicular components of the electron velocity with respect to the static magnetic field, $U_\parallel = (2\kappa T_\parallel/M)^{1/2}$, $U_\perp = (2\kappa T_\perp/M)^{1/2}$ the mean parallel and perpendicular velocities, and κ the Boltzmann constant. For the above case, A is independent of the resonant velocity(V_R) and is given by

$$A(V_R) = T_\perp/T_\parallel - 1.$$

Hence, the growth rate(γ) can be written as

$$\gamma = \sqrt{\pi}\frac{n_1}{n_0}\Omega_e(1-x)(A-(A+1)x)\sqrt{y}\exp(-y),$$

where n_0 is the total electron density, $x = \omega/\Omega_e$, and y the ratio of the resonant energy to the mean parallel energy of electrons. For the conjugate propagation of LF transmitter signals at 85.725 kHz, the wave frequency(ω) is normalized by the equatorial electron cyclotron frequency(Ω_e) by means of the IGRF(1980) model, and the normalized value(x) is 0.32. Assumming that A=2.0 and y=1, the growth rate is roughly estimated as

$$\gamma = 0.46\times10^{-3}\Omega_e.$$

The phase path of a whistler-mode signal in wave lengths is 2.93×10^4 for 85.725 kHz with the average dispersion value 50 sec for whistlers at Moshiri(L=1.56), as is estimated in section 3 [9]. Also, assumming the interaction on a small portion (1%) of the phase path around the equator, the interaction time(Δt) is

$$\Delta t = 10^{-2}\times\phi\times\frac{1}{f} = 2.93\times10^2\times\frac{1}{f}.$$

Correspondingly,the intensity enhancement of the signals may amount to

$$20\log_{10}\exp(\gamma\Delta t) = 23.0\,(\text{dB}).$$

In the same conditions except for a smaller anisotropy(A=1.0), the enhancement decreases to be 8 dB.

References

1. Andrews MK. Thomson NR. (1977) Doppler shifts on whistler mode signals received at Siple station,Antarctica. Geophys. Res.Lett. 10: 399-401
2. Andrews MK. Knox FB. Thomson NR. (1978) Magnetospheric electric fields and protonospheric coupling fluxes inferred from simultaneous phase and group path merasurements on whistler-mode signals. Planet. Space Sci. 26: 171-183
3. Andrews MK. (1980) Night-time radial plasma drifts and coupling fluxes at L=2.3 from whistler-mode measurements. Planet.Space Sci. 28: 407-417
4. Carpenter DL. Stone K. Siren JC. Crystal TL. (1972) Magnetospheric electric fields deduced from drifting whistler path. J. Geophys. Res. 77: 2819-2834
5. Cayton TT. Belian RD. Gary SP. Fritz TA. Baker DN.(1989) Energetic electron components at geosynchronous orbit. Geophys.Res.Lett. 16: 147-150
6. Gendrin R. Lacourly S. Roux A. Solomon J. Feigin FZ. Gokhberg MV. Troitskaya VA. Yakimenko VL. (1971) Wave packet propagation in an amplifying medium and its application to the dispersion characteristics and to the generation mechanisms of Pc 1 events. Planet.Space Sci. 19: 165-194
7. Gonzalez WD. Pinto Jr O. Mendes Jr O. Mozer FS .(1986) Large plasmaspheric electric fields at L=2 measured by the S3-3 satellite during strong geomagnetic activity. Geophys. Res.Lett. 13: 363-365
8. Hayakawa M. Tanaka Y. Ohtsu J. (1975) Satellite and ground observations of magnetospheric VLF hiss associated with the severe magnetic storm on May 25-27,1967. J.Geophys.Res. 80: 86-92
9. Hayakawa M. Tanaka Y. (1978) On the propagation of low-latitude whistlers. Rev. Geophys. Space Phys. 16: 111-123
10. Hayakawa M. Tanaka Y. Shimakura S. Iizuka A. (1986) Statistical characteristics of medium latitude VLF emissions (unstructured and structured):local time dependence and the association with geomagnetic disturbances. Planet. Space Sci. 34: 1361-1372
11. Helliwell RA. (1965) Whistlers and Related Ionospheric Phenomena. Stanford Univ.Press, Stanford Calif.
12. Helliwell RA.(1979) Siple station experiments on wave-particle interactions in the magnetosphere, Wave instabilities in space plasmas. Reidel Pub. Company: 191-203
13. Imhof WL. Reagan JB. Gaines EE. Datlowe DW. (1984) The L shell region of importances for waves emitted at ground level as a loss mechanism for trapped electrons > 68 keV. J.Geophys.Res.89: 10,827-10,835
14. Iwai A. Katoh Y. Nishino M. Okada T. Hayakawa M. Tanaka Y. (1985) Ground-based reception of the whistler-mode Decca signals. Proc.Res.Inst.Atmos.,Nagoya Univ. 32: 29-44
15. Kennel CF. Petschek HE. (1966) Limit of stably trapped particle fluxes J.Geophys.Res. 71: 1-28

16. Kikuchi T. Evans DS (1989) Energetic electrons observed by NOAA-6 over Japan(L=1.3) at the time of geomagnetic storm on February 8-9,1986. Proc.Res.Inst.Atmos.,Nagoya Univ. 36,No.2: 137-150
17. Matsushita S.(1971) Interactions between the ionosphere and magnetosphere for Sq and L variations. Radio Sci. 6: 279-294
18. McNeil FA. Andrews NK. (1975) Quiet-time characteristics of middle latitude whistler-mode signals during an 8-yr period. J.Atmos.Terr.Phys. 37: 531-543
19. Mozer FS. (1970) Electric field mapping in the ionosphere at the equatorial planes. Planet. Space Sci. 18: 259-263
20. Nagata K. Kohno T. Murakami H. Nakamoto A. Hasebe N. Kikuchi J. Doke T. (1988) Electron(0.19-3.2 MeV) and proton(0.58-35 MeV) precipitation observed by OHZORA satellite at low latitude zones L=1.6-1.8. Planet.Space Sci. 36: 591-606
21. Nakata Y. (1968) Variations of total electron content in the ionospheric F region(in Japanese). Rev. Radio Res.Lab. 14, 70: 169-174
22. Nishino M. Tanaka Y. Lynn KJW. (1989) Doppler shifts of LF whistler-mode signal observed at a low latitude(L=1.54). Planet.Space Sci. 37: 825-836
23. Okada T. Iwai A. Hayakawa M. (1981) A new whistler direction finder. J.Atmos.Terr.Phys. 43: 679-691
24. Ohtsu J. Kashiwagi M. (1979) An unusual VLF noise event observed at Sakushima on September 29,1978. Proc.Res.Inst. Atmos.,Nagoya Univ. 26: 133-143
25. Park CG. (1970) Whistler observations of the interchange of ionization between the ionosphere and the protonosphere. J.Geophys. Res. 75: 4249-4260
26. Pitteway MLV. Jespersen JL. (1966) A numerical study of the excitation,internal reflection and limiting polarization of whistler waves in the low ionosphere. J.Atmos.Terr.Phys. 28: 17-43
27. Smith AJ. Yearby KH. Bullough K. Saxton JM. Strangeways HJ. Thomson NR. (1987) Whistler mode signals from VLF transmitters observed at Faraday,Antarctica. Memo. Natl Inst. Polar Res., Spec. Issue, 48: 183-195
28. Tanaka Y. Hayakawa M. Ohtsu J. (1974) VLF hiss observed at a low-latitude ground station and its relation to drifting ring current electrons. Rep.Iono.Space Res.Japan 28: 168-172
29. Tanaka Y. Cairo L. (1980) Propagation of VLF waves through the equatorial anomaly. Ann.Geophys. 36: 555-575
30. Tanaka Y. Hayakawa M. (1985) On the propagation of daytime whistlers at low latitudes. J.Geophys.Res. 90: 3457-3464
31. Tanaka Y. Nishino M. Hayakawa M. (1987) Conjugate measurements of VLF transmitter signals at middle latitudes (L=1.93). Planet.Space Sci. 35: 1053-1059
32. Tanaka Y. Nishino M. Lynn KJW. (1989) On the propagation of LF whistler-mode waves deduced from conjugate measurements at low latitudes. Planet.Space Sci. 37: 1215-1226
33. Tanaka Y. Nishino M. Lynn KJW. (1990) Magnetic storm-related energetic electrons and magnetospheric electric fields penetrating into the low-latitude magnetosphere (L-1.5). Planet.Space Sci. 38: 1051-1059
34. Thomson NR. (1976) Causes of the frequency shift of whistler-mode signals. Planet.Space Sci. 24: 447-458
35. Thomson NR. (1976) Electric fields from whistler-mode Doppler shifts. Planet. Space Sci. 24: 455-458

5.4 Ray Focussing of Whistler-Mode Waves in a Magnetoplasma

K. Ishikawa, K. Hattori, and M. Hayakawa

Research Institute of Atmospherics, Nagoya University, Aichi, 442 Japan

ABSTRACT The purpose of this paper is to investigate the ray focussing of whistler-mode waves which results in an enhanced wave-particle interaction. The critical frequency in a slightly inhomogeneous magnetoplasma is studied, at which the refractive index surface of whistler-mode waves indicates a zero curvature at a longitudinal wave normal angle. This critical frequency is also found to be consistent with the zero diffraction coefficient in the full-wave theory. It is further suggested that the critical frequency is also important in the wave-particle interaction for the inhomogeneous outer magnetosphere outside the plasmapause and in the vicinity of the plasmapause.

1. INTRODUCTION

Ray focussing of whistler-mode waves along the magnetic field line is very important in the effective phase bunching of gyroresonant electrons by the counterpropagating whistler-mode waves which leads to the generation of VLF/ELF emissions in the magnetosphere. The critical frequency in this ray focussing is derived based on the consideration of the topological change in whistler-mode refractive index surface. Then, we discuss this critical frequency by using the wave equation for a slightly inhomogeneous magnetoplasma, and compare this result with the previous ray-theory result. Finally,the importance of this critical frequency is discussed even for the inhomogeneous realistic outer magnetosphere outside the plasmapause and also for the region in the vicinity of the plasmapause.

2. CRITICAL FREQUENCY IN WHISTLER-MODE RAY FOCUSSING BASED ON RAY THEORY [1,2] AND FULL-WAVE THEORY [2,3]

The propagation of whistler-mode waves in a homogeneous magnetoplasma can be studied with the aid of refractive index surfaces. For dense plasmas such that the electron plasma frequency f_p is much larger than the electron gyrofrequency f_H ($f_p \gg f_H$),we have plotted the change in morphology of the whistler-mode refractive index surface above and below the critical frequency of $f_H/2$. Figure 1 illustrates this situation, and Fig.1(a) refers to the frequency below the critical frequency and Fig.1 (b) above the critical frequency. As seen from the figure, the surface below the critical frequency is convex at small wave normal angles, while above the critical frequency we find the surface simply concave. Hence,we can anticipate that the curvature of the surface vanishes at zero wave normal angle just at this critical frequency. Because the ray direction is perpendicular to the refractive index surface, this suggests a possibility of enhanced phase-bunching of incoming electrons caused by focussing of the radiation in the direction of the Earth's magnetic field $\mathbf{B}_0$.

In the outer magnetosphere the f_p becomes, on some occasions, smaller than the f_H, and so we will derive the general equation for this critical frequency for different plasma conditions. By using the full, but collisionless, Appleton-Hartree's equation for the whistler-mode refractive index and by putting $d^2(n \cos\theta)/ d\theta^2|_{\theta=0} =0$ (n:refractive index), the critical frequency f_c is found to satify the following equation,

$$2 \Lambda_c^2 (1 - \Lambda_c)/ (1 - 2 \Lambda_c) = (f_p/f_H)^2 \tag{1}$$

where Λ_c is the critical frequency (f_c) normalized by f_H ($\Lambda_c=f_c/f_H$). The relationship of Λ_c with the plasma parameter, f_p/f_H is presented in Fig.2. This figure indicates that the critical frequency Λ_c decreases considerably below 0.5 for the teneous plasma ($f_p < f_H$). The variation of the curvature of the refractive index surface at $\theta=0°$ is plotted in Fig.3 for varying the normalized frequency $\Lambda(=f/f_H)$ along the broken line in Fig.2 (i.e. $f_p/f_H=1.0$). When the frequency Λ is increased, the curvature is negative (i.e. concave), then crosses zero at Λ_c and we have a positive curvature (convex surface).

The importance of the above-obtained critical frequency derived from the concept of refractive index surfaces (in other words, the ray theory) is investigated by means of a full-wave concept. We consider that the wave propagation is nearly along the Earth's magnetic field Bo (z direction) and that the plasma is weakly inhomogeneous across the magnetic field line (or in the x direction). The reductive perturbation method is applied to solve the Maxwell's equations, in which the time (t) and the space (x,z) are stretched in multi-time and -space, respectively, and we finally obtain a general wave equation [3],

$$-\frac{1}{2}\frac{\partial^2 k}{\partial \omega^2}\frac{\partial^2 \phi}{\partial t^2} + i\left(\frac{1}{v_g}\frac{\partial \phi}{\partial t} + \frac{\partial \phi}{\partial z}\right) + p\frac{\partial^2 \phi}{\partial x^2} - U\phi - i\frac{4}{\omega}\frac{\partial U}{\partial t}\phi = 0 \tag{2}$$

where ϕ is the wave electric field, k, the wave number, ω, the angular wave frequency and vg, the group velocity. The potential U(x) is related with the plasma inhomogeneity and expressed by,

$$U(x) = -(k/2)(\Delta \varepsilon(x)/ \varepsilon_0) \tag{3}$$

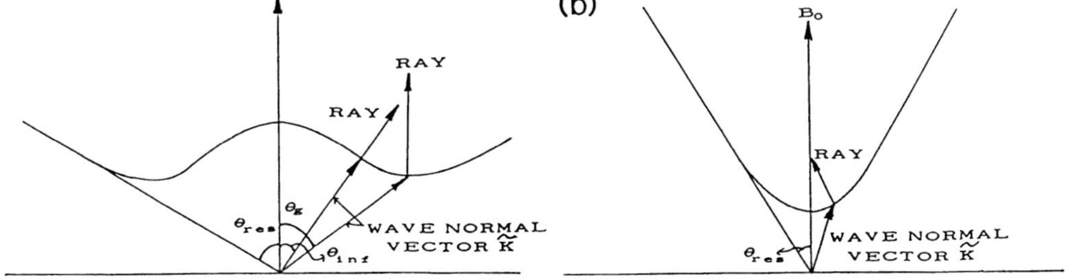

Fig.1 Change in morphology of whistler-mode refractive index surface (a) below and (b) above the critical frequency.

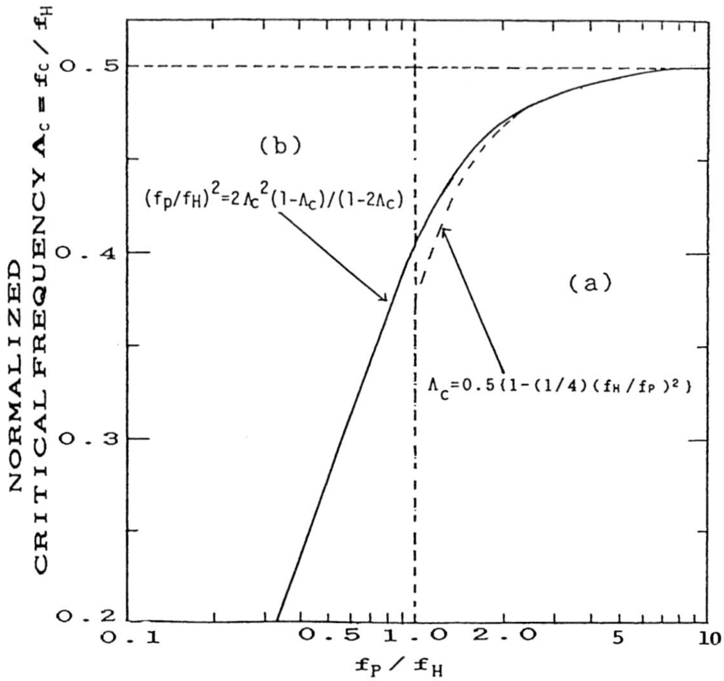

Fig.2 Critical normalized frequency (in a full line) as a function of of the plasma parameter, f_P/f_H. An approximate curve for the critical frequency on the condition of $f_P/f_H > 1$ is illustrated in a broken line.

where ε_0 and $\Delta\varepsilon$ are the homogeneous and inhomogeneous parts of the dielectric constant ε, respectively ($\varepsilon = \varepsilon_0 + \Delta\varepsilon$). The coefficient of the diffraction term, p is given by

$$p = \nu / 2k \tag{4}$$

and

$$\nu = 1 + \frac{1}{2}\left(- \frac{\omega_P^2}{\omega^2} \frac{\omega_H}{\omega - \omega_H} \right) / \left(1 - \frac{\omega_P^2}{\omega^2} \right) \tag{5}$$

where $\omega_P = 2\pi f_P$ and $\omega_H = 2\pi f_H$.

If a time-stationary state is assumed, i.e. $\partial/\partial t = 0$, the wave equation (2) reduces to the following two-dimensional Schrodinger equation.

$$i \frac{\partial \phi}{\partial z} + p \frac{\partial^2 \phi}{\partial x^2} - U(x,z)\phi = 0 \tag{6}$$

The important coefficient p represents the degree of wave focussing and defocussing. Figure 4 illustrates the diffraction p value as a function of the normalized frequency Λ for the same fixed plasma parameter, $f_P/f_H = 1$ as in Fig.3. The figure indicates that the p value decreases with

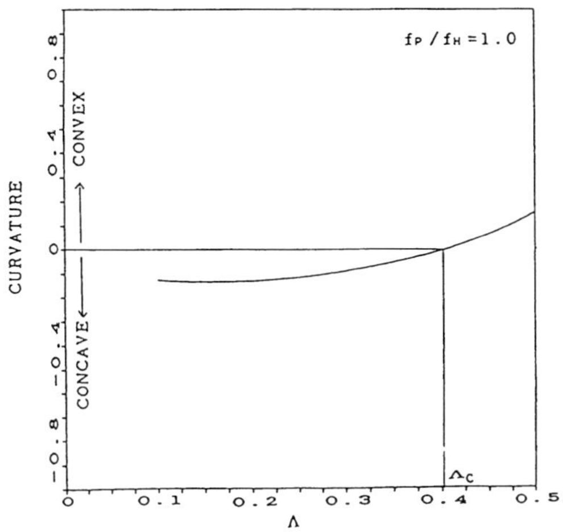

Fig.3 Variation with the normalized frequency of the curvature of refractive index surface for a longitudinal wave normal direction for the case of $f_p/f_H=1.0$.

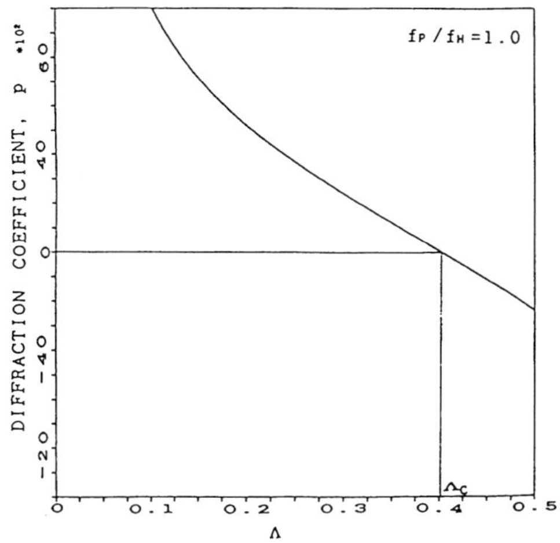

Fig.4 Variation with the normalized frequency of the diffraction p value for the same plasma condition of $f_p/f_H=1.0$.

increasing wave frequency and changes its sign from positive to negative at a specific frequency. This specific frequency can be obtained by putting p (in Eq.(4)) =0, which is found to yield exactly the same relationship with Eq.(1). In other words, the condition of no diffraction (p=0 in Eq.(4)) indicates no ray defocussing and correspondingly the strongest ray focussing. And, this critical frequency Λ_c is well approximated for higher fp/fH by the following equation.

$$\Lambda_c \simeq 0.5\ (\ 1 - (1/4)(f_H/f_p)^2)\quad (f_p > f_H) \qquad (7)$$

This approximation is also plotted in Fig.2 in a broken line, which indicates the validity of Eq.(7) for $f_p/f_H > 2.0$.

3. CONCLUDING REMARKS

Both from the refractive index surface study and full-wave concept, the critical frequency is found to be important for having a strong ray focussing in a slightly inhomogeneous magnetoplasma and correspondingly we can anticipate a strong phase-bunching of incoming electrons, leading to the enhancement of wave-particle interactions.

Although not shown in this report, we have performed the two-dimensional ray tracing computations for varying the wave frequency and initial wave normal direction in an inhomogeneous realistic model of the outer magnetosphere outside the plasmapause, which have yielded that although the critical frequency for the homogeneous case has its importance even in the inhomogeneous plasma,the strongest ray focussing seems to occur at a frequency slightly below the critical frequency. See the details in Ishikawa et al.[2]. Further, the ray focussing of whistler-mode waves in the vicinity of the plasmapause is also discussed and will be published elsewhere [4].

REFERENCES

1. Burtis WJ. Helliwell RA (1976) Magnetospheric chorus: Occurrence frequency and normalized frequency. Planet. Space Sci. 24: 1007-1024
2. Ishikawa K. Hattori K. Hayakawa M (1990) A study of ray focussing of whistler-mode waves in the outer magnetosphere. Trans. Inst. Electr. Inform. Comm. Engrs. Japan E73: 149-154
3. Washimi H (1988) Theory of wave trapping in space plasmas. Proc. Res. Inst. Atmosph. Nagoya Univ. 35: 39-57
4. Hattori K. Ishikawa K. Hayakawa M (1990) The effect of the plasmapause on the whistler-mode wave propagation, To be submitted.

5.5 Dispersion Characteristics and Waveform Analysis of Tweek Atmospherics

S. YANO[1], T. OGAWA[2], and H. HAGINO[3]

Department of Electrical Engineering, Kochi National College of Technology, Kochi, 783 Japan[1], Department of Physics, Kochi University, Kochi, 780 Japan (Now at Science Laboratory International, Kochi, 780 Japan)[2], and Kagawa Technical College, Kagawa, 763 Japan[3]

ABSTRACT

The frequency characteristics and the waveforms of tweeks are analyzed theoretically. In the waveform analysis the mode current method is used: a method based upon the source current expansion and the multi-reflection. In the theory the source current is Fourier-expanded by using the electric images. The calculated frequency dispersion and the first mode waveform are compared with the observed ones and discussions follow.

1. INTRODUCTION

Lightning produces impulse noises which propagate under multiple reflections between the ground and the ionosphere. The frequencies are dispersed along the pass and the higher frequency components arrive at the observation point earlier than the lower frequency components. This phenomenon is called "tweek" after the "tweet" of a bird [1] and has been reported by a number of authors [2][3][4][5][12]. The tweek atmospherics of the duration of several tens ms to 160 ms are observed. In the previous paper [6], we noticed the waveform characteristics and analyzed the propagation mechanism of this phenomenon. In this paper we discuss the dispersion characteristics of the tweeks in detail and try to explain an amplitude modulation mechanism of the first mode waveform.

2. MEASURED TWEEK WAVES

Figure 1 shows an example of sound spectrogram of a multi-mode tweek wave. Waves up to the 7th mode are received. As the mode number gets higher, the duration of the wave gets shorter. The duration of the nth mode is about one nth of the duration of the first mode. Figures 2(a),(b) show the waveforms of the first mode filtered out from the observed two tweek waves, respectively. When only the first-mode waveform is extracted, the amplitude of the waveform is greatly modulated. The modulated amplitude shows a guitar-shaped outline with a neck near 10 ms. Figure 3 shows an example of measured frequency to time curve in log-log scales. There is a rise and fall of the frequency at the beginning of a tweek. A peak of the curve is situated near 0.5 ms.

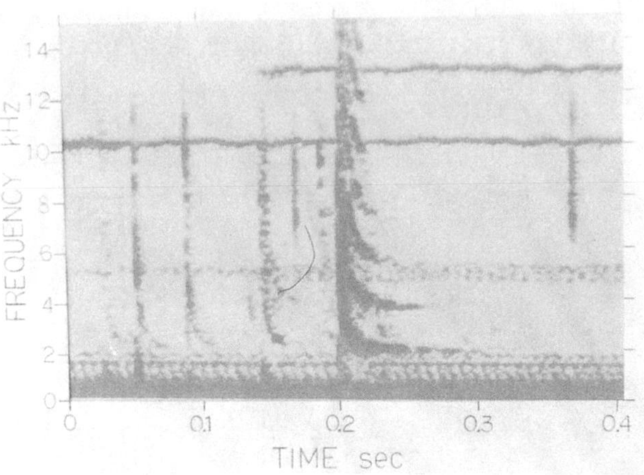

Fig.1. An example of multi-mode tweek wave upto the 7th modes displayed in sonagram. It was measured on Jan. 6, 1986.

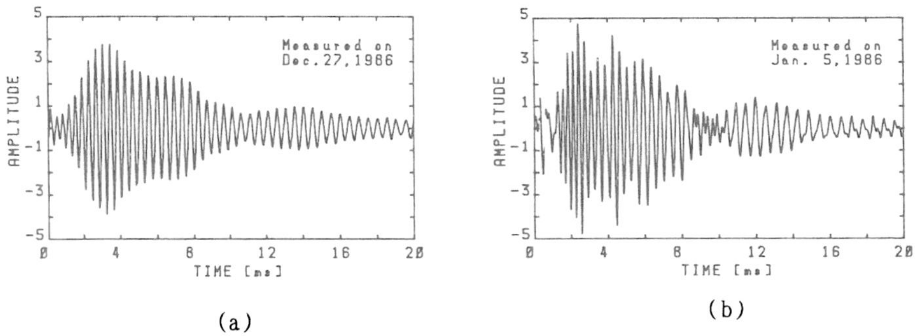

Fig.2. The first mode waveform filtered out with a pass-band from 1 kHz to 3 kHz from an observed tweek. (a) An example measured on Dec. 27, 1986. (b) An example measured on Jan. 5, 1986.

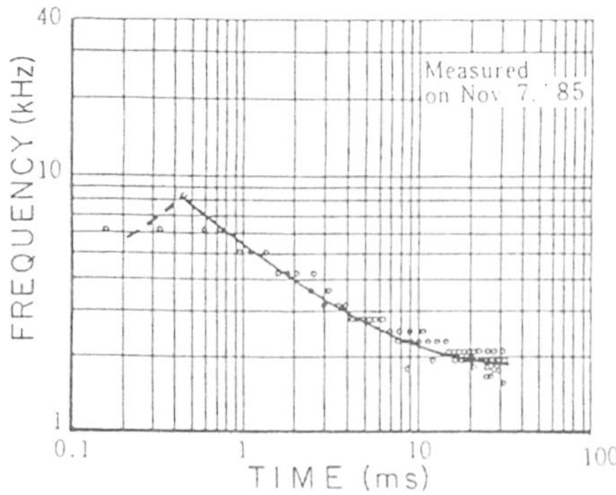

Fig.3. An example of measured frequency-time curve of the first mode of a tweek. A rise and fall of the frequency-time curve can be seen at the beginning.

It is a purpose of this paper to explain under what mechanism these characteristics result in.

3. DERIVATION OF THEORETICAL EXPRESSION

1) Frequency Dispersion

In the flat Earth approximation, the frequency dispersion is derived from the Inverse Laplace operation of the Hankel function [6][7]. It may also be possible to use the same method in the spherical Earth approximation. But the frequency may tend to infinity at the beginning of the occurrence of a tweek. Besides it is somewhat difficult to find a physical reason of the frequency dispersion. Alternatively the following method is recommended for understanding the dispersion intuitively.

The path element l_m in Fig.4 is given by the following equations according to the second cosine formula:

$$l_m = \{a^2+(a+h)^2 - 2a(a+h)\cos\theta_m\}^{1/2} \tag{1}$$

where a is the Earth's radius, h is the ionospheric height and θ_m is given by

$$\theta_m = \frac{d}{2ma} \tag{2}$$

where d is the propagation distance. The total path length is 2m times the path element. The propagation time t is given by this total path length divided by the speed of light c:

$$t = \frac{2ml_m}{c}, \qquad m = 1, 2, 3, \ldots \tag{3}$$

Differentiate t by m and obtain the reciprocal. Then set dm as unity. Thus the frequency of the tweek wave is obtained. We can derive the following expressions:

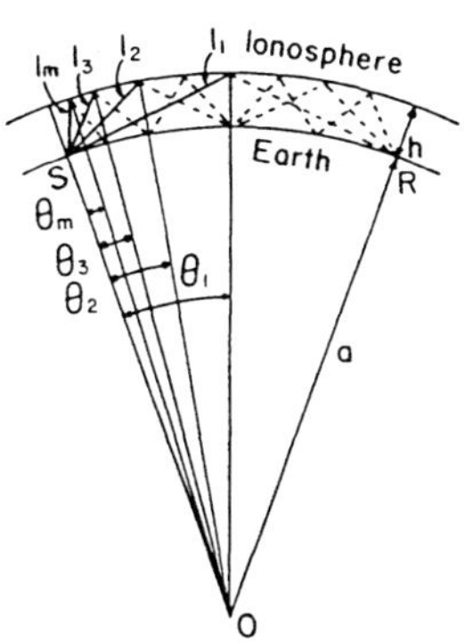

Fig.4. Definition of path element l_m in the spherical waveguide.

$$\frac{dt}{dm} = \frac{1}{cl_m}\{2l_m^2 - \frac{(a+h)d}{m}\sin\theta_m\} \tag{4}$$

$$f = \frac{cl_m}{2l_m^2 - \frac{(a+h)d}{m}\sin\theta_m} \tag{5}$$

For the higher mode of number n, h/n can be substituted for the height of the ionosphere h in Eqs.(1) and (5) due to the next character of the electric field: The vertical z distribution of the electric field of the nth mode matches the distribution provided when conductor plates are placed where the height h can be divided into n equal parts.

2) Image Source Array

In Fig.5, the elevation angle α is derived from the second cosine formula as follows.

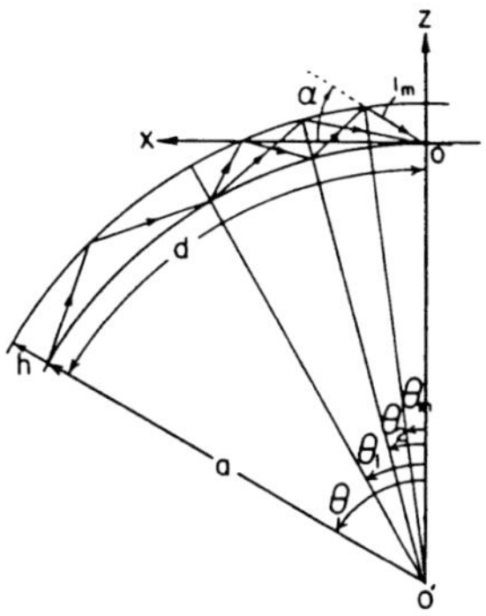

Fig.5. Definition of elevation angle α.

$$\cos(\alpha + \frac{\pi}{2}) = \frac{a^2 + l_m^2 - (a+h)^2}{2al_m} \tag{6}$$

$$\sin\alpha = \frac{(1+h/a)^2 - 1 - (l_m/a)^2}{2l_m/a} \tag{7}$$

y and z coordinates are given by this angle α:

$$y = 2ml_m(1-\sin^2\alpha)^{1/2}, \qquad z = 2ml_m\sin\alpha \tag{8}$$

3) Derivation of a Field Expression by the Mode Current Method

Since lightning current has a dominant vertical component and its channel length is comparatively shorter than the wavelength of the VLF band, the lightning source can be assumed to be a short dipole antenna [8]. If both the ground and the ionosphere have sharp boundaries, each reflected wave would be replaced by a radiation field from the electric images as shown in Fig.6. Namely, if the charges $\pm Q$ indicated in Fig.6 are neutralized by discharge, the current is typically shown by a series of uniformly distributed currents labeled as 'Source Current' in Fig.7.

The Fourier expansion of this current with respect to z produces each mode current shown in Fig.7. Numbers assigned to each current, n=0,1,2,... correspond to the mode numbers of the propagating electromagnetic field. The reason is as follows. Even if the boundaries were removed after introducing the electric images, the radiation field from each current satisfies the boundary condition at the both planes z=0 and z=h. That is to say the electromagnetic field is distributed as the current is distributed in the waveguide. We will call this kind of current by the term 'mode current'.

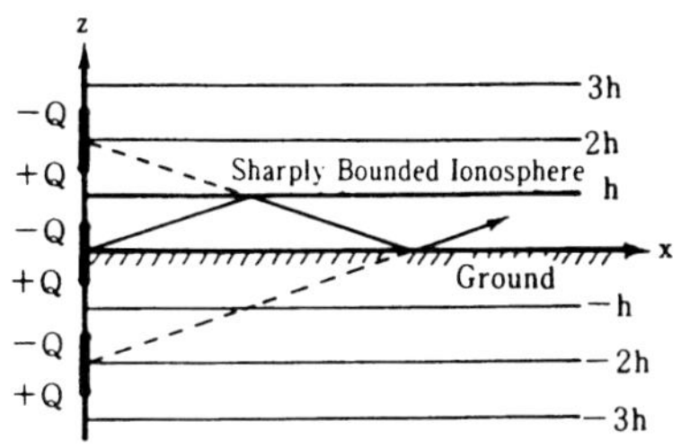

Fig.6. Array antenna approximation and Earth-ionosphere waveguide.

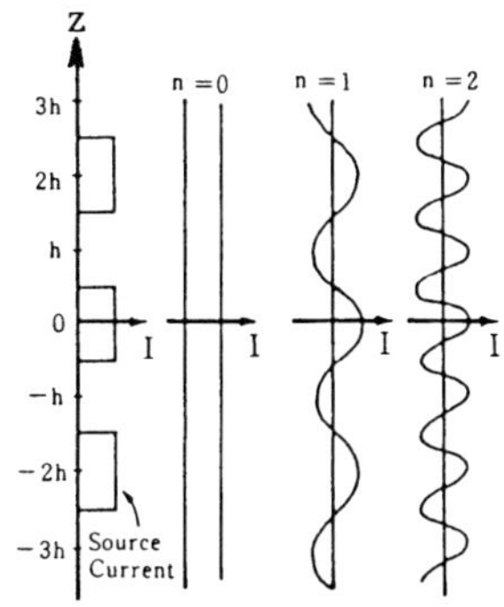

Fig.7. Fourier expansion of the source current and mode numbers.

We integrate the mode currents to get the tweek waveforms. As shown in Fig.8, z component of the electric field dE_z radiated from a segment dz of the mode current is given by Eq.(9)[9].

$$e(t)=\frac{\eta}{2\pi c}\int_0^S \left\{\frac{d}{dt}\, i\left(t-\frac{r}{c}\right)\right\}\{R(z)\}^m \frac{d^2}{r^3}\cos\left(\frac{n\pi}{h}z\right)dz \qquad (9)$$

where

$$i(t)=\frac{2I_0}{1+ue^{-2\gamma t}}\left(e^{-\alpha t}-e^{-\beta t}\right)\cos\left(\frac{n\pi}{h}z\right) \qquad (10)$$

$$\alpha = 2\times 10^4\ [1/s],\quad \beta = 2\times 10^5\ [1/s],\quad \gamma = 5\times 10^4\ [1/s] \qquad (11)$$

$$u = e^{-2\gamma t_0},\quad t_0 = 1\times 10^{-4}\ [s],\quad I_0 = 3\times 10^4\ [A] \qquad (12)$$

and

$$R(z)=\frac{1+\left(1-\sqrt{2}\,N\,\frac{d}{r}\right)^2}{1+\left(1+\sqrt{2}\,N\,\frac{d}{r}\right)^2},\quad r^2 = z^2+d^2 \qquad (13)$$

$$N=\frac{\omega_p}{(\nu\,\omega)^{1/2}},\quad \omega=\frac{2\pi c}{\frac{2hr}{z}-\frac{2(a+h)d}{r}\sin\left(\frac{hd}{za}\right)} \qquad (14)$$

$$m=\frac{z}{2h},\quad S^2=(ct)^2-d^2,\quad \eta=(\mu_0/\varepsilon_0)^{1/2} \qquad (15)$$

m denotes the number of reflections at the ionosphere. N indicates the magnitude of the refraction coefficient of the ionosphere. ω_p and ν are the plasma frequency and the electronic collision frequency in the ionosphere, respectively. Local angular frequency ω is derived from Eq.(5) by substituting the next three equations.

$$l_m = \frac{hr}{z}, \quad \theta_m = \frac{hd}{za}, \quad m = \frac{d}{2h} \tag{16}$$

Integration limit S in Eq.(9) also can be obtained from the fact that no signal reaches in the range of $t<r/c$.

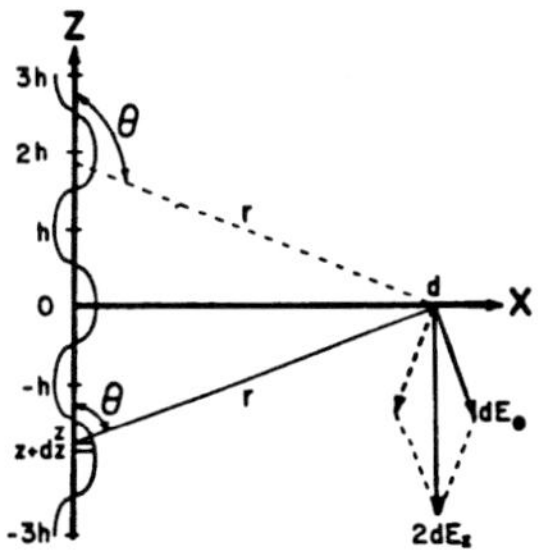

Fig.8. Radiation field from the mode currents.

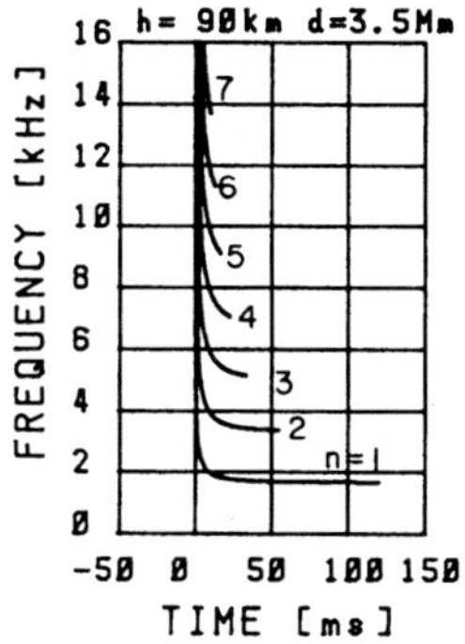

Fig.9. Calculated frequency-time curves by Eq.(5).

4. CALCULATION RESULTS

1) Frequency Dispersion

Figure 9 shows curves of frequency versus time obtained from Eq.(5) substituting 90 km to the ionospheric height h and 3.5 Mm to the propagation distance d, assuming the number of reflections be 220 for each mode. The numbers by each curve are the orders of mode. The number of the reflections was adjusted as the duration of each mode agrees with those in Fig.1. 90 km and 3.5 Mm were chosen for h and d, respectively, so as to fit the frequency intervals between the successive mode curves and to fit the curvature of each curve to the experimental data in Fig.1.

Figure 10 shows the frequency to time characteristics of the tweek as displayed in log-log scales. Figure 10(a) shows the frequency dispersion for d=3.5 Mm and Fig.10(b) for d=5 Mm. It is interesting to see in these figures that the frequency-time curves of each mode of a tweek have maxima between 0.01 and 0.1 ms.

The reason to have rises in the initial parts of the frequency-time curves is follows. The image source array obtained from Eq.(8) is shown in Fig.11. We can see some sources located at the negative heights: one source for d=3 Mm, two sources for d=5 Mm and three sources for d=7 Mm. The radiation field from shallow sources in the negative region may propagate to the receiving point by diffraction, but that from deep sources may give little contribution to the observed field by large attenuation. Image sources located in smaller heights in the positive region are separated from each successive others greater than in larger heights. This is a reason why the frequency-time curves show the chevrons at the initial part of the tweek. The mode number is higher the effect is clearer.

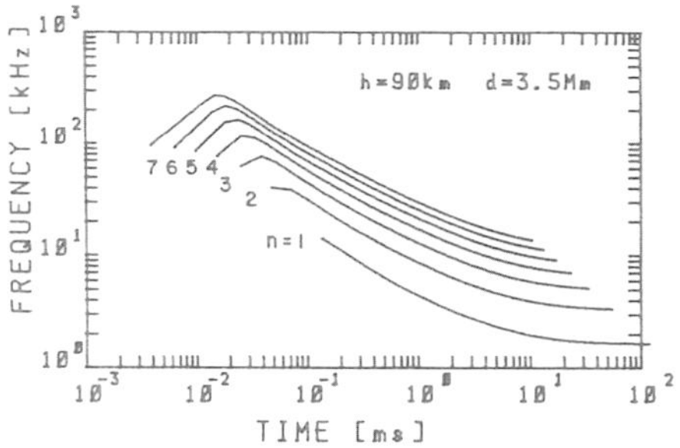

(a)

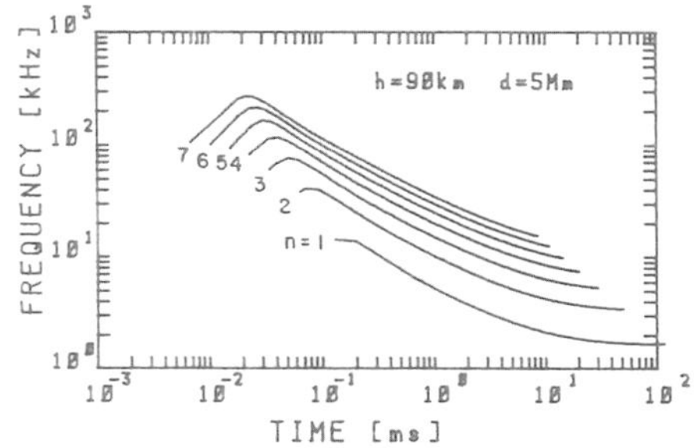

(b)

Fig.10. Calculated frequency dispersion for h=90 km. (a) Propagation distance is 3.5 Mm. (b) Propagation distance is 5 Mm.

2) Waveforms

Figure 12 shows a calculated result from Eq.(9) with h=90 km, d=5 Mm, $\omega_p = 2\pi$ MHz, ν =100 kHz, and with the parameters given by Eqs.(11) and (12). Conductivity of the ground is assumed as infinity. $e(t) \times 2\pi c/\eta$ and $\tau = t - d/c$ [ms] are used for the vertical and horizontal axes, respectively. Figures 13 and 14 show the radiation field which are given by the time differential of the source current waveform. Numbers by the curves give the order of image sources which correspond to the number of reflections m. Each field comes from each image source. In Fig.13, the field from the image sources of m=1 to 18 are displayed. We can see that the first swing in the field of m=10 just overlaps with the second swing of m=9. This is a reason why the amplitude is emphasized at the initial part of the occurrence of the tweek. In Fig.14 the field from the

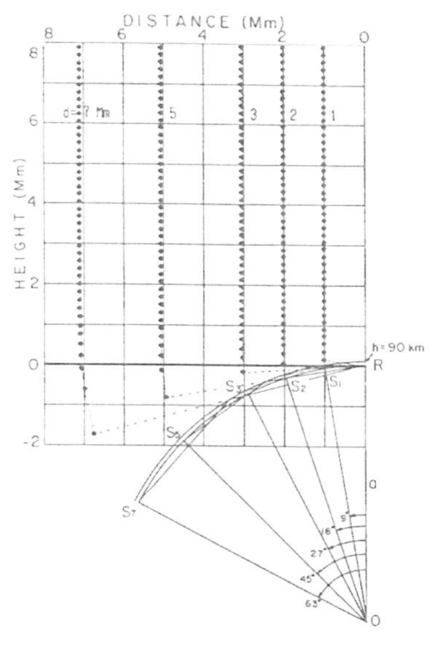

Fig.11. Image source array for spherical Earth approximation.

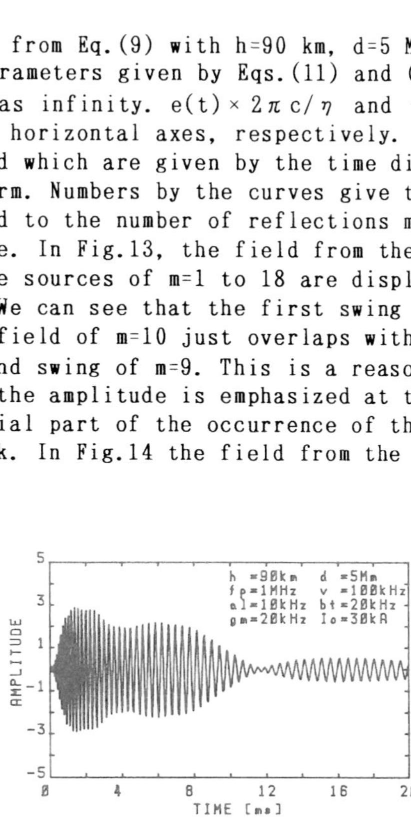

Fig.12. Calculated tweek waveform for the first mode by using Eq.(9).

image sources of m=65 to 72 are displayed. In this range of m, the overlapping of the successive number fields do not occur so that the amplitude is deemphasized at around τ =10 ms (around m=69), thus the waveforms seem as a guitar.

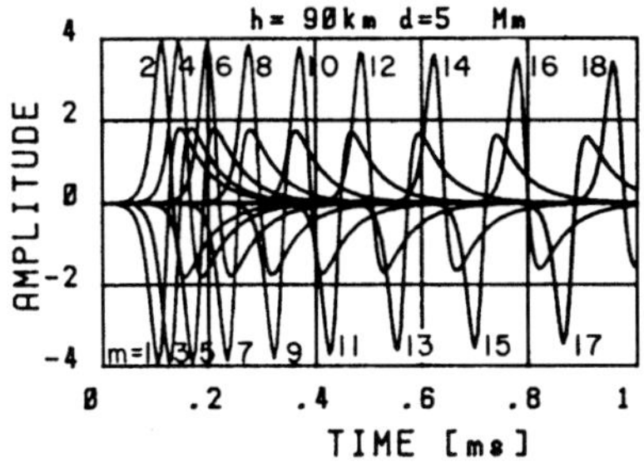

Fig.13. Radiation fields from images of m=1 to 18.

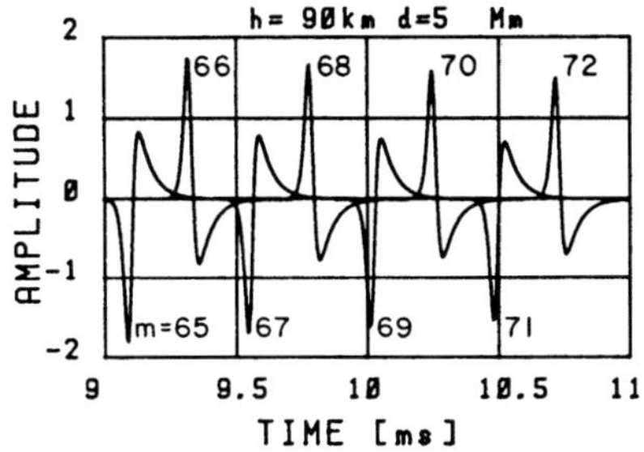

Fig.14. Radiation fields from images of m=65 to 72.

5. DISCUSSIONS

1) Frequency Dispersion

In comparison of Fig.9 with Fig.1, the frequency-time curves agree well with each other, provided the ionosphere height and the propagation distance were determined skillfully. We plotted curves in Fig.9 on several transparent sheets for each ionospheric height with each propagation distance, then put them over the sonagram as shown in Fig.1 one by one. Then we determined the propagation distance by a better agreement between both curves. In other words, we first determined the ionospheric height h by observing the cutoff frequency of the lower mode, then determined the propagation distance d by observing the gradient of the curves of higher modes. The ionospheric height and the propagation distance of the measured tweek in Fig.1 were estimated to be 86 km and 3.5 Mm, respectively. The frequency spaces between modes are also useful to determine the ionospheric height. Pay attention to the frequency separation between the modes in a comparison of Fig.1 with Fig.9. It can be said that the ionospheric height tends to be larger for the higher mode waves than for the lower modes.

In comparison of the curve of the first mode in Fig.10(b) with Fig.3, the chevron of the curve at the initial part is similar with each other. This characteristic is not well explained by the method using the group velocity in the mode theory [6][10].

2) Amplitude modulation

The calculated waveform in Fig.12 is similar to the observed waveform in Fig.2 in the following points;
(1) The waveform slowly rises from zero, producing a sector at the origin.
(2) The amplitude of the waveform is modulated so that the outline of the

amplitude is shaped as a guitar.

These characteristics are neither explained by the method using the source current waveform as given by exp($-\alpha t$)-exp($-\beta t$), nor by the method using the approximate solution of the modal equation [10][11]. In the waves in Fig.2(a),(b) the higher frequency components over 3 kHz were filtered out, then the beginning of the waveforms are distorted so as not to agree with the calculated one in Fig.12. We will discuss this problem in the future.

6. CONCLUSIONS

1) We derived the expression which gave the electric field of the tweek wave by using the multiple reflection method which was based upon the mode current method. In this process, the effects of the Earth's magnetic field was neglected and the ground was assumed to be a perfect conductor.
2) We showed that the effect of the Earth's curvature appeared in the chevron of the frequency-time curves at the beginning of a tweek.
3) We compared the calculated result with the measured data. As a result, good agreements were shown in the initial rising part and the guitar-shaped section in the first mode waveform.
4) We proposed the expression for the source current waveform which gives both positive and negative swings of radiation field. As a source current waveform in both swings becomes more symmetrical, a guitar-neck becomes deeper in the outline of the amplitude.
5) Number of reflections of the typical tweek wave is a few hundreds. A constant number of reflections should be applied throughout all modes.

ACKNOWLEDGMENTS

The authors would like to thank Dr. Akira Ishimaru of University of Washington for his kind discussions and useful suggestions. Tweek data analyzed in this paper were obtained with the Stanford University ELF/VLF Radiometer which was set at the Earth Observatory, Kochi University. The authors thank Dr. A. C. Fraser-Smith and Prof. R. A. Helliwell for their cooperation in this project. This research is partly supported by the Grant-in-Aid for Scientific Research, The Ministry of Education, Science and Culture, and partly by Hitachi Communication Systems, Incorporated.

REFERENCES

1. Helliwell RA.(1965) Whistlers and related ionospheric phenomena. Stanford University Press. California. pp 349
2. Liebermann L.(1956) Extremely low-frequency electromagnetic waves. I. Reception from lightning. J Appl Phys 27: 1473-1476
3. Lynn KJW. Crouchley J.(1967) Night-time sferic propagation at frequency below 10 kHz. Aust J Phys 20: 101-108
4. Yamashita M.(1978) Propagation of tweek atmospherics. J Atmos Terr Phys 40: 151-156

5. Likhter JI. Gul'elmi AV. Erukhimov LM. Mikhailova GA.(1988) Wave diagnostics of the near-Earth plasma. Nauka Moscow. pp 215 (in Russian)
6. Yano S. Ogawa T. Hagino H.(1989) Waveform analysis of tweek atmospherics. Res Lett Atmos Electr 9: 31-42
7. Budden KG.(1961) The wave-guide mode theory of wave propagation. Logos Press. London pp 325
8. Wolff EA.(1966) Antenna analysis. John Wiley & Sons. New York. pp 514
9. Uman MA.(1987) The lightning discharge. Academic Press. New York. pp 377
10. Wait JR.(1962) Electromagnetic waves in stratified media, Pergamon Press Oxford. pp 372
11. Galejs J.(1972) Terrestrial propagation of long electromagnetic waves. Pergamon Press. New York. pp 362
12. Tixier M. and Rivault R.(1974) Study of typical night time atmospheric waveforms observed at Poitiers (France). ELF-VLF Radio Wave Propagation. Edited by Holtet J A. D.Reidel Publishing Co. Dordrecht: 245-249

6 Terrestrial and Extraterrestrial Noise Environment

6.1 Natural Noise Above 50 MHZ from Terrestrial and Extraterrestrial Sources

E.K. SMITH and W.L. FLOCK

NASA Propagation Information Center, Department of Electrical and Computer Engineering, University of Colorado, Boulder, CO 80309, USA

ABSTRACT

This paper offers a brief overview of natural radio noise for frequencies above 50 MHz in terms of brightness temperature as observed from two vantage points. The first is from an Earth station located at 40 degrees north latitude and observing at elevation angles from 0 to 90 degrees with an ideal antenna. The second is a satellite in geostationary orbit communicating with the Earth.

Earth station noise at VHF and UHF is dominated by galactic and solar noise. Emission from the atmosphere, gases and hydrometeors, are dominant at EHF and SHF. Radiative transfer theory is invoked in the calculation of brightness temperature from the atmosphere.

The situation is not vastly different from geostationary orbit if communications is with the Earth. Emission from the land and sea, even under idealized conditions, enters significantly. Land is a much more effective emitter than sea water, but at frequencies above 30 GHz the differential becomes much less due to the increasing significance of atmospheric emission.

INTRODUCTION

Brightness temperature, which relates comfortably to antenna temperature in communications, is derived from Planck's blackbody radiation law for brightness, b(f), a measure of radiant energy in terms of frequency.

$$b(f) = \frac{2hf^3}{c^2}\,\frac{1}{\exp\left(\frac{hf}{kT}\right) - 1} \quad \text{Watts m}^{-2}\,\text{Hz}^{-1}\,\text{rad}^{-2} \qquad (1)$$

where

h is Planck's constant = 6.625×10^{-34} Joule-sec.

f is frequency in Hertz

c is the velocity of light = 299,792,458 m sec^{-1}

k is Boltzmann's constant = 1.381×10^{-23} J K^{-1}

T is Temperature in Kelvin

The Rayleigh-Jeans approximation to Planck's law is given by

$$b(f) = 22.2\ kT\,(f_{GHz})^{-2} \ \text{watts m}^{-2}\,\text{Hz}^{-1}\,\text{rad}^{-2} \qquad (2)$$

A measure of the goodness of this approximation can be obtained by solving Eqs. (1) and (2) for T = 300 K, f = 1 THz. The result is that the Rayleigh-Jeans approximation is high by 8.3%. This error also holds for T = 30 K, f = 100 GHz, or T = 3 K, f = 10 GHz. If we solve both relations for T = 300 K, f = 100 GHz the error in the Rayleigh-Jeans approximation is down to 0.8 %. The Rayleigh-Jeans relation is seen to be a good approximation for almost all radio-frequency problems.

The overall operating noise factor, f, of a receiving system as used in the CCIR is given by

$$f = f_a + (\ell_c - 1)\frac{T_c}{T_o} + \ell_c(\ell_c - 1)\frac{T_c}{T_o} + \ell_c\ell_t\,(f_r - 1) \tag{3}$$

where

$$f_a = \frac{p_n}{k\,T_o\,b}$$

p_n is the available noise power in watts from a lossless antenna.
T_o is the reference temperature taken as 288 K
b is the noise power bandwidth in Hz
ℓ_c is the antenna circuit loss (available input power to available output power)
T_c is the actual temperature, in K, of the antenna and nearby ground
ℓ_t is the transmission line loss
T_t is the actual temperature, in K, of the transmission line
f_r is the noise factor of the receiver

In CCIR usage noise figure is the decibel expression of noise factor and is shown with capital letters (CCIR 1986b). Hence the overall noise figure F is given by

$$F = 10 \log f \tag{4}$$

and the external noise factor, F_a, is given by

$$F_a = 10 \log f_a \tag{5}$$

Similarly the available noise power p_n can now be written in decibel form

$$P_n = F_a + B - 204 \quad \text{dBW} \tag{6}$$

where B = 10 log b

The effective aperture of an isotropic radiator, A_i is given by

$$A_i = \frac{\lambda^2}{4\pi} \tag{7}$$

where

λ is the wavelength in m

An arbitrary antenna of gain, g, relative to an isotropic radiator then has an aperture, A, of

$$A = g\,A_i = g\frac{\lambda^2}{4\pi} \tag{8}$$

The available noise power, p_n, is related to the the noise power flux, s, of a plane wave in the direction of maximum gain by

$$p_n = s\,A = s\,g\frac{\lambda^2}{4\pi} \tag{9}$$

The field strength, e, is related to power flux by

$$e^2 = 120\,\pi s \tag{10}$$

which, when written in decibel form, becomes

$$E\{dB(1\mu V/m)\} = S\{dB(W/m)\} + 145.8 \tag{11}$$

where

$$E\{dB(1\mu V/m)\} = 20 \log e\,(V/m) + 120$$

and

$$S\{dB(W/m^2)\} = 10 \log s(W/m^2)$$

Combining (3) through (11) we arrive at

$$E\,\{dB\,(1\mu V/m)\} = 20 \log f_{MHz} + F_a + B - G - 96.8 \tag{12}$$

The concept of antenna temperature is analogous to that of f_a as defined following Eq. (3). As $e_n = (4\,R\,k\,T\,b\,)^{1/2}$ is the mean square noise voltage in a resistor of R ohms, it follows that the noise power p_n transferred to a matched load is

$$P_n = \frac{e_n^2}{4R} = \frac{4RkTb}{4R} = kTb$$

where

T is the ambient temperature of the resistor in K,
R is its resistance (13)

This expression unfortunately yields an infinite power if bandwidth is allowed to expand from 0 to ∞. The remedy is given by equipartition theory in quantum mechanics.

$$P_n\,(f) = \frac{hf}{\exp\,(\frac{hf}{kT}) - 1}$$

where

$$P_n\,(f) = kT \quad \text{for } hf << kT \tag{14}$$

This expression gives a finite value when integrated from 0 to ∞. At room temperature kT = hf occurs at 6 THz so the problem is fairly academic. Note however that if the resistor is cooled to 3 K then hf = kT occurs at 60 GHz.

Let us now rewrite the expression for the external noise factor, f_a, in the light of Eq. (13).

$$f_a = \frac{P_n}{k\,T_o\,b} = \frac{T_a}{T_o} \tag{15}$$

or

$$T_a = \frac{P_n}{k\,b} \tag{16}$$

where

T_a is antenna temperature due to external noise in Kelvin

Antenna temperature is seen to be directly proportional to noise power. It appears to differ from brightness temperature of equations (1) and (2) in that the former has a frequency squared term. This problem may be rationalized by observing that there is a frequency squared term in the receiving antenna aperture as in relations (7) and (8).

NATURAL NOISE AT AN EARTH STATION

A receiving antenna on the Earth's surface will have a total antenna temperature, T_A, given by [1] as

$$T_A = \frac{1}{\Omega_A} \int_0^{\pi} \int_0^{2\pi} T_S\,(\theta, \phi)\, P_n\,(\theta, \phi)\, d\Omega \tag{17}$$

Where

T_A is the total antenna temperature in Kelvin

T_s (θ,ϕ) is the brightness temperature of sources as a function of angle, K

P_n (θ,ϕ) is the normalized antenna power pattern

Ω_A is the antenna beam solid angle sr

$d\Omega = \sin\theta\, d\theta\, d\phi$ is an infinitesimal element of solid angle, sr

The value of T_A is influenced by all noise sources entering the antenna's main beam and side lobes. However in this section attention will be limited to values of T_s due to sky noise.

Shown in Fig. 1 are the maximum and minimum levels of noise power portrayed as a function of frequency, of the various noise sources which enter into sky noise at frequencies above 50 MHz. This figure is modeled after one originally prepared by Spaulding and Hagn [2] and now found in CCIR Report 670 [3]. It illustrates that noise due to lightning is no longer dominant above 50 MHz, that extraterrestrial noise will be the primary source of sky noise at VHF and UHF, after which noise from the gaseous atmosphere, rain, and clouds will dominate. These will be reviewed in that order.

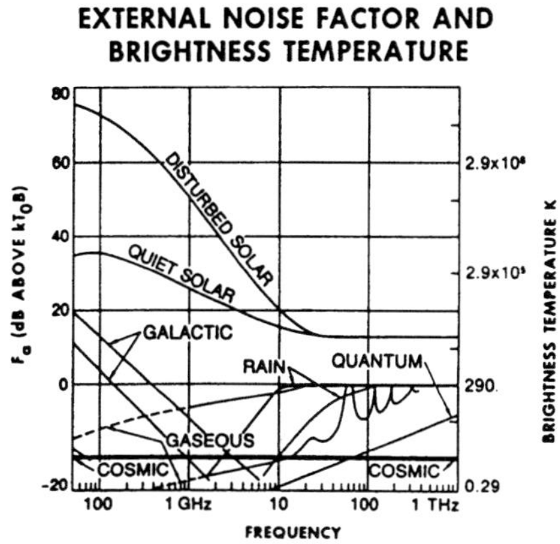

Fig. 1. Approximate maximum and minimum values of natural noise levels in the radio spectrum, 50 MHz to 1 THz [2, 4],

The radio astronomy literature abounds with surveys of galactic noise in the frequency range 100 MHz to 3 GHz. A particularly nice set of four charts is found in CCIR Report 720 for 1986 [5] and 1990 [6] courtesy of the Max Planck Institute at Bonn, FRG..

Another expression for the Rayleigh-Jeans law, applicable to an extended source is

$$S = \frac{2k}{\lambda^2} \int T \, d\Omega \quad \text{Watts m}^{-2}\text{Hz}^{-1} \tag{18}$$

where

λ is the wavelength, m

T is the equivalent blackbody temperature, K

S is the flux density of the source.

If T is uniform over the source, the (18) reduces to

$$S = \frac{2k}{\lambda^2} \, T \, \Omega_s \tag{19}$$

where

Ω_s is the solid angle subtended by the source

The variation of the flux density with frequency is normally given as

$$S \, \alpha \, \lambda^{\alpha} \tag{20}$$

where

α is the spectral index, and is dimensionless

If the blackbody temperature T is constant with frequency, as in the case of a thermal source, then, from (19) the spectral index is -2. The spectral index for radio stars such as Cassiopeia A or Cygnus A is approximatel +0.8 and is attributed to sychrotron emission [1]. The actual range of spectral indices for objects or regions beyond our solar system is -0.3 to +1.3 indicating a mix of emission from ionized gases (free-free transitions) and synchrotron radiation. Extrapolation of values from a radio sky chart to a different frequency than the one for which it is drawn must be done with caution as the spectral index tends to vary with location on the celestial sphere. In addition the spectral index can vary with frequency. However, as illustrated in Fig. 2, the range of galactic noise, as a as measured by brightness temperature, decreases steadilly with frequency to the extent that in the EHF it has essentially decayed into the 2.7 K cosmic background. Certain radio stars are still discernable well above 3 GHz.

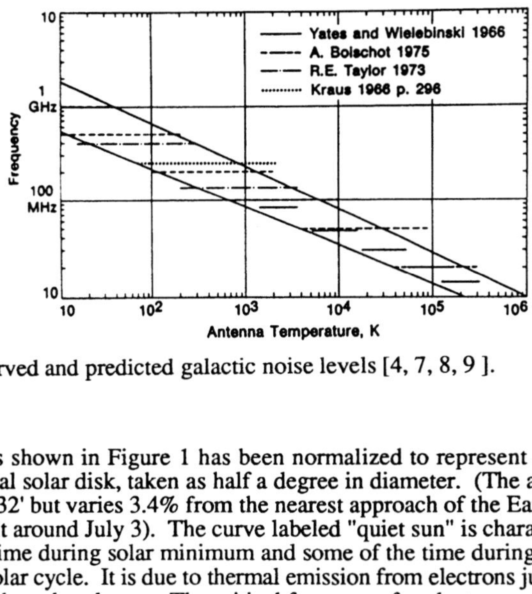

Fig. 2. Observed and predicted galactic noise levels [4, 7, 8, 9].

Solar noise as shown in Figure 1 has been normalized to represent uniform temperature over the optical solar disk, taken as half a degree in diameter. (The actual diameter of the sun averages 32' but varies 3.4% from the nearest approach of the Earth around January 5, to most distant around July 3). The curve labeled "quiet sun" is characteristic of almost all of the of the time during solar minimum and some of the time during the maximum of the eleven-year solar cycle. It is due to thermal emission from electrons just beyond the critical frequency of the solar plasma. The critical frequency for electrons at vertical incidence is given by

$$f_N = [80.5 \times 10^{-12} N(m^{-3})]^{1/2} \text{ MHz} \qquad (21)$$

where

f_N is the critical or plasma frequency of electrons in MHz

N is the electron density, m^{-3}

The Allen-Baumbach formula [10] modeled the electron density N(R) out to a few solar radii

$$N(R) = \left[\frac{1.55}{R^6} + \frac{2.99}{R^{16}} \right] 10^{14} \text{ electrons m}^{-3} \qquad (22)$$

where

N(R) is the electron density, m^{-3}

R is distance from the center of the sun in solar radii (= 1 at photosphere)

Recent models of the electron density, since spacecraft measurements, extend the solar ionosphere out to planetary distances. Berman [11] proposes the following formula for electron density

$$N(R) = \frac{2.21 \times 10^{14}}{R^6} + \frac{1.55 \times 10^{12}}{R^{2.3}} \quad \text{for } R > 1.1\ R_o \qquad (23)$$

Expression (22) gives higher electron densities up to 1.4 solar radii, R_o while expression (23) yields slightly higher values from 1.5 R_o.to 3.8 R_o and much higher values beyond that point.

The fact that the Sun produces higher noise fluxes than would be expected from a 6000 K blackbody was first noticed by radio amateurs in 1936 [12]. Then during World War II a

significant source of radar interference was identified by Hey [13]. Smerd [14] published the picture of the radio sun which is still in vogue today. He showed the radio cross-section to be substantially larger than the visual disk and the brightness temperature to approach one million degrees for frequencies below 1 GHz with substantial limb brightening. Above 1 GHz the temperature monotonically decreases to 6000 K by 100 GHz and the area reduces to that of the visual disk.

The disturbed radio sun correlates with periods of sunspot activity and can be characterized by a slowly varying component (days to months) for frequencies of 0.5 to 10 GHz, and a rapidly varying component (seconds to hours). The Ottawa 10.7 cm flux [15] has become a standard measure of the former [1].

The Moon can be a factor in space communications but primarily when it blocks a radio path. Its brightness temperature is 200 ± 50 K. This temperature varies with frequency and the phase of the Moon and has been summarized by Mayer [16].

Emission from Atmospheric Gases

From Kirchoff's law one can relate the emission coefficient of a gas in thermodynamic equilibrium to its absorption coefficient. This leads to the radiative transfer expression for the brightness temperature T_{B1} of the gaseous atmosphere [17].

$$T_{B1} = T_{sky} e^{-\tau_o} + \int_{surf}^{\infty} T(r)\, \alpha(r)\, e^{-\tau_1(r)}\, dr \text{ K} \qquad (24)$$

where

T_{sky} is the brightness temperature external to the atmosphere in the direction of interest
T_{B1} is the brightness temperature seen by an upward looking antenna at the Earth's surface.
T(r) is the ambient temperature of the atmosphere, K
$\alpha(r)$ is the absorption coefficient of the atmosphere at r, dB/km

$\tau_o = \int_{surf}^{\infty} \alpha(r)\, dr$ is the optical depth from the surface through the entire atmosphere in the direc

$\tau_1(r) = \int_{surf}^{r} \alpha(r')\, dr'$ is the optical depth from the surface to point r

Note that scattering does not enter in this formulation. In the case where the absorbing atmosphere is isothermal a simplification to equation (24) is

$$T_{B1} = T_{sky} e^{-\tau_o} + T_m (1 - e^{-\tau_o}) \quad K \qquad (25)$$

where

T_m is the isothermal value chosen for the atmosphere

This is the relation used in remote sensing to estimate atmospheric absorption. It is critically dependent on the choice of T_m for estimating values of absorption higher than three or four dB.

Curves of brightness temperature for frequencies of 1 - 50 GHz, derlved from a JPL radiative transfer code similar to (24) prepared by Joe Waters, are shown in Fig. 3. The curve parameter is elevation angle. All curves are for an average atmosphere, i.e. 7.5 g/m^3 of water vapor at the Earth's surface with a scale height of 2 km. Other frequency ranges and water vapor contents are available in the literature [5, 18. 19].

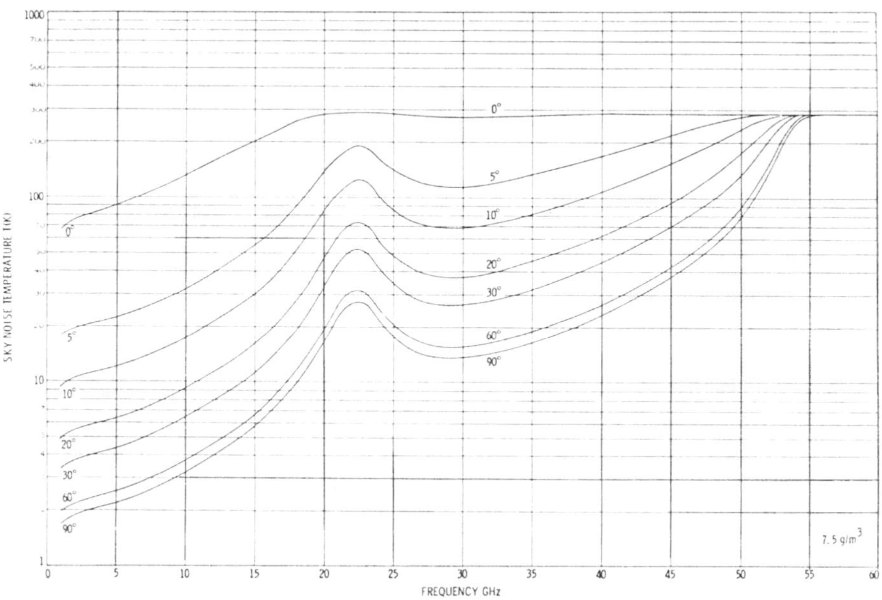

Fig. 3. Clear air brightness temperature for an average atmosphere (7.5 g/m3 of water vapor at the surface, 2 km scale height) 1 - 60 GHz [18]

Brightness temperature due to rain and cloud is a complicated problem but approximate solutions can be obtained simply by using Eq. (25). Care must be taken to separate the absorption and scattering components of the observed attenuation and only enter the absorption component [20, 21]. The effect of clouds on noise temperature based on meteorological data has been studied by Slobin [22].

BRIGHTNESS TEMPERATURE OF THE EARTH FROM GEOSTATIONARY ORBIT

The brightness temperature T_{B2} of the Earth as viewed from geostationary orbit involves several source terms

- extraterrestrial noise and atmospheric emission reflected from the surface of the Earth,
- emission from the land and sea surface,
- emission from the atmosphere.

Njoku and Smith [23] ran the JPL radiative transfer program for mean land and mean sea, for a variety of frequencies, geostationary longitudes, and incident angles. The resultant brightness temperatures were then combined with the polar diagram of an antenna with a global gaussian beam at geostationary orbit to produce Fig. 4. The brightness temperature of the Earth is much greater from land areas than from the oceans so the curves show maxima over Africa at 30 degrees east longitude and minima over the Pacific Ocean at 160 degrees west longitude. However these geographic features are smoothed out above 40 GHz due to the dominance of atmospheric effects.

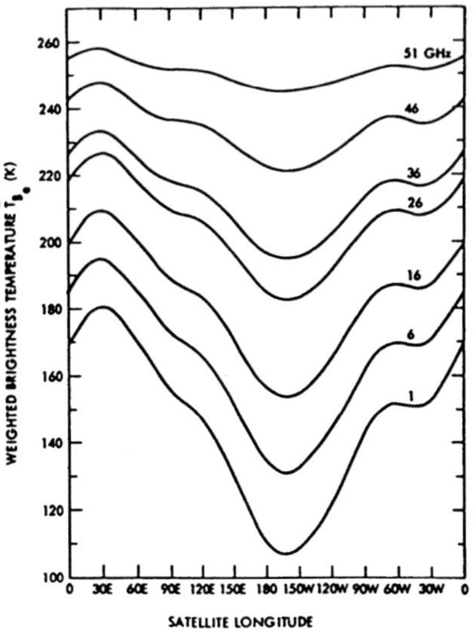

Fig.4. Weighted brightness temperature as a function of longitude viewed from geostationary orbit at frequencies between 1 and 51 GHz. Curves are for U.S. Standard Atmosphere with 2.5 g/m^2 integrated water vapor and 50% cloud cover [23].

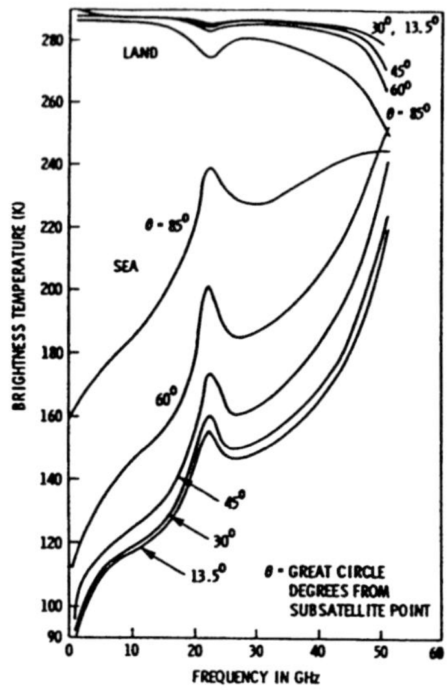

Fig. 5. Mean brightness temperature from geostationary orbit of selected radial distances in degrees (1 degree = 111.11 km) from subsatellite point [24].

Modern communications satellites are more likely to use shaped regional beams and spot beams than global ones. Hence it is helpful to separate the brightness contributions from land and sea so as to calculate the noise temperature of the non-global satellite antenna. This calculation can be done using the curves in Fig. 5 plus knowledge of the satellite antenna.

REFERENCES

1. Kraus, J. D. (1988) *Radio Astronomy, 2nd Edition*, Cygnus .Quasar Books, Powell, Ohio.

2. Spaulding, A. D. & Hagn, G. H. (1978) Worldwide minimum radio noise levels (0.1 Hz to 100 Ghz), *Effect of the Ionosphere on Space and Terrestrial Systems*, J. M. Goodwin, Ed., ONR/NRL, Arlington, VA, Jan. 24-26, 1978.

3. CCIR (1986b) Report 670, Worldwide external noise levels, 0.1 Hz to 100 GHz,*Vol. 1, Spectrum Utilization and Monitoring, Recommendations and Reports of the CCIR, XVIth Plenary Assembly, Dubrovnik, 1986*, ITU, Geneva.

4. Smith, E. K. (1982) The natural radio noise environment, *Proc. IEEE Int'l Symp.on EMC,* Santa Clara, CA, Sept. 6-8, 1982., pp 266-277

5. CCIR (1986a) Report 720-2, Radio emission from natural sources in the frequency range above about 50 MHz, Vol. 5, *XVI Plenary Assembly , Dubrovnik, 1986,* ITU Geneva.

6. CCIR (1990) *Recommendations and Reports of the CCIR, XVIIth Plenary Assembly, Dusseldorf, 1990,* ITU, Geneva.

7. Boischot, A. (1975) Les Bruits Cosmiques, *AGARD Conference on Electromagnetic Noise, Interference, and Compatibility,* NATO, November, 1975.

8. Taylor, R. E. (1973) 136 MHz/400 MHz radio-sky maps, *Proc. IEEE,* Vol. 61, No. 4, pp 469-472.

9. Yates, K. W. & Wielebinski, R. (1966) Intensity-frequency dependence of the radio sky background, *Australian J. Phys.,* Vol. 19, pp 389-407.

10. Allen, C. W. (1947) Interpretation of electron densities from coronal brightness, *Mon. Not. Roy. Astron. Soc.,* vol. 107, p 426.

11. Berman, A. L. (1979) A unified observational theory for solar wind columnar turbulence, Deep Space Network Progress Report 42-50, pp 743-808, NASA-JPL, Caltech Jet Propulsion Laboratory, Pasadena, CA 91109.

12. Shimabukuro, F. I. (1988) The solar spectrum above 1 mm. *The Solar Output and Its Variation,* O. R. White, Ed. , Colorado Associated University Press, Boulder, Colorado.

13. Hey, J. S. (1946) Solar radiation in the 4 and 6 meter radio wavelength band, *Nature,* Vol. 157, p 47.

14. Smerd, S. F. (1950) Radio frequency radiation from the quiet sun, *Australian J. Sci. Res.* , Vol. 3A, pp 34-59.

15. Covington, A. E. (1979) Historical background of the 1970 absolute calibration of solar flux density, *Herzberg Institute of Astrophys.,* Rept. ARO-6, NRC No. 17686.

16. Mayer, C. H. (1964) Thermal radiation from the Moon and planets, *IEEE Trans. Ant. & Propag.* Vol. AP-12, pp 902-913.

17. Waters, J. W. (1976) Absorption and Emission by atmospheric gases, *Methods of Experimental Physics, Vol. 12B, Radio Telescopes,* Ed. M. L. Meeks, pp 142-176, Academic Press, New York, N.Y.

18. Smith E. K., & Waters, J. W. (1981) Microwave attenuation and brightness temperature due to atmospheric oxygen and water vapor, *JPL Publication 81-81,* Aug. 15, 1981.

19. Smith, E. K. (1982) Centimeter and millimeter wave attenuation and brightness temperature due to atmospheric oxygen and water vapor, *Radio Science,* Vol. 17, No. 6, pp 1455-1464.

20. Zavody, A. M. (1974) Effect of scattering by rain on radiometer measurements at millimeter wavelengths, *Proc. IEE* , Vol. 121, pp 257-263.

21. Tsang, L., Long, J. A, Njoku, E. G. Staelin, D. H. & Waters, J. W. (1977) Theory of microwave thermal emission from a layer of cloud or rain, *IEEE Trans. Ant. & Propag.* Vol. AP-25, No. 5, pp 650-657.

22. Slobin, S. D. (1982) Microwave noise temperature and attenuation of clouds: statistics of these effects at various sites in the United States, Hawaii and Alaska, *Radio Science*, Vol. 17, No. 6, pp 1443-1454.

23. Njoku, E. G. & Smith, E. K. (1985) Microwave antenna temperature of the Earth from geostationary orbit, *Radio Science*, Vol. 20, No. 3, pp 591-599.

24. Smith, E. K. and Njoku, E. G. (1985) The microwave noise environment at a geostationary satellite caused by the brightness of the Earth, *Proc. IEEE Int'l Symp. on EMC*, Wakefield MA, Aug. 20-22, 1985, pp 318-323.

6.2 Amplification and Propagation of Geomagnetic Pulsations Pc1 in the Plasmapause Region

Y. HIGUCHI

Department of Electrical and Information Engineering, Yamagata University, Yamagata, 992 Japan

ABSTRACT

An electromagnetic ion cyclotron instability caused by proton temperature anisotropy of ring current particles during the recovery phase of geomagnetic storms is investigated. It is confirmed that the unstable frequency bands occur mainly below the helium ion gyrofrequency. There is an optimum value of the cold helium ion density to hot proton density for the maximum growth rate. The total wave amplification values are calculated along a single propagation path through the unstable equatorial region. It is shown that the maximum amplification value occurs just inside the plasmapause region.

INTRODUCTION

A classification of geomagnetic micropulsations based on a morphological description has been adopted after the IGY(Jacobs et al.,1964). The Pc 1 pulsations are continuous sinusoidal oscillations with periods mainly from 0.2 to 5 sec. During the IMS, the observational knowledge of Pc 1 pulsations has been advanced by the multi-station experiments in the middle and high latitudes(Hayashi et al.,1981; Fukunishi,1984). The maximum amplitude of Pc 1 pulsations with periodic structure on the dynamic spectrum was distributed along the plasmapause latitude. It is now generally accepted that the Pc 1 pulsations are generated in the equatorial region of the plasmapause by the electromagnetic ion cyclotron instability caused by proton temperature anisotropy(Kennel and Petschek,1966). The particle kinetic energy of stormtime ring current protons can be transfered into the hydromagnetic wave energy through the electromagnetic ion cyclotron instability process. It has been also known that the occurrences of Pc 1 pulsations with periodic structure are concentrated generally in the recovery phase of geomagnetic storms. The relaxation of pitch angle anisotropy can produce the Pc 1 pulsations in the plasmapause region accompaning with the particle precipitation into the ionosphere.

The Pc 1 pulsations have been used as a diagnostic tool to obtain an information about plasma parameters in the magnetosphere. Higuchi(1985) improved a diagnostic method to estimate the plasma density along the propagation path including the effect of helium ions. It was demonstrated that the generation region of the Pc 1 pulsations moved close to the plasmapause region from the outer magnetosphere, when the helium ion density to hot proton density is approximately 0.1. The Pc 1 pulsations observed at auroral latitudes have been classified into eight subgroups; Pc 1-2 band emissions, hydromagnetic burst, hydromagnetic whistlers, periodic hydromagnetic emissions, hydromagnetic chorus, irregular hydromagnetic emissions, morning side IPDP, and IPDP(Fukunishi,1981). Sato et al(1987) have introduced a new sort of Pc 1 pulsations called the fingerprint type because of their distinctive appearance in frequency-time spectrograms. The new pulsations were observed at high latitude geomganetic field lines (L-6) and they suggested that the considerable number of spectral lines and the constant frequency separation (16-20 mHz) might provide evidence for non-linear effects in the Pc 1 pulsation generation region. On the contrary, Fraser-smith(1987) opposed that the fine frequency structure was not a non-linear effect but was simply a manifestation of field line resonances of the kind suggested by Jacobs and Watanabe(1962). In spite of many research works mentioned above several important problems remain unresolved. One of the unresolved problems ,in my view, is what the role of heavy ions is in the propagation process and the amplification mechanism for the Pc 1 pulsations in the plasmapause region. It will be shown that the maximum amplification occurs just inside the plasmapause region provided that a typical distribution of Alfven wave velocity in the magnetosphere is assumed.

TEMPORAL GROWTH RATE and WAVE AMPLIFICATION

In Figure 1 we show how the temporal growth rate depends on the increase of cold helium ions to hot proton density. Note that the maximum growth rate is increased with increasing the density of cold helium ions. However, there is an optimum value of the cold helium ion density to hot proton density for the maximum growth rate. Furthermore, the unstable frequency region occurs mainly below the helium ion gyrofrequency because of the expansion of the stop band into the higher frequency unstable region(Higuchi,1991).

In Figure 2 a-c we show how the temporal growth rate depends on the increase of cold oxygen ions to hot proton density. Similarly, there is an optimum value of the cold oxygen ion density to hot proton density for the maximum growth rate. However, the growth rate of unstable frequency region just below the helium ion gyrofrequency is increased with increasing the cold helium ion density to hot proton density.

For the sake of simplification we have approximated the geomagnetic field by a dipole magnetic field. Since the Pc 1 pulsations are observed in the dayside of the magnetosphere, the dipole field remains a good approximation. We have a model of plasma density variation, the Alfven wave velocity variation and the distribution of the proton temperature anisotropy in the magnetosphere which are shown in Figure 3.

In Figure 4 we show the wave amplification integrated along a single propagation path through the unstable equatorial region. Note that the maximum wave amplification occurs just inside the plasmapause. This is in agreement with the recent ground-satellite observation of Pc 1 pulsations (Fraser, Kemp and Webster,1989).

DISCUSSION and CONCLUSIONS

The presence of heavy ions modifies the propagation characteristics and generation process for Pc 1 pulsations in the magnetosphere giving rise to polarization reversals, wave reflection and tunnelling effects. As a consequence the observation of Pc 1 pulsations provides an information on the generation mechanism and propagation characteristics in the magnetospheric plasma. Particulary, in the generation process, it was shown that there was an optimum value of the cold helium ion density to hot proton density for the maximum growth rate. Furthermore, it was demonstrated that the source of Pc 1 pulsations was just inside the plasmapause region for a simple model of geomagnetic field, cold plasma density distribution and proton temperature anisotropy variation.

REFERENCES

Fraser,B.J., W.J.Kemp and D.J.Webster(1989) J.Geophys.Res., 94, 11855-11863.

Fraser-smith,A.C.(1987) Geophys.Res.Lett., 14, 1178-1179.

Fukunishi,H.(1981) J.Geophys.Res., 87, 9029-9039.

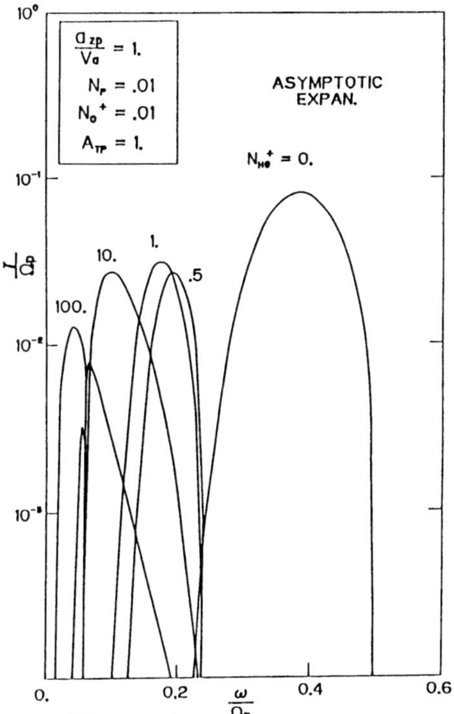

Figure 1

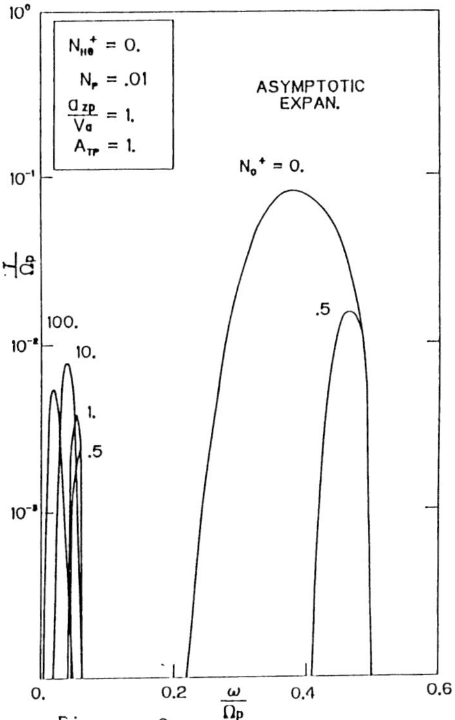

Figure 2-a

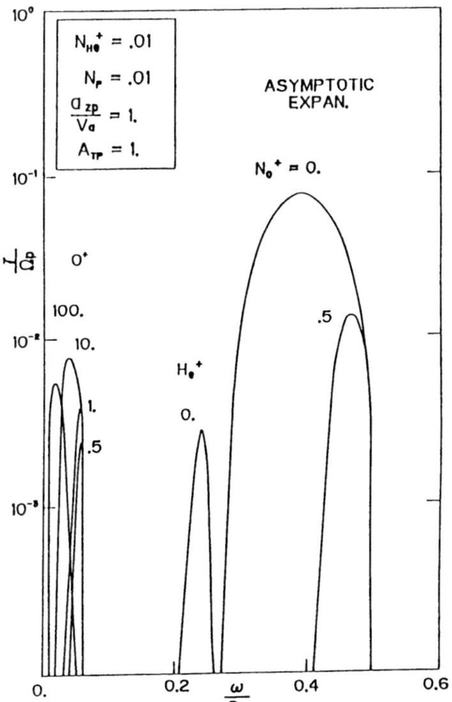

Figure 2-b

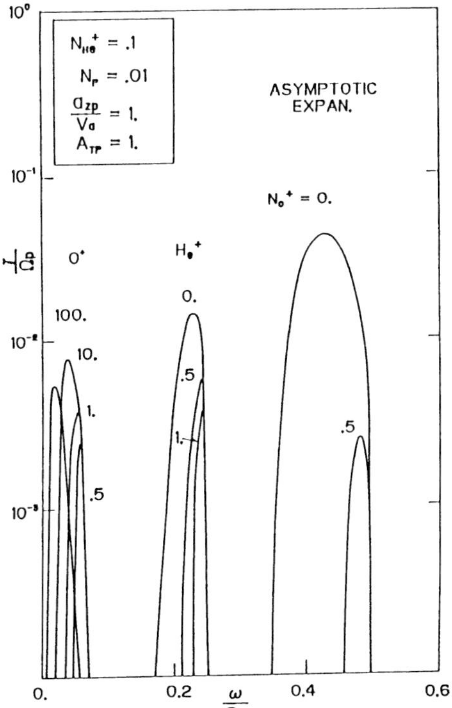

Figure 2-c

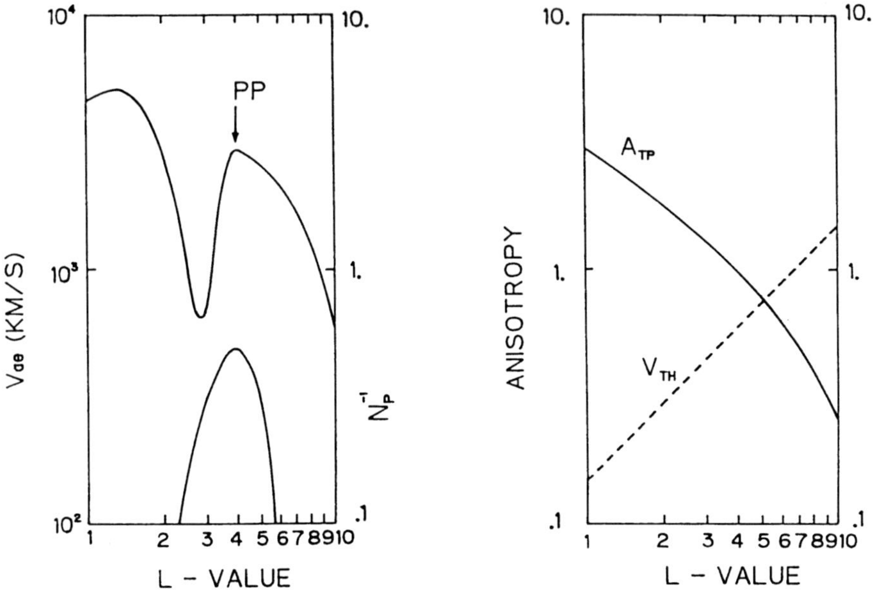

Figure 3

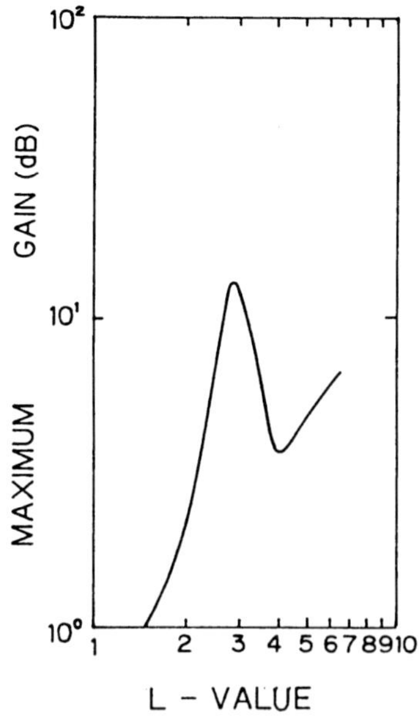

Figure 4

Fukunishi,H.(1984) Proc.Conf.Achievements of the IMS, ESA SP-217, 26-28.

Hayashi,K., S.Kokubun, T.Oguti, K.Tsuruda, S.Machida, T.Kitamura, O.Saka, and T.Watanabe(1981) Canad.J.Phys., 59, 1097-1105.

Higuchi,Y.(1985) J.Geomag.Geoelec., 37, 999-1015.

Higuchi,Y.(1991) J.Plasma Phys., to be published.

Jacobs,J.A. and T.Watanabe(1962) J.Atmos.Terr.Phys.,24, 413-434.

Jacobs,J.A., Y.Kato, S.Matsushita and V.A.Troitskaya(1964) J.Geophys.Res., 69, 180-181.

Kennel,C.F. and H.E.Petschek(1966) J.Geophys.Res., 71, 1-28.

Sato,N., Hirasawa,T. and Shirokura,Y.(1987) Geophys.Res.Lett., 14, 664-667.

6.3 On the Experimental Results of Hiss-Triggered Chorus Observed Onboard GEOS-1 Satellite in the Outer Magnetosphere

K. HATTORI[1], M. HAYAKAWA[1], D. LAGOUTTE[2], F. LEFEUVRE[2], and M. PARROT[2]

Research Institute of Atmospherics, Nagoya University, Aichi, 442 Japan[1] and Laboratoire de Physique et Chimie de l'Environnement, CNRS, Orleans, France[2]

ABSTRACT

The co-existence of hiss and chorus is frequently observed onboard satellites in the outer magnetosphere. The purpose of this paper is to investigate experimentally the association between hiss and chorus and to clarify the role of hiss in chorus generation. The paper is based on the detailed spectral analyses and direction finding measurements for the simultaneous occurrence of hiss and chorus observed onboard GEOS-1 satellite in the outer magnetosphere. The event analyzed here is observed near the equator (geomagnetic latitude, 6.8-8.5 degrees). Then, we propose the mechanism of hiss-triggered chorus based on our experimental findings.

1. INTRODUCTION

Naturally occurring magnetospheric VLF/ELF emissions are known to be basically classified into two different forms; (1) unstructured hiss (Helliwell[1], Dowden[2], Hayakawa et al.[3,4,5,6]) and (2) structured and discrete emissions collectively called "chorus" (Helliwell[1], Tsurutani and Smith[7], Burtis and Helliwell[8], Hayakawa et al.[3,6,9]) . However, the fundamental problems such as whether those two types of emissions are essentially different or not and the link between hiss and chorus, are quite uncertain. The phenomenon suitable for studying these unsolved problems is so-called hiss-triggered chorus emissions. The ground and satellite VLF measurements have indicated that chorus is often accompanied by a background of hiss (Burtis and Helliwell[10], Cornilleau-Wehrlin et al.[11], Koons[12]), although chorus is, on many occasions, spontaneously generated (e.g. Helliwell[1]). Another possible stimulus to trigger a chorus is pointed out by Luette et al. [13,14] to be power line harmonic radiation (PLHR), and they have concluded that man-made VLF noise such as PLHR plays an important role in triggering chorus. But, this hypothesis has been questioned by Tsurutani et al. [15]. Hence, two opposite possibilities have been suggested; one is hiss considered to be incoherent and turbulent, and the other is monochromatic and coherent PLHR. By using the events of hiss-triggered chorus observed onboard the SCATHA satellite, Koons [12] has arrived at the conclusionthat some structures or large-amplitude spectral components existingin the hiss band, are able to phase-bunch the electrons, then leading to the excitation of chorus emissions. However, it seems that the experimental data presented by Koons[12] are not so persuasive as to indicate that hiss is the origin of chorus. Hence, more extensive experimental studies are highly required on his implication concerning the mechanism of hiss-triggered chorus, i.e. whether hiss is actuallya source of chorus or not. For this purpose, the application of detailed signal analyses is of essential use.

The purpose of this paper is to present detailed experimental results and to clarify the relationship between hiss and chorus, that is, to exhibit the role of hiss in chorus generation. The study is based on the detailed spectral analyses and direction finding measurements for the VLF emission events indicating the simultaneous presence of hiss and chorus seeming to be closely correlated with each other. In this paper we make full use of a complete set of wave data observed onboard the GEOS-1 satellite. The application of detailed spectral analyses to the waveform data enables us to investigate the fine structures existing in the hiss band, and the simultaneous measurements of multiple field components make it possible for us to determine the wave normal directions of both hiss and chorus.

In Section 2 we describe the wave data observed onboard the GEOS-1 satellite. In Section 3 we show the wave characteristics of the present event. Section 4 indicates the investigation of fine structures within the hiss band. In Section 5 we present direction finding results of both hiss and chorus. Finally, we summarize the experimental findings obtained from this investigation, discuss the physical implications, and propose an idea of the chorus generation in Section 6.

2. WAVE DATA

The field data are the signals obtained by the so-called S-300 experiment onboard GEOS-1 satellite, which measures continuously the electric and magnetic field components of the wave field by six antennas. The observed signals are subjected to an onboard analysis; swept frequency analyzers (SFA's) and a correlator. Six SFA's, which can be operated with any sensor combination, have a bandwidth of 300Hz and are swept in frequency in the range of 150Hz-77kHz. Before being telemetered onto ground, the signals are transposed in frequency, passed through identical low pass filters at 450Hz and sampled 1488 times per second. More detailed explanation of these wave experiments onboard the GEOS-1 satellite is given in S-300 experimenters [16].

The spectral analyses to find out the fine structures of the hiss emissions are based on the waveform data of SFA's. And the spectral matrix composed of the mean auto-power and cross-power spectra among the multiple field components at each Fourier frequency component is estimated by using SFA data in the VLF range, and these are used for the subsequent direction finding measurement to determine the wave normal direction.

Wave data used in the present study are obtained on the GEOS-1 satellite during the period of about 30 minutes from 12h 12m to 12h 41m U.T. on 21st July, 1977 at L value of 6.5-6.7 at geomagnetic latitude from 6.8-8.5 deg. at L.T. 13h. The Kp index at the time of observation was 2, but the occurrence of the object event is likely to be closely correlated with the severe magnetic disturbance about one day before.

3. WAVE CHARACTERISTICS

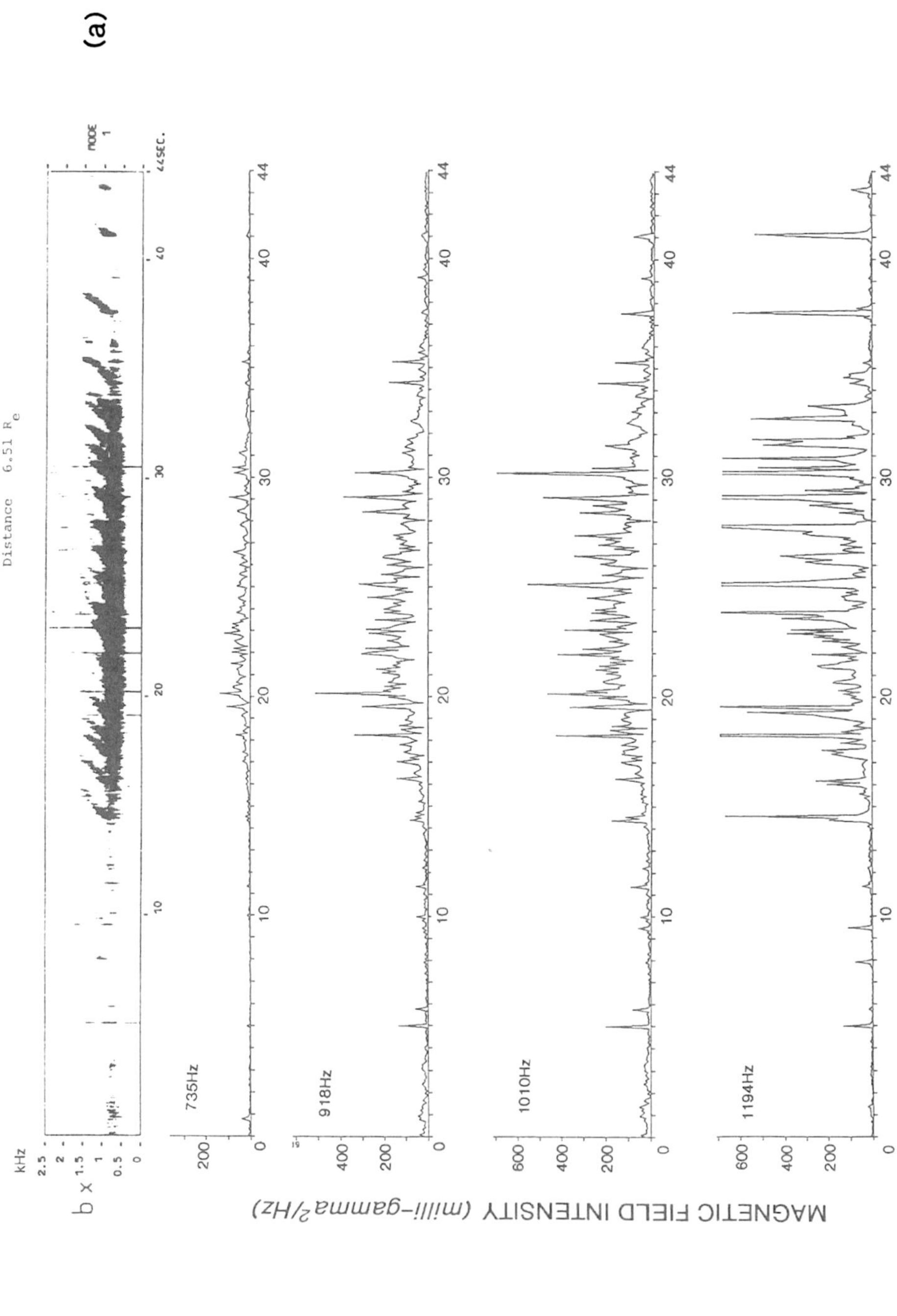
(a)
21/ 7/77 12h 15m 38s U.T.
Latitude 6.75 deg
Longitude 19.38 deg
Distance 6.51 Re
fH = 3.063 kHz
kHz
MODE 1
44SEC.
735Hz
918Hz
1010Hz
1194Hz
MAGNETIC FIELD INTENSITY (milli-gamma²/Hz)

Fig.1(a)

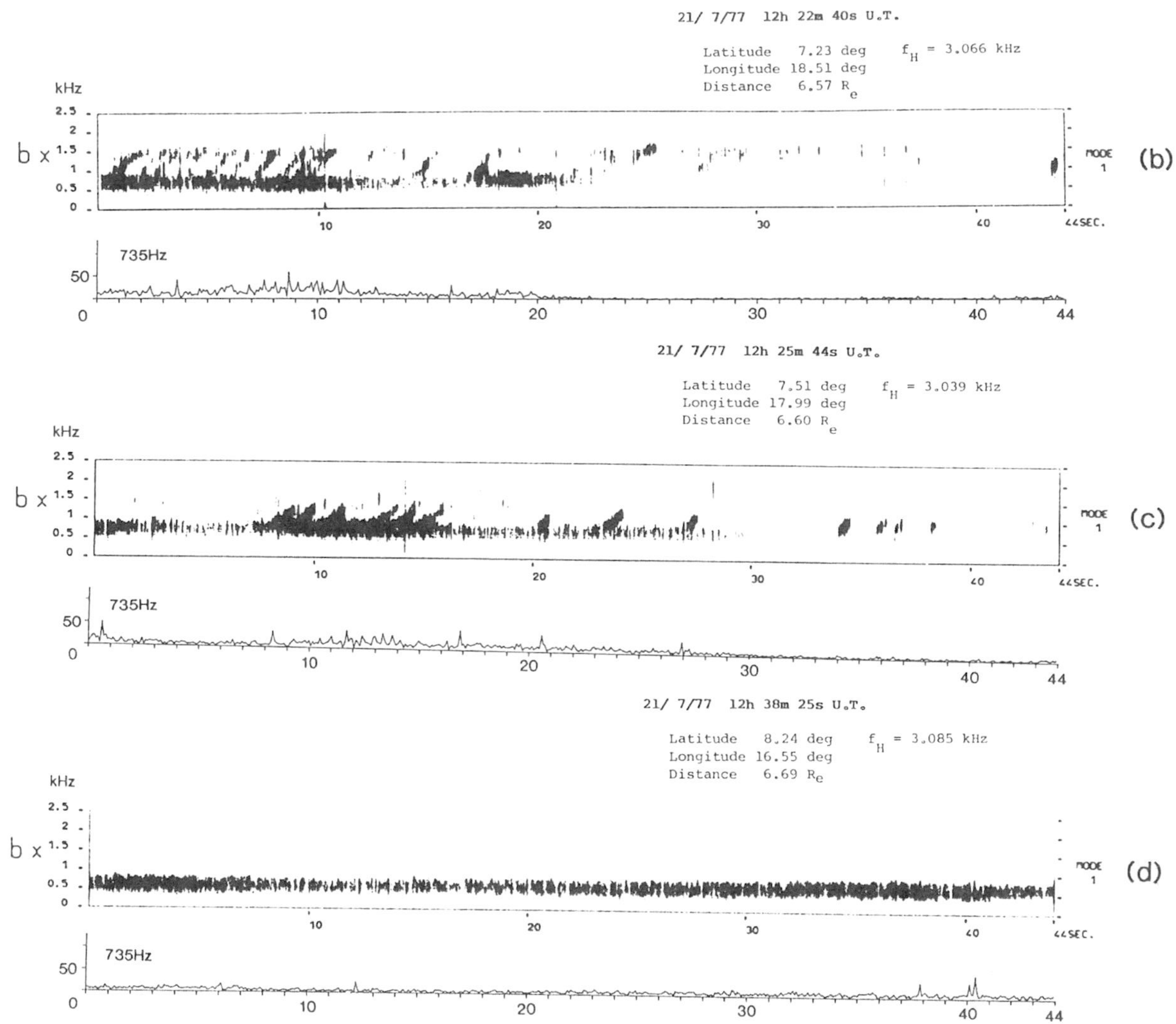
21/ 7/77 12h 22m 40s U.T.
Latitude 7.23 deg
Longitude 18.51 deg
Distance 6.57 R_e
f_H = 3.066 kHz
kHz
b x
MODE 1
(b)
44SEC.
735Hz
21/ 7/77 12h 25m 44s U.T.
Latitude 7.51 deg
Longitude 17.99 deg
Distance 6.60 R_e
f_H = 3.039 kHz
(c)
21/ 7/77 12h 38m 25s U.T.
Latitude 8.24 deg
Longitude 16.55 deg
Distance 6.69 R_e
f_H = 3.085 kHz
(d)

Fig.1(b)-(d)

Fig.1 The spectrograms of magnetic field (bx component) observed onboard GEOS-1satellite in the near-equatorial region on 21/ 7/1977. The alphabetical order (a)-(d) corresponds to the time progress. The vertical stripe structure is owing to the are illustrated the intensities at the four frequencies of 735, 918, 1010 and 1194Hz. The component at 735Hz refers to the intensity of hiss, that at 918Hz means the intensity of the boundary between hiss and chorus, and the other two frequencies correspond to the intensity of chorus. Below the figures (b)-(d) is shown the temporal variation of only the hiss component at 735Hz. From these spectrograms, we can find that each chorus element is asymptotic to the hiss band its slope (df/dt) is about 0.7kHz/s with linear fit. Furthermore, when hiss intensity is high the occurrence rate of chorus is also enhanced.

A few sonagrams in the ELF band (0-2.5kHz) observed by bx antenna are selected and illustrated in Fig.1. The spectrograms in the figures are results of Fourier transforms applied on the ground to the onboard-computed correlograms, and the alphabetical order (a-d) means the time sequence. Especially, there are indicated four temporal evolutions (time resolution=86ms) of magnetic field (bx) intensity of four specific frequencies (735, 918, 1010, and 1194Hz, all with the same bandwidth of 92Hz) in the same linear scale below the the spectrogram of Fig.1(a). The intensity at the lowest frequency of 735Hz is considered to be a contribution of hiss as may be understood from the above spectrogram. The intensity at the higher two frequencies (1010 and 1194kHz) is mainly owing to the chorus. While, the frequency of 918Hz seems to correspond to the boundary between hiss and chorus. In other three figures, Figs.1(b)-1(d), there is only plotted the corresponding temporal evolution of the emission intensity at 735Hz which may reflect the hiss intensity as in Fig.1(a) (note the difference of the scale) so that we can compare the hiss intensity in the different figures. We here comment on the identification of the phenomenon. We have called the signal below 900Hz as hiss, but someone may wonder that at least from Fig.1 it is not clear why this part of the signal be called hiss. Someone might point out this phenomenon as "banded chorus" (Burtis and Helliwell[8]) because of vertical bars. But this is clearly an artifact due to the analysis technique. The spectra of Figure 1 have been obtained by Fourier transform of the correlation function produced by the onboard correlator. The correlator being operated in time sharing (autocorrelogram on one component, autocorrelogram on the other component, etc.) one get time intervals with zero values. This is easily checked by performing a spectral analysis on the waveforms transmitted to the ground. Spectrograms obtained over the same time intervals have not the vertical bar structures. A double-check has been made by performing stationary tests (Lefeuvre and Parrot[17]) on the data of Figure 1. With a high confidence level, the signal is stationary in time. The emissions are definitely hiss emissions.

These figures provide us with general characteristics of the present event. It is likely that chorus element in every spectrogram looks to be triggered from the top of the underlying hiss band. Some of risers are likely to be asymptotic to the hiss band (i.e. initially they have very small df/dt values). A comparison of Fig.1 implies that chorus occurrence is enhanced only when underlying hiss becomes intense; that is, a close correlation between chorus and hiss which is also quantitatively investigated later.

The emission band around 1.5kHz in Fig.1(b) is identified as the half-gyrofrequency VLF emissions which are generated near the equator at frequency above one half the electron gyrofrequency (Hayakawa et al.[3], Muto et al.[18]), and this is not the subject of the present study.

4. SPECTRAL ANALYSES FOR HISS AND CHORUS AND FINE STRUCTURES OF HISS BAND

The detailed spectral analyses have been performed for the waveforms taken at the output of the SFA's. Two steps of frequencies are swept; step 2 ($592 \le f < 1336$Hz) for which the transfer function is constant between 742 and 1042Hz and step 3 ($888 \le f < 1632$Hz) for which the transfer function is nearly constant between 1038 and 1338Hz.

We adopt the periodogram method (Welch[19], Lefeuvre et al.[20]) for investigating the fine structure within the hiss band. This method is a direct application of FFT (fast Fourier transform) on the signal, and the resolution of this spectral analysis is taken as 23.25Hz in frequency and 43ms in time. The results are presented in the form of intensity contour maps, and several examples will be illustrated in this paper.

Koons[12] has suggested the presence of wavelets in the hiss structure based on the simple FFT analysis for very limited data, and recently Tsuji et al.[21] have been made the spectral analyses for VLF/ELF hiss emissions observed on the ground by means of different methods as adopted in this paper, and have supported the idea of Koons that there are some wavelets within the hiss band. This point will be treated in the following for the hiss and chorus observed in Figs.1.

Figure 2 illustrates an example of fine structure analyses for an interval (three successive time intervals with each interval of 688ms) selected from Fig.1(b), and similar results for the two time intervals (starting at 12h 25m 52s 010ms and 12h 25m 56s 138ms) in Fig.1(c) are presented in Fig.3. In the contour maps in Figs.2 and 3, the amplitudes in each 688ms time interval are normalized by the maximum intensity. Then, we plot the contour lines between a minimum value chosen equal to 0.02 and maximum value of 1.0, and the increment between successive contours is linear and equal to 0.04. Especially, the middle of the bottom panels of Fig.2 indicates a very random nature of the hiss band. However, on some occasions, we can find out some structures in the hiss band, and a few examples can be selected in Figs.2 and 3. We notice good examples of a monochromatic wave component (or a wavelet) with duration of about 100-200ms at a frequency in the hiss band. For example, we can find out a wavelet with duration of about 100ms at a frequency of about 833Hz, which is indicated by A in the middle of the bottom panel of Fig.2. Similar wavelets with the same order of duration in hiss band are also found in other figures. Although we have investigated from the spectrograms that the hiss has occupied the frequencies below 900Hz, these contour maps suggest that hiss band extends even up to 1000Hz. Hence, we cannot point out the boundary between hiss and chorus clearly, but we can safely say or confirm again from these contour maps that chorus elements have a

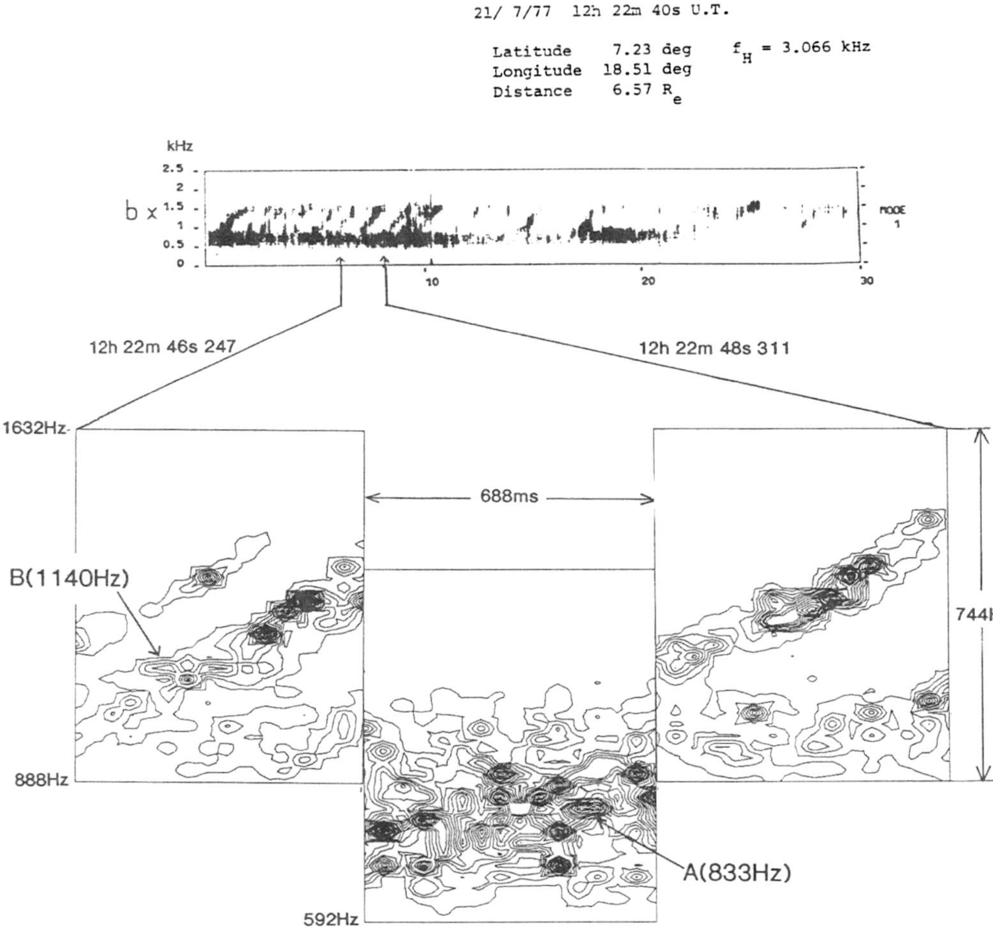

Fig.2 The results of the detailed spectral analyses during 12h 22m 46s 247ms-12h 22m 48s 311ms U.T. in Fig.1(b). This analysis was done with **Δ**t=43ms and **Δ**f=23.25Hz. Coherent wavelets with duration of >100ms can be seen in the hiss band in the middle of the bottom contour map. An example is marked by A. Also, we can find that chorus elements are asymptotic to the hiss band. Mark B is considered to be the causative wavelet for a chorus element. ***(upper figure)***

Fig.3 The intensity contour maps during 12h 25m 52s 010ms-12h 25m 54s 074ms U.T. and 12h 25m56s 138ms-12h 25m 58s 202ms U.T. in Fig.1(c). There are also some wavelets in the hiss band. Among them we select the wavelets which exist at the foot of the chorus elements and labeled C-E. These wavelets may be considered to be the origins of the relevant chorus emissions. ***(next page)***

21/ 7/77 12h 25m 44s U.T.

Latitude 7.51 deg f_H = 3.039 kHz
Longitude 17.99 deg
Distance 6.60 R_e

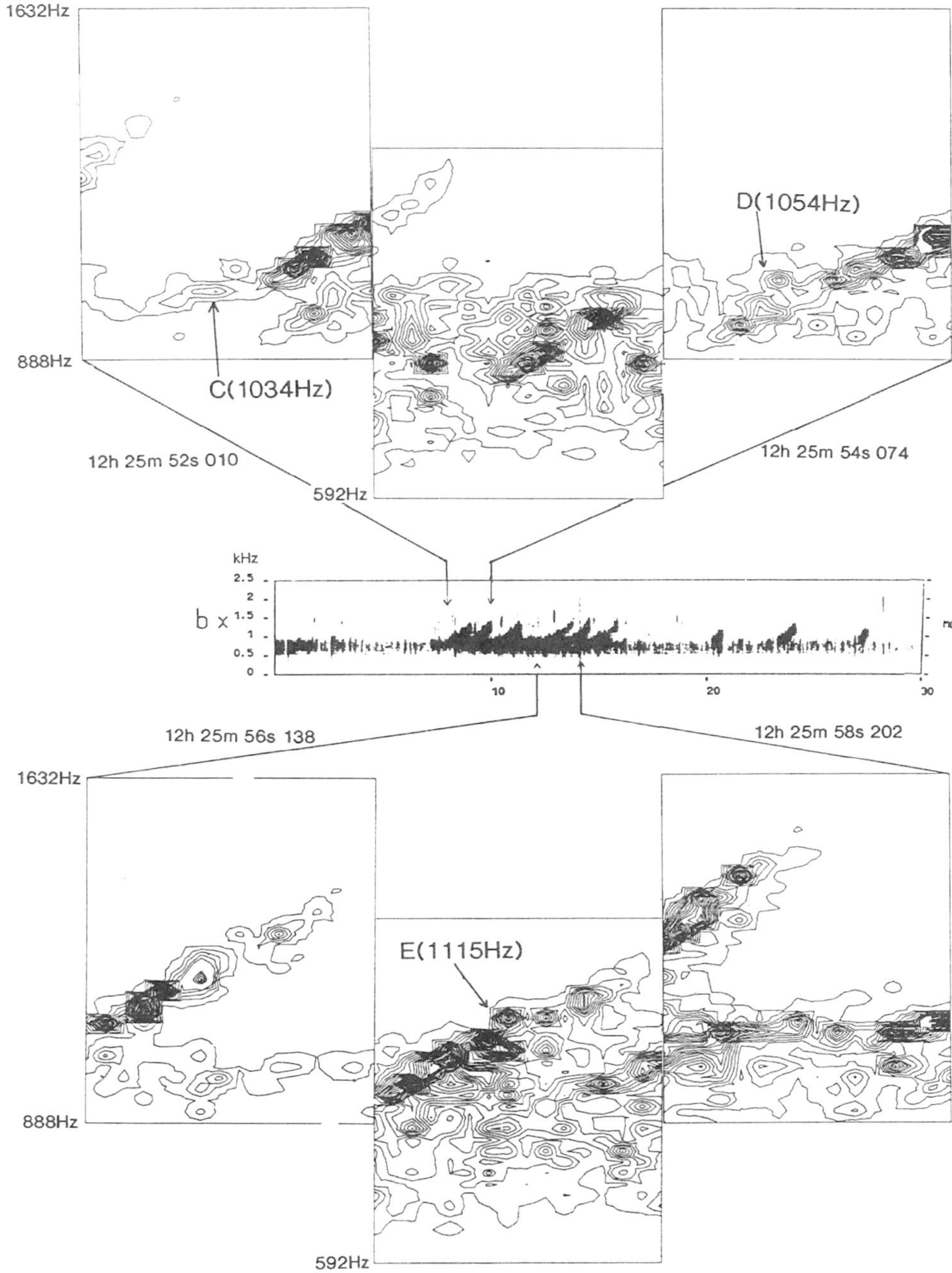

Fig.3

tendency to be asymptotic to the upper edge of the hiss band. Therefore, the suggestion by Koons[12] and Tsuji et al.[21] on the presence of wavelets within the hiss band has been definitely confirmed by this study.

Next, we have to investigate experimentally that those wavelets near the upper edge of the hiss band are hypothesized to be responsible for triggering chorus emissions. The spectrograms in Fig.1 have suggested that chorus elements seem to be triggered from the hiss band, and we look at the detailed spectral analysis results in Figs.2 and 3 with paying a special attention to the part of the foot of each chorus element. We notice examples of a wavelet with duration about 200ms at a frequency around 1140Hz indicated by B in the first bottom panel in Fig.2 and similar wavelets at a frequency of 1034Hz indicated by C in the upper first panel, at a frequency of 1054Hz indicated by D in the upper last panel and at a frequency of 1115Hz indicated by E in the middle of the bottom panel in Fig.3. In these examples of the wavelet associated with a chorus element, the best one is in the first panel in Fig.3. A wavelet at the frequency of 1034Hz (indicated by C) is very probable to be the origin of the chorus, because it is located exactly at the foot of the relevant chorus element. Then, in the middle bottom panel of Fig.3, it is clear that there are a few structures at the part of the hiss where the chorus element is asymptotic to, but it seems difficult to say which wavelet is the origin of the chorus element.

The frequency drift rate of chorus elements (df/dt) can be investigated by using the spectral analyses as in Figs.2 and 3. It is found that the df/dt's of the chorus elements are about 0.7kHz/s when we approximate the chorus element above the hiss band by a linear line fitting.

A close correlation between the hiss intensity and chorus occurrence as inferred in Fig.1 is more quantitatively investigated by comparing the intensity and occurrence rate of chorus elements with respect to those of hiss. The intensities at the four frequencies of 735, 918, 1010 and 1194Hz measured every 86ms are plotted in Fig.1(a), and we use the magnetic field intensities at these frequencies. It is safe to say that the signal at the frequency of 735Hz is attributed to the hiss and this frequency is always below the foot of the chorus elements. While, the higher two frequencies of 1010 and 1194Hz correspond to the intensity of chorus from the sonagram. Fig.4 shows an example of the scatter plot between the hiss intensity at 735Hz and the chorus intensity at 1010Hz at every 86ms at the same time for the interval of Fig.1(a) measured by bx antenna. If we hypothesize as mentioned before that a wavelet near the upper edge of the hiss band is responsible for a chorus element, we have to investigate the correlation of chorus intensity with that of hiss slightly ahead of the chorus, but this kind of study seems to be unimportant because it is difficult to identify the association of the wavelet with the chorus element on some occasions though some wavelets are likely to be the origin of the corresponding chorus elements, and furthermore the temporal resolution of 86ms used in this paper seems to be insufficient. Nevertheless, Fig.4 based on the comparison of the intensities of hiss and chorus at the same time, is still useful to show the association of hiss with chorus. This figure implies clearly a good correlation between the chorus and hiss intensities and the correlation coefficient is estimated to be significantly high, 0.74. This fact indicates strongly that the chorus intensity is very much influenced by the hiss intensity. The intensity ratios at these two frequencies are studied for all events (a)-(d) in Fig.1, and it is then found that this ratio is always around 2.0, indicating the chorus intensity is about 6dB higher than the hiss intensity. However, intensities at the higher two frequencies of 1010 and 1194Hz are nearly of the same magnitude. If we assume that the upper edge of the hiss band is at 900Hz and we adopt 0.7kHz/s for the normal frequency drift of chorus, the time difference between the two frequencies of 900Hz and 1010Hz will be 0.157sec. Hence, we will have the apparent growth rate of 38dB/s during the initial 0.16sec and the subsequent saturation.

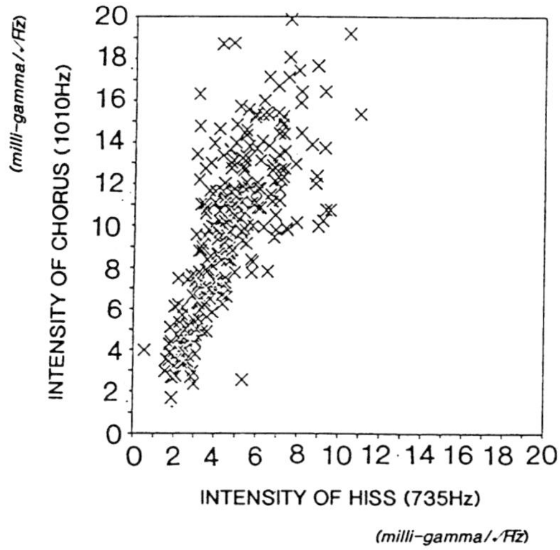

Fig.4 The relationship of magnetic field intensity of hiss (735Hz) and that of chorus (1010Hz) for the event of Fig.1(a).

It is easily thought that very low intensities at 1010Hz in Fig.4 imply the absence of chorus or the background noise. Therefore, the occurrence histogram of the hiss intensity (735Hz) is shown in Fig.5(a). The full line indicates the period of high chorus activity and the broken line is corresponding to the inactive chorus period. The criterion distinguishing between the active chorus period and period of no chorus is a little bit subjectively estimated from the spectrogram of Fig.1(a). For example, the main active period is taken as the interval from 12h 15m 52s 751ms to 12h 16m 13s 821ms, and other interval corresponds to that of no chorus activity. The magnetic field hiss intensity during the period of no chorus indicated in a broken line, is found to be well peaked from 1 milli-gamma/ $\sqrt{Hz}$ to 3 milli-gamma/ $\sqrt{Hz}$, with a weak tail up to 7 milli-gammma/ $\sqrt{Hz}$. On the other hand, the hiss intensity at 735Hz during active chorus period exhibits a broad maximum with its low intensity cutoff of 1.5 milli-gamma/ $\sqrt{Hz}$ and its high intensity cutoff of 12 milli-gamma/ $\sqrt{Hz}$. By comparing these two intensities and by considering our crude criterion on the chorus activity, the hiss intensity is likely to be about 3-5 milli-gamma/ $\sqrt{Hz}$ in order to have an active occurrence, this value being considered to be a threshold value of the hiss intensity for the excitation of chorus. A similar histogram of the hiss intensity for the case of Fig.1(d) in which no chorus takes place, is illustrated in figure 5(b). The histogram in a broken line extends up to about 4 milli-gamma/ $\sqrt{Hz}$, and this lends a further support to the previous threshold hiss intensity some milli-gamma/$\sqrt{Hz}$ for exciting a chorus.

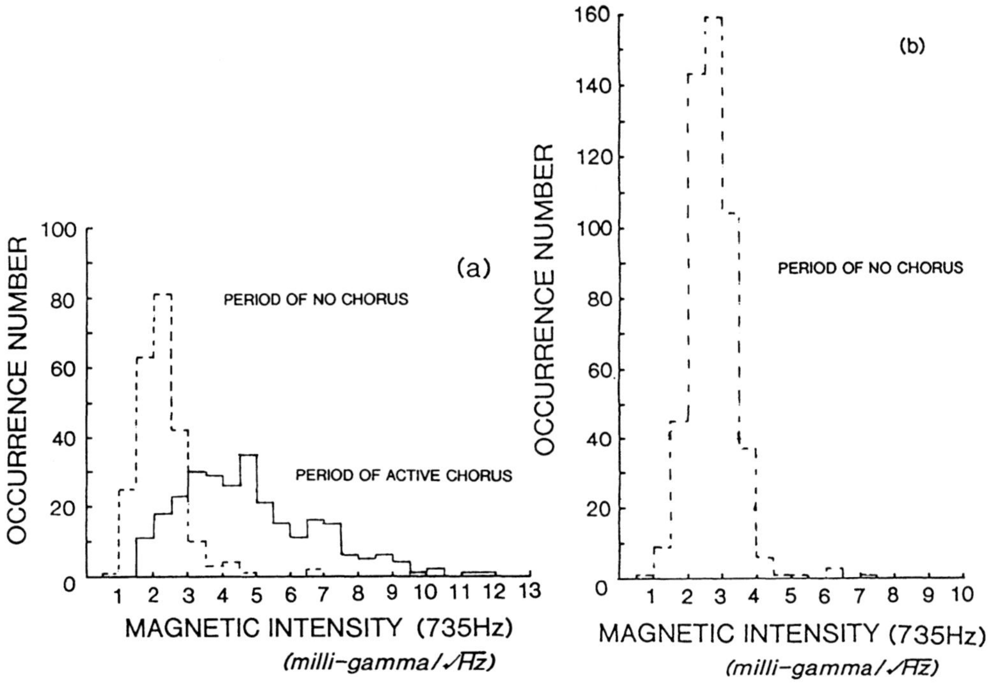

Fig.5 The occurrence histogram of the magnetic field intensity of hiss (735Hz). (a) indicates the result of Fig.1(a) and a full line refers to the period of active chorus period and a broken line corresponds to the period of inactive chorus period. (b) shows the case of no chorus period of Fig.1(d).

Above experiments are statistical treatments and now we adopt the intensity of wavelets indicated in alphabets as shown in Figs.2 and 3. All wavelets indicated in alphabets have an intensity beyond 5 milli-gamma/$\sqrt{Hz}$, so the statistical consideration is available. The wavelet B is about 5.1 milli-gamma/$\sqrt{Hz}$ at its peek intensity. The wavelet C is the best example for the association of the hiss and chorus and has a peek intensity about 6.6 milli-gamma/$\sqrt{Hz}$ and 5.9 milli-gamma/$\sqrt{Hz}$ as the average over the wavelet. The wavelets D and E have an intensity of about 9.3 and 15.5 milli-gamma/ $\sqrt{Hz}$, respectively.

5. WAVE NORMAL DIRECTIONS OF HISS AND CHORUS

Figure 6 illustrates an example of the computer display on the measurement of wave normal directions (θ,Φ) of VLF emissions of both hiss and chorus, where θ is the angle between the wave normal and Earth's magnetic field and Φ is the azimuthal angle measured from the magnetic meridian plane as defined in Lefeuvre et al.[20] and Hayakawa

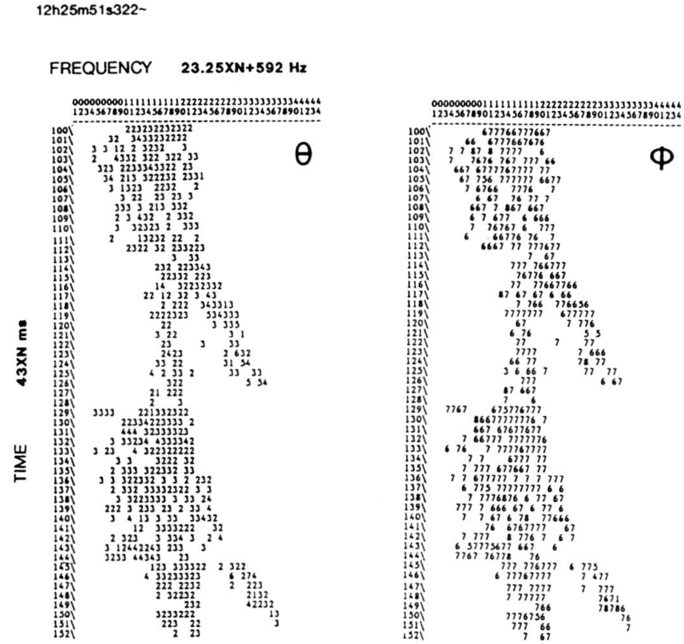

Fig.6 An example of the direction finding results by Means' method. This figure is part of Fig.1(c) and the wave normal direction θ, corresponding to the polar angle, and the azimuthal angle Φ are shown as the range numbers. As for θ, the range number 1 denotes $0^\circ<\theta<10^\circ$, 2 means $10^\circ<\theta<20^\circ$, and so on. As for Φ, the range number denotes the value in every 45°, for example, the range number 1 indicates $0^\circ<\Phi<45^\circ$.

et al.[3,4,5]. 43ms is taken as a time resolution and one frequency division means 23.25Hz in Fig.6. The same analyses have been done for successive many time intervals. The determination of wave normal directions of emissions is based on Means' method [22] assuming a single plane wave. The numerics in the θ plot on the left half indicate the following values; 1 denotes $0<\theta<10$, $10<\theta<20$, and so on in degree. Similarly, in the Φ plot on the right side, 1 indicates $0<\Phi<45$, $45<\Phi<90$, and so on in degree.

Based on the computer plots as in Fig.6, Fig.7 summarizes the occurrence histograms of θ and Φ for the 11 seconds starting at 12h 22m 43s 495ms U.T. in Fig.1(b), and the similar figures are shown in Fig.8 for 11 seconds starting at 12h 25m 51s 322ms U.T. as in Fig.1(c). In both figures the full lines refer to the frequency range below 800Hz and the broken lines, above 1000Hz. As seen from the spectrograms in Fig.1, 900Hz is considered to be the upper edge of the hiss band and the signal around this frequency is a combination of hiss and chorus. Actually, taking account of the contour maps (Figs.2 and 3), there is an ambiguity for the boundary between hiss and chorus. In order to avoid this ambiguity, we have selected the above two frequencies, and so we can consider the frequencies below 800Hz as mainly hiss and above 1000Hz as mainly chorus. Fig.8 implies that the θ value for the hiss is considerably distributed, but smaller θ values less than 40 degree are common, and that the chorus elements tend to have θ values slightly larger by about 10 degrees. Very similar distributions in Φ,

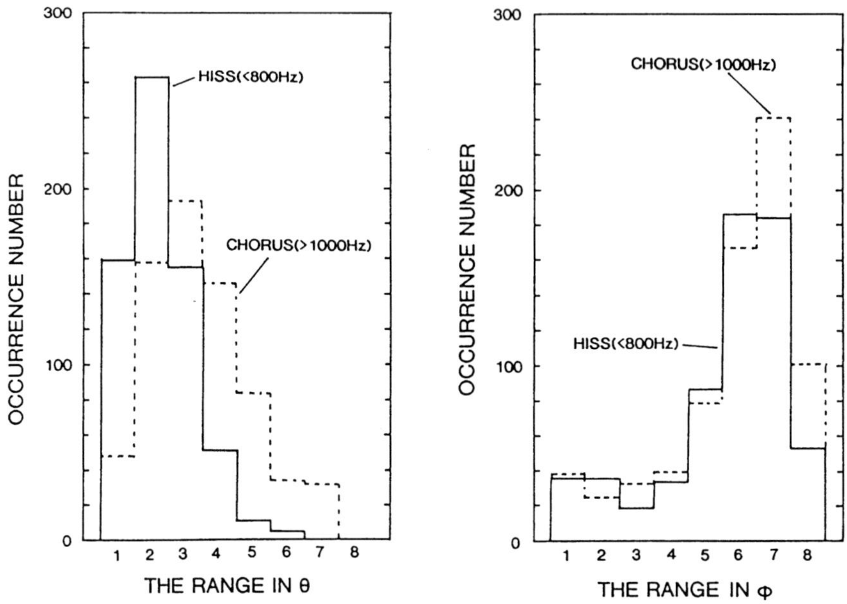

Fig.7 The occurrence histograms of wave normal directions **θ** and **Φ** values, for 11 seconds starting at 12h 22m 43s 495ms U.T. in Fig.1(b). Full lines refer to the frequency range below 800Hz (hiss) and broken lines, the frequency range above 1000Hz (mainly chorus).

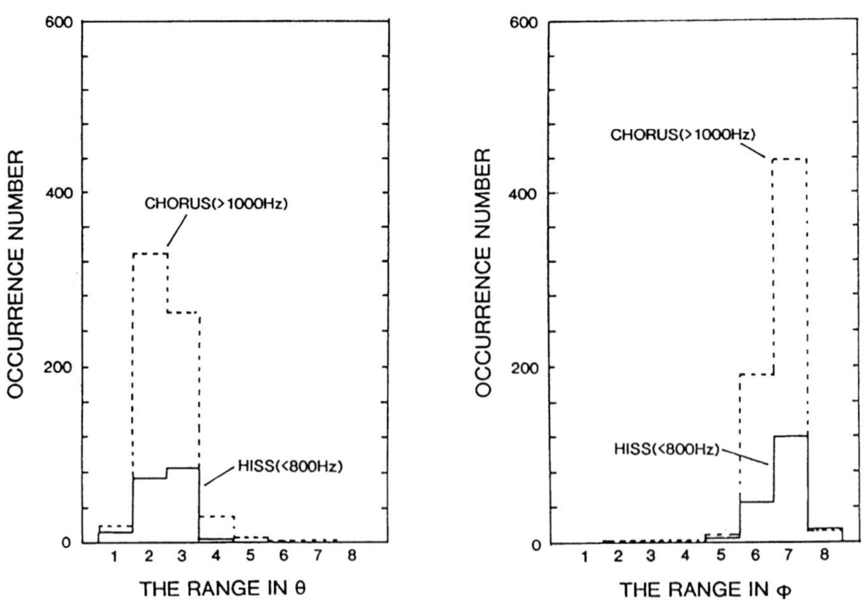

Fig.8 The occurrence histograms of wave normal direction **θ** and **Φ** values, for 11 seconds starting at 12h 25m 51s 322ms U.T. in Fig.1(c). Full lines refer to the frequency range below 800Hz (hiss) and broken lines, the frequency range above 1000Hz (mainly chorus).

being different from the θ behaviours, are noticed for both hiss and chorus.

The same analyses are performed for the event in Fig.1(c) and the corresponding results are presented in Fig.8. The θ's for the hiss illustrated in a full line are well concentrated to θ=10-30 degrees, and a very similar distribution in θ is noticed even for the chorus in a broken line. This point is a little bit different from the previous case. However, we have indicated the very same distributions in Φ for hiss and chorus.

The above information on the the wave normal directions of VLF emissions is statistical, and we cannot study the frequency dependence of the wave normal direction for each chorus element. Fig.9 illustrates the frequency dependence of wave normal angle (θ) of chorus elements during the period of Fig.8. Five elements are indicated by different marks. This figure indicates that there seems to exist no specific dependence of the θ value for each chorus elements on frequency. Then, Fig.11 corresponds to Fig.8, and four chorus elements are analyzed. It appears that there is, generally, a tendency of the Φ value to increase with increasing frequency for this analysis period, except for a chorus element indicated by a box. Although not shown in this paper, we have to add that the Φ value for each chorus element is nearly the same. From these direction finding results for Φ, we can conclude that hiss and chorus came from the same source region.

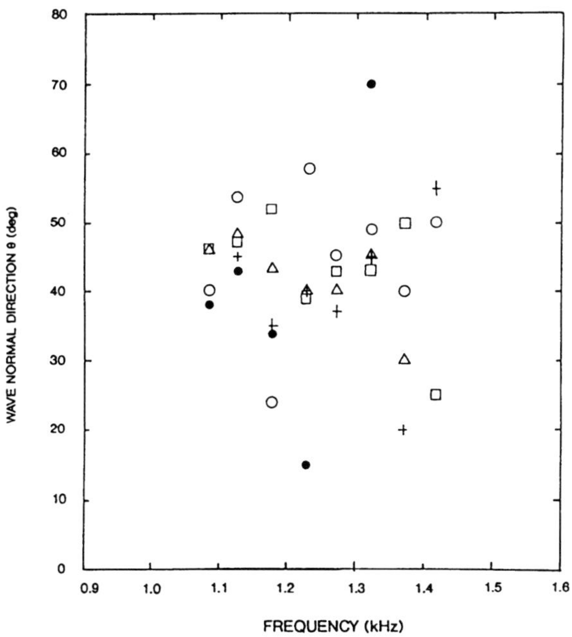

Fig.9 The frequency dependence of wave normal angle θ of chorus elements during the period of Fig.7. There are five elements indicated by different symbols.

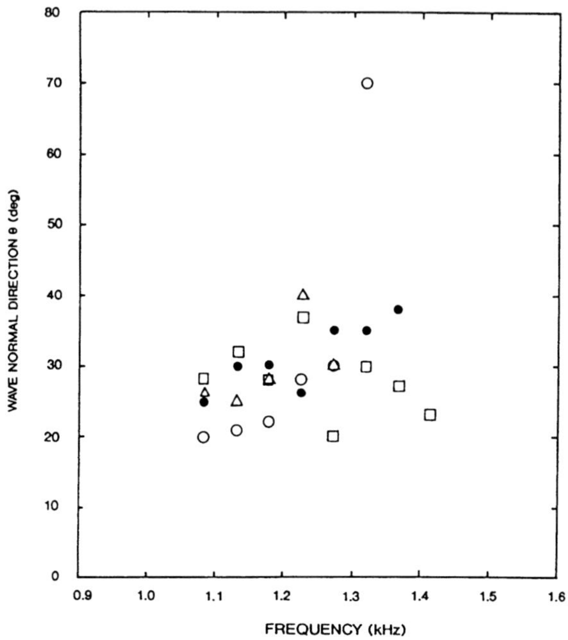

Fig.10 The frequency dependence of wave normal angle θ of chorus elements during the period of Fig.8. There are four elements indicated by different symbols.

6. SUMMARY AND DISCUSSION

Based on the spectral analyses and direction finding measurements for VLF emissions with the simultaneous occurrence of hiss and chorus observed near the geomagnetic equator in the outer magnetosphere by the GEOS-1 satellite, the followings are the summary of the observational facts which are important for the study of the role of hiss in triggering chorus.

(1) The spectrogram by means of the correlators and also spectral analyses have yielded that each chorus element is very likely to be originated from the underlying hiss band. Some of chorus elements are asymptotic to the hiss band and the df/dt at the foot of them are nearly zero, and the frequency drift rate of chorus elements near the equator is about 0.7kHz/s.
(2) The intensity and occurrence of chorus emissions are found to be closely correlated with the intensity of the underlying hiss band. When the intensity of the underlying hiss exceeds a threshold value--some milligamma/$\sqrt{Hz}$, the excitation of chorus emission is extremely enhanced.
(3) The spectral analyses indicate that hiss is, on some occasions, very random and incoherent as has long been considered, but that some parts of hiss exhibit some structures or wavelets (monochromatic wave component with duration of 100ms or more). We can identify, on some occasions, wavelets at the foot of some chorus elements with the intensity of 5 milligamma/$\sqrt{Hz}$ or more, which are considered to have an ability to generate chorus emissions; that is, they are the causative to the relevant chorus. But in other cases, the association of the

wavelet with the chorus emission is not identified clearly.
(4) The hiss emissions are found to be propagating with small wave normal angles, less than 30 degrees, but the wave normal directions(θ) of chorus emissions exhibit different behaviours for different events. Some chorus emissions are propagating with small wave normal angles, but other chorus emissions are found to have large θ values, The more important direction finding fact about the association between hiss and chorus is that same Φ values are obtained for hiss and chorus, which indicates that both phenomena of hiss and chorus come from the same source region.
(5) The θ values for some chorus elements exhibit a tendency to increase with increasing frequency, but other chorus elements have no definite relationship of θ with frequency.

Based on these experimental results, we briefly discuss the possible association between hiss and chorus. The most important observational finding is that wavelets existing near the upper edge of the hiss band are very plausible to be responsible for the associated chorus element. We have obtained a few good examples of this association between the causative wavelet and the chorus. However, this association was not always persuasive, and even in this case there were always a few possible wavelets which may be causative to the relevant chorus element. Such an unclear association may, in fact, be due to the propagation effect such that even if the association between the causative wavelet and chorus element is definite at the source region, it may be destroyed during the course of propagation to the spacecraft location. Of course, the hiss intensity, or to be more exact, the intensity of wavelets within the hiss band exceeds a threshold value for triggering a chorus. The contour maps of intensity with the frequency resolution of 23.25Hz are presented in this paper in order to demonstrate the presence of wavelets, but the presence of wavelets is also checked by the vectorial A.R. method (Lagoutte and Lefeuvre[23]) with frequency resolution of 6Hz, although not shown in this paper. Hence, we can conclude that emissions are triggered by the wavelets existing in the hiss band and that the present experimental study has given a convincing support to terminology of hiss-triggered chorus. This work has supported experimentally very much the work by Koons[12], and the mechanism of hiss-triggered chorus is just coherent wave-particle interactions as in the case of Siple wave injection experiments. Helliwell et al.[24] have used the Siple transmitter to simulate the magnetospheric hiss, and have presented the experimental results in which relatively short duration wavelets of a hiss spectrum is converted into the longer, semicoherent discrete emissions that are typical of chorus. The experimental findings obtained in the present paper have furthermore supported the conclusion by Helliwell et al. [24]. In these coherent wave instabilities, the most serious problem is the effect of other wavelets at different frequencies on the coherent wave-particle interactions. Matsumoto and Omura [25] have made the test particle simulation on phase bunching in the presence of two waves whose frequencies are slightly different from each other. Then, when the frequency difference $\delta\omega$ is greater than $2\omega_t$ (ω_t :trapping frequency), electrons which are trapped by the first wave are not detrapped by the second wave on the assumption that the two waves are of equal amplitude. If we take the wave amplitude as 5 milli-gamma, $2\omega_t$ becomes about 30Hz for the plasma parameters observed. The similar "control frequency" fc is derived by Dowden et al.[26], and this fc is estimated by means of their expression. Then, we obtain fc~35Hz, which is in good agreement with the previous value. Hence, the wavelets near the upper edge of the hiss band are able to phase-bunch the resonant electrons near the equatorial plane, and these phase-bunched electrons are not detrapped by another wavelet whose frequency is spaced more than 30Hz.

REFERENCES

1. Helliwell RA (1965) Whistlers and Related Ionospheric Phenomena, Stanford Univ. Press, Stanford, CA
2. Dowden RL (1971) Distinctions between mid-latitude VLF hiss and discrete emissions, Planet.Space Sci. 19: 374-376
3. Hayakawa M. Yamanaka Y. Parrot M. Lefeuvre F (1984) The wave normals of magnetospheric chorus emissions observed onboard GEOS-2, J.Geophys.Res. 89: 2811-2821
4. Hayakawa M. Ohmi N. Parrot M. Lefeuvre F (1986) Direction finding of ELF hiss in a detached plasma region of the magnetosphere, J.Geophys.Res. 91: 135-141
5. Hayakawa M. Tanaka Y. Sazhin SS. Okada T. Kurita K (1986) Characteristics of dawnside mid-latitude VLF emissions associated with substorms as deduced from the two-stationed direction finding measurement, Planet.Space Sci. 34: 225-243
6. Hayakawa M. Tanaka Y. Shimakura S. Iizuka A (1986) Statistical characteristics of mid-latitude VLF emissions (unstructured and structured): The local time dependence and the association with geomagnetic disturbances, Planet.Space Sci. 34: 1361-1372
7. Tsurutani BT. Smith EJ (1974) Postmidnight chorus: A substorm phenomenon, J.Geophys.Res. 79: 118-127
8. Burtis WJ. Helliwell RA (1969) Banded chorus-A new type VLF radiation observed in the magnetosphere by OGO 1 and OGO 3, J.Geophys.Res. 74: 3202-3010
9. Hayakawa M. Muto H. Shimakura S. Hattori K. Parrot M. Lefeuvre F (1989) The wave normals of chorus emissions in the outer magnetosphere, Proc.Natl Inst.Polar Res. 2: 62-73
10. Burtis WJ. Helliwell RA (1976) Magnetospheric chorus: Occurrence patterns and normalized frequency, Planet.Space Sci. 24: 1007-1024
11. Cornilleau-Wehrlin N. Gendrin R. Lefeuvre F. Parrot M. Grard R. Jones D. Bahnsen A. Ungstrup E. Gibbons WG (1978) VLF waves observed onboard GEOS-1, Space Sci.Rev. 22: 371-382
12. Koons HC (1981) The role of hiss in magnetospheric chorus emissions, J.Geophys.Res. 86: 6745-6754
13. Luette JP. Park CG. Helliwell RA (1977) Longitudinal variations of very low frequency chorus activity in the magnetosphere: Evidence of excitation by electrical power transmission lines, Geophys.Res.Lett. 4: 275-278
14. Luette JP. Park CG. Helliwell RA (1979) The control of the magnetosphere by power line radiation, J.Geophys.Res. 84: 2567-2660
15. Tsurutani BT. Smith EJ. Church JR. Thorne RM. Holzer RE (1979) Does ELF chorus show evidence of power line stimulation?, in "Wave Instabilities in Space Plasma", Edited by P.J.Palmadesso and K.Papadoulos, Reidel, Holland.
16. S-300 experimenters (1979) Measurements of electric and magnetic wave fields and of cold plasma parameters onboard GEOS-1, Preliminary results, Planet.Space Sci. 27: 317-339
17. Lefeuvre F. Parrot M (1979) The use of coherence function for the automatic recognition of chorus and hiss observed by GEOS, J.Atmos.Terr.Phys. 41: 143-152
18. Muto H. Hayakawa M. Parrot M. Lefeuvre F (1987) Direction finding of half-gyrofrequency VLF emissions in the off-equatorial region of the magnetosphere and their generation and propagation, J.Geophys.Res. 92: 7538-7550
19. Welch PD (1967) The use of fast Fourier transform for the estimation of power spectra: A method based on time averaging over short,

modified periodograms, IEEE Trans. Audio and Electroacoust. 15: 70-74
20. Lefeuvre F. Parrot M. Delannoy C (1981) Wave distribution function estimation of VLF electromagnetic waves, J.Geophys.Res. 86: 2539-2375
21. Tsuji S. Hayakawa M Shimakura S. Hattori K (1989) On the statistical properties of magnetospheric VLF/ELF hiss, Proc.Natl Inst.Polar Res. Tokyo, 2: 74-83
22. Means JD (1972) Use of the three dimensional covariance matrix in analyzing the polarization properties of plane waves, J.Geophys.Res. 77: 5551-5559
23. Lagoutte D. Lefeuvre F (1985) Multispectral analysis for electromagnetic wave field component in a magnetoplasma: Application to narrow-band VLF emission, J.Geophys.Res. 90: 4117-4127
24. Helliwell RA. Carpenter DL. Inan US. Katsufrakis JP (1986) Generation of band-limited VLF noise using the Siple transmitter: A model for magnetospheric hiss, J.Geophys.Res. 91: 4381-4392
25. Matsumoto H. Omura Y (1981) Cluster and channel effect phase bunchings by whistler waves in the nonuniform geomagnetic field, J.Geophys.Res. 86: 779-791
26. Dowden RL. McKay AD. Amon LES. Koons HC. Dazey MH (1978) Linear and nonlinear amplification in the magnetosphere during a 6.6kHz transmission, J.Geophys.Res. 83: 169-181

6.4 Geomagnetic Induction in the Finnish 400 KV Power System

R.J. PIRJOLA and A.T. VILJANEN

Finnish Meteorological Institute, Geophysics Department, SF-00101 Helsinki, Finland

The physical principle of geomagnetically induced currents (GICs) in power transmission grids during geomagnetic disturbances is directly based on Faraday's law of induction. By saturating transformers GICs may disturb power systems, and the risk is great in particular at auroral latitudes where geomagnetic disturbances are most intense. However until now, practically no inconveniences have been observed in the Finnish power system, but studies on GICs in the 400 kV grid including both measurements and theoretical calculations are going on. This paper deals with the research work done in Finland. The largest recorded GIC flowing through a 400 kV transformer into the earth is 165 A, but theoretical estimates indicate the possibility of even several hundreds of amperes in the Finnish system during very great geomagnetic storms.

1. INTRODUCTION

The solar wind, which is a stream of charged particles emitted by the sun, deforms the earth's geomagnetic dipole field giving it the shape of a comet. Resulting from intense disruptions in the sun, there are sometimes "gusts" in the solar wind which, when hitting the geomagnetic field, disturb the electric current system in the magnetosphere and ionosphere. On the earth's surface such a disturbance is observed in the time behaviour of the geomagnetic field. The most violent geomagnetic disturbances are known as geomagnetic storms. Another geophysical phenomenon that is closely related to geomagnetic disturbances is the production of the aurora.

Intense disruptions occur in the sun especially when the sunspot number is large. It follows an approximate eleven-year periodicity, and so there is also a statistical periodicity in the occurrence of geomagnetic disturbances. Now we are close to a sunspot maximum which will probably be reached in 1990. Hence questions connected with geomagnetic disturbances are particularly significant at present and in the near future. A great magnetic storm occurred in March 1989 [1].

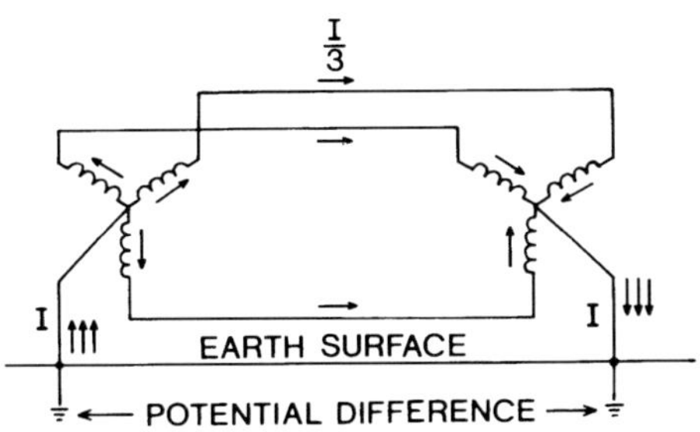

Fig. 1: Flow of a geomagnetically induced current (GIC) in an isolated power transmission system section consisting of two earthed transformers and three phase conductors between them [2].

According to Faraday's law of induction a geomagnetic disturbance is always accompanied by a geoelectric field. The latter in turn drives currents in the earth. Corresponding currents also flow in other conductors, as in power transmission networks, oil and gas pipelines, telecommunication lines and railway systems, and then they are called geomagnetically induced currents (GICs). Figure 1 schematically shows the flow of a GIC in an isolated three-phase power system section earthed at both ends through transformers.

Geomagnetic induction in man-made technical equipment has been widely discussed in the literature, e.g. [3]–[10]. The first observations of the phenomenon were made already more than 100 years ago, e.g. [11]. The induced currents and voltages may disturb or damage the technical system in question. In a.c. power transmission networks, GICs, which are (quasi-)d.c. currents compared to the 50 or 60 Hz frequency, may cause saturation of transformers leading to production of harmonics, false relay trippings etc., and the saturation may cause overheating of transformers and so even damage them. For example during the March 1989 storm, a part of Canada was without electricity for hours because of GICs [1].

Due to the location at auroral latitudes where geomagnetic disturbances are most frequent and intense, the GIC phenomenon is of particular importance in Finland (as well as in Sweden, Norway and North America). Therefore the Imatran Voima Oy power company and the Finnish Meteorological Institute have studied geomagnetic induction in the Finnish 400 kV power grid since 1977 making both measurements and theoretical studies. However, until now no noteworthy inconveniences have been observed in the Finnish system. An important reason for this is the relay equipment used in Finland: no tripping occurs if the second harmonic is present preventing false trippings by GICs. Also neutral point reactors commonly used in the Finnish 400 kV grid decrease GIC magnitudes (cf. Section 3). On the other hand, it should be noted that several changes have been made in the Finnish 400 kV system since 1977 affecting the GICs, and statistically definite conclusions of GIC effects require observations during at least a sunspot cycle.

2. THEORETICAL CALCULATION OF GICS

It seems to be convenient to divide theoretical estimation of GICs in a power system (or any earthed network of conductors) into two steps:

1° The horizontal electric field occurring at the earth's surface at the particular network during a geomagnetic disturbance is first determined assuming the conductors to be absent. The whole earthing system is considered present.
2° This electric field is regarded as externally impressed on the conductors, and the resulting GICs are calculated.

The first purely geophysical problem requires assumptions about the primary sources of the geomagnetic disturbance located in the magnetosphere and ionosphere and about the conductivity structure of the earth. If the primary source is an infinite horizontal sheet current giving rise to a plane wave electromagnetic field propagating vertically downwards and the earth is homogeneous the relation between a horizontal electric field component $E = E(t)$ (t = time) and the time derivative $g = g(t)$ of the perpendicular horizontal magnetic field component $B = B(t)$ on the earth's surface is

$$E(t) = -\frac{1}{\sqrt{\pi\mu_0\sigma}}\int_0^{\infty}\frac{g(t-u)}{\sqrt{u}}\,du = -\frac{1}{\sqrt{\pi\mu_0\sigma}}\int_{-\infty}^{t}\frac{g(u)}{\sqrt{t-u}}\,du \qquad (1)$$

where σ is the conductivity of the earth, and μ_0 is the vacuum permeability valid also for the earth [12, p. 611]. In formula (1) the displacement current in the earth is neglected which is acceptable for geomagnetic frequencies [13, pp. 23–25]. It is necessary to note that $E(t)$ depends on all earlier values of the magnetic time derivative. As stated, equation (1) is obtained when the primary field is a vertically propagating plane wave. However, for formula (1) to be applicable, it is sufficient to assume that the primary field does not vary much in the horizontal direction. This plane wave assumption, is usually true except at auroral latitudes.

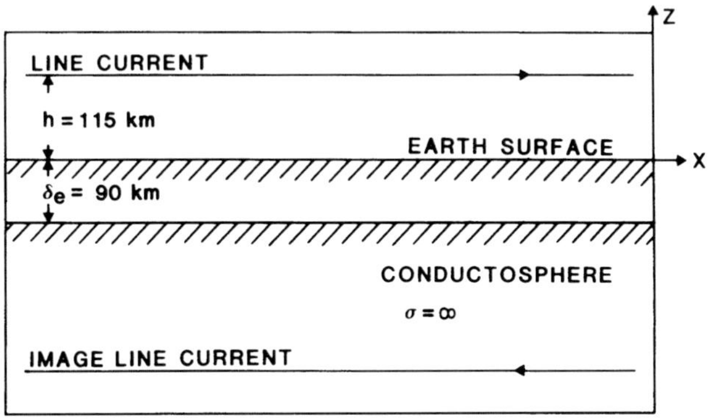

Fig. 2: Line current model of an ionospheric electrojet above the earth's surface. The earth contains a non-conducting layer above a perfect conductor, and so the induction in the earth is described by a mirror image current. The model including the values of h and δ_e shown in the figure was adopted from [15].

A different model for calculating the electric field associated with a geomagnetic disturbance is depicted in Fig. 2. The primary source is a horizontal line current of infinite length describing an ionospheric auroral (or equatorial) electrojet, and the earth is composed of two regions: a non-conducting surface layer above a perfect conductor. The earth model permits the determination of the electromagnetic field using the method of images. Formulas expressing the electromagnetic field on the earth's surface are given in [14]. Now the electric field is proportional to the time derivative of the magnetic field with a space-dependent coefficient. The proportionality, which is thus contrary to the case in equation (1), is a general property of image models (assuming low frequencies). The values of the height h of the electrojet and of the thickness δ_e of the surface earth layer are taken from [15], but as indicated in [14], δ_e should obviously be larger to be more suitable for modelling the situation in Finland.

The image model has been further developed to a "backward method of images" by one of the authors of this paper (ATV), and a different approach also containing a perfect conductor is presented in [17] based on conformal mapping. These extensions will make the first step of theoretical GIC calculation more accurate in the future.

The second step is straightforward circuit theory based on Ohm's and Kirchhoff's laws, and because frequencies associated with geomagnetic phenomena are very small, a d.c. treatment can be used. The following equation has been derived in [17] for the column matrix $\mathbb{I}^e$ whose elements are the currents flowing from the network into the earth at earthing points:

$$\mathbb{I}^e = (\mathbb{1} + \mathbb{Y}^n \mathbb{Z}^e)^{-1} \mathbb{J}^e \qquad (2)$$

Here $\mathbb{1}$ is the unit matrix, and $\mathbb{Y}^n$ and $\mathbb{Z}^e$ are called the network admittance matrix and the earthing impedance matrix, respectively. In a d.c. treatment $\mathbb{Y}^n$ only depends on the resistances of the conductors of the network, and $\mathbb{Z}^e$ contains the earthing resistances of the network. The elements of the column matrix $\mathbb{J}^e$ are called the perfect-earthing currents. They can easily be calculated from the external geomagnetically induced electric field (the first step) and from the resistances of the conductors, and they express the earthing currents in the case that the earthings are perfect, i.e. $\mathbb{Z}^e = \mathbb{O}$. A formula for the currents flowing in the conductors is neglected here because the earthing currents give the GICs flowing through transformers and are so more important in practice.

3. GIC CALCULATION IN THE FINNISH POWER SYSTEM

The value of the induced horizontal electric field occurring at the earth's surface during a geomagnetic storm may be in the order of 1 V/km, and the largest values we have found in the literature are about 10 ... 50 V/km , e.g. [18].

Let us now consider the isolated power line shown in Fig. 1 and assume that its length equals 200 km. Let further a constant external horizontal electric field of 1 V/km be parallel to it (the first step in Section 2). The (total) current I indicated in Fig. 1 and caused by the (external) voltage $U = 200$ V between the two transformers is given by

$$I = \frac{U}{R + R_{e1} + R_{e2}} \qquad (3)$$

where R is the total resistance of the conductors (three phase conductors in parallel) and the earthing resistances of the transformers are denoted by R_{e1} and R_{e2}. In GIC investigations of power lines, we mean by the earthing resistance the sum of the total resistance of the transformer (three phases in parallel), the resistance of a possible neutral point reactor and the actual earthing resistance. The interpretation is not completely correct, since the total conductor resistance R is not connected in series to the total transformer resistance, as assumed in equation (3), but each phase conductor is connected individually to the corresponding transformer phase. However, no error is obviously caused due to "symmetry" between the phases, and so the interpretation is applicable.

The presence or absence of a neutral point reactor is essential to the values of R_{e1} and R_{e2}. For conditions in the Finnish 400 kV power grid, let us now assume that $R_{e1} + R_{e2}$ equals 2.0 Ω, and let R be 1.8 Ω. Equation (3) then gives $I = 53$ A, i.e. 18 A per phase.

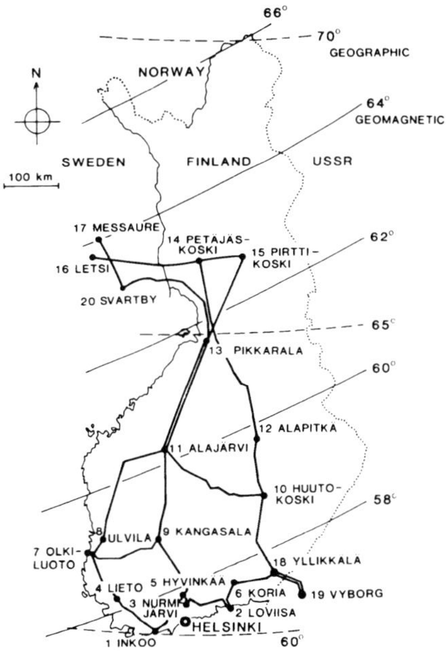

Fig. 3: Finnish 400 kV power transmission system in 1986.

Figure 3 presents the Finnish 400 kV power transmission system in the 1986 situation, which is almost valid also today. A description of the grid by isolated sections as in Fig. 1 is a rough approximation, and a calculation of GICs based on formula (3) only gives estimates of orders of magnitude of the currents.

The dashed curve in Fig. 4 shows an example of the estimation of GICs in the Finnish 400 kV power system using the two-step method outlined in Section 2. The figure concerns the current flowing into the earth through the Huutokoski transformer (number 10 in Fig. 3) during an hour on September 22, 1982. The calculation was based on geomagnetic data from the Nurmijärvi Geophysical Observatory and on formula (1). For comparison, the measured current is also shown (see Section 5). The agreement between the calculated and measured GICs is qualitatively good, but quantitative differences exist.

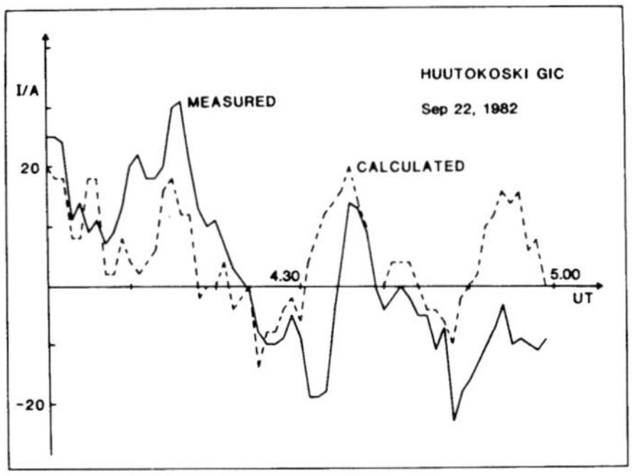

Fig. 4: Geomagnetically induced earthing current (positive into the earth) at the Huutokoski 400 kV transformer on September 22, 1982, at 04.00 ... 05.00 UT [19]. The solid and dashed lines show the recorded and calculated currents, respectively. The calculation was based on geomagnetic data of the Nurmijärvi Geophysical Observatory assuming a homogeneous earth (conductivity = $2.5 \cdot 10^{-4}\ \Omega^{-1}m^{-1}$) and using the plane wave formula (1) for the determination of the external electric field.

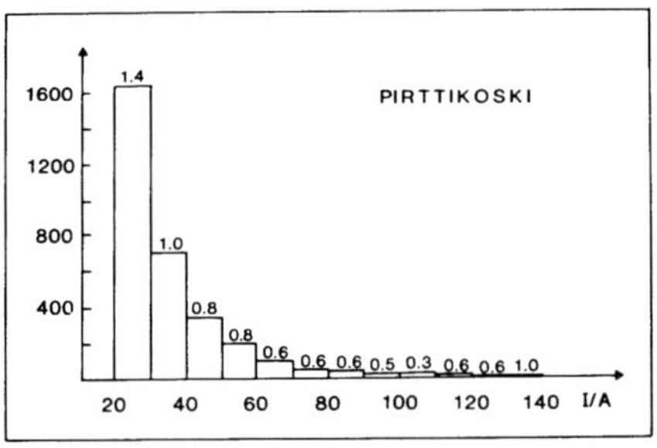

Fig. 5: Statistics of the occurrence of geomagnetically induced earthing currents at the Pirttikoski 400 kV transformer determined theoretically based on geomagnetic data of the Nurmijärvi Geophysical Observatory and using the 1986 situation of the Finnish power system (Fig. 3) [19], [21]. The vertical axis gives the number of events per year when the absolute value of the current has the value expressed on the horizontal axis. The numbers above the columns are average durations of the events.

A critical parameter in a GIC estimation using equation (1) is the conductivity σ of the earth. We have showed that letting the conductivity change with the intensity of the geomagnetic disturbance, i.e. with the (Nurmijärvi) K-index, shortcomings of the simple model can be partly compensated [19]. The conductivity, which thus represents some kind of an effective conductivity, seems to increase with increasing K. The change of the earth's conductivity is, of course, unphysical, but improves calculated GIC results compared to the use of a fixed value of the conductivity. The fitting of the earth's conductivity in a theoretical GIC calculation is also discussed in [20].

For engineering planning purposes, it is important to have some statistical estimates of the occurrence of GICs in a power system. Therefore we have made such studies for the Imatran Voima Oy power company concerning the Finnish 400 kV grid [19], [21]. Our investigations have mainly been based on formula (1), on Nurmijärvi geomagnetic recordings and on Huutokoski GIC data. Figure 5 gives an example of estimated statistics of GICs flowing into the earth through the Pirttikoski transformer (number 15 in Fig. 3). The figure shows, e.g., that a GIC event of 30 – 40 A occurs about 700 times per year. Each of these events lasts for a minute on average, which means that the total annual duration of the GIC earthing current of the particular magnitude is about twelve hours. A detailed definition of an "event" in this connection is given in [21, pp. 18–19]. Large GICs may be expected to occur especially at Pirttikoski, because it is situated at a "corner" of the 400 kV grid and in the auroral region. Our theoretical estimates indicate the possibility of even 700 A there during a very great geomagnetic storm. This value approximately equals the 50 Hz rated current of the Pirttikoski transformer. It should be noted that the rated current refers to the current flowing in a single phase conductor while a GIC earthing current discussed in this paper represents the sum of the currents in each phase, i.e. the GIC is three times the current in a single phase.

Table 1: Calculated geomagnetically induced currents (GICs) at different sites of the Finnish and Scandinavian power transmission system. Two models for the external horizontal electric field are considered: a constant field of 1 V/km to the east (*C*) and the field caused by the line current with a perfect conductor in the earth as shown in Fig. 2 (*L*). Three different cases (*1*, *2* and *3*) concerning the grid and the area where the external field occurs are included. The 1991 situation (only slightly different from the 1986 situation, Fig. 3) was assumed to be valid for the Finnish grid, while an older configuration was assumed for the Scandinavian grid. Rough estimates were used for the resistance values of the Scandinavian grid. Series capacitors used in some Swedish conductors and preventing the flow of GICs were neglected in these calculations.

THEORETICALLY CALCULATED GIC–CURRENT IN THE NORDIC 400 kV POWER SYSTEM

Stations:

symbol	*station*	*country*	*geogr.lat.*
HAR	Harsprånget	Sweden	66.9° N
LET	Letsi	Sweden	66.5° N
PIR	Pirttikoski	Finland	66.4° N
PET	Petäjäskoski	Finland	66.3° N
HUU	Huutokoski	Finland	62.3° N
LIE	Lieto	Finland	60.5° N
MAL	Malmö	Sweden	55.4° N

C = constant external electric field to the east 1 V/km
L = external electric field caused by a west-east ionospheric line current at geogr.lat. 67.2° N ; average electric field at the Finnish 400 kV grid 1 V/km pointing to the east

1 = the whole Nordic 400 kV grid included
2 = the grid outside Finland neglected
3 = the external field equal to zero in Finland

GIC earthing currents (*ea*) and conductor current (*co*) (in amperes):

	C1	C2	C3	L1	L2	L3
HAR (*ea*)	0.4	–	12.9	–10.9	–	22.8
PIR (*ea*)	92.3	92.1	3.8	231.7	234.2	7.0
HUU (*ea*)	24.4	24.4	0.2	12.3	12.3	0.3
LIE (*ea*)	–17.6	–17.6	0.0	–4.9	–4.9	0.0
MAL (*ea*)	–13.3	–	–13.3	–1.5	–	–1.5
PET-LET (*co*)	–95.7	–97.0	–14.7	–255.4	–270.6	–29.1

The Finnish 400 kV power transmission grid continues in galvanic connection from Letsi (number 16 in Fig. 3) and Messaure (17) over whole Scandinavia, but towards the Soviet Union the galvanic connection is interrupted at Vyborg (number 19). Also at some stations there is a galvanic connection to the 220 kV system through autotransformers. In a theoretical calculation of GICs the whole network that is galvanically connected should be included. In the estimations described above, we have, however, neglected the 220 kV grid and "stopped" the 400 kV grid at Letsi and Messaure and taken the Scandinavian network into account only by assuming the earthing resistances of Letsi and Messaure to be equal to zero. It is indicated in [22] that the neglect of the lower voltage system is obviously acceptable as the first approximation.

The effect of the galvanic connection between the Scandinavian and Finnish power systems on GICs is studied in Table 1, which concerns the current in the power line (three phases in parallel) between Petäjäskoski (number 14 in Fig. 3) and Letsi (16) and the earthing currents at five stations: Pirttikoski (15), Huutokoski (10), Lieto (4), Harsprånget (in northern Sweden near the connection to the Finnish grid) and Malmö (in southern Sweden). Two different horizontal electric fields external to the conductors are considered: a constant eastward field of 1 V/km (denoted by *C*) and the field caused by the line current model shown in Fig. 2 (denoted by *L*). In the latter case the current has a west-east orientation and is located about 100 km north of Pirttikoski and the parameters are so chosen that the average of the eastward field at the Finnish grid is 1 V/km. Case *1* refers to the situation in which the whole galvanically connected Nordic 400 kV grid is taken into account. Case *2* corresponds to the computations discussed above, i.e. the Scandinavian grid is included only in the zero earthing resistances of Letsi and Messaure. In case *3* the external electric field is (artificially) set equal to zero at the Finnish grid.

There are three facts seen in Table 1 which indicate that GICs do not flow long distances in a power network:

1. The GIC is the same at Malmö in cases *1* and *3* for both *C* and *L* showing that the situation in Finland does not affect southern Sweden.
2. Cases *1* and *2* yield practically the same GICs in Finland for both *C* and *L* justifying the above-mentioned neglect of the galvanic connection to the Scandinavian grid when the Finnish side is concerned.
3. The GICs are small in Finland in case *3*, i.e. when they are generated outside, for both *C* and *L*.

Hence only the parts of the grid that lie in the vicinity of a particular site are significant to the GIC at the site. This is a conclusion that greatly simplifies theoretical GIC estimation in a complicated network in practice; the matrices included in formula (2) may be kept small.

4. EFFECT OF THE CONDUCTIVITY OF THE EARTH ON GICS

We have claimed several times previously that the low conductivity of the earth in Scandinavia increases the risk of geomagnetic induction in technical systems, e.g. [20]. This conclusion would seem to be supported by equation (1) which indicates that a decrease of σ increases the external horizontal electric field. However, the conclusion need not be quite true since a decrease of the conductivity also increases the actual earthing resistances of a network thus tending to decrease GICs. Hence there are two opposite effects, and we will now consider the matter in greater detail.

For simplicity, let us discuss an isolated power line shown in Fig. 1 earthed at both ends to a homogeneous earth (with σ). We consider the plane wave formula (1) valid for the external horizontal electric field. The magnetic field on the earth's surface is practically independent of σ being twice the primary field, e.g. [23, p. 50]. Thus the electric field can be written as $E = \alpha/\sqrt{\sigma}$ where α does not depend on σ, and if the length of the line is d, U included in equation (3) equals $\alpha d/\sqrt{\sigma}$. Let us now rewrite the denominator of the right-hand side of formula (3) as $R_1 + R_m$ where R_m is the sum of the actual earthing resistances of the two transformers and R_1 is the sum of the other resistances. It is obvious that R_m is proportional to the inverse of the conductivity of the earth: $R_m = \beta/\sigma$, e.g. [24, pp. 66–97]. Now the GIC given by equation (3) can be written as

$$I = \frac{\alpha d}{\beta} \frac{\sqrt{\sigma}}{1 + \frac{R_1}{\beta}\sigma} \qquad (4)$$

The GIC I regarded as a function of σ is zero if $\sigma = 0$ (which is unrealistic and makes formula (1) invalid), and I also goes to zero when σ approaches infinity. A maximum (or a minimum if $\alpha < 0$) occurs at $\sigma = \beta/R_1$, i.e. $R_1 = R_m$. Hence the above-mentioned statement of an increase of GICs with the decrease of σ is only true when $\sigma > \beta/R_1$, i.e. $R_m < R_1$. In practice for a power transmission system, this probably seems to be the usual case. So the conclusion is correct, but as indicated, the reasoning is not straightforward based only on the external electric field. Figure 6 illustrates the behaviour of I (in units $\alpha d/\beta$) as a function of σ with $\beta = 1/300\pi$ m^{-1} and $R_1 = 3$ Ω. Typically the conductivity of the earth is in the decreasing part of the curve.

An obviously equivalent way to say the above-mentioned is the following: When the current induced in the earth by a geomagnetic disturbance meets a particular section of a power system (or any network), it is divided into two parts depending on the values of the resistances, one part flowing through the network as a GIC, and the other continuing in the earth. If the conductivity of the earth decreases the former part relatively to the original total earth current increases as $R_m/(R_1 + R_m) = 1/(1 + R_1\sigma/\beta)$. The point is, however, that the original current also decreases with decreasing σ being proportional to $\sqrt{\sigma}$. These facts exactly lead to formula (4).

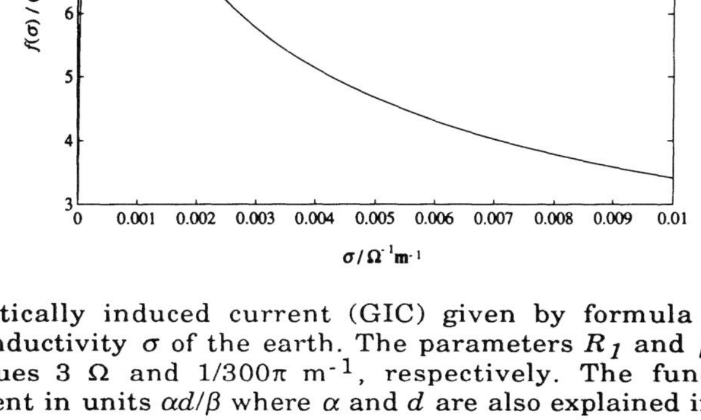

Fig. 6: Geomagnetically induced current (GIC) given by formula (4) depicted as a function of the conductivity σ of the earth. The parameters R_1 and β explained in the text have the values 3 Ω and $1/300\pi$ m^{-1}, respectively. The function $f(\sigma)$ plotted expresses the current in units $\alpha d/\beta$ where α and d are also explained in the text.

We have here discussed a homogeneous earth. However, another case is evidently obtained if the conductivity of the earth is locally different from its large-scale value at a particular section of a power system. Such a local inhomogeneity probably has no essential effect on the magnitude of the original total current induced in the earth. But as above, it affects the division of the current between the network and the ground. Thus the GIC increases with decreasing local earth conductivity. This conclusion is in accordance with the statement given in [25] about large GICs at regions of highly resistive igneous rocks in North America.

5. GIC RECORDINGS IN FINLAND

The geomagnetically induced current flowing in the earthing lead of the neutral(s) of the Huutokoski 400 kV transformer(s) (number 10 in Fig. 3) has been monitored since 1977. The choice of Huutokoski was somewhat arbitrary based on the structure of the 400 kV grid at that time, e.g. [6, Fig. 3]. The recordings are mainly digital 10 second mean values taken for one to two hours after the current exceeds a certain threshold value (about 40 A today). The recording system is shown in Fig. 7, which concerns the situation since 1985 when the second main transformer and the neutral point reactor were installed. Analogue recordings of GICs were also performed at Pikkarala (number 13 in Fig. 3), Ulvila (8) and Pirttikoski (15) for a short time about ten years ago. They, however, did not yield as useful data as the Huutokoski recordings.

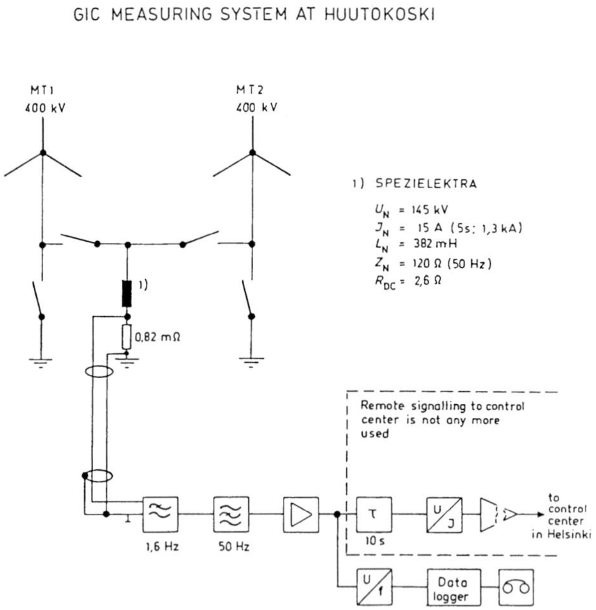

Fig. 7: Recording system of the geomagnetically induced current (GIC) in the earthing lead of the neutrals of the Huutokoski 400 kV transformers (number 10 in Fig. 3). The figure was obtained from the Imatran Voima Oy power company.

An example of the measured GIC at Huutokoski is shown in Fig. 4. The largest GIC observed at Huutokoski is 165 A recorded on January 4, 1979, as a 10 second mean value. The simultaneous current at Pirttikoski was only 34 A. It is interesting to note that the geomagnetic disturbance occurring then was not at all of the most intense kind. It means that geomagnetic induction effects on a power system cannot be determined by considering only the strength of magnetic disturbances, but the spatial distribution of the disturbance over the network is obviously also essential.

Unfortunately, the Huutokoski recordings were out of operation during the geomagnetic storm of July 1982, which is the greatest storm since starting the Huutokoski measurements. Theoretical calculations indicate that the GIC certainly exceeded 165 A and was probably a few hundreds of amperes then. Theoretical estimates also show that the Huutokoski currents do not at all represent the highest GICs in the Finnish 400 kV system, which is mostly due to the neutral point reactor (cf. Section 3). For example during the large geomagnetic storm in March 1989 mentioned in Section 1, the Huutokoski GIC was only 45 A.

To find out the occurrence of geomagnetically induced currents in the Finnish power system in greater detail and to verify and improve previous theoretical calculations, the Imatran Voima Oy power company and the Finnish Meteorological Institute are planning a GIC measuring campaign to be performed in 1991–92 when great geomagnetic disturbances are expectable due to the sunspot maximum. Induced currents would then be recorded in a manner similar to the present Huutokoski measurement at about four sites in Finland simultaneously. We will also try to record the GIC in a line by observing the magnetic field in its vicinity. It is further planned that some GIC recording sites could be situated in Sweden because of the galvanic connection between the 400 kV power systems of the two countries. On the other hand it was concluded in Section 3 that the flow of GICs from one part of the grid to another is not very significant.

6. CONCLUDING REMARKS

By Faraday's law of induction, an electric field is always connected to a time variation of a magnetic field. This is the simple physical basis of the occurrence of geomagnetically induced currents (GICs) in power transmission grids and other man-made conductors during geomagnetic disturbances. From the viewpoint of power systems, GICs must be regarded as d.c. currents. When flowing through transformers, GICs may saturate them and so cause harm to the power system. Risk areas in this respect are auroral latitudes, as for example North America and the Nordic countries, and disturbances have occurred during magnetic storms.

However until now, no significant GIC inconveniences have been observed in the Finnish power system (yet). Research work on GICs in the Finnish 400 kV grid containing both measurements and theoretical calculations has been done since 1977. The measurements mainly concern the current flowing through the Huutokoski transformer(s) into the earth (number 10 in Fig. 3), and the highest GIC recorded there is 165 A. Theoretical estimates indicate that currents of even several hundreds of amperes are possible in the Finnish 400 kV system during great geomagnetic storms.

We are planning to arrange a special GIC measuring campaign in the Finnish and Swedish 400 kV power system in 1991–92 , and record GICs at different sites simultaneously then. The purpose of the measurements combined to geomagnetic data and theoretical studies is to find out reliably the possible GIC magnitudes, which will be the basis for the engineering estimation of induction risk in the power system and for the planning of countermeasures if necessary. The campaign will also yield significant geophysical information applicable to studies of currents in the ionosphere and magnetosphere and to studies of the structure of the earth.

7. ACKNOWLEDGEMENT

We would like to express our gratitude to the Imatran Voima Oy power company for collaboration and support in studies of geomagnetically induced currents in the Finnish power transmission system for many years. We also wish to thank Imatran Voima Oy for providing us with network data necessary in theoretical GIC estimation.

8. REFERENCES

1. *Joselyn JA (1989)* The Great Geomagnetic Storm of 13 March 1989. *Geomagnetic Applications Bulletin*, International Association of Geomagnetism and Aeronomy, No. 3: 2

2. *Albertson VD. Van Baelen JA (1970)* Electric and Magnetic Fields at the Earth's Surface Due to Auroral Currents. *IEEE Trans. on Power Apparatus and Systems* **PAS-89**, 4: 578–584

3. *Albertson VD. Thorson JM. Jr. Clayton RE. Tripathy SC (1973)* Solar-induced-currents in power systems: cause and effects. *IEEE Trans. on Power Apparatus and Systems* **PAS-92**, 2: 471–477

4. *Lanzerotti LJ* (ed) *(1979)* Impacts of ionospheric/magnetospheric processes on terrestrial science and technology. In: Lanzerotti LJ. Kennel CF. Parker EN (eds) *Solar System Plasma Physics* **Vol. III**, North-Holland Publishing Company, Chapter III.2.1: 319–363

5. *Campbell WH (1980)* Observation of electric currents in the Alaska oil pipeline resulting from auroral electrojet current sources. *Geophys. J. R. astr. Soc.* **61**: 437–449

6. *Pirjola R (1983)* Induction in power transmission lines during geomagnetic disturbances. *Space Sci. Rev.* **35**, 2: 185–193

7. *Lundby S. Chapel BE. Boteler DH. Watanabe T. Horita RE (1985)* Occurrence Frequency of Geomagnetically Induced Currents: A Case Study on a B. C. Hydro 500 kV Power Line. *J. Geomag. Geoelectr.* **37**: 1097–1114

8. *Boteler DH. Cookson MJ (1986)* Telluric currents and their effects on pipelines in the Cook Strait region of New Zealand. *Materials Performance*, March 1986: 27–32

9. *Markström A. Enbom A (1989)* The effect of solar flares on geomagnetic disturbances. *Pre-graduate thesis*, School of Electrical and Computer Engineering, Chalmers University of Technology, Gothenburg, Sweden, 41 pp.

10. *Medford LV. Lanzerotti LJ. Kraus JS. Maclennan CG (1989)* Transatlantic earth potential variations during the March 1989 magnetic storms. *Geophys. Res. Lett.* **16**, 10: 1145–1148

11. *Prescott GB (1860)* Observations made at Boston, Mass., and its vicinity. Art. XII. – The Great Auroral Exhibition of August 28th to September 4th, 1859. – 2d Article, *Am. J. Sci. Arts* **29**: 92–95

12. *Cagniard L (1953)* Basic theory of the magneto-telluric method of geophysical prospecting. *Geophysics* **18**, 3: 605–635

13. *Pirjola R (1982)* Electromagnetic induction in the earth by a plane wave or by fields of line currents harmonic in time and space. *Geophysica* **18**, 1-2: 1–161

14. *Pirjola R. Viljanen A (1989)* On geomagnetically-induced currents in the Finnish 400 kV power system by an auroral electrojet current. *IEEE Trans. on Power Delivery* **4**, 2: 1239–1245

15. *Lühr H. Klöcker N. Thürey S (1984)* Ground-based observations of a very intense substorm-related pulsation event. *J. Geophys.* **55**, 1: 41–53

16. *Janhunen P. Viljanen A (*submitted*)* Application of conformal mapping to two-dimensional conductivity structures with nonuniform primary sources. *Geophysical Journal*, 7 pp. (manuscript)

17. *Lehtinen M. Pirjola R (1985)* Currents produced in earthed conductor networks by geomagnetically-induced electric fields. *Annales Geophysicae* **3**, 4: 479–484

18. *Sanders R (1961)* Effects of Terrestrial Electromagnetic Storms on Wireline Com-

munications. *IRE Trans. on Communications Systems* **V.C.S.-9**, 4: 367–377

19. *Viljanen A. Pirjola R (1989)* Statistics on Geomagnetically-Induced Currents in the Finnish 400 kV Power System Based on Recordings of Geomagnetic Variations. *J. Geomag. Geoelectr.* **41**: 411–420

20. *Pirjola R (1989)* Geomagnetically induced currents in the Finnish 400 kV power transmission system. *Phys. Earth Planet. Inter.* **53,** 3-4: 214–220

21. *Viljanen A (1987)* Geomagneettisesti indusoituvat virrat Suomen 400 kV sähköverkossa (= Geomagnetically-induced currents (GICs) in the Finnish 400 kV power grid). (Finnish with an English abstract), *R&D-notes*, Imatran Voima Oy, IVO–B–07/87, 75 pp.

22. *Pirjola R. Lehtinen M (1985)* Currents produced in the Finnish 400 kV power transmission grid and in the Finnish natural gas pipeline by geomagnetically-induced electric fields. *Annales Geophysicae* **3**, 4: 485–491

23. *Kaufman AA. Keller GV (1981)* The magnetotelluric sounding method. *Methods in Geochemistry and Geophysics* **15**, Elsevier Scientific Publishing Company, 595 pp.

24. *Sunde ED (1949) Earth Conduction Effects in Transmission Systems.* D. Van Nostrand Company, Inc.

25. *Douglas J (1989)* A Storm From the Sun. *EPRI Journal*, Electric Power Research Institute, July/August 1989: 14–21

6.5 Electromagnetic Field Caused by an Auroral Electrojet Current System Model

R.J. PIRJOLA and L.V.T. HÄKKINEN

Finnish Meteorological Institute, Geophysics Department, SF-00101 Helsinki, Finland

A theoretical model of the auroral electrojet current system is discussed. The model contains a horizontal sheet electrojet of a finite length with a Gaussian transverse distribution of the current, and geomagnetic-field-aligned currents that make the divergence of the total current vanish. The time dependence is harmonic, and a harmonic spatial dependence along the electrojet is assumed. Induction in the earth is included by applying a homogeneous or four-layer model. Exact formulas expressing the electromagnetic field on the earth's surface are used. Examples of the magnitudes of the field components are given. Particular attention is paid to the magnetotelluric apparent resistivity near an electrojet system, and it is indicated that incorrect results of the structure of the earth may be obtained if source effects are neglected in a magnetotelluric study.

1. INTRODUCTION

The most simple theoretical model of an electrojet current flowing in the auroral or equatorial ionosphere and causing time variations of the geomagnetic field is a horizontal infinitely long straight line current situated above the flat surface of the earth considered a half-space, e.g. [1], [2]. The line current model has further been developed to sheet currents in [3]–[6].

A different generalization of the basic line current was presented in [7] and [8] when a space dependence in the direction of the current flow was considered. In fact, a similar model was also included in [9]. Although the space dependence in the direction of the current makes an extension of the original line current, it also means that the current is divergent and hence implies a geophysical drawback: a divergent current is located in a poorly-conducting medium (the air) causing significant accumulation of electric charge. In reality, all accumulation is practically prevented due to the high (anisotropic) conductivity of the ionosphere. An improved model in which additional vertical currents, i.e. almost geomagnetic-field-aligned currents at auroral latitudes, start upwards from the electrojet and make the total current divergence-free was discussed in [10] and [11]. The formulation was not limited to a line current electrojet in [10], but a sheet current was considered.

Finally, an even more general theoretical model of an electrojet system was presented in [12]. There the intensity and direction of the sheet current electrojet are arbitrary functions of space and time, and the additional field-aligned currents, which are located above the electrojet and make the divergence of the total current vanish, have any direction. Formulas giving the total electromagnetic field (primary + induced) on the surface of a layered-half-space earth were derived. They are rigorous including even the effects of displacement currents, which, on the other hand, are insignificant at geomagnetic frequencies.

Numerical results of the electromagnetic field on the surface of a homogeneous earth caused by a line current electrojet of a finite length with field-aligned currents were shown in [13]. The model is a special case of the general model of [12]. Particular attention was paid to the apparent resistivity parameter commonly determined in electromagnetic induction studies of the earth's structure. It was demonstrated that its straightforward use may lead to incorrect conclusions in the vicinity of an electrojet. Electromagnetic investigations of the earth thus require modelling of the electrojet system.

In this paper besides a homogeneous earth, a four-layer earth model approximately valid for northern Scandinavia is also considered, and numerical calculations are extended to sheet current electrojets. Special discussion of the apparent resistivity is included again.

A final aim of electrojet system modelling works is to find out real (and not only equivalent) currents flowing in the ionosphere and magnetosphere. Theoretical results shown e.g. in this paper may be compared to electric and magnetic measurements on the earth's surface. Magnetic data usable then have been collected by the so-called EISCAT magnetometer cross in northern Scandinavia since 1982 [13], [14]. In this paper, also all three electric components are studied, so the possibility of using recorded electric data, too, would be valuable. No comparison of measured electric or magnetic data to theoretical calculations is, however, included in this paper, which only contains modelling.

2. MODEL

The earth is described by a half-space with a flat surface, which implies a local nature of the study instead of a global one. Let the earth's surface be the xy-plane of a right-handed Cartesian coordinate system fixed to the earth with the z-axis pointing downwards. The electrojet is a horizontal infinitely thin sheet current of a finite length L at a height h. We further assume that the current flow is parallel to the y-axis (unit vector $\hat{\mathbf{e}}_y$) and that the current has a harmonic time dependence (angular freguency ω), a harmonic y-dependence along the electrojet (wave number q) and a Gaussian spatial distribution in the x-direction (width parameter s). If the (complex) coefficient implying the magnitude and phase of the electrojet current is denoted by J, the current density can thus be mathematically expressed as follows:

$$\mathbf{j}(x,y,z,t) = \frac{J}{s\sqrt{2\pi}} e^{-\frac{x^2}{2s^2}} e^{i(\omega t - qy)} \delta(z+h)\left(\Theta\left(y+\frac{L}{2}\right) - \Theta\left(y-\frac{L}{2}\right)\right)\hat{\mathbf{e}}_y \qquad (1)$$

where δ and Θ denote the Dirac delta and Heaviside step function, respectively. The use of a harmonic time dependence should, of course, be understood as a Fourier transform of an arbitrary time dependence, and the parameters appearing in formula (1) may be regarded as functions of ω. When s goes to zero a delta function dependence on x is obtained, and so equation (1) reduces to the line current electrojet discussed in [13].

Equation (1) means that the z-axis goes through the centre of the electrojet and that the electrojet has a fixed location and position with respect to the earth. The latter fact implies that the rotation of the earth below the electrojet current system is not taken into account. Hence the model is applicable only to phenomena sufficiently rapid, i.e. the current given by equation (1) shall represent the Fourier transform of a time variation fast enough. To be more precise and assuming that the electrojet has a geographic east-west orientation, the requirement of sufficient rapidity is obviously satisfied if the duration of the current at each point of the electrojet is much smaller than L/v where v is the rotation velocity of the earth at the area investigated. The requirement given in [13] and concerning the frequencies thus seems to be too restrictive. At auroral latitudes L/v is roughly obtained in seconds when L in kilometres is multiplied by five.

The electric charge is assumed to be zero at the electrojet system, so the sheet current of formula (1) is accompanied by field-aligned currents which start upwards from the sheet and make the total current non-divergent. The direction of the field-aligned currents is given by the unit vector

$$\hat{\mathbf{e}}_{\mathbf{B}} = \cos I \cos D \; \hat{\mathbf{e}}_{\mathbf{x}} + \cos I \sin D \; \hat{\mathbf{e}}_{\mathbf{y}} + \sin I \; \hat{\mathbf{e}}_{\mathbf{z}} \qquad (2)$$

where I and D are the inclination and declination of the geomagnetic field. For brevity, we here neglect the mathematical expression of the field-aligned currents and just refer to [12] or, if the field-aligned currents are vertical, to [10].

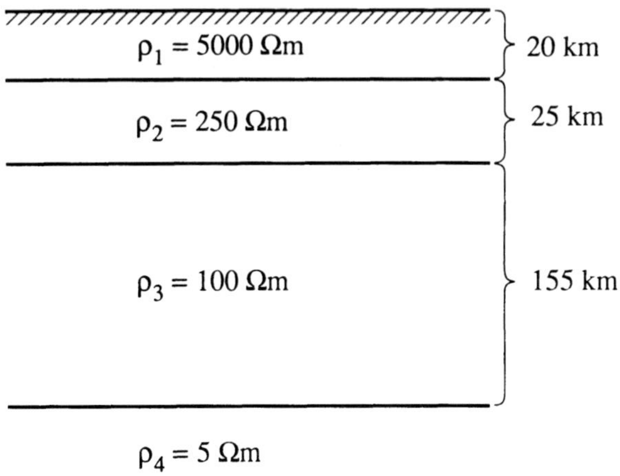

Fig. 1: Earth model consisting of four horizontal layers with different resistivities and thicknesses [15]. The model is approximately valid for northern Scandinavia.

In this paper, we assume that the earth is either homogeneous with a resistivity ρ = 100 Ωm or has the four-layer structure shown in Fig. 1 and approximately valid for northern Scandinavia [15]. The value 100 Ωm mentioned may be considered as representative if the model of Fig. 1 is described by a single resistivity. On the other hand however, a homogeneous earth model is just a very rough approximation, and it may possibly always yield quantitatively unreliable results, so an exact choice of the value of a homogeneous resistivity is not extremely important.

Several parameters have constant values throughout this paper as follows: J = 100 kA, h = 100 km, $\sigma_0 = 2\cdot 10^{-14}\ \Omega^{-1}\mathrm{m}^{-1}$, $\mu_0 = 4\pi\cdot 10^{-7}\ \mathrm{VsA^{-1}m^{-1}}$, $\varepsilon_0 = 8.854\cdot 10^{-12}\ \mathrm{AsV^{-1}m^{-1}}$, $\mu = \mu_0$, $\varepsilon = 5\varepsilon_0$. The symbols σ_0, μ_0 and ε_0 refer to the conductivity (i.e. the inverse of the resistivity), the permeability and the permittivity of the half-space z<0 (the air), and as usual μ_0 and ε_0 have their vacuum values. The permeability and permittivity of the earth are denoted by μ and ε. It should be noted that ε and ε_0 clearly do not have any effect in practice at geomagnetic frequencies since displacement currents are negligible. Also σ_0 is obviously unessential. The fact that J is real and positive simply fixes the origin of the time axis at the particular frequency considered, and of course, it does not mean any restriction. Choices of the values of the parameters are discussed in [8].

To avoid complicated mathematical formulas in this paper, we ignore the expressions of the electromagnetic field on the earth's surface and refer to the general theory given in [12]. If the field-aligned currents are vertical and the earth is homogeneous (instead of the four-layer model) [10] may also be referred to. The expressions are double integrals from $-\infty$ to $+\infty$ over two wave numbers q_1 and q_2, i.e. inverse Fourier transforms from the q_1q_2-plane to the xy-plane.

3. *NUMERICAL CALCULATION OF THE INTEGRALS*

Numerical calculation of the double integrals yielding the electromagnetic field on the earth's surface is laborious and requires an efficient computer.

We adopted the inverse h^{-1} of the height of the electrojet and the inverse L^{-1} of its length as scale parameters in the q_1 and q_2 integrations, respectively. These scale intervals were then divided into a certain number of subintervals (~4 being a good choice). The Gauss' Integration Formula of the eighth order was applied to each subinterval [16, pp. 887–888 and 916]. The q_1 and q_2 integrations were extended over a sufficient number (~90 for q_1 and ~130 for q_2 being good choices) of subintervals in the regions where the integrals get their largest contributions.

We also used a different numerical method based on the Fast Hankel Transform [17] in the q_1 integration, see also [11], and on the Extended Simpson's Rule [16, p. 886] in the q_2 integration.

Results obtained by the two integration methods seem to be practically equal, which justifies sufficient numerical accuracy in the computations.

4. ELECTROMAGNETIC FIELD ON THE EARTH'S SURFACE

Although several parameters are assumed to have fixed values in the present calculations (Section 2), there are still many free variables in the expressions of the electromagnetic field on the earth's surface caused by the electrojet current system: the length L and width s of the electrojet, the period T $(= 2\pi/\omega)$, the wave number q, the direction of the field-aligned currents (the inclination I and the declination D), the point of observation (x,y). Also two different earth models are considered. A detailed investigation of the effects of all parameters not yet fixed would be laborious, and so we just discuss examples.

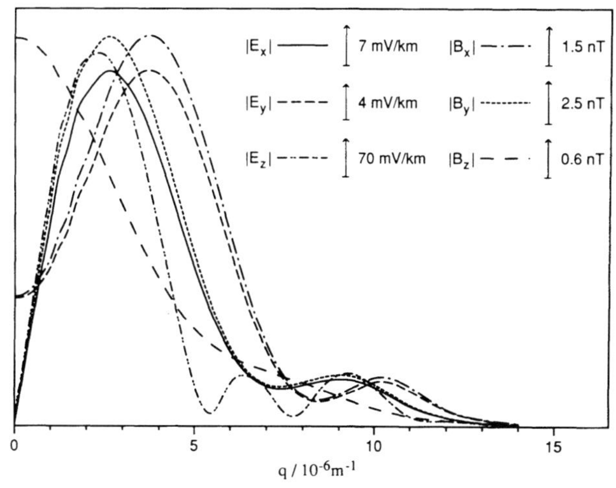

Fig. 2: Amplitudes of the total electromagnetic field on the earth's surface caused by an electrojet current system consisting of a line current of a finite length (= 2000 km) and of geomagnetic-field-aligned currents (inclination 77.5°, declination 6.0°). The point of observation has a horizontal distance coordinate x = 500 km measured from the electrojet in the perpendicular direction, and in the parallel direction, the point is located at the centre of the electrojet. The field amplitudes are given as functions of the wave number q expressing a harmonic space dependence of the current along the electrojet. The harmonic time period considered is 3 min, and the earth has the structure shown in Fig. 1.

Let T and L be equal to 3 min and 2000 km and consider a line current, i.e. $s = 0$. We further assume that I and D are 77.5° and 6.0°, respectively, which corresponds to the situation in northern Scandinavia [18]. Figure 2 then shows the absolute values (i.e. the amplitudes) of the electromagnetic field components as functions of q at the point x = 500 km, y = 0 on the surface of the four-layer earth of Fig. 1. As in all examples presented in [8], [11] and [13], the components also now seem to approach zero at large values of q.

Figure 2 is comparable to Fig. 2 of [11]. Essential differences also exist in the models, since an electrojet of an infinite length with vertical currents (i.e. I = 90°) above a homogeneous earth (ρ = 100 Ωm) is discussed in [11]. Anyway, the two figures are generally speaking rather similar. But in Fig. 2 of this paper the maxima occur at larger values of q than in Fig. 2 of [11], and the decrease of the amplitudes with q shows some oscillation now. Although the accuracy of Fig. 2 of this paper is not sufficient, it should be noted that $|E_x|$, $|E_z|$ and $|B_y|$ are not exactly zero when q = 0 in the figure, which is due to the non-symmetric field-aligned currents ($I \neq 90°$, $D \neq 0°$).

The largest differences between Figs 2 of this paper and of [11] concern $|E_z|$ whose maximum value is now much smaller and is achieved at a clearly higher value of q. Further, it is interesting to note that $|E_z|$ of Fig. 2 of this paper has local maxima approximately at $q = 4\pi/L$ and $6\pi/L$ and a minimum at $q = 5\pi/L$. This suggests that the differences between the present results and those in [11] are essentially due to the finite length of the electrojet, and the layered structure of the earth and the non-vertical inclination are less significant. It may be pointed out here that an "oscillation" is also seen in $|E_z|$ in Fig. 5 of [13] for a different electrojet of a finite length.

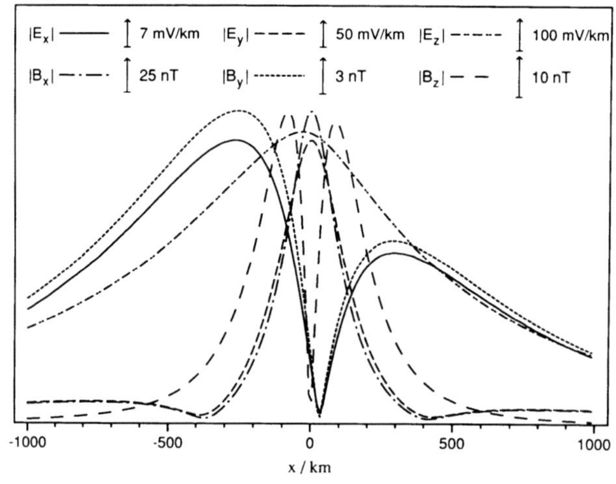

Fig. 3: This figure exactly corresponds to Fig. 2, but now the amplitudes of the electromagnetic field components are presented as functions of the horizontal distance coordinate x of the point of observation. The wave number is fixed to $q = 10^{-6}\mathrm{m}^{-1}$.

Figure 3 exactly corresponds to Fig. 2, but in Fig. 3 the amplitudes of the field components are depicted as functions of x with a constant value of q (= 10^{-6} m^{-1}). The fact that the components are not symmetric with respect to x = 0 is obviously a result of the field-aligned currents that are also not symmetric.

The effect of the width of the electrojet sheet is demonstrated in Table 1 in which T, L and q are equal to 20 s, 2000 km and $10^{-6}\mathrm{m}^{-1}$, and the point of observation is located at x = 100 km, y = 0 on the surface of the homogeneous earth. The field-aligned currents are assumed to be vertical. Generally speaking, the amplitudes of the field components seem to decrease as the width parameter s of the electrojet increases. In some components, however, a slight increase is observed when a line current is replaced by a sheet current of s = 50 km.

Table 1: Amplitudes of the total electromagnetic field on the earth's surface caused by an electrojet current system consisting of a sheet current electrojet of a finite length (= 2000 km) and of vertical currents. The point of observation has a horizontal distance coordinate x = 100 km measured from the centre of the electrojet in the perpendicular direction, and in the parallel direction, the point is located at the centre of the electrojet. The width parameter s having four different values gives the Gaussian transverse distribution of the current in the electrojet sheet. The harmonic time period considered is 20 s, and the wave number expressing a harmonic space dependence of the current along the electrojet equals $10^{-6}m^{-1}$. The earth is homogeneous with a resistivity 100 Ωm.

s (km)	0 (line)	50	200	500
$\lvert E_x \rvert$ (mV/km)	106	96	46	14
$\lvert E_y \rvert$ (mV/km)	756	800	553	230
$\lvert E_z \rvert$ (mV/km)	7792	7741	7075	5534
$\lvert B_x \rvert$ (nT)	150	160	111	46
$\lvert B_y \rvert$ (nT)	21	19	9	3
$\lvert B_z \rvert$ (nT)	28	24	4	0.4

5. *APPARENT RESISTIVITY*

The apparent resistivity is a parameter commonly used in electromagnetic induction studies of the structure of the earth, e.g. [19, pp.75–104]. It expresses an equivalent homogeneous-earth resistivity for a more complex real earth being a function of the angular frequency ω of the harmonic time dependence. Normally the so-called plane wave assumption is made, which is satisfied when the variations of the primary field caused by sources in the ionosphere and magnetosphere can be neglected in the horizontal direction, i.e. the primary field may be considered laterally uniform. Expressed more quantitatively, the spatial variation of the primary field should be insignificant at a horizontal distance equal to the skin depth of the earth. In this respect, a vertically propagating plane wave, of course, constitutes the ideal case. For the plane wave assumption, the apparent resistivity has the formula

$$\rho_a = \frac{\mu_0^2}{\mu\omega}\left(\frac{E}{B}\right)^2 \tag{3}$$

where E and B are the amplitudes of two perpendicular horizontal components of the electric and magnetic fields, respectively, at a point on the earth's surface and at the particular angular frequency ω. (Note that μ was assumed to be equal to μ_0 in Section 2 making formula (3) simpler.)

Possible source effects violating the plane wave assumption and making equation (3) questionable have been largely discussed since the basic magnetotelluric paper [20], e.g. [2], [22]–[26]. Now we will also demonstrate that the use of formula (3) may lead to incorrect conclusions of the structure of the earth in the presence of the electrojet system discussed in this paper. Besides a homogeneous earth, the earth model presented in [15] and shown in Fig. 1 will again be used below, but we want to emphasize that the purpose is not to make the particular Scandinavian four-layer model questionable. It has certainly been deduced carefully by selecting suitable events in the electromagnetic analysis. Here the model is just adopted as an example.

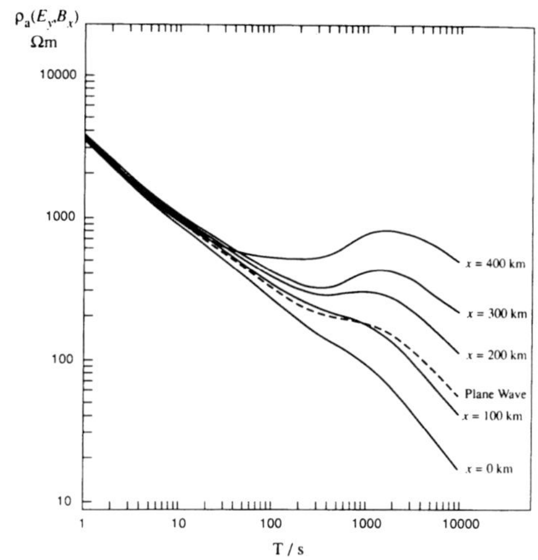

Fig. 4: Apparent resistivity calculated from the plane wave formula (3) using the parallel horizontal electric component E_y and perpendicular horizontal magnetic component B_x produced by an electrojet current system on the surface of the four-layer earth of Fig. 1. The electrojet system consists of a line current electrojet of a finite length (= 2000 km) and of geomagnetic-field-aligned currents (inclination 77.5°, declination 6.0°). The wave number expressing a harmonic space dependence of the current along the electrojet equals 10^{-6}m^{-1}. The apparent resistivity is depicted as a function of the period T of the harmonic time oscillation at different values of the horizontal distance coordinate x of the point of observation. The coordinate x is measured from the electrojet in the perpendicular direction, and in the parallel direction the point of observation is located at the centre of the electrojet.

As in Fig. 2, let us consider a line current electrojet with L = 2000 km supplemented by field-aligned currents having the inclination I = 77.5° and the declination D = 6.0°. The wave number q now equals 10^{-6}m^{-1}. The structure of the earth is shown in Fig. 1, and the points of observation are always located on the earth's surface at y = 0 while x is variable. The apparent resistivity calculated from formula (3) with $E = |E_y|$ and $B = |B_x|$ is then depicted as a function of the period of the harmonic time dependence at different x-values in Fig. 4. The curve labelled by "Plane Wave" shows the correct apparent resistivity obtained for the model of Fig. 1 when the plane wave assumption is satisfied. Deviations from the "Plane Wave" curve are caused by source effects, which according to Fig. 4 become significant at periods larger than ~100 s, implying the possibility of incorrect interpretation of the earth's conductivity. The behaviour of ρ_a as a function of x seems to be more systematic in Fig. 4 than in Fig. 6 of [13]. This is probably due to the fact that x remains far below the "resonance" value q^{-1} = 1000 km in Fig. 4.

In Fig. 5, the apparent resistivities obtained from E_y and B_x using a line current (s = 0) or a Gaussian sheet current (s = 50 km) electrojet models are compared as functions of the period of the harmonic time oscillation. The length of the electrojet equals L = 500 km, and the field-aligned currents are vertical. The wave number has the value q =

$5 \cdot 10^{-6}$ m^{-1}, and the point of observation lies at x = 200 km, y = 0 on the earth's surface. The earth is assumed to be homogeneous with ρ = 100 Ωm, so naturally, the correct plane wave apparent resistivity equals this value as shown in Fig. 5. We observe that deviations of the apparent resistivity from the plane wave value start at lower periods in the sheet current than in the line current case, but at larger periods the line current curve becomes worse while at even larger periods the sheet current curve again differs more. The apparent resistivities seem to be above the correct plane wave value in Fig. 5. It should be noted that the figure obviously represents an exceptional case due to the "resonance" $x = q^{-1}$.

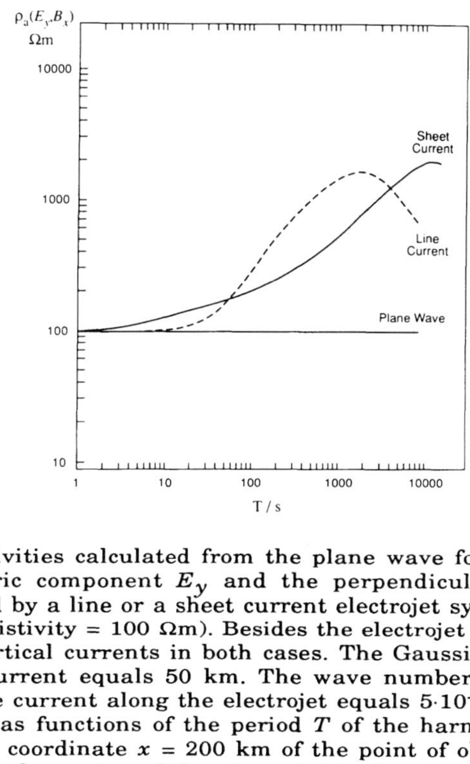

Fig. 5: Apparent resistivities calculated from the plane wave formula (3) using the parallel horizontal electric component E_y and the perpendicular horizontal magnetic component B_x produced by a line or a sheet current electrojet system on the surface of a homogeneous earth (resistivity = 100 Ωm). Besides the electrojet whose length is 500 km, the system contains vertical currents in both cases. The Gaussian transverse width parameter of the sheet current equals 50 km. The wave number expressing a harmonic space dependence of the current along the electrojet equals $5 \cdot 10^{-6}$m^{-1}.The apparent resistivities are depicted as functions of the period T of the harmonic time oscillation at the horizontal distance coordinate x = 200 km of the point of observation. The coordinate x is measured from the centre of the electrojet in the perpendicular direction, and in the parallel direction, the point of observation is located at the centre of the electrojet.

Figure 6 exactly corresponds to Fig.5, but in Fig. 6 the apparent resistivities are depicted as functions of x with a constant value of T (= 1000 s). At small values of x both apparent resistivities are too small compared to the correct plane wave resistivity. Between x = ~200 to 300 km they are clearly too large but when $x \geq$ ~350 km they approach the correct value. The latter fact is due to the decrease of source effects at long distances from the electrojet system.

The apparent resistivities calculated from E_y and B_x and from E_x and B_y are in general not equal. It is demonstrated by an example in Fig. 7 which corresponds to the sheet current case of Fig. 5 except for the value of the x-coordinate of the point of observation being now 250 km. We see that $\rho_a(E_x,B_y)$ differs less than $\rho_a(E_y,B_x)$ from the plane wave apparent resistivity for periods smaller than ~5000 s. This observation seems to be in agreement with a conclusion given in [25].

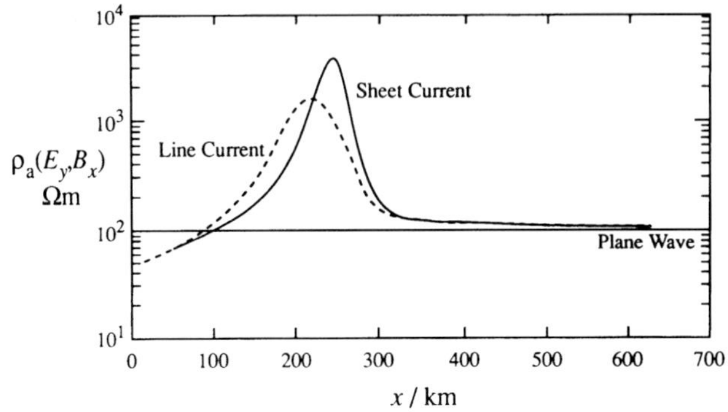

Fig. 6: This figure exactly corresponds to Fig. 5, but now the apparent resistivities are presented as functions of the horizontal distance coordinate x of the point of observation. The period is fixed to $T = 1000$ s.

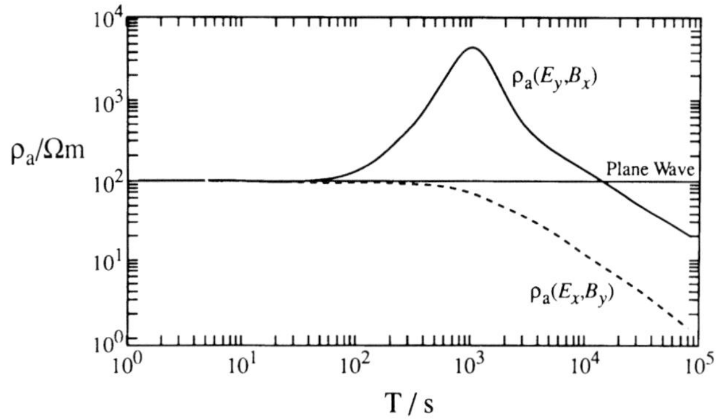

Fig. 7: Apparent resistivities calculated from the plane wave formula (3) using the parallel horizontal electric component E_y and the perpendicular horizontal magnetic component B_x or using the perpendicular horizontal electric component E_x and the parallel horizontal magnetic component B_y. In both cases the situation corresponds to the sheet current case of Fig. 5 except for the value of the horizontal distance coordinate of the point of observation being now $x = 250$ km. The apparent resistivities are depicted as functions of the period T of the harmonic time oscillation.

6. CONCLUDING REMARKS

This paper deals with the electromagnetic field produced by a complex electrojet current system on the surface of a homogeneous or four-layer earth. The system consists of a horizontal sheet electrojet of a finite length and having a Gaussian transverse distribution of the current, and of geomagnetic-field-aligned currents which make the total current of the electrojet system non-divergent. The results presented here were calculated utilizing general rigorous formulas.

The electrojet system model discussed is a basis in studies of real (and not only equivalent) currents flowing in the ionosphere and magnetosphere. Suitable magnetic data which may be compared with model results in the future have been obtained by the EISCAT magnetometer cross in northern Norway and Finland since 1982.

In this paper, the magnetotelluric apparent resistivity is discussed in particular. It is shown that the vicinity of an electrojet current system can seriously distort the resistivity values calculated from the usual formula based on the plane wave assumption.

7. REFERENCES

1. *Albertson VD. Van Baelen JA (1970)* Electric and Magnetic Fields at the Earth's Surface Due to Auroral Currents. *IEEE Trans. on Power Apparatus and Systems* **PAS-89**, 4: 578–584

2. *Hermance JF. Peltier WR (1970)* Magnetotelluric Fields of a Line Current. *J. Geophys. Res.* **75**, 17: 3351–3356

3. *Peltier WR. Hermance JF (1971)* Magnetotelluric Fields of a Gaussian Electrojet. *Canadian Journal of Earth Sciences* **8**: 338–346

4. *Hibbs RD. Jr. Jones FW (1973)* Electromagnetic Induction in the Earth by a Non-Symmetric Non-Uniform Source *J. Geomag. Geoelectr.* **25**: 75–86

5. *Hibbs RD. Jones FW (1976)* The Calculation of the Electromagnetic Fields of a Sheet Current Source with Arbitrary Spatial Intensity Distribution over a Layered Half Space – I. *Geophys. J. R. astr. Soc.* **46**: 433–452

6. *Hibbs RD. Jones FW (1976)* The Calculation of the Electromagnetic Fields of a Sheet Current Source with Arbitrary Spatial Intensity Distribution over a Layered Half Space – II. The Computer Program and its Application. *Geophys. J. R. astr. Soc.* **46**: 453–465

7. *Pirjola R (1982)* Electromagnetic induction in the earth by a plane wave or by fields of line currents harmonic in time and space. *Geophysica* **18**, 1-2: 1–161

8. *Pirjola R (1985)* Electromagnetic induction in the earth by a line current harmonic in time and space. *Geophysica* **21**, 2: 127–143

9. *Wait JR (1980)* Electromagnetic surface impedance for a layered earth for general excitation. *Radio Science* **15**, 1: 129–134

10. *Lehto K (1984)* Electromagnetic field caused by a three-dimensional time- and space-dependent electrojet current system. *Geophysica* **20**, 2: 105–121

11. *Pirjola R (1985)* Electromagnetic induction in the earth by an electrojet current system harmonic in time and space. *Geophysica* **21**, 2: 145–159

12. *Häkkinen L. Pirjola R (1986)* Calculation of electric and magnetic fields due to an electrojet current system above a layered earth. *Geophysica* **22**, 1-2: 31–44

13. *Häkkinen L. Pirjola R. Sucksdorff C (1990)* EISCAT magnetometer cross and

theoretical studies connected with the electrojet current system. To be published in *Geophysica* **25**, 1, 15 pp. (manuscript)

14. *Lühr H. Klöcker N. Thürey S (1984)* Ground-based observations of a very intense substorm-related pulsation event. *J. Geophys.* **55**, 1: 41–53

15. *Jones AG. Olafsdottir B. Tiikkainen J (1983)* Geomagnetic induction studies in Scandinavia. III. Magnetotelluric observations. *J. Geophys.* **54**: 35–50

16. *Abramowitz M. Stegun IA* (eds) *(1970) Handbook of mathematical functions with formulas, graphs and mathematical tables.* Ninth printing, Dover Publications, Inc., New York, 1046 pp.

17. *Johansen HK. Sørensen K (1979)* Fast Hankel Transforms. *Geophysical Prospecting* **27**: 876–901

18. *Sucksdorff C. Eerola K. Sallinen T (1986)* Magnetic charts of Finland for 1985.0. *Geophysical Publications*, Finnish Meteorological Institute, No. 1, 58 pp.

19. *Kaufman AA. Keller GV (1981)* The magnetotelluric sounding method. *Methods in Geochemistry and Geophysics* **15**, Elsevier Scientific Publishing Company, 595 pp.

20. *Cagniard L. (1953)* Basic theory of the magneto-telluric method of geophysical prospecting. *Geophysics* **18**, 3: 605–635

21. *Wait JR (1954)* On the relation between telluric currents and the earth's magnetic field. *Geophysics* **19**, 2: 281–289

22. *Price AT (1962)* The Theory of Magnetotelluric Methods When the Source Field Is Considered. *J. Geophys. Res.*, **67**, 5: 1907–1918

23. *Dmitriev VI. Berdichevsky MN (1979)* The Fundamental Model of Magnetotelluric Sounding. *Proc. IEEE* **67**, 7: 1034–1044

24. *Mareschal M (1986)* Modelling of natural sources of magnetospheric origin in the interpretation of regional induction studies: a review. *Surveys in Geophysics* **8**, 3: 261–300

25. *Osipova IL (1983)* Consideration of the influence of ionospheric source field structure on deep electromagnetic sounding results. In: Hjelt SE. Vanyan LL (eds) *The development of the deep geoelectric model of the Baltic Shield. Part 1. Numerical methods*, University of Oulu, Finland, Department of Geophysics, Report no. 7: 8–38

26. *Osipova IL. Hjelt SE. Vanyan LL (1989)* Source field problems in northern parts of the Baltic Shield. *Phys. Earth Planet. Inter.* **53**, 3-4: 337–342

6.6 On the Estimation of the Ionospheric Exit Regions of Magnetospheric VLF Radio Waves by the Use of Wave Energy Distribution in Wave Number Space

S. SHIMAKURA[1] and M. HAYAKAWA[2]

Department of Electrical Engineering, Chiba University, Chiba, 260 Japan[1] and Research Institute of Atmospherics, Nagoya University, Aichi, 442 Japan[2]

Abstract. A new ground-based direction finding technique is proposed to estimate the ionospheric exit regions of magnetospheric VLF waves on the basis of wave energy distribution of those waves in wave number space which is evaluated by means of maximum entropy concept. This method is essentially different from the previous direction finding techniques by which the wave normal direction of an equivalent plane wave can be estimated. The solution by maximum entropy method, however, depends on the inversion model, or on the wave polarization included in the integration kernels, because TE- and TM-mode waves can independently propagate in free space. In this paper, the effectiveness and dependence of maximum entropy solutions on the inversion model are discussed for the investigation of the ionospheric transmission and magnetospheric propagation mechanisms of magnetospheric VLF waves.

1. Introduction

The ground-based direction finding (DF) to locate the ionospheric exit regions of magnetospheric VLF/ELF radio waves is extensively used to investigate the generation and propagation mechanisms of those waves and also to study the dynamics of magnetospheric plasma [1,2,3,4,5,7, 11,16,17]. Several DF systems have so far been proposed; e. g., (1) goniometer triangulation [1], (2) field-analysis method [12,14,15] and (3) Poynting vector method [8,23]. In the principle of all these systems, it is assumed that a single plane wave with a definite wave normal direction is incident on the direction finder. Hence, each of these DF systems yields only the average direction of magnetospheric VLF waves, even though the wave energy of magnetospheric VLF waves is distributed in the lower ionosphere. For this situation, it is highly required to derive the wave energy distribution at the base of the ionosphere by using the ground-based multiple field components. In this paper a new direction finding technique for the magnetospheric VLF radio waves observed on the ground is proposed on the basis of the wave energy distribution of those waves in wave number space. This method is essentially different from the previous methods influenced by the propagation characteristics of the waves and/or the source effects on the lower edge of the ionosphere. So, it is possible for us to estimate the ionospheric exit-region of magnetospheric VLF waves by using this method.

We utilize three electromagnetic field components observable on the ground; two horizontal magnetic field components (B_x and B_y) and a vertical electric field component (E_z), which are used to construct the auto- and cross-power spectra, or the elements of spectral matrix[20]. Those measured field components are understood as being the sum of a lot of independent elementary plane waves, and the integration kernels have been estimated. They are significantly different from the kernels in space plasma because two characteristic waves (TE and TM mode waves) can independently propagate in free space. Then the polarization (the ratio of magnetic field components of TE and TM modes) is included in the kernels. The inverse problem to deduce the wave energy distribution function at the base of the ionosphere is to apply the maximum entropy concept to the observed spectral matrix.

We use the simulated data to make certain of the usefulness of maximum entropy solution; we simulate a single wave source which is constructed by a large number of independent elementary plane waves whose central polarization is right-handed circular, with its amplitude and argument having Gaussian distribution. The maximum entropy solution on the basis of the right-handed circular polarization model for the above simulated data have yielded that the present method is very effective to deduce the wave energy distribution function. Further simulation studies have been done for the wave sources with central polarization deviated from circular polarization, in order to make sure that the stable and smooth solution with small prediction error can be obtained by using right-handed elliptical polarization model, and the dependence of solution on polarization model is shown by changing the number of eigenvalues used in the inversion. And the application of the present method to actually observed data (narrow band whistler signals around 5 kHz) has yielded that the results obtained are in good agreement with those deduced from our previous field-analysis DF method. It is possible for us to discuss the details of the ionospheric transmission and magnetospheric propagation mechanisms of VLF waves by the use of the present method.

2. Estimation of wave energy distribution of VLF waves observed on the ground

The ground can be considered as a perfect conductor for electromagnetic waves at VLF [14,15], and hence the observable field components on the ground are two horizontal magnetic field components (B_x, B_y) and a vertical electric field component (E_z), which are used to construct the auto- and cross-power spectra, or the elements of the spectral matrix [20]. Those measured field components are understood as being the sum of a lot of plane waves with appropriate amplitude, with different propagation direction and with their phase relationship being quite random. The properties of such an incoherent and random wave fields can be described statistically. Storey and Lefeuvre [21] have proposed to characterize it in terms of a function called the wave distribution function (WDF) which specifies how the wave energy density is distributed with respect to wave number space $\vec{k}$, or to the space of angular frequency ω and direction (θ, ϕ) of propagation. When we normalize the field components as $B_1 = B_x/2$, $B_2 = B_y/2$ and $B_3 = E_z/2c$ (c is light velocity of free space), the element of spectral matrix $S_{ij}(\omega)$ at an angular frequency ω, which is given by ensemble mean $< B_i B_j >$ composed of the observable field components on the ground, is related to the wave distribution function $F(\omega, \theta, \phi)$ by the following relation [18,20],

$$S_{ij}(\omega) = \frac{\pi}{2} \int_{-1}^{1} \int_{0}^{2\pi} a_{ij}(\omega, \theta, \phi; p) \, F(\omega, \theta, \phi) \, d(\cos\theta) \, d\phi \quad (i, j = 1, 2, 3) \tag{1}$$

where $a_{ij}(\omega, \theta, \phi; p)$ is the element of integration kernels, $p = B_{\|}/B_{\perp}$ the polarization of the down-coming wave, which is defined by the ratio of the magnetic field components corresponding to TE $(B_{\|}, E_{\perp})$ and TM $(B_{\perp}, E_{\|})$ modes, respectively (see Fig. 1), and $F(\omega, \theta, \phi)$ is the WDF to be determined. This is an inversion problem in which we will estimate this WDF by means of the maximum entropy concept.

The wave mode in a magnetoplasma can be specified by magneto-ionic theory such that whistler mode is known to be only possible propagation mode if we consider VLF waves, but in free space TE and TM modes are two possible propagation modes and they can propagate independently of each other. So, the element of kernels $a_{ij}(\omega, \theta, \phi; p)$ depends on the inversion model or on the polarization p. The element of the kernels $a_{ij}(\omega, \theta, \phi; p)$ can be described as follows [18,20]:

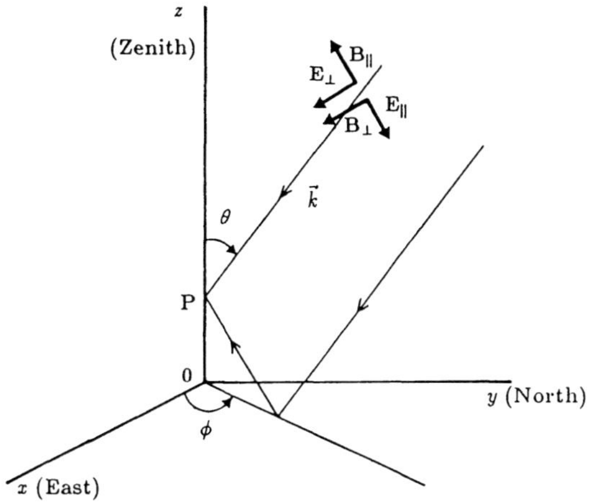

Fig. 1 The electromagnetic field components at the observing point P.

$$\left.\begin{aligned}
a_{11} &= A(p)\,[\sin^2\phi + pp^*\cos^2\theta\,\cos^2\phi - (p+p^*)\cos\theta\,\sin\phi\,\cos\phi] \\
a_{12} &= a_{21}^* = A(p)\,[-(1-pp^*\cos^2\theta)\sin\phi\,\cos\phi + (p\,\cos^2\phi - p^*\,\sin^2\phi)\cos\theta] \\
a_{13} &= a_{31}^* = A(p)\,[-(\sin\phi - p\,\cos\theta\,\cos\phi)\sin\theta] \\
a_{22} &= A(p)\,[\cos^2\phi + pp^*\cos^2\theta\,\sin^2\phi + (p+p^*)\cos\theta\,\sin\phi\,\cos\phi] \\
a_{23} &= a_{32}^* = A(p)\,[(\cos\phi + p\,\cos\theta\,\sin\phi)\sin\theta] \\
a_{33} &= A(p)\,\sin^2\theta
\end{aligned}\right\} \qquad (2)$$

where $A(p) = 2\mu_0/(1+pp*)$. In the derivation of these kernels, it is considered only the fact such that the waves with polarization p arrive at a direction finder. However, the observable waves on the ground would be constituted not only by the direct waves from the ionospheric exit regions but also by the multiply-reflected waves in the Earth-ionosphere waveguide. And each set of waves may have the proper polarization, corresponding to the direct waves and to the multiply-reflected waves. So, it may be considered that when the spectral matrix consist of only the direct waves, the polarization p of kernels is applied to be approximately right-handed circular, and when the multiply-reflected waves are dominant, the polarization is to be right-handed elliptical. The effect of multiple ray is evidenced by Strangeways [22] based on the point source assumption, to be negligible for the range where the field-analysis DF is effective. According to his result, the observable field components on the ground may be considered to be composed mainly of a set of direct waves radiated from the ionospheric exit region, though his point source assumption is a problem to be investigated.

3. Application of the maximum entropy solution to simulated data

3.1. Maximum entropy solution based on right-handed circular polarization model

The most important interest is whether we can obtain a stable maximum entropy solution with small prediction error for the ground-based data, and so we have considered the simplest model: a single wave source with the central polarization being right-handed circular which is composed of the sum of random plane waves with some deviations of polarization. Then it has been found that the inversion on the basis of the kernels with the exact right-handed circular polarization yields very satisfactory results for a single source model [19,20]. An example of the results is shown in

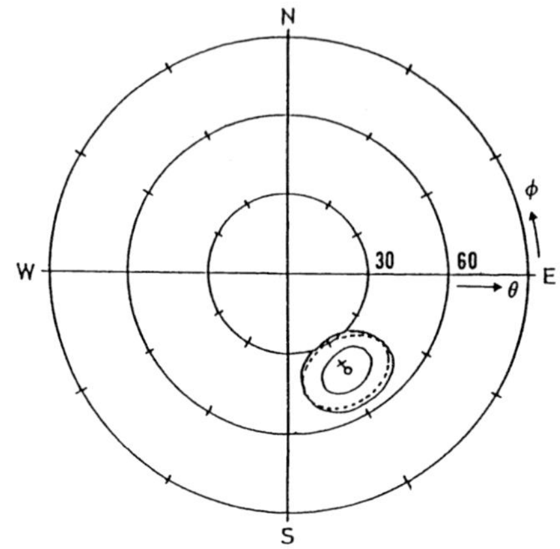

Fig. 2 An example of WDF solution estimated with right-handed circular polarization model: $Q = 0.11$ and $P_r = 0.039$. The broken line shows the level of assumed standard deviation of the source and the full line the levels of 75% and 50% of the maximum WDF indicated by the small circle. × is the direction of the equivalent plane wave.

Fig. 2: The broken line indicates the level of the assumed standard deviation of the source, and the full lines the level of 75% and 50% of the maximum of the WDF. The stability parameter Q and prediction parameter P_r are the quantities which indicate the acceptability of the WDF solution, and the solution with $Q \leq 1$ and $P_r \leq 9$ is stable one with small prediction error and can be considered to be acceptable for the ground observation [9,20]. And even if the source is composed of two peaks, it has been found that reasonable WDF solutions with small prediction error can be obtained by the present method [18,20]: The WDF becomes a reasonable ellipse if two sources are close to each other, and also is completely separated as two peaks when the sources are well-detached from each other, though the assumption of a single (equivalent) plane wave yields only the ineffective information in the simultaneous presence of a few ducts[24].

And also we have very useful relationship between the assumed source region and the scale of 50% level of the estimated WDF with the correlation coefficient of 0.94. When we define S_R as the scale of the estimated exit region and S_E as the source dimension, the following relationship are obtained [18,20]:

$$S_R = 1.54 S_E + 13.7 \tag{3}$$

where S_R indicates the distance of 50% level measured at the ionospheric base of 80 km in θ and ϕ directions, and S_E the standard deviation of the source. This high correlation implies that the WDF solutions will enable us to infer, with sufficient accuracy, the wave energy distribution of magnetospheric waves at the base of the ionosphere.

3.2. The dependence of WDF solution on inversion model

When whistler-mode wave is radiated into free space after leaving the ionosphere, the average polarization of its equivalent plane wave was found experimentally to be right-handed circular on the statistical sense [12,13,14]. Hence, we have so far considered the determination of WDF solutions based on the kernels with "right-handed circular polarization"[20]. However, it is not certain whether the right-handed circular polarization is of the most appropriate for the actually observed wave because the magnetospheric VLF waves radiated in free space can propagate with any polarization. It is plausible that the polarization of waves below the ionosphere may deviate from right-handed circular, depending on the conditions of the lower ionosphere and magnetospheric duct.

For a wave source constituted by the ensemble of elementary plane waves whose central polarization is somehow deviated from right-handed circular, i. e. elliptical, it is necessary to study the following four items; (1) the WDF inversions by using the right-handed circular polarization kernels, (2) the effect of larger standard deviations for the polarization of elementary plane waves as constituting the wave source, (3) the WDF inversions based on the elliptical polarization kernels, and (4) the effect of limiting the number, M, of eigenvalues. In this section, we study the WDF solutions estimated by using the inversion model with right-handed circular and elliptical polarizations. The $\ln|p|$ and $\arg(p)$ of elementary plane waves are all taken as Gaussian, with standard deviation of 0.5 and 30° ($\Delta p = (0.5, 30°)$), respectively. These quantities are acceptable from the observational results [12,13,14]. And the center of the source is taken as $(\bar{x}, \bar{y}) = (40\,\text{km}, -70\,\text{km})$, indicating $\theta = 45°$ and $\phi = -60°$. The spread of the source is 40 km (standard deviation = 20 km). Furthermore, the convergence constraint ε, defined as the ratio of the prediction error to the measuring error of the element of spectral matrix, is taken as being less than 2 [20].

We have used the following five wave sources for which the average of the argument of polarization of elementary plane waves constituting the source is −70°, −80°, −90°, −100° and −110°, with the average of the polarization ratio being fixed as 1.0. The WDF solutions are examined for the spectral matrices having the above polarization characteristics, by means of the maximum entropy concept. On some occasions when we derive the maximum entropy solution with using all the eigenvalues ($M = 9$), we cannot obtain the stable solution with small prediction error. Hence, in this sense, it is worthwhile to compare the WDF solutions for different M values, $M = 9$, 8, 7 and 6.

When adopting the right-handed circular polarization model, we could not obtain the stable solutions for two cases when the average of polarization argument of elementary plane waves is deviated by more than 20° from the inversion polarization model (−90°) such as −70° and −110°. The obtained solution for these cases is found to be a point, which is not a smooth distribution as expected from the maximum entropy concept [18]. This may suggest a tool of inferring an invaluable information on the polarization of observed magnetospheric waves with special reference to its inversion polarization model. For the deviation of about ±10° in the polarizations, we should expect an estimating error of about 10° both in θ and ϕ, and we obtain elongated, elliptical distributions, though smooth and stable. It must be recalled that such elliptical WDF solutions with sufficient stability and smoothness are already obtained for the case when a few neighbouring sources are present, but such elliptical WDF solutions can be resulted also from the difference in polarization between the source and the inversion model as in the present situation. On the contrary, even when the standard deviation of the polarization is relatively large, we can predict the original wave source distribution by the maximum entropy solution when the polarization of inversion model is the same as the source. This is shown in Fig. 3, the energy distribution tends to be estimated as being larger with decreasing the value of M, and we find no difference in the WDF solutions for $M = 9$ and 8.

In estimating the WDF from the estimated spectral matrix, the properties of the kernels a_{ij}'s are of great importance. For example, the kernels for the magnetospheric plasma are descried by the magneto-ionic theory, and when we use all the eigenvalues, the solution will be unstable because 3 among 36 are slightly linearly dependent [10,21]. On the other hand, the elements of kernels for free space are the function of polarization. For the circular polarization, the real part of the wave polarization is zero, resulting in the behaviour completely different from the elliptical polarization. The WDF solutions as inversions based on the model of the same elliptical polarization with the average polarization of the source, have been derived by changing the number of eigenvalues (M), in order to consider the effectiveness of elliptical polarization model. Figures 4, 5, 6 and 7 refer to the WDF solutions for the elliptical polarization with its argument, −70°, −80°, −100° and −110°, respectively. The convergence constraint ε is less than 2. The simulation studies have shown that even for elliptical polarization model, we are able to obtain the stable solution with small prediction

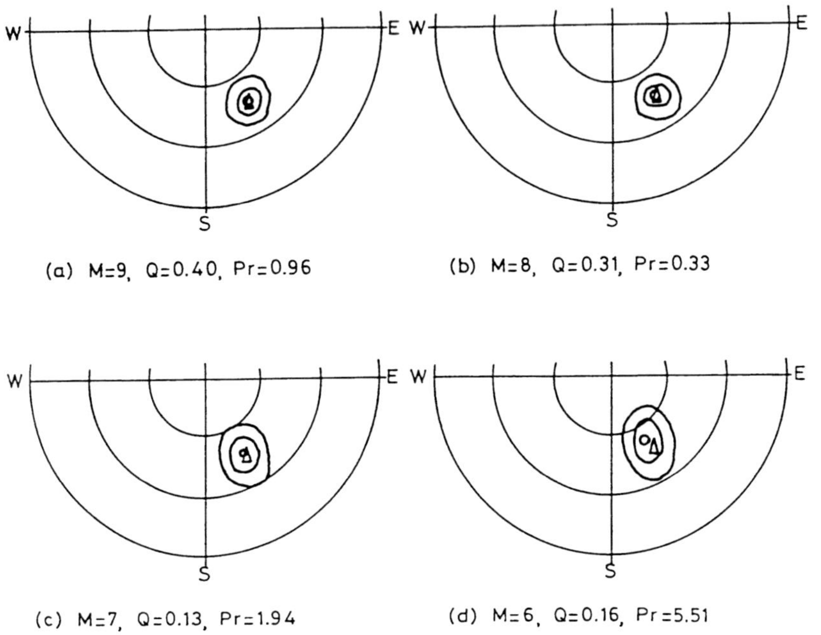

Fig. 3 The WDF solutions with right-handed polarization model for the wave source with the average of the polarization argument of elementary plane waves being $-90°$. The number, M, of eigenvalues is changed; (a) $M = 9$, (b) $M = 8$, (c) $M = 7$, (d) $M = 6$.

error for $M = 9$. On the other hand, we can obtain the stable solutions with smaller prediction errors even for $M = 8$, 7 and 6, but the center of WDF is deviated from the source center by about 10° both in incident and azimuthal angles except for $M = 8$ when the polarization angle is $-70°$ and $-80°$.

Considering that the polarization of actual whistler data is not so much departed from right-handed circular one and taking into account the simulation results for the elliptical polarization model, the wave distribution of magnetospheric radio waves whose exit region is relatively close to the observing station, can be estimated with sufficient accuracy by using the maximum entropy method. In the present simulations, we have changed only the argument of the polarization, and the characteristics of integration kernels and WDF solutions, when the real part of the polarization is not zero, are of the greatest concern. In this sense, it is verified that we can obtain the stable solutions with small prediction errors in the free space problem.

4. Application to actual data

We have applied the method to the actual data. Unfortunately, the data of VLF/ELF hiss, to which the present method may be most effective, are not available, but we use the data of whistlers observed at Yamaoka (geomag. lat. 25° N), Japan. In this experiment, Ohta *et al.* [12] have recorded the wave forms of three field components (B_x, B_y, E_z) whose center frequency is 5.0 kHz ($\Delta f = \pm 200$ Hz), which have been used for the automatic DF on the basis of field-analysis principle [12]. The enhanced daytime events on two successive days (15*th* and 16*th* January, 1982) have been studied based on the right-handed circular polarization model [20]. According to the analyzing results, the WDF solution for all events observed at 16:30 – 16:40, on 16*th* January, is much more stable compared with the case of 15*th*.

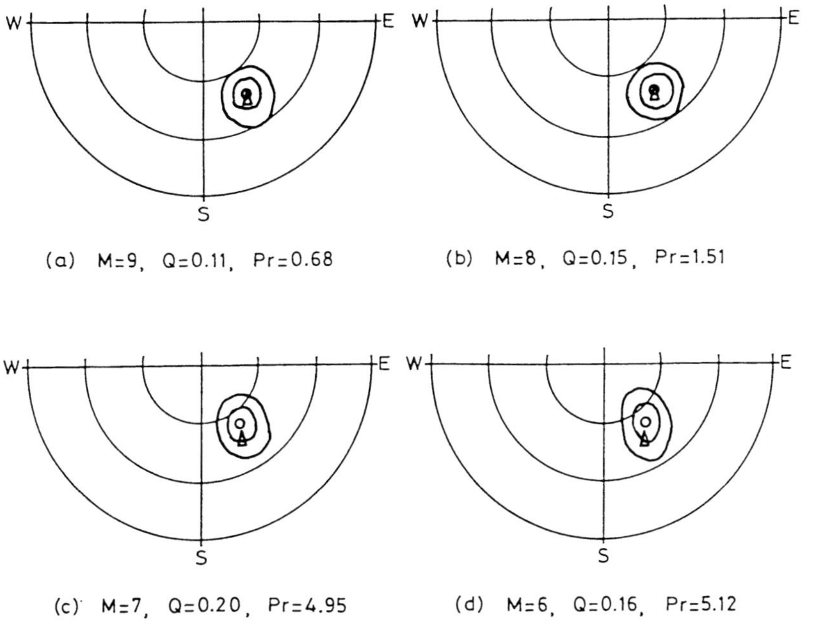

Fig. 4 The WDF solutions by using right-handed elliptical polarization model with the same polarization as the average of the argument being $-70°$: (a) $M = 9$, (b) $M = 8$, (c) $M = 7$, (d) $M = 6$.

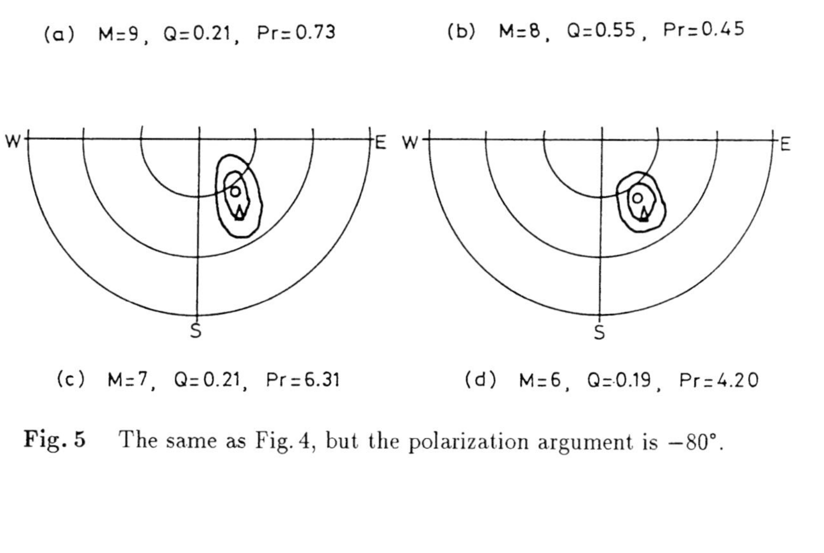

Fig. 5 The same as Fig. 4, but the polarization argument is $-80°$.

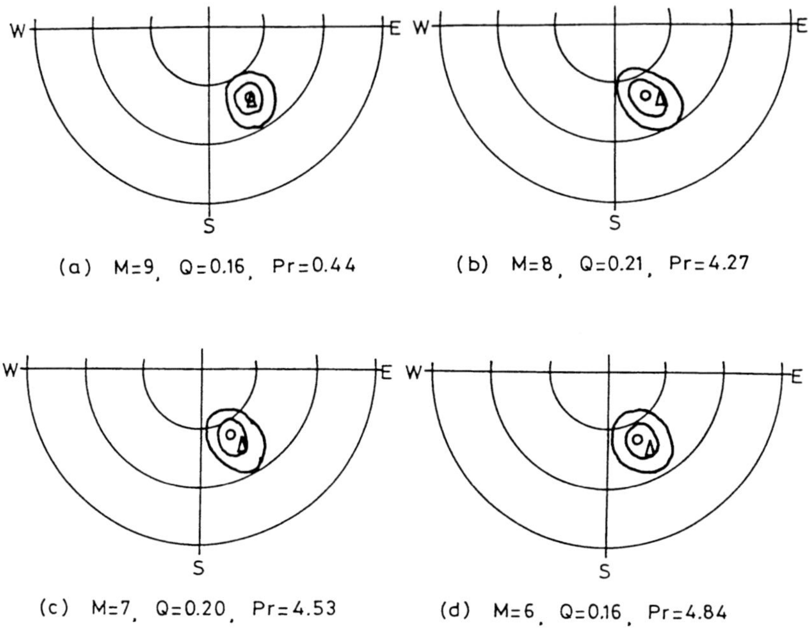

Fig. 6 The same as Fig. 4, but the polarization argument is $-100°$.

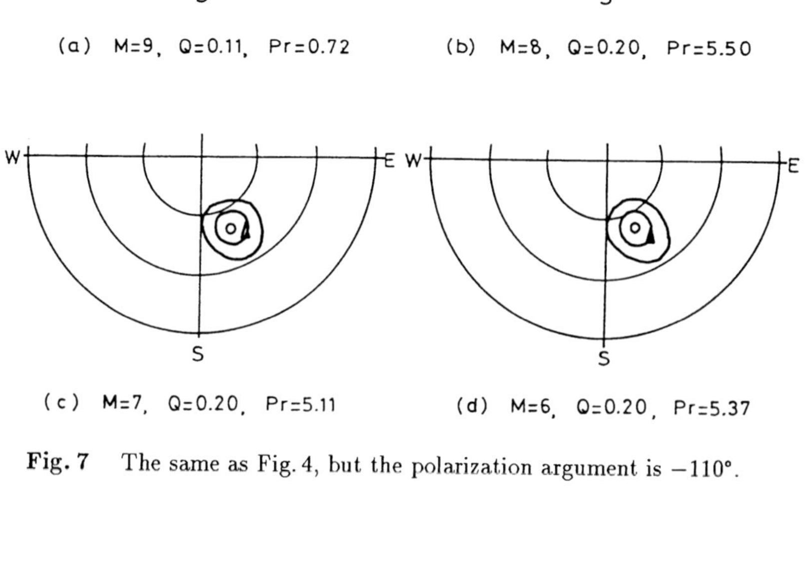

Fig. 7 The same as Fig. 4, but the polarization argument is $-110°$.

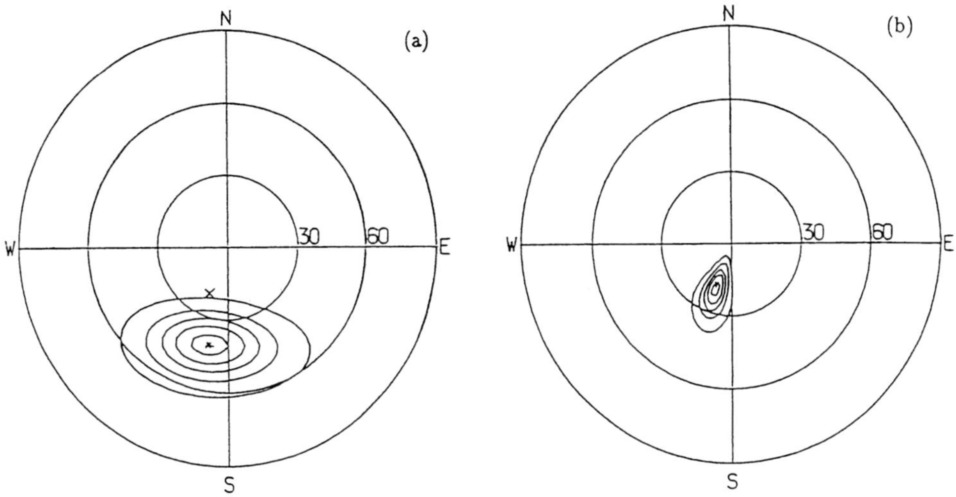

Fig. 8 WDF solutions of a daytime whistler observed at Yamaoka on 15*th* January, 1982. Panel (a) indicates the WDF solution obtained by right-handed circular polarization model; $Q = 2.34$ and $P_r = 5.72$. Panel (b) is the WDF of the same whistler by the right-handed elliptical model with the polarization of equivalent plane wave; $Q = 1.70$ and $P_r = 1.02$.

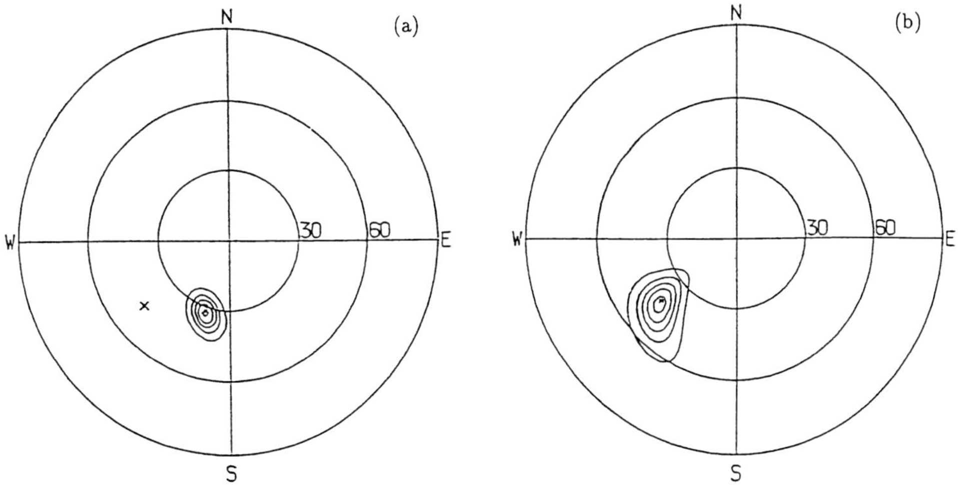

Fig. 9 WDF solutions of another whistler. Panel (a) indicates the WDF solution obtained by right-handed circular polarization model; $Q = 3.67$ and $P_r = 4.91$. (b) the WDF of the same whistler by right-handed elliptical polarization model; $Q = 0.626$ and $P_r = 0.985$.

Two examples of the WDF results on 15*th* January, 1982, are illustrated in Figs. 8 and 9. In these cases, we can not obtain the sufficiently accurate WDF solutions by right-handed circular polarization model. Figures 8(a) and 9(a) are the WDF results obtained by the right-handed circular polarization model, and panel (b) of these figures the WDF by the right-handed elliptic polarization model based on the polarization of equivalent plane wave determined by using only the imaginary part of spectral matrix. As both stability parameter Q and prediction parameter P_r, which indicate the acceptability of the WDF solution, are improved well enough, the WDF solutions is considered to be acceptable for the ground observation, though Q exceeds unity in Fig. 8(b), because this threshold, unity, has physical meaning only if the order of the error in the data is well estimated [9]. Comparing (a) with (b) in each figure, the changes of the shape of WDF, stability and prediction parameters might lead us to image the effect of the multiply reflected waves, with small incident angle, in the Earth-ionosphere waveguide. In conclusion, The present WDF method is more effective compared to the previous DF methods and also will provide us with further information concerned with the magnetospheric propagation (such as duct characteristics) and ionospheric transmission mechanism.

5. Conclusions

We have shown how to estimate the ionospheric exit regions of magnetospheric VLF waves, on the basis of two horizontal magnetic and a vertical electric field components, by the wave distribution function obtained as the maximum entropy solution which gives the wave energy distribution in wave number space. The wave distribution function depends on the inversion model because the integration kernels are the function of wave polarization. However, it is found that we can get a stable and smooth WDF solution with small prediction error by using suitable polarization model in the inversion with $M = 9$ or 8, and that the solution with direction error less than 10° may be obtained, except for its shape, even when $M < 8$. And it is shown that quite acceptable solution can be derived, for the actual data, by using the right-handed elliptical model with inferable polarization from equivalent plane wave.

In this paper we have not included the effect of multiple rays, but it may be possible for us to obtain some useful information on this effect by using the present method in the range where field-analysis DF method is effective, i. e., where the energy of direct waves is dominant.

Acknowledgement. The authors would like to express to Dr. K. Ohta of Chubu University for his useful discussion and supplying whistler data observed at Yamaoka.

References

[1] Bullough, K., and J. L. Sagredo, VLF goniometer observations at Halley Bay, Antarctica, I. The equipment and the measurement of signal bearing, Planet. Space Sci., **21**, 899, 1973.

[2] Carpenter, D. L., Fast fluctuations in the arrival bearing of magnetospherically propagating signals from the Siple, Antarctica, J. Geophys. Res., **85**, 4157, 1980.

[3] Hayakawa, M., Y. Tanaka, T. Okada and A. Iwai, Goniometric direction finding for low-latitude whistlers and their propagation mechanism, J. Geophys. Res., **86**, 6781, 1981.

[4] Hayakawa, M., T. Okada and A. Iwai, Direction finding of medium latitude whistlers and their propagation characteristics, J. Geophys. Res., **86**, 6939, 1981.

[5] Hayakawa, M., Y. Tanaka, A. Iwai, J. Ohtsu, L. R. O. Storey, C. Beghin and T. S. Jorgensen, Simultaneous spaced direction finding measurements for medium-latitude VLF/ELF emissions, Planet. Space Sci., **29**, 505, 1981.

[6] Hayakawa, M., Y. Tanaka, T. Okada and J. Ohtsu, Time scales for the formation, lifetime and decay of low latitude whistler ducts, Ann. Geophysicae, **1**, 515, 1983.

[7] Hayakawa, M., Y. Tanaka, S. S. Sazhin, T. Okada and K. Kurita, Characteristics of dawnside mid-latitude VLF emissions associated with substorms as deduced from the two-stationed direction finding measurements, Planet. Space Sci., **34**, 225, 1986.

[8] Leavitt, M. K., D. L. Carpenter, N. T. Seely, R. R. Padden and J. H. Doolittle, Initial results from a tracking direction finding receiver for whistler mode signals, J. Geophys. Res., **83**, 1601, 1978.

[9] Lefeuvre, F. and C. Delannoy, Analysis of random electromagnetic wave field by a maximum entropy method, Ann. Telecomm., **34**, 204, 1979.

[10] Lefeuvre, F., M. Parrot and C. Delannoy, Wave distribution functions estimation of VLF electromagnetic waves observed onboard GEOS 2, J. Geophys. Res., **86**, 2359, 1981.

[11] Lester, M. and A. J. Smith, Whistler duct structure and formation, Planet. Space Sci., **28**, 645, 1980.

[12] Ohta, K., M. Hayakawa and Y. Tanaka, Ducted propagation of daytime whistlers at low latitudes as deduced from the ground-based direction findings, J. Geophys. Res., **89**, 7557, 1984.

[13] Ohta, K. and H. Eguchi, The automatic observation of the polarization of whistlers at low latitudes, Trans. Inst. Electr. Comm. Engrs. Japan, **J69-B**, 967, 1986.

[14] Okada, T., A. Iwai and M. Hayakawa, Measurement of incident and azimuthal angles and the polarization of whistlers at low latitudes, Planet. Space Sci., **25**, 233, 1977.

[15] Okada, T., A. Iwai and M. Hayakawa, A new whistler direction finder, J. Atmos. Terr. Phys., **43**, 679, 1981.

[16] Rycroft, M. J., Whistlers and discrete VLF/ELF emissions, in "ELF-VLF Radio Wave Propagation" ed. by J. A. Holtet, p.317, D. Reidel, 1974.

[17] Sagredo, J. L. and K. Bullough, VLF goniometer observation at Halley Bay, Antarctica, II. Magnetospheric structure deduced from whistler observations, Planet. Space Sci., **21**, 913, 1973.

[18] Shimakura, S., Study on signal analysis and estimation of the ionospheric exit regions of whistler waves, Doctoral thesis, Nagoya University, 1987.

[19] Shimakura, S. and M. Hayakawa, On the estimation of wave energy distribution of magnetospheric VLF waves at the ionospheric base with ground-based multiple electromagnetic field components, Proc. Chapman Conf. on Plasma Waves and Instabilities in Magnetospheres and at Comets, Sendai, p.238, 1987.

[20] Shimakura, S., H. Suzuki, M. Hayakawa, F. Lefeuvre and D. Lagoutte, On the estimation of wave energy distribution of magnetospheric VLF waves at the ionospheric base with ground-based multiple electromagnetic field components, to be submitted to Planet. Space Sci., 1990.

[21] Storey, L. R. O. and F. Lefeuvre, Analysis of a wave field in a magnetoplasma, 1. The direct problem, Geophys. J. Roy. Astron. Soc., **56**, 255, 1979.

[22] Strangeways, H. J., Systematic errors in VLF direction finding of whistler ducts-I, J. Atmos. Terr. Phys., **42**, 995, 1980.

[23] Tsuruda, K. and K. Hayashi, Direction finding technique for elliptically polarized whistlers, J. Atmos. Terr. Phys., **37**, 1193, 1975.

[24] Tsuruda, K. and M. Ikeda, Comparison of three different types of VLF direction finding techniques, J. Geophys. Res., **84**, 5325, 1979.

6.7 Radio Sky Mapping from Satellites at Very Low Frequencies

L.R.O. Storey

National Space Science Data Center (Code 630.2), NASA Goddard Space Flight Center, Greenbelt, MD 20771, USA

Wave Distribution Function (WDF) Analysis is a procedure for making sky maps of the sources of natural electromagnetic waves in space plasmas, given local measurements of some or all of the three magnetic and three electric field components. The work that still needs to be done on this subject includes solving basic methodological problems, translating the solutions into efficient algorithms, and embodying the algorithms in computer software. One important scientific use of WDF analysis is to identify the mode of origin of plasmaspheric hiss. Suitable data for this purpose are likely to come from the Japanese satellite Akebono (EXOS-D).

1. INTRODUCTION

Wave Distribution Function (WDF) analysis is a generalized form of radio direction finding. As input, it accepts *in situ* data on electromagnetic (EM) plasma waves in space, specifically data in the form of simultaneous measurements of several or all of their six (three electric and three magnetic) field components. Its output consists of *sky maps,* showing, at a given frequency and for a given EM plasma wave mode, how the intensity of the waves varies as a function of their direction of arrival at the point of observation; this function is the WDF. Its methodology was developed originally in France, in the early 1970s, by the present author in association with F. Lefeuvre, who had it as a thesis assignment [1–3]. Lefeuvre received his State Doctorate degree from the University of Orleans in 1977, and then continued working in the same area. With his French colleagues, and later in association with some Japanese workers, he analyzed data from the European GEOS-1 and GEOS-2 satellites, from the NASA satellite ISEE-1, and from the Soviet satellite Aureol-3. The outcome of their sustained efforts is a large body of published work, referenced in §6.1, which has established the value of WDF analysis for the experimental study of EM wave phenomena in space plasmas.

Plasmaspheric extra-low-frequency or very-low-frequency (ELF/VLF) hiss is a phenomenon that has been much studied in this way, because of its importance in the dynamics of Earth's radiation belts. It occurs widely and frequently, and has been observed off and on for a quarter of a century, yet its mode of origin is still not fully understood. Its random, continuous character makes it well suited for WDF analysis.

Section 2 of this paper describes the present state of published knowledge about WDF analysis (§2.1) and plasmaspheric hiss (§2.2), explaining why further research is needed on both fronts. Briefly, the methodology of WDF analysis needs improvement because, at present, too many data are being wasted. As for plasmaspheric hiss, the problem of its origin remains unsolved because the right data are not yet available, though some progress is being made with the data in hand.

These topics were investigated by the author, at Stanford University, during the period from February 1984 to November 1986. Some possible causes of the waste of data were identified, and solutions to this problem found, but time did not then permit the development of new software based upon them; this work , which is described in §3.1, is now being continued at the National Space Science Data Center (NSSDC). Meanwhile, in a collaboration with Lefeuvre and others, a study has been made of plasmaspheric hiss data from the ISEE-1 satellite, analyzed with existing software. The results from this study, and also from some related work performed by R.A. Helliwell and his colleagues at Stanford, are described in §3.2. Section 3 also lists the improvements that are still needed in the methodology of WDF analysis, and explains how it could then be applied to solving the persistent problem of the origin of plasmaspheric hiss.

2. PRESENT KNOWLEDGE

2.1. WDF Analysis

The aim of methodological research on WDF analysis is to devise procedures for making radio sky maps at VLF with the best possible directional resolution: the difficulty is that the resolution offered by a single antenna is very poor indeed. The resolution is determined by the beam width, and an antenna of characteristic dimension d receiving EM radiation of wavelength λ has a beam width of the order of λ/d, provided that $\lambda \ll d$. However, this condition cannot be met at VLF. At 10 kHz, for instance, the free-space wavelength is 30 km, and although the wavelengths in the magnetospheric plasma are substantially smaller, they are still much greater than the dimensions of any antenna that can be deployed from a satellite. Hence all such antennas, whether electric or magnetic, behave like Hertzian dipoles and have the same simple cosine-squared reception pattern, offering little in the way of resolution.

The way around this difficulty is to collect all the available relevant information and make the best possible use of it. This means that, to begin with, accurate measurements should be made of as many as possible of the six components of the EM wave field, namely the three electric components and the three magnetic ones. The static magnetic field vector should be measured as well, together with the properties of the ambient plasma: the electron concentration is essential, the ion composition highly desirable. Theoretical knowledge about the waves, in particular as to how their polarization depends on their direction of propagation, should also be brought to bear. A simple but very helpful fact is that the WDF cannot be negative. Finding out how best to exploit this set of information is the task of methodology.

At the start of any WDF analysis, there is a preliminary step of data reduction. Since the electric and magnetic field data are random, or assumed so, only their statistical properties are of interest. Hence the first step is to use the waveforms of the N (≤ 6) measured components of the EM field to estimate the N^2 power spectra, *i.e.*, the N purely real auto-spectra together with the real and imaginary parts of the $N(N-1)/2$ complex cross-spectra, that characterize their second-order statistics; this operation greatly reduces the amount of data. It is convenient to regard these spectra as the elements of a Hermitian $N \times N$ *spectral matrix*. At any given frequency the N^2 values of the spectral matrix elements are the actual input data for the WDF analysis.

The earliest work on the methodology of WDF analysis solved the *direct problem* of how to calculate the spectral matrix, given the WDF [1, 6, 7]. There then remained the *inverse problem* of estimating the WDF from the spectral matrix, which is much more difficult.

The initial approach was by model-fitting, which consists of adopting, for the WDF, some analytic model governed by a finite number of adjustable parameters [2]. The unique set of parameter values that fits the data best can be found by nonlinear optimization, using the principle of maximum likelihood. Unfortunately this simple procedure is unsatisfactory because, in general, there is no way of telling from the data whether or not the model is appropriate.

There is, however, an exception to this rule, which occurs when all of the waves observed at the spacecraft appear to be arriving from just a few discrete directions, typically one or two. Moreover, it is possible to tell from the data whether this state of affairs exists. Only in this special case, which is quite common, has the model-fitting method of WDF analysis so far found practical use [4, 9].

In the general case, the inverse problem is more difficult to solve. The solution favored by Lefeuvre and his colleagues was based on the principle of maximum entropy, which had been used previously to solve similar problems in signal theory, image processing, geophysics, and many other fields. Development of the maximum-entropy method for WDF analysis, and some initial application of it to multi-component ELF/VLF wave data from the GEOS-1 satellite, provided Lefeuvre with the material for his thesis [3]. For its large-scale application he was assisted by C. Delannoy, a professional software engineer and author of several books on computer programming. The first description of their method in the open literature was published in 1979 by Lefeuvre and Delannoy jointly [5], and their GEOS-1 data analyses were published two years later [8]. At the same time they undertook to improve the method by experimenting with the solution of analogous inverse problems in one dimension instead of two [10, 11]. Finally they embodied it in a large Fortran 77 software package called MAXENTWDF, which is now in the public domain [24].

Figure 1 indicates succinctly how the analysis proceeds. The maximum-entropy model is fitted to the input data by an iterative process of nonlinear optimization. Sometimes the iteration fails to converge to a solution, in which case the data are rejected. Even when a solution is obtained, it may have to be rejected because it is too unstable, *i.e.*, too sensitive to small errors in the data. Solutions for the WDF that pass the test for stability can be plotted as sky maps.

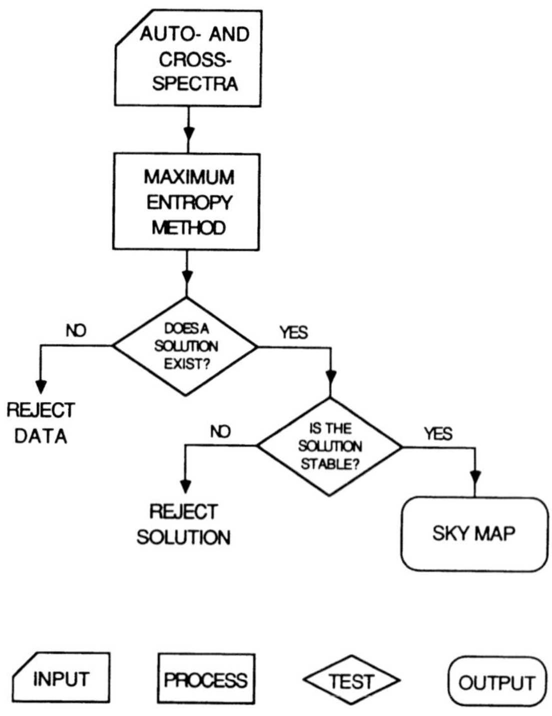

Fig. 1. Existing maximum-entropy method of WDF analysis.

Other persons have contributed to the development of WDF analysis. Recently a Swedish physicist, K. Rönnmark, and his collaborators have refined the definition of the WDF, and have shown that its spatio-temporal variations in a weakly inhomogeneous magnetoplasma are governed by a transport equation analogous to the Vlasov equation for particle distribution functions [29, 30]. Turning their attention to the methodology, they have developed a variant of the maximum-entropy method in which the maximization is performed by "simulated annealing," a process used in statistical mechanics and in operations research [31, 58]. This work has been undertaken with a view to analyzing data from satellites of the Viking series, in the Swedish national space program.

Of all the satellites designed to make multi-component measurements of ELF/VLF wave fields, the one best equipped for this purpose is the Japanese satellite Akebono (EXOS-D), launched in February 1989. The ELF/VLF wave instrument for Akebono was supplied by I. Kimura. It measures three components of the wave magnetic field and two components of the wave electric field, in the frequency range from 100 Hz to 12.7 kHz. The electric components are sensed by two crossed dipole antennas, each 60 m tip-to-tip. A unique feature of the instrument is that it can measure the vector impedances of the electric antennas; these data are used to correct for the significant attenuation occurring between the antennas and the preamplifiers, thus improving the accuracy of the electric field measurements.

A number of satellites now being built will also be equipped for multi-component ELF/VLF wave measurements. In scheduled order of launch, they include the Auroral Probe satellite in the Soviet INTERBALL Program, the four Cluster satellites in the International Solar-Terrestrial Physics (ISTP) program, and the Polar satellite in the NASA Global Geospace Science (GGS) program. The persons responsible for the multi-component wave measurements in these three programs are F. Lefeuvre, N. Cornilleau-Wehrlin, and D.A. Gurnett respectively.

Maximum-entropy WDF analysis of wave data from past or current satellites has already provided the subject matter for a number of papers, of which Lefeuvre and his colleagues are the principal authors [3, 8, 12, 14, 17–19, 22, 23]. They have also written several papers in collaboration with a Japanese group at the University of Nagoya [16, 20, 21, 25–27], and have collaborated briefly with US [14, 18] and Soviet [19] workers. The next sub-section reports the impact that some of this work has had on current understanding of plasmaspheric hiss.

2.2. Plasmaspheric Hiss

The ELF/VLF emission known as plasmaspheric hiss is a widespread and almost permanent feature of the natural EM wave field in Earth's plasmasphere [38, 39]. It is noise-like, seeming structureless when displayed on a frequency-time spectrogram, although at times it is associated with structured wave phenomena such as chorus. It covers a band of frequencies from a few hundred hertz to a few kilohertz. The upper cut-off frequency is always well below the local electron gyrofrequency [36, 37, 42]. On the other hand, most of the hiss band is above the local proton gyrofrequency, in a frequency range where only the whistler mode exists. Thus plasmaspheric hiss consists of EM waves propagated in the whistler mode.

Plasmaspheric hiss is widely believed to be produced by the Doppler-shifted electron gyroresonance instability (or *whistler instability* for short), the immediate source of free energy being the energetic (tens of keV) electrons trapped in Earth's radiation belts; the anisotropy of their distribution in pitch angle makes the whistler mode unstable [34]. The theoretical growth rate of the waves is largest in the magnetic equatorial plane, just below the plasmapause, so this region is often presumed to be their source, from which they spread out to fill the plasmasphere.

The first comprehensive theory of the origin of plasmaspheric hiss was put forward by Kennel and Petschek (K & P) in a classic paper that has influenced all subsequent thinking in the field [35, 40]. They recognized that the whistler instability is convective, not absolute, which means that although the unstable region around the magnetic equator can amplify whistler-mode waves that traverse it, such waves cannot be generated there spontaneously. Moreover, they estimated that the amplification resulting from a single pass through this region is insufficient to raise the waves from their thermal level to the levels observed for the hiss. Multiple passes are needed, which would occur by propagation back and forth between the two hemispheres along the magnetic lines of force, with reflections from the ionosphere or lower magnetosphere at either end. This process of feedback by propagation would convert the amplifying region into an oscillatory one, generating waves at levels limited by quasilinear modification of the distribution function of the energetic electrons. In sum, according to K & P, the hiss is produced by the plasmasphere acting as a laser.

The main difficulty with this theory concerns the direction of propagation of the waves (*i.e.*, the wave normal direction), which governs their spatial growth rate in the amplifying region around the magnetic equator. K & P found that the waves propagating parallel to the magnetic field ("longitudinally") are amplified the most, but that the amplification decreases as the angle between the wave normal and the field increases, and beyond a certain angle the waves are damped. Now, the wave normal angle varies along the propagation path of a wave, and the way in which it does so can be discovered by ray tracing. Such studies show that, in the plasmasphere, waves starting from the equator with their normals parallel to the field, and following paths of the type pictured by K & P, usually return to the equator with their normals oblique. For these waves, at any rate, there can be no further amplification, so no laser action.

Two consequences follow. Firstly, if K & P's theory is to survive, it must be supplemented by some mechanism for bringing the waves back repeatedly to their source region at the equator with their normals always more or less parallel to the field. Secondly, for testing this or any other theory, it is important to be able to measure wave normal directions in situations where several sets of waves are present at once, all propagating in different directions; hence the value of WDF analysis.

As regards the first of these two points, various mechanisms have been proposed, including guidance of the waves by trapping inside field-aligned ducts [19, 20, 41, 52, 55], guidance by propagation just below the plasmapause [22, 56], and propagation on closed "cyclic trajectories" with one reflection from the plasmapause in each hemisphere [43]. Of the papers that deal with trapping or guidance, [55] and [56] concern theories of these modes of propagation, while the others present experimental evidence for their roles in producing some kinds of hiss. In [19, 20], and also in [22], the evidence comes from WDF analysis: hiss waves have been observed in ducts, or just below the plasmapause, with their normals more or less parallel to the magnetic field. Hiss produced as guided waves can also be observed on the ground, because such waves are able to escape from the ionosphere. However, ELF hiss is detected more often in the plasmasphere than it is on the ground, which suggests that its waves are not always guided; this suggestion is borne out by the fact that WDF analyses of plasmaspheric hiss generally show it to be propagating obliquely [21]. The paper [43] on cyclic trajectories is purely theoretical, but some relevant experimental results are described in §3.2. In these three mechanisms, the waves are all supposed to propagate on closed paths.

Several variants of the K & P theory have been suggested in which the requirement for closed paths is abandoned. Thus Church and Thorne [45], and also Huang and Goertz [46, 47], have traced ray paths for non-ducted waves, and have found cases where the waves make several successive passes through the equator with their normals remaining fairly close to the magnetic field, though becoming oblique on later passes. Thus the waves gain more energy than they could on a single pass, but they are not amplified indefinitely, *i.e.*, there is no laser action. Church and Thorne suggest that plasmaspheric hiss is produced on such trajectories, by amplification of waves from some non-thermal "embryonic source" which they have not yet clearly identified. Huang and Goertz disagree, having made less favorable assumptions about the distribution function of the energetic electrons responsible for the whistler instability, and there, for the moment, this matter rests.

Recently Thorne and Summers have considered energetic electron distribution functions that permit the growth of obliquely propagating whistler-mode waves, as well as of longitudinally propagating ones [60]. Their findings may help to support Church and Thorne's theory of the origin of plasmaspheric hiss, but the application has not yet been made, at least, not to the present author's knowledge.

Using particle as well as wave data from the GEOS-1 and GEOS-2 satellites, Solomon *et al.* [50] have found cases in which the measured distribution functions of the energetic electrons implied growth rates sufficient to raise ELF waves from thermal level to the observed hiss levels on a single pass through the magnetic equator, as had already been suggested by Cornilleau-Wehrlin *et al.* [48]. Their data were taken when the satellite was near the equator and just below the plasmapause. They cite WDF analyses by Parrot and Lefeuvre [22], which showed that waves observed at such points were propagating almost parallel to the magnetic field, in agreement with the theory. They recognize, however, that their findings in these cases need not apply to others, and in particular, not to waves observed at places well below the plasmapause; GEOS-1 did not descend below $L = 4.5$, while GEOS-2 was in a geostationary orbit. Sky maps of plasmaspheric hiss at lower altitudes show that oblique propagation is more common than parallel propagation, even at points close to the magnetic equator (see §3.2).

Another common feature of plasmaspheric hiss is that the sky maps often show one – or, more commonly, two – distinct peaks. The occurrence of a peak suggests that the satellite is observing waves from a spatially localized but remote source. The presence of two peaks can be explained by the peculiar relationship between the wave normal and the ray directions that applies to the whistler mode. Two different wave normal directions may correspond to the same ray direction, the latter being the direction of propagation of the energy. Hence, between a localized source and a remote point of observation there may be two propagation paths, corresponding to different wave normal directions. The observed directions are generally consistent with this interpretation.

The current status of the various theories of the origin of plasmaspheric hiss will be summarized in §3.2, after the presentation of results from some recent research at Stanford University.

3. RECENT PROGRESS

3.1. WDF Analysis

§2.1 described the maximum-entropy method of WDF analysis in its standard form, which was used for the wave data analyses referred to in §2.2. It was also used for those studied in the course of the recent work that will be discussed briefly in §3.2 below.

Unfortunately, practical experience with WDF analysis has revealed one serious drawback, namely that of wasting data. Not all of the data encountered in practice can be exploited by this method. With some data sets, when their analysis is attempted, either no solution for the WDF can be found, or the solution is unstable. For instance, in a collection of ELF hiss data from GEOS-

1, comprising 262 sets of 4-component wave field measurements, 171 (65% of the total) were exploitable if only the three magnetic components were analyzed, but the number fell to 22 (8.4%) if the single electric field component was included in the analysis. Now, it is important to be able to include at least one electric component, because with only the magnetic components the results are ambiguous: there is no way to distinguish between two waves traveling in opposite directions. On the other hand, a wastage rate exceeding 90% is obviously unacceptable.

Some effort had gone previously towards solving this problem, and one cause of instability had been recognized, namely redundancy in the data. The elements of the spectral matrix are related to the WDF by a set of linear integral equations, and the kernels of these various equations are not altogether linearly independent of one another. The resulting instability can be cured, at the price of some loss of data, by orthogonalizing the kernels and then eliminating some of the smallest ones [3, 5, 24]. The cause of the frequent failures to find any solution at all for the WDF was not known, however.

Therefore, in the methodological work performed at Stanford in 1984–1986, one of the first issues examined was whether solutions for the WDF always exist. Firstly, considering that the WDF cannot be negative, it was found that measurement errors can indeed lead to situations where there are no non-negative solutions. Secondly, when the data are on the border of the range of values within which such solutions do exist, these solutions are usually both singular and unique, in which case they represent waves arriving from a few discrete directions only. Thirdly, when the data are just inside this border, the maximum-entropy solution is highly unstable in some respects, but not in others; this is an additional cause of instability, distinct from the one just mentioned. These findings, most of which have been published [32], are the bases for the new method of WDF analysis now under development at NSSDC.

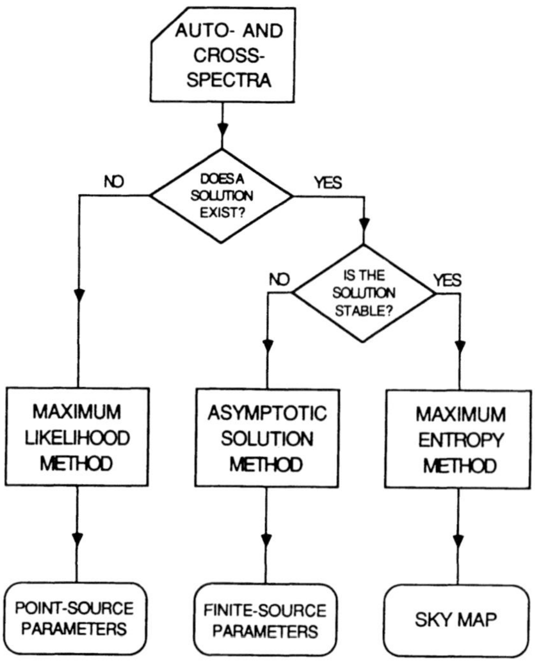

Fig. 2. Proposed new method of WDF analysis.

Figure 2, on the previous page, illustrates this method, in outline. The main difference from the current method (Fig. 1) is that tests for the existence and stability of solutions are applied to the data before these are analyzed, and the route that the analysis follows depends on the outcome.

Two algorithms for for performing existence tests have already been devised, the first of which is systematic and yields unequivocal answers, though at the price of much computation [32]. The second, based on work by Agmon *et al.* [57], involves a random search and is more efficient computationally, but the answers that it yields are only highly probable, not absolutely certain. Future efforts will be devoted to finding an even better algorithm for this purpose.

The role of the existence test is to determine whether any solution for the WDF exists, when the data are taken at their face value. If the answer is no, however, this means that the data are erroneous rather than that no solution can be found. In this case, they must be corrected to the most likely set of feasible values, which usually corresponds to a singular solution. Thus, in fact, this test determines whether the solution to be found is singular or non-singular.

The role of the stability test is to determine, in cases where a solution exists, how close the data are to the border of the region of "data space" within which there is no solution. If they are quite close, then the solution is likely to be unstable.

In fig. 2, the right-hand branch represents the procedure to be followed when the tests indicate that that a stable non-singular solution exists. The solution is found by using the principle of maximum entropy. Here the new method differs from the existing one only in points of detail, such as the exact form of the entropy functional.

The central branch in fig. 2 represents the procedure in the case where the data lie close to the border referred to above, and hence are unstable, at least when judged by the stability criteria that previous workers have used. In point of fact, these solutions consist of one or more narrow peaks, and though their heights are unstable (*i.e.*, sensitive to slight changes in the data), their positions and the volume under each of them are stable. Moreover, the sub-region of data space within which there is instability, close to the border, is also the domain of validity of certain analytic approximations to the true maximum-entropy solutions; these approximations become asymptotically exact as the data set approaches the border. Hence the proper tactic for dealing with unstable maximum-entropy solutions, which cause trouble for any iterative procedures, is to replace them by the corresponding asymptotic approximations; it is hoped that the data needed for calculating these approximations can be obtained as by-products of the existence test.

Finally, the left-hand branch in fig. 2 represents the procedure to be followed in the case where the outcome of this test is negative, meaning that the data are erroneous and should be replaced by their most likely true values. These are the nearest values, in the maximum-likelihood sense, for which a solution for the WDF exists. Provided that the errors have reasonable statistics, this set of data must necessarily lie on the border, which, as already stated, means that the solution must be singular, consisting of one or several Dirac distributions (or "delta functions"). In physical terms, this means that the wave field consists of just a few discrete sets of waves propagating in well-defined directions, a special case that was discussed previously in §2.1. Although such singular solutions can be found as limiting cases of the asymptotic approximations mentioned above, a much simpler and more efficient way of finding them is model-fitting: the solution for the WDF is modeled as the sum of the appropriate number of delta functions, the strengths and positions of which are the parameters of the model. Again, preliminary estimates of the values of these parameters should be obtainable from the existence test. With these features based on the principle of maximum likelihood, the new method of WDF analysis will unify the existing maximum-entropy and model-fitting methods [8, 9].

The question might be asked as to why no method of this kind has yet been developed in any other of the many fields where linear inverse problems are solved by invoking the maximum-entropy principle. The answer lies in a peculiarity of natural whistler-mode waves in Earth's magnetosphere, namely that WDFs consisting of one two fairly narrow peaks are quite common, as was mentioned in §2.2. Such data lie close to the border of the region of data space in which solutions for the WDF exist, so the solutions that they yield are unstable, and moreover, measurement errors can readily displace the data into the region where no solutions exist. In other fields, similar difficulties do not appear because the unknown functions to be solved for rarely have this peaky character. One field in which the unknown functions are sometimes peaky, and where the new method may therefore be needed, is the mapping of starspots [59].

3.2. Plasmaspheric Hiss

In the course of the research at Stanford, no analyses of satellite data on plasmaspheric hiss were made by the new method described above, since its development had only just begun. However, in collaboration with Lefeuvre and others, efforts were made to interpret some data from ISEE-1 that he had analyzed by his existing method. A brief account of this work is given in a report prepared for the sponsor [28]. Some of its broader implications, which were not covered in this report, are discussed below.

Since some theories of the origin of plasmaspheric hiss assign an important role to the plasmapause, as a guide or as a reflector for the whistler-mode waves, it was decided to analyze data taken at times following long periods of magnetic quiet, when either there was no plasmapause or it was located unusually far from the Earth, at $L > 7$. None the less, strong hiss was observed on all of the four satellite passes from which the data were taken. On two of them, during which the satellite covered large ranges of L while remaining close to the magnetic equatorial plane, the upper cut-off frequency of the hiss increased steadily with decreasing L, as several previous authors have reported; at the lowest L-values, it was about 10 kHz. Waves of these frequencies could not have originated in the region just below the plasmapause, since the electron gyrofrequency, which is the upper limit for propagation in the whistler mode, is only about 2.5 kHz at L = 7. These observations support the view that plasmaspheric hiss is generated over a wide range of L-values, and not only just below the plasmapause.

Confirmation of this view was forthcoming from the WDF analyses, which, on one of the two equatorial passes, yielded evidence for a broad source region extending over the range $2.4 < L < 4.3$. Though near to the equator, most of the waves observed on these passes were propagating obliquely; similar results have been reported recently by Sonwalkar and Inan, based on data from the Dynamics Explorer 1 satellite [51] In our ISEE-1 data, however, some waves were also observed at relatively small wave normal angles, and a ray-tracing study suggested that they might have been propagating on cyclic trajectories as proposed by Thorne *et al.* [43].

These results, taken with those published by previous experimenters, have led us to the following conclusions concerning the mode of origin of plasmaspheric hiss. The generation mechanisms proposed by Kennel and Petschek [35], by Thorne *et al.* [43], and by Solomon *et al.* [50], among others (see §2.2), are all physically plausible and can act whenever the necessary conditions exist, in which case they give rise to waves that cross the magnetic equator with their normals at small angles to the magnetic field. However, hiss occurs even when the conditions for none of these mechanisms exist, and then it appears to be generated near the equatorial plane over a wide range of L-values, with the wave normals at large angles to the field. It would be premature to speculate as to what the generation mechanism may be in such cases. Suffice it to say that we share the opinions of certain other workers who, having noted that hiss often occurs in association with the more coherent emission known as "chorus," have suggested that the two may be produced by the same basic mechanism, which must be a nonlinear rather than a quasilinear one [44, 49, 51, 53]; the recent observations by Sonwalkar and Inan [54] of hiss emissions triggered by lightning-generated whistlers are supporting evidence.

It is now clear what experimental data are required for identifying this unknown mechanism. The need is for multi-component ELF/VLF wave data, including accurate measurements of at least two and preferably all three electric field components so that the WDFs can be derived without ambiguity. Moreover, these data should be taken by a satellite in a polar orbit with its perigee in the plasmasphere. Then any equatorial region in which waves are being generated should be readily detectable, from the reversal that would occur in the predominant direction of the wave energy flux as the satellite passed through it. Other aspects of the variation of the WDF through such a region should help to identify the generation mechanism: *e.g.*, for those waves that have their Poynting vectors parallel or anti-parallel to the spacecraft orbit at the equator, it should be possible to determine the spatial growth rates. Suitable data are already being taken by the Japanese satellite Akebono, and several other satellites will be doing the same in a few years from now.

4. ACKNOWLEDGEMENTS

The work at Stanford University described in §3.1 was supported by Grant No. ATM-8318186 from the National Science Foundation. This paper was written while the author held a National Research Council Senior Research Associateship at NASA Goddard Space Flight Center. The study of plasmaspheric hiss summarized in §3.2 was made in association with F. Lefeuvre, M. Parrot, L. Cairó, and R.R. Anderson, and will be published fully elsewhere.

5. REFERENCES

In each of the three following sub-sections, the references are given in approximate chronological order.

5.1. WDF Analysis

1. Storey, L.R.O., and Lefeuvre, F. (1974). Theory for the interpretation of measurements of a random electromagnetic wave field in space. In: Rycroft, M.J. and Reasenberg, R.D. (eds), Space Research, vol. 14, pp. 381–386, Akademie-Verlag, Berlin.

2. Lefeuvre, F., and Storey, L.R.O. (1977). The analysis of 6-component measurements of a random electromagnetic wave field in a magnetoplasma: – Model identification, CNRS Center for Research in Physics of the Environment, Orléans, France, Technical Report No. CRPE/41.

3. Lefeuvre, F. (1977). Analyse de champs d'ondes électromagnétiques aléatoires observés dans la magnétosphère à partir de la mesure simultanée de leurs six composantes (Analysis of random electromagnetic wave fields in the magnetosphere, based on simultaneous measurements of their six components), State Doctorate Thesis, University of Orléans, France.

4. Buchalet, L.J. (1979). Analyse de champs d'ondes électromagnétiques à partir de modèles à une ou deux directions privilegiées (Analysis of electromagnetic wave fields, based on models with one or two privileged directions), Third Cycle Thesis, University of Orléans, France.

5. Lefeuvre, F., and Delannoy, C. (1979). Analysis of a random electromagnetic wave field by a maximum entropy method, Ann. Télécomm. 34: 204–213.

6. Storey, L.R.O., and Lefeuvre, F. (1979). The analysis of 6-component measurements of a random electromagnetic wave field in a magnetoplasma: – I. The direct problem, Geophys. J. R. astr. Soc. 56: 255–270.

7. Storey, L.R.O., and Lefeuvre F. (1980). The analysis of 6-component measurements of a random electromagnetic wave field in a magnetoplasma: – 2. The integration kernels, Geophys. J. R. astr. Soc. 62: 173–194.

8. Lefeuvre, F., Parrot, M., and Delannoy, C. (1981). Wave distribution functions estimation of VLF electromagnetic waves observed onboard GEOS 1, J. Geophys. Res. 86: 2359–2375.

9. Buchalet, L.J., and Lefeuvre, F. (1981). One and two direction models for VLF electromagnetic waves observed on-board GEOS 1, J. Geophys. Res. 86: 2377–2383.

10. Delannoy, C., and Lefeuvre, F. (1981). Logiciel de résolution d'un problème inverse à 1 variable: – 1. Exposé des méthodes, (Software for the solution of an inverse problem with 1 variable: – 1. Description of the methods), CNRS Center for Research in Physics of the Environment, Orléans, France, Technical Report No. CRPE/80.

11. Delannoy, C., and Lefeuvre, F. (1981). Logiciel de résolution d'un problème inverse à 1 variable: – 2. Documentation d'utilisation, (Software for the solution of an inverse problem with 1 variable: – 2. Users' guide), CNRS Center for Research in Physics of the Environment, Orléans, France, Technical Report No. CRPE/81.

12. Lefeuvre, F., Neubert, T., and Parrot, M. (1982). Wave normal directions and wave distribution functions for ground-based transmitter signals observed on GEOS-1, J. Geophys. Res. 87: 6203–6217.

13. Lefeuvre, F., and Lagoutte, D. (1983). Bias for spectral density estimates of electromagnetic wave field components in a magnetoplasma, Ann. Geophys. 1: 265–270.

14. Lefeuvre, F., Parrot, M., Storey, L.R.O., and Anderson, R.R. (1983). Wave distribution functions for plasmaspheric hiss observed on board ISEE-1, CNRS Laboratory for Physics and Chemistry of the Environment, Orléans, France, Technical Report No. LPCE/6.

15. Storey, L.R.O. (1984). The energy density of electromagnetic waves in a cold magnetoplasma, J. Plasma Phys. 32: 309–318.

16. Hayakawa, M., Yamanaka, Y., Parrot, M., and Lefeuvre, F. (1984), The wave normals of magnetospheric chorus emissions observed on board GEOS 2, J. Geophys. Res. 89: 2811–2821.

17. Lagoutte, D., and Lefeuvre, F. (1985). Multispectral analysis for electromagnetic wave field components in a magnetoplasma: application to narrow-band VLF emissions, J. Geophys. Res. 90: 4117–4127.

18. Lefeuvre, F., and Helliwell, R.A. (1985). Characterization of the sources of VLF hiss and chorus observed on GEOS 1, J. Geophys. Res. 90: 6419–6438.

19. Beghin, C., Cerisier, J.C., Rauch, J.L., Berthelier, J.J., Lefeuvre, F., Debrie, R., Maltseva, O.A., and Massevitch, N.I. (1985). Experimental evidence of field-aligned ELF plasma ducts in the ionospheric trough and in the auroral zone, In: Résultats du Projet ARCAD 3 et des Programmes Récents en Physique de la Magnétosphère et de l'Ionosphère (Results from the ARCAD 3 Project and Recent Programs in Magnetospheric and Ionospheric Physics), Cepadues Editions, Toulouse, France. pp. 517–527.

20. Hayakawa, M., Ohmi, N., Parrot, M., and Lefeuvre, F. (1986). Direction finding of ELF hiss emissions in a detached plasma region of the magnetosphere, J. Geophys. Res. 91: 135–141.

21. Hayakawa, M., Parrot, M., and Lefeuvre, F. (1986). The wave normals of ELF hiss emissions observed onboard GEOS 1 at the equatorial and off-equatorial regions of the plasmasphere, J. Geophys. Res. 91: 7899–7999.

22. Parrot, M., and Lefeuvre, F. (1986). Statistical study of the propagation characteristics of ELF hiss observed on GEOS-1, inside and outside the plasmapause, Ann. Geophys. 4, 363–384.

23. Cairo, L., and Lefeuvre, F. (1986). Localization of sources of ELF/VLF hiss observed in the magnetosphere: three-dimensional ray-tracing, J. Geophys. Res. 91: 4352–4364.

24. Delannoy, C., and Lefeuvre, F. (1986). MAXENTWDF – A computer program for the maximum entropy estimation of a wave distribution function, Comput. Phys. Comm. 40: 389–419.

25. Hayakawa, M., Parrot, M., and Lefeuvre, F. (1987). The wave distribution functions of plasmaspheric ELF hiss: GEOS 1 observation in the equatorial region, Mem. Natl. Inst. Polar Res. (Japan), Spec. Issue, 47: 157–172.

26. Hayakawa, M. (1987). The generation mechanism of ELF hiss in detached plasma regions of the magnetosphere, as based on the direction finding results, Mem. Natl. Inst. Polar Res. (Japan), Spec. Issue, 47: 173–182.

27. Muto, H., Hayakawa, M., Parrot, M., and Lefeuvre, F. (1987). Direction finding of half-gyro-frequency VLF emissions in the off-equatorial region of the magnetosphere and their generation and propagation, J. Geophys. Res. 92: 7538-7550.

28. Storey, L.R.O. (1987). Final project report on wave distribution function analysis of plasmaspheric hiss (Prepared under NSF Grant No. ATM-8318186), Space, Telecommunications, and Radioscience Laboratory, Stanford University, Technical Report No. D-711-2.

29. Rönnmark, K. (1987). The evolution of spectral densities in weakly inhomogeneous plasmas, J. Geophys. Res. 92: 11,053–11,058.

30. Rönnmark, K., and Larsson, J. (1988). Local spectra and wave distribution functions, J. Geophys. Res. 93: 1809–1815.

31. Oscarsson, T.E., and Rönnmark, K.G. (1989). Reconstruction of wave distribution functions in warm plasmas, J. Geophys. Res. 94: 2417–2428.

32. Storey, L.R.O., and Chapron, S. (1988). Conditions for the existence and uniqueness of solutions to a constrained linear inverse problem, Inverse Problems, 4: 249–271.

33. Hayakawa, M., Hattori, K., Shimakura, S., Parrot, M., and Lefeuvre, F. (1990). Direction finding of chorus emissions in the outer magnetosphere and their generation and propagation, Planet. Space Sci., 38: 135–143.

5.2. Plasmaspheric hiss

34. Cornwall, J.M. (1965). Cyclotron instabilities and electromagnetic emission in the ultra-low and very low frequency ranges, J. Geophys. Res. 70: 61–69.

35. Kennel, C.F., and Petschek, H.E. (1966). Limit on stably trapped particle fluxes, J. Geophys. Res. 71: 1–28.

36. Dunckel, N., and Helliwell, R.A. (1969). Whistler-mode emissions on the OGO-1 satellite, J. Geophys. Res. 74: 6371–6385.

37. Russell, C.T., R.E. Holzer, and E.J. Smith, OGO-3 observations of ELF noise in the magnetosphere: 1 - Spatial extent and frequency of occurrence, J. Geophys. Res., 74, 755-777, 1969.

38. Russell, C.T., McPherron, R.L., and Coleman, P.J. (1972). Fluctuating magnetic fields in the magnetosphere: 1 – ELF and VLF fluctuations, Space Sci. Rev. 12: 810–856.

39. Thorne, R.M., Smith, E.J., Burton, R.J., and Holzer, R.E. (1973). Plasmaspheric hiss, J. Geophys. Res. 78: 1581–1596.

40. Etcheto, J., Gendrin, R., Solomon, J., and Roux, A. (1973). A self-consistent theory of magnetospheric ELF hiss, J. Geophys. Res. 78: 8150–8166.

41. Chan, K.W., and Holzer, R.E., (1974). ELF hiss associated with plasma density enhancements in the outer magnetosphere, J. Geophys. Res. 81: 2267–2274.

42. Parady, B.K., Eberlein, D.D., Marvin, J.A., Taylor, W.W.L., and Cahill, L.J. Jr. (1975). Plasmaspheric hiss observations in the evening and afternoon quadrants, J. Geophys. Res. 80: 2183–2198.

43. Thorne, R.M., Church, S.R., and Gorney, D.J. (1979). On the origin of plasmaspheric hiss: the importance of wave propagation and the plasmapause, J. Geophys. Res. 84: 5241–5247.

44. Koons, H.C., The role of hiss in magnetospheric chorus emissions, J. Geophys. Res. 86: 6745–6754.

45. Church, S.R., and Thorne, R.M. (1983). On the origin of plasmaspheric hiss: ray path integrated amplification, J. Geophys. Res. 88: 7941–7957.

46. Huang, C.Y., and Goertz, C.K. (1983). Ray-tracing studies and path-integrated gains of ELF unducted whistler-mode waves in the earth's magnetosphere, J. Geophys. Res. 88: 6181–6187.

47. Huang, C.Y., Goertz, C.K., and Anderson, R.R. (1983). A theoretical study of plasmaspheric hiss generation, J. Geophys. Res. 88: 7927–7940.

48. Cornilleau-Wehrlin, N., Solomon, J., Korth, A., and Kremser, G. (1985). Experimental study

of the relationship between energetic electrons and ELF waves observed on-board GEOS: a support to quasilinear theory, J. Geophys. Res. 90: 4141–4154.

49. Helliwell, R.A., Carpenter, D.L., Inan, U.S., and Katsufrakis, J.P. (1986). Generation of band-limited VLF noise using the Siple transmitter: a model for magnetospheric hiss, J. Geophys. Res. 91: 4381–4392.

50. Solomon, J., Cornilleau-Wehrlin, N., Korth, A., and Kremser, G. (1988). An experimental study of ELF/VLF hiss generation in the Earth's magnetosphere, J. Geophys. Res. 93: 1839–1847.

51. Sonwalkar, V.S., and Inan, U.S. (1988). Wave normal direction and spectral properties of whistler mode hiss observed on the DE-1 satellite, J. Geophys. Res. 93: 7493–7514.

52. Koons, H.C. (1989). Observations of large-amplitude, whistler mode wave ducts in the outer plasmasphere, J. Geophys. Res. 94: 15,393–15,397.

53. Sonwalkar, V.S., and Inan, U.S. (1989). Lightning as an embryonic source of VLF hiss, J. Geophys. Res. 94: 6986–6994.

5.3. Related Subjects

54. Smith, R.L. (1961). Propagation characteristics of whistlers trapped in field-aligned columns of enhanced ionization, J. Geophys. Res. 66: 3669–3707.

55. Inan, U.S., and Bell, T.F. (1977). The plasmapause as a wave guide, J. Geophys. Res. 82: 2819–2827.

56. Agmon, N., Alhassid, Y., and Levine, R.D. (1979). An algorithm for finding the distribution of maximal entropy, J. Comput. Phys. 30: 250–258.

57. Kirkpatrick, S., Gelatt, C.D. Jr., and Vecchi, M.P. (1983). Optimization by simulated annealing, Science, 220: 671–680.

58. Sonwalkar, V.S., and Inan, U.S. (1986). Measurements of Siple transmitter signals on the DE 1 satellite: wave normal direction and antenna effective length, J. Geophys. Res. 91: 154–164.

59. Cameron, A.C., and Horne, K.D. (1986). Maximum entropy reconstruction of starspot distributions. In: Zeilik, M. and Gibson, D.M. (eds), Cool Stars, Stellar Systems, and the Sun, pp. 205–208, Springer, Berlin.

60. Thorne, R.M., and Summers, D. (1988). On the marginal stability criterion for a loss-cone distribution, Ann. Geophys. 6: 275-286.

7 *Planetary Lightning and Cosmic Plasma Noise*

7.1 Voyager Neptune Encounter

J.W. Warwick
Radiophysics, Incorporated, 5475 Western Avenue, Boulder, CO 80301, USA

ABSTRACT

At 0500 UT on 25 August 1989 the spacecraft Voyager 2 will fly over the north pole of the planet Neptune whose clouds will then be only 5000 kilometers away. Previously the Voyager spacecraft detected strong electrostatic discharges from Saturn as well as many kinds of radio emissions from the other giant planets. No nonthermal emission has yet unambiguously been seen from Neptune (as of the time of my writing this abstract, 27 July 1989). Since, however, each of the other giants has been an extraordinary radio source, we believe that Neptune will also be one. I will present a review of the principal results that we will have obtained during the fly-by.

It has been my great privilege to participate in the Voyager program for the exploration of the outer solar system for 18 years, culminating in the flyby of Neptune only 12 days ago. Many of the glorious pictures of Neptune, his satellite Triton, and his rings, have no doubt been publicized both here in Japan and throughout the world.

In addition to those results, Voyager gave us data in many other ways. Although these have also been brought to the attention of the press, by their very nature they have not received as wide attention throughout the world.

The Voyager program leaders have enjoined us scientists from presenting or publishing papers on the Voyager results until after the first major scientific presentation is given in the scientific journal Science, of the American Association for the Advancement of Science.

After that (submission for publication) Voyager results will be presented at a half day session of the Division of Planetary Sciences of the American Astronomical Society, October meeeting, and to the San Francisco meeting of the American Geophysical Union in December.

I will honor the injunction to the extent that I will present to results of Voyager at Neptune as they have already been released to the press, but not widely publicized and therefore not in all probability familiar to you.

Also I am prepared to answer individual questions in an informal way, as any of us is always happy to do for colleagues and friends.

Because of the subject of our Symposium, and because of my own familiarity and interest in Voyager radio astronomy data, those are what I will emphasize.

To present general background, I will first show you some other material.

Both Voyagers have now passed through the entire solar system, from Earth to beyond Neptune, at this moment the outermost planet of the nine our system comprises. Both spacecraft are now on trajectories that will intersect the heliopause- the boundary between plasma flowing out from Sun and the plasma of interstellar space- at some unknown and essentially unknowable time in the next several decades. After that the spacecraft will move within interstellar space. This final phase of the Voyager program, beginning with the next few months, is known as "VIM", the Voyager Interstellar Mission. Given the excellent condition of both space craft, it will continue for at least the next 20 to 30 years.

No substantial evidence has yet demonstrated that either spacecraft has observed the heliopause.

As Voyager 2 approached Neptune, it continually captured images of the planet, with better and better resolution, exceeding that of the most powerful Earth-or-space-based telescopes already many months before closest approach on 25 August 1989.

An early image of Neptune already shows a large dark spot associated with white clouds to one side. The center of the image lies at about 30° south latitude. The Great Dark Spot (GDS) is about the size of Earth and the material within it circulates counterclockwise as it is observed from space. Counterclockwise means that the circulation is like that around a high pressure center in Earth's southern hemisphere. The rotation of Neptune brings GDS into view about once every 18 hours.

Further imaging revealed a small white spot at high southern temperate latitudes. The imaging team named this "scooter": on time lapse movies of Neptune (what for Jupiter were called "zoom" movies, but what are now called "approach" movies) with GDS fixed in the middle of the frame scooter whips by rapidly to the east, passing GDS at a speed corresponding to a rotation period of only about 16 hours [later revised to about 16.7 hours].

The relative velocity of these two objects is about 1000 km hr^{-1}. This enormous shear corresponds to winds of about 300 m s^{-1} which must somehow be rationalized between scooter and GDS.

Voyager carried an Imaging Infrared Spectrometer (IRIS), which made maps of thermal emission from Neptune, despite the extreme coldness of the atmosphere. They demonstrated that the equatorial regions of Neptune are hot (relatively speaking), that temperate regions are cool, and that polar regions were hot, like the equator. These were interpreted by the IRIS team in terms of a Hadley cell model of Neptune's atmospheric circulation, in which gas descends at the equator and poles, but rises in temperate latitudes.

They showed that this temperature and circulation are consistent with winds corresponding to the relative motion of GDS and scooter, where scooter is at high levels (white and bright) and GDS is at lower levels.

The radio astronomy experiment (PRA) is my team's responsibility. We detected radio emission first in our data from 17 August 1989, but were able to discover emission after the fact as early as 13 and 14 August 1989. Remember that closest approach occurred on 25 August 1989, about 12 days after our earliest data. Although other Voyager experiments, for example the imaging team (!), had detected Neptune earlier than that, the other wave and particles and fields experiments had not. The importance of our early observations lies in the fact that radio astronomy has been able to provide the program with accurate information on the rotation of Neptune observed over an interval of several weeks duration.

Why, you may well ask, do we need a radio rotation period distinct from optical ones?

Non-thermal radio waves, such as we observed, surely arise in sources closely tied to planetary magnetic fields. These, in turn, depend, for their genesis, upon electrical currents flowing deep within planetary interiors.

In short, observations of planetary magnetic fields can determine the rotation rate of the bulk of the planetary mass, within the interior and distinctly below the observed planetary atmosphere.

I will return to this point in a moment.

We detected Neptune's radio emissions first in a dynamic spectral mode, in which the bursts lay as much as 18 decibels above the noise background (essentially the cosmic radio noise background). For the most part, a given burst appeared in only one frequency channel of the PRA, which has channels spaced at intervals of 19.2 kHz. In early episodes of these bursts, they drifted upwards in frequency from burst to burst in some kind of average sense, from perhaps 700 kHz to perhaps 850 kHz.

These bursts are extremely short and are strongly polarized. In early cases they were virtually without exception left-handed.

They may be produced in regions near the south pole of Neptune and at radio frequencies close to the electron cyclotron frequency in that region. In that case, we estimated that Neptune's magnetic pole nearest the southern rotational pole has polarity opposite that of Earth's magnetic pole near Earth's southern rotational pole. The sense of Neptune's magnetic field therefore has the same relation to Neptune's rotation axis as do the magnetic fields of Jupiter, Saturn, and Uranus, but opposite that of Earth. This orientation is in agreement with that determined later by the magnetometer team.

The remarkable feature of these bursts (other than their bandwidth and polarization) is the precision with which they define a period of 16 hours, to within +/- 0.1 or 0.2 hours, just on the basis of these limited data.

About four days before closest approach we began to detect "continuum" radio emission that is decidedly non-bursty and lasting for several hours during each episode of occurence. Its polarization was opposite that of the bursts and it occurred half way between their episodes. We continued to detect this emission through closest approach and, now, after closest approach as well. Its radio frequency lies in the range generally about 300 to 600 kHz, distinctly below that of most of the bursts.

The power in the continuum emission can be plotted as a function of longitude of the sub-spacecraft position on Neptune, where longitude is given as a function of time modulo the 16 hour rotation period defined by the bursts. This kind of diagram shows that the continuum is emitted in a rotating searchlight pattern; we observe this pattern from a spacecraft which is itself moving around the planet. The period of rotation is $16^{h}3^{m}$+/-4^{m} [later revised to be 16.11 +/- 0.05 hours].

The bursts do not follow the same pattern as the continuum. They occur instead at strictly repetitive intervals throughout the times before and after closest approach. In other words, they are fixed in terms of Sun (or perhaps better, an inertial frame).

As we approached Neptune, the burst source directed radio energy in a broad beam toward Sun (and toward Voyager, incident from that direction) once each time it rotated past that direction. As we orbited around Neptune, their source continued to emit each time it faced Sun, and therefore shifted in phase relative to the continuum sources distributed broadly in longitude.

In conclusion, I mention that Neptune's rotation is very unusual in relation to the rotation of his atmosphere. Apparently, the interior rotates more rapidly than, or as rapidly as, any point in the atmosphere, being equalled only by scooter. In this sense, it may be more appropriate to call scooter "floater"!

7.2 Venus Plasma Noise: An Important Interdisciplinary Issue?

H.A. TAYLOR, JR.
Taylor Enterprises, Beaver River, Nova Scotia B5A 4A5 Canada

ABSTRACT

Low frequency electric field noise in the Venus night ionosphere has been persistently interpreted as evidence of almost continuous lightning discharges, and inferred to originate from explosive volcanism. If correct, these assertions would indeed be an important interdisciplinary issue. If present, these phenomena would contradict current understanding of both atmospheric electrification and geological processes at Earth and published interpretations of conditions at Venus. However, independent analysis has revealed that much of the noise reported is either linked with natural ionospheric disturbances, or the result of interference generated at or near the spacecraft. In no case have the advocates of the lightning-volcanism scenario presented analyses of the dispersion and polarization characteristics of the vlf signals, and we note that the data do not exhibit characteristics expected for whistler waves propagating up to the spacecraft. Thus we conclude that the plasma noise is totally unrelated to either lighting or volcanism, and is certainly not an important interdisciplinary issue.

BACKGROUND

After initial encounter in December, 1978, the Pioneer Venus Orbiter (PVO) has continued to the present to provide the most complete in-situ observations of any planet other than Earth. Daily orbits, with measurements of the atmospheric and plasma environment and radar observations of the surface have, over more than a decade, provided a wealth of information on our sister planet.

Much of the justification for the detailed exploration of Venus was derived from the premise that comparative planetology, featuring tests of conditions on a planet similar to, yet distinct from, the Earth would provide new insights into problems of Earth physics, ranging from plate tectonics and volcanism to atmospheric circulation and the details of the ionosphere and solar wind interaction.

Indeed, much has been learned about interesting processes and conditions, ranging from the detection of unexpected traces of atmospheric gases to the confirmation of a lack of an intrinsic magnetic field and the details of a super-rotating upper atmosphere. These findings have stimulated theories of similar processes in on-going studies of Earth physics.

Such accomplishments notwithstanding, there has evolved a significantly misleading scenario of published results which, if credible, would constitute a most serious and far-reaching interdisciplinary challenge to our understanding of Earth processes. This is the asserted case for atmospheric lightning and associated massive explosive volcanism. These concepts are based almost entirely upon the vast body of PVO observations, accumulated over a number of years.

EARLIER CLAIMS FOR LIGHTNING

The lightning portion of this scenario was first hesitently suggested by studies of Soviet radio noise measurements from probes descending through the lower atmosphere. In these very limited observations, which amounted to little more than "snapshot" glimpses over a number of hours in each case, patches of unexplained noise initially were puzzling. Lacking a credible interpretation, and since lightning was unexpected, the Soviet researchers suggested that a "Venus dragon" might be the source of the noise [1]. Later, however, it was suggested that lightning might possibly be a source for the noise [2].

Closely following the Soviet observations, the electric field measurement on the PVO began to report plasma noise along many of the nightside orbits. Impulsive noise most prominent at 100 Hz was noted at altitudes usually above about 160 km, and ranging to much higher heights in the ionosphere. The researchers interpreting these data were well aware of the Soviet measurements and also began to suggest that the noise must be linked to atmospheric lightning. In their first comprehensive paper on the subject, Scarf and Russell [3] examined a large body of noise events detected exclusively at 100 Hz, and interpreted these narrowband impulses as whistler waves originating in lightning discharges, and propagating upward into the ionosphere to the PVO. Scarf and Russell presented plots suggesting that the noise did not occur randomly, but rather was encountered most often as the satellite passed over mountainous terrain, considered by some geologists to be of possible volcanic origin. These authors insisted that the vlf noise was topographically clustered, with the majority of multiple impulse events detected as the PVO passes over the highest elevations.

LINKAGE OF LIGHTNING AND VOLCANISM CLAIMS

This asserted topographic ordering of noise was critical to the proposed credibility of the "lightning" interpretation, since comparative planetological arguments discounted the possibility of the formation of lightning discharges at Venus. For example, Williams, et al. [4] and others argued that the lack of aerosols needed as charge-carriers and the oppressive atmospheric pressure would severely limit the possibility of lightning formation. Thus the notion that the noise might be clustered over exploding volcanoes presented itself as a rationale for providing the particles and the strong updrafts otherwise lacking. In this way, it was apparently perceived that the assertion of lightning could be justified.

It is important to realize that at the time of the Scarf and Russell paper, and to this time, there is no physical evidence of any explosive volcanic activity. Indeed, there is no hard evidence for any form of volcanism during the years of the PVO observations, and no geologist has published such claims, from any of the detailed radar observations of the planet. Although Exposito [5] speculated that increases in the Venus sulfur dioxide content in late 1978-early 1979 arose from massive exploding volcanoes, there is no supporting evidence, and clearly such increases could be caused by changes in atmospheric circulation, with upward convection raising sulfur dioxide from the large reservoir below. After an extensive analysis, Head and Wilson [6] state that the occurrence of explosive volcanism would be "extremely rare" and make no claim for any volcanic activity during the progress of the PVO observations following encounter in 1978. These results are consistent with earlier reports of geological studies available at the time of the "lightning" interpretation.

The PVO data base used for the "lightning" studies includes observations taken throughout the period 1978-1984, and includes thousands of noise bursts detected on hundreds of individual nightside orbits. The volume and extent of data base totally dominates any other form of evidence, making the Soviet snapshots almost negligible by comparison. In this base, the narrowband 100 Hz noise attributed to lightning and indirectly also to volcanism does not decrease in the years following the 1978 encounter, but rather actually increases significantly in the years 1979, 80, 81 and 82. Thus, since the interpretation was linked to topography and active volcanism, the clear implication must be that these processes were not only active at the time of the encounter but actually increased through the following years. This fact alone serves to negate the lightning-volcanism scenario, since the sulfur dioxide concentration was observed to drop dramatically shortly after encounter in early 1979, and to remain very low thereafter. This would imply a cessation of volcanism, and thus is totally inconsistent with the "lightning" noise linkage concept.

PROBLEMS WITH ASSERTED EVIDENCE OF LIGHTNING

Published opposition to the initial interpretation of the lightning-volcanism scenario has been provided in a number of papers, beginning with the work of Taylor et al. [7], in which caution was urged for this interpretation, and followed by studies in which the very credibility of the data and its analysis was questioned on several very basic grounds. In a comprehensive study of the total body of the electric field data available, Taylor and Cloutier [8] showed that (1) the noise was restricted to regions in which the ionosphere was distinctly and significantly disturbed, i.e. in ionospheric troughs, and (2) when the complete data set was examined, the noise was distributed at random, with no evidence whatsoever of the asserted topographic clustering. By linking the noise directly with ionospheric events, and by showing that the claimed topographic clustering could not be replicated, Taylor and Cloutier rejected the most basic concepts of the published "lightning" results.

Taylor and Cloutier argued that the noise was surely linked to the plasma instabilities inevitably occurring within the ion troughs. The density gradients marking the ion troughs assure that ion-electron instabilities, capable of generating a variety of in-situ noise types, such as ion acoustic noise, were the probable source of the 100 Hz noise impulses. These authors also argued that the altitude distribution of the noise simply resulted from the solar wind interaction believed to generate the ion troughs. The troughs, like the high-latitude ionospheric troughs of the Earth ionosphere, are understood to result from the interaction of the solar wind with the magnetic field configuration (the draped interplanetary magnetic field in the case of Venus). As in the case of the Earth troughs, field aligned particle precipitation and associated plasma noise is quite predictable, as discussed by Grebowsky and Curtis [9].

The fundamental criticisms of the data and its interpretation have never been properly addressed in subsequent papers clinging to the lightning-volcanism scenario. Yet, interestingly, after these faults were documented, a totally new and revised definition of the specification for "lightning" signals in the PVO electric field data was brought forward. Singh and Russell [10] reported that the earlier definition by Scarf of exclusively narrowband signals was not adequate, and that indeed <u>wideband</u> signals were the best evidence for lightning. This assertion was, in itself, remarkable, in that the acknowledged expert on vlf measurements, F. Scarf, insisted that any wideband whistlers emanating from lightning could not propagate up to the PVO. This position is underscored by the fact that all of the events selected in the studies of Scarf were not accompanied by any signals at higher frequencies. He was clearly devoted to the concept that the wideband signals would not reach the satellite as stated in his publications. Still, in the face of this, Singh and Russell insisted that they had found "wideband" evidence of lightning. These authors proceeded to develop a new data base, including impulses measured on the 730 Hz, 5.4kHz, and 30kHz channels of the detector. Singh and Russell asserted that the new wideband noise exhibited low altitude dominance, and further that it showed year to year variations indicative of possible seasonal variation of the "lightning".

Soon after, Taylor and Cloutier [11] showed that not only were the new data proffered by Singh and Russell not linked in any way with either lightning or volcanism, but in fact were not even scientific results. Specifically, they showed that the "signals" were actually telemetry interference spikes and not vlf signals of any sort. Incredibly, the telemetry interference was read as impulsive vlf whistlers, indicating that there was no tangible analysis of the character of the raw data, but rather simply a tabulation of impulsive activity, whatever the source. Taylor and Cloutier showed that the results of Singh and Russell could only be replicated by the use of telemetry noise, and thus that no significant amount of actual electric field signals were included. Thus these authors showed that the published results were irrelevant and totally unrelated to the lower atmosphere or the planet.

MOST RECENT EVIDENCE ALSO REJECTED

It is emphasized that Singh and Russell examined and reported on an extensive body of data, collected from hundreds of orbits over a number of years. In so doing, they clearly intended to present the best and most representative of the implied "broadband" data. Thus the impact of the revelation that these data were largely telemetry noise would normally be expected to significantly detract from the asserted scientific arguments from this area. Yet, following this revelation, Russell and Singh subsequently introduced still another data set, insisting that the broadband nature of the data were even more evident in this new (the third in the series) data set. Using this data, Russell and Singh [12] continue to insist that there is clear evidence of upward propagating broadband electromagnetic waves, and that these waves originate from lightning discharges.

We have now examined this newest data set and find that the so-called broadband signals do not exhibit the characteristics of upward propagating whistler waves. To illustrate, we consider several examples of the same signals specifically identified by Russell et al. [13] as key samples of the "broadband" data. First, in Figure 1 we note the several types of impulsive noise evident in the electric field data. In this example from orbit 531, we designate three distinct types of noise, as (a), (b) and (c). In segment (a) a precisely periodic train of impulses is seen, with spike-like rises occurring regularly at 6-second intervals. The PVO spin rate about its axis is 12-seconds, and thus the field measurement experiences an interference burst two times each spin cycle. These responses are clearly not evidence of ambient electric field activity, but rather a form of spacecraft-plasma interaction, indicating that the detector is quite sensitive to local plasma perturbations. Close inspection of this form of response shows that it must at times be difficult to distinguish from the second type of response indicated in interval (b). Interval (b) is typical of the 100 Hz noise bursts originally asserted by Scarf to be the narrowband evidence of "lightning". Note that in this interval, the 100 Hz impulses are not accompanied by any activity on the higher frequency channels, a characteristic repeatedly specified by Scarf and colleagues as key to the association with lightning. Note also some evidence of 6 and 12-second timing of pulses.

As we have shown in several publications [7, 8], the clusters of 100 Hz noise seen in interval (b) are consistently coincident with ion troughs. Very often, the onset and cessation of the noise is abrupt, coinciding with the sharp density gradients which mark the boundaries of the trough. We are quite confident that this noise is also interference, arising from the plasma instabilities encountered within these troughs. Thus, we assert that the signals of interval (b) are also not indicative of the ambient electric field, but rather artifacts of the solar wind induced disturbance of the nightside ionosphere.

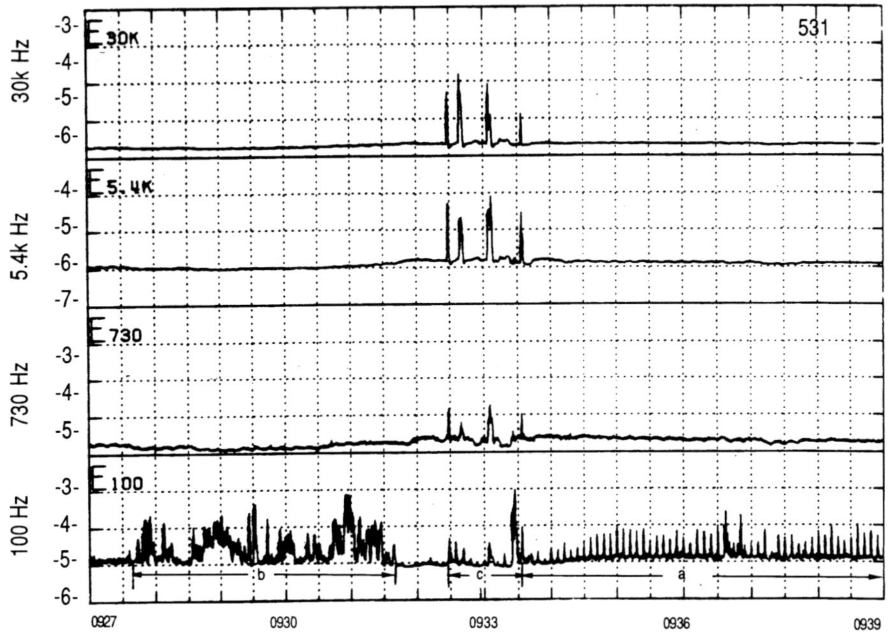

Fig. 1 Electric field data from PVO orbit 531 with impulsive noise on all four frequency channels. On the 100 Hz channel, three segments of impulses are marked to indicate distinct types of noise often appearing on that channel. Segment (a) is spacecraft-plasma interference with spikes occurring two times each 12-second spin period. Interval (b) is 100 Hz noise attributed by Scarf and Russell to "lightning", Interval (c) indicates so-called "broadband" signals later attributed by Russell and Scarf to "lightning". Note that the interference in (a) occurs at rates similar to portions of the noise in (b). Also note that the signals of (c) show no dispersion with frequency. Ordinate values are in negative decade units of $V/m\text{-}Hz^{\frac{1}{2}}$.

The third type of interference, indicated in interval (c) of Figure 1, is that most recently identified by Russell and colleagues as the new "broadband" evidence of lightning. If indeed the impulses of the type illustrated in interval (c) were broadband whistler mode waves emanating from lightning, they would exhibit key characteristics consistent with this hypothesis. However, they do not.

First, we note that it is not difficult to see that there is no evidence of spectral dispersion with frequency. A close look at the figure will show that the impulses occurring on one channel rise and fall simultaneously with their counterpart impulses on the other channels. This discounts the assertion that these are whistler waves.

In separate work, we have also examined the polarization of examples of these signals, by relating their amplitude variation relative to the orientation of the magnetic field. (Note that by the term "magnetic field" we are referring to the draped interplanetary magnetic field, since Venus appears to exhibit little or no detectable intrinsic magnetic field.) Our comparisons showed that there is no convincing evidence of any coherent polarization of the so-called broadband impulses evident in interval (c).

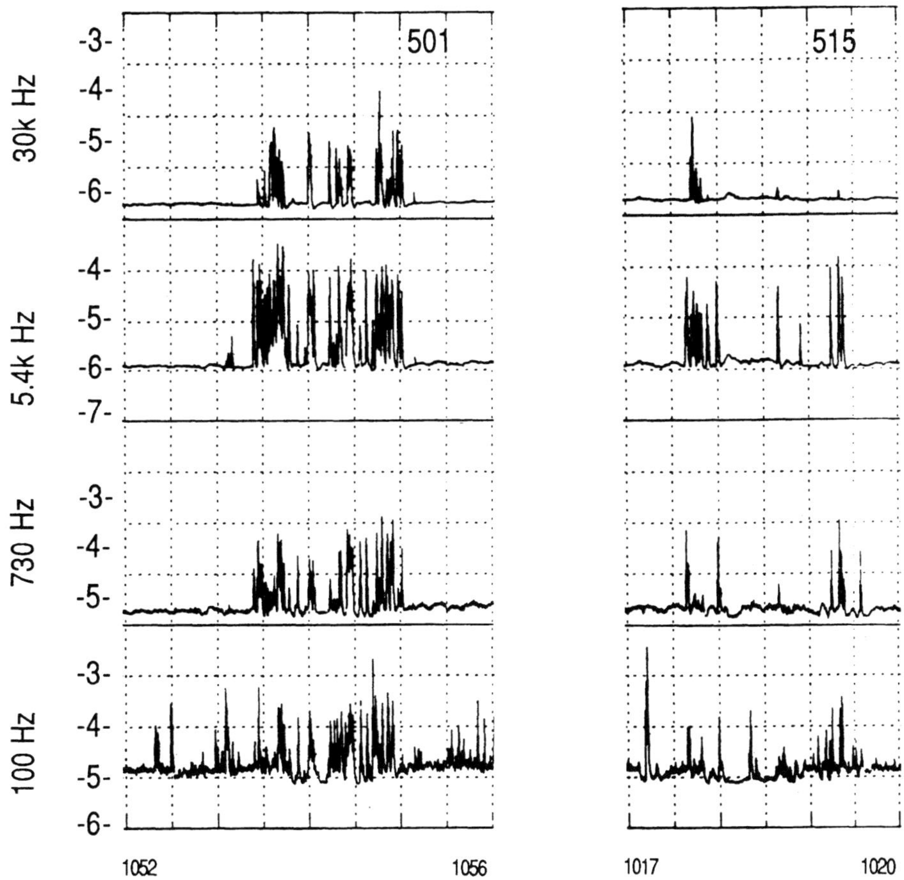

Fig. 2 Additional examples of the "broadband" impulsive noise like that shown in interval (c) of Fig. 1. As in the earlier example, the impulsive events show no evidence of expected dispersion, nor the amplitude behavior indicative of upward propagating whistler waves. Units same as for Fig. 1.

The lack of dispersion evident in this new "lightning" data set is further certified in Figure 2 which gives two additional examples of the data published as broadband evidence. In these data from two orbits, the impulses again exhibit simultaneous relationships across the frequency channels. There is no evidence of the expected dispersion.

We call attention to several additional features of this new data, which further convince us that this response, too, is simply noise generated in the ionosphere, near or at the PVO. First, we note that there is no tendency for a decrease in signal amplitude with frequency. If, as originally argued by Scarf and Russell, higher frequency signals emanating from lightning would have difficulty propagating up to the PVO, we should expect that even if some signals were to "leak" into the ionosphere their amplitudes, if detectable, would be lowest at the highest frequencies, and vice versa. This is not at all indicated by the results.

Second, we note that the bursts of signals at the three higher frequency channels do not exhibit a gradual onset and cessation as might be expected in a naturally occurring situation. Rather, we observe that these signals occur and end abruptly, with initial and final impulses being as strong in amplitude as any within the interval. This characteristic is suggestive of the likelihood that some form of spacecraft-plasma interaction is "turned-on" and off at some critical point, dependent perhaps upon reaching a critical atmospheric pressure. For example, it has been shown that the electric field measurement is dominated by impact ionization interference as the PVO plunges abruptly into the dense lower neutral atmosphere [14].

SUMMARY AND CONCLUSIONS

Summarizing our views of the several data types asserted sequentially to be evidence of lightning and thus volcanism, we offer the following comments. First, we emphasize that in response to our published criticisms of a given type of data, a new and distinct set of data is introduced, with new criteria and definitions, and without adequate responses to the criticisms of the prior data set and its interpretation. Second, we find no evidence whatsoever of any published analyses of any of these data sets which would constitute reliable evidence that these are naturally occurring electric field signals, and certainly no evidence to support the contention that they are upward propagating whistler mode waves.

Since much of the encouragement for the interpretation of currently active explosive volcanism has arisen from the claims for "topographic clustering" of the field noise, we also assert that our published evidence that the noise claimed to be clustered is instead randomly distributed removes the support for the interpretation. In the absence of any hard evidence for explosive volcanism, we can only surmise that the notion of massive eruptions is unsupported speculation, inappropriately encouraged by the claims for "lightning".

We note that Ksanfomaliti [2], musing upon the unproven nature of the "lightning" evidence, was strongly persuaded by the claimed clustering of the noise over mountains, stating "... there would seem to be no other way of explaining why the electromagnetic pulses recorded by Venera 11-14 and PVO were so highly concentrated." In response, we firmly state that indeed there is another way to understand the PVO electric field data. First, it is not at all topographically clustered, contrary to claims otherwise, and second, the data does not exhibit any evidence of upward propagating electromagnetic waves. Any particular distribution of these signals arises from pecularities of the orbit and associated spacecraft relationships to regions of the ionosphere and the neutral exobase where densities rise abruptly.

In conclusion, we remain convinced that unfortunately the Venus plasma measurements do not constitute an important interdisciplinary issue. Due to the fact that these observations have been widely and repeatedly promoted as evidence for totally unexpected and unpredicted physical processes at Venus, they have naturally excited the community and encouraged some to speculate beyond reliable evidence, about discoveries in comparative planetology. We would wish that such revelations might be valid, but however the realities of the data do not support the conjectures. Lightning, like explosive volcanism, may or may not have been occurring during the investigations of the PVO, but surely the electric field data are unrelated to these issues, and provide no such evidence.

REFERENCES

1. Taylor, H. A., Jr. (1987) Auroras at Venus: taming the Venus dragon, The Planetary Report, 7, 4-6

2. Ksanfomaliti, L. V. (1985) Signs of possible volcanism on Venus, Sol. Sys. Res., 18, 200-207

3. Scarf, F. L., and Russell, C. T. (1983) Lightning measurements from Pioneer Venus Orbiter, Geophy. Res. Lett., 10, 1192-1195

4. Williams, M. A., Krider, E. P., and Hunten, D. M. (1983) Planetary lightning: Earth, Jupiter and Venus, Rev. Geophy., 21, 892-902

5. Esposito, L. W. (1984) Sulfur dioxide: episodic injection shows evidence for active Venus volcanism, Sci., 223, 1072-1074

6. Head, J. W., and Wilson, L. (1986) Volcanic processes and landforms on Venus: theory, predictions, and observations, J. Geophys Res. 91, 9407-9446

7. Taylor, H. A., Jr., Grebowsky, J. M., and Cloutier, P. A. (1985) Venus nightside ionospheric troughs: implications for evidence of lightning and volcanism, J. Geophys. Res., 90, 7415-7426

8. Taylor, H. A., Jr., and Cloutier, P. A. (1986) Venus: dead or alive? Sci., 234, 1087-1093

9. Grebowsky, J. M., and Curtis, S. A. (1981) Venus nightside ionospheric holes: the signatures of parallel electric field acceleration regions? Geophys. Res. Lett., 8, 1273-1276

10. Singh, R. N., and Russell, C. T. (1986) Further evidence for lightning on Venus, Geophys. Res. Lett. 13, 1051-1054

11. Taylor, H. A., Jr., and Cloutier, P. A. (1988) Telemetry interference incorrectly interpreted as evidence for lightning and present-day volcanism at Venus, Geophys. Res. Lett., 15, 729-732

12. Russell, C. T., and Singh, R. N. (1989) A re-examination of impulsive vlf signals in the night ionosphere of Venus, Geophys. Res. Lett. 16, 1481-1484

13. Russell, C. T., vonDornum, M., and Scarf, F. L. (1989) Source locations for impulsive electric signals seen in the night ionosphere of Venus, Icarus, 80, 390-415

14. Curtis, S. A., Brace, L. H., Niemann, H. B., and Scarf, F. L. (1985) CO2 impact ionization-driven plasma instability observed by Pioneer Venus Orbiter at periapsis, J. Geophys. Res. 90, 6631-6636

7.3 Selfgravitational Space Plasma

P.V. Bliokh and V.V. Yaroshenko

Institute of Radio Astronomy of the Ukrainian Academy of Sciences, Kharkov, 310002 USSR

Collective processes in a plasma with charged dust particles with account of gravitational interaction between the particles are considered. Formulae for the dielectric constant are derived in the cases of an isotropic and magnetized plasma. The secular equations are investigated and energy losses of a probe particle are considered.

The authors wish to discuss some aspects of wave propagation in selfgravitational plasma, assuming first that the medium is homogeneous and boundless. Then the central problem of the linear theory is to establish dispersion laws. Dispersion manifests itself as a frequency dependence of the dielectric constnat $\mathcal{E}(\omega)$ or the refractive index $n(\omega)$. It is conditioned by the existence of eigenfrequencies in the medium, which are determined by the interaction between particles. Let us consider some typical examples (restricting the analysis, for the sake of simplicity, to one sort of particles).

If there is no interaction between the particles, then no eigenfrequencies arise in the medium, and hence $\mathcal{E}$ does not depend on ω. Interaction by means of electric forces gives rise to one characteristic frequency, namely the plasma frequency $\omega_p = \sqrt{4\pi q^2 n_o/m}$ (n_o - density, q and m - the particle charge and mass, respectively). As a result, we arrive at the well-known formula for the $\mathcal{E}$ of an isotropic plasma,

$$\mathcal{E} = 1 - \frac{\omega_p{}^2}{\omega^2}. \tag{1}$$

With application of an external magnetic field $\vec{H}$, an additional magnetic interaction arises, bringing forth a second eigenfrequency ω_L = qH/mc, i.e. the Larmor frequency. The formula for $\mathcal{E}$ becomes more complicated and takes a tensorial form. The tensor components depend on two eigenfrequencies, ω_p and ω_L. For example the refractive indices for the ordinary and the extraordinary wave under the longitudinal propagation conditions are

$$n^2_{o,e} = 1 - \frac{\omega_p{}^2}{\omega(\omega \mp \omega_L)} . \tag{2}$$

A third eigenfrequency arises if we take into account the gravitation forces between the particles (selfgravitation). This is the Geans frequency, $\omega_G = \sqrt{4\pi G n_0 m}$ (G is the gravitation constant). Even in the simplest case of an isotropic plasma ($\omega_L = 0$) the formula for ε differs in principle from (1), i.e.

$$\varepsilon = 1 - \frac{\omega_p{}^2}{\omega^2 + \omega_G{}^2} \tag{3}$$

One can see that the cut-off frequency decreases from ω_p to $\sqrt{\omega_p{}^2 - \omega_G{}^2}$ and the plasma becomes transparent at lower frequencies.

If the interactions of all three kinds act together (in a selfgravitational magnetized plasma), then the formula for the tensor $\hat{\varepsilon}$ becomes so complex that we do not present it here. The pattern of calculations for the $\hat{\varepsilon}$ of a selfgravitational plasma does not differ from the ordinary scheme. Within the hydrodynamics approximation, the equation of motion should be supplemented with the term $f_G = -m\nabla\psi_G$ representing the gravitation force, where ψ_G is the gravitation potential of the entire ensemble of particles. It can be determined from Poisson's equation.

$$\Delta \psi_G = 4\pi G m n. \tag{4}$$

Thus, one extra function, ψ_G, and, accordingly, an extra equation (4) are added to the ordinary set of equations.

Our examples show that selfgravitation can essentially change the properties of a plasma. Naturally, the question arises, why does not anyone take into account the gravitation forces between plasma particles? To give an answer, let us compare the frequencies ω_p and ω_G for the electrons. The ratio ω_p/ω_G is about $\sim 10^{21}$, thus the error in determining ε without account of ω_G is $\sim (\omega_G/\omega_p)^2 \sim 10^{-42}$. Obviously, there is no reason in considering selfgravitation effects for electrons or ions. But the situation is quite different in the case of macroscopic particles (like dust b articles or bigger fragments). Using the simplest formulas relating the charge q, mass m and the potential φ of one particle with its dimension a ($q = \varphi a$, $m \simeq \rho a^3$) we obtain $\omega_p/\omega_G \sim 10\,\varphi\,[V]/\,a^2\,[cm]$.

With $\varphi \sim$ IV and $a \sim 3$ cm we have $\omega_p/\omega_G \sim 1$, and hence neither electrical nor gravitation forces can be neglected.

Charged macroparticles are often encountered in space (e.g. in planetary rings, comet tails, interstellar dust clouds, etc). The idea of representing such a medium as a special plasma was first expressed by Alfven (Alfven, 1981), but collective processes were only discussed for the case of planetary rings by the present authors (Bliokh and Yaroshenko, 1985).

Now we shall describe some effects of interest, relating to waves in a selfgravitational plasma. Generally, there is a broad spectrum of particle sizes in space (at the same time it is the spectrum of charges and masses). The ω_p/ω_G

ratio may be $\omega_p/\omega_G >> 1$ for small particles and $\omega_p/\omega_G << 1$ for big ones. For the case of a multicomponent plasma, the calculation of ε becomes more complicated and the results are different from the single-component medium. The essential effects may be illustrated for the simple example of two sorts of particles. We shall assume the particle dimensions to satisfy the inequality $R_1 << R_2$, and also $\omega_{G1} << << \omega_{G2}$ and $\omega_{p1} >> \omega_{p2}$. Let us neglect for a time the lowest frequency $\omega_{G1} \simeq 0$, as well as the thermal velocity of the bigger particles $v_{T2} \simeq 0$. Then we obtain some approximate formulas for ε, which are not too complicated, yet demonstrate distinctive features of the selfgravitational plasma.

In the conventional plasma the ion sound wave shows no dispersion, i.e. the phase velocity is not frequency dependent, $v_{ph}^2 = v_s^2 = v_{T1}^2\ \omega_{p2}^2/\omega_{p1}^2 = T_1/m_2$. If the gravitational interaction is taken into account ($\omega_{G2} \neq 0$), then the phase velocity of the analog to the ionacoustic wave becomes

$$v_{ph}^2 \simeq v_s^2 \cdot \frac{1-(\omega_{G2}^2/\omega_{p1}^2)}{1-(\omega_{G2}^2/\omega_{p1}^2)(1-\omega_{p1}^2/\omega^2)} . \qquad (5)$$

Essential dispersion appears in the vicinity of $\omega \sim \omega_{G2}$

The phase velocity of magnetoacoustic waves under the transverse propagation conditions in an electron-ion plasma coincides with the Alfven velocity, $v_{ph}^2 = v_A^2 = H/4\pi n_{02} m_2$, and shows no dispersion either. If $\omega_{G2} \neq 0$, then

$$v_{ph}^2 \simeq \frac{v_A^2}{1+\omega_{G2}^2/\omega^2} , \qquad (6)$$

and dispersion apears again near $\omega \sim \omega_{G2}$.

We have considered only some simpler manifectations of selfgravitation. Generally, the selfgraviational plasma is characterized by a very complex spectrum of eigenwaves $\omega = f(\omega_{p\alpha}, \omega_{G\alpha}, kv_{T\alpha}, \omega_{L\alpha})$. Besides, the criteria of instability and values of hybrid resonance frequencies are markedly different from the corresponding characteristics of the conventional plasma.

In order to illustrate applications of this theory, let us analyze some plasma processes in planetary rings. A ring may be considered as a system interpenetrating streams of charged particles. Neglecting the stream curvature, we can arrive at the following modal equation

$$\left[1-\sum_\alpha \frac{\omega_{p\alpha}^2}{(\omega-kv_{0\alpha})^2}\right]\left[1+\sum_\alpha \frac{\omega_{G\alpha}^2}{(\omega-kv_{0\alpha})^2}\right] + \left[\sum_\alpha \frac{\omega_{p\alpha}\omega_{G\alpha}}{(\omega-kv_{0\alpha})^2}\right]^2 = 0, \qquad (7)$$

where α specifies the sort of particles.

This has a form typical for the selfgravitational plasma. The main part of (7) consists of two factors, with one involving only gravitation forces and the other only electrical. Besides, there is an additional term to couple the two processes. As a rule, the magnitude of this latter is much less then unity and hence equation (7) has solutions of two classes with corresponding characteristic frequencies $\omega^{(I)} \sim \omega_p$ and $\omega^{(II)} \sim \omega_G$. If $\omega_G \ll \omega_p$, then the two kinds of waves can be analyzed separately. For the Saturn rings, the characteristic times are $\tau_G \sim \omega_G^{-1} \sim 10$ years and $\tau_p \sim \omega_p^{-1} \sim 10$ minutes, which values differ very much. τ_G is the characteristic time for the principal ring to become subdivided into smaller ringlets which was estimated in a theory without electrical interaction. The value τ_p corresponds to the life time of the radial irregularities known as "spokes". The discovery of "spokes" by Voyager 1 and Voyager 2 was quite unexpected, and many theories were developed to explain the phenomenon. We have also made some calculations in an attempt to describe "spokes" as envelope solitons of longitudinal density waves in the multicomponent streams of charged particles.

REFERENCES:

1. Alfven, H., 1981. Cosmic plasma. D. Reidel Publishing Company, Dordrecht, Holland.

2. Bliokh, P.V. and V.V. Yaroshenko, Electrostatic waves in Saturn rings, Astron. J., 62, 569-579, 1985 (Russian).

7.4 Microwave Observations of the Sun During the 22nd Solar Activity Cycle – 17 GHz Radio Heliograph

K. SHIBASAKI[1], K. KAI[1], S. ENOME[1], H. NAKAJIMA[1], M. NISHIO[1], T. TAKANO[1], C. TORII[1], Y. SHIOMI[1], H. SEKIGUCHI[1], M. SAWA[1], T. BUSHIMATA[1], S. KAWASHIMA[1], N. SHINOHARA[1], and T. KOSUGI[2]
Nobeyama Radio Observatory, National Astronomical Observatory, Nagano, 384-13 Japan[1] and Institute of Astronomy, Faculty of Science, University of Tokyo, Tokyo, 181 Japan[2]

ABSTRACT

For the study of high energy phenomena on the Sun during the maximum period of the 22-nd solar activity cycle, we started to construct a radio heliograph at Nobeyama Radio Observatory. The operating frequency was selected to be 17GHz to catch the emission from the accelerated electrons in solar flares. The 17GHz Radio Heliograph consists of "T" shaped array of 490 m in E-W direction and 220 m in N-S direction with 84 antenna elements. The heliograph can map the whole Sun with 10 arc second spatial resolution and with 50 millisecond temporal resolution.

1. INTRODUCTION

Particle accelerations in solar flares and build-up process of flare energy are key issues in solar physics. Accelerated electrons emit microwave when they interact with active region magnetic fields. High density and high temperature plasma trapped by magnetic loops also emit microwave. Microwave study of the Sun gives us useful information about the solar flare processes such as build-up of the flare energy and the high energy particle production.

So far, high resolution microwave observations of the Sun have been done using large interferometers designed mainly for the study of cosmic radio emission such as the Very Large Array [1] and the Westerbork Synthesis Radio Telescope [2]. Element antennas of these arrays have large diameter so that the field of view is small and can cover only one part of the solar disk at a time. Forecasting technique of solar flares at particular active regions at particular time is not accurate enough and allocation of telescope time is not flexible. It is very hard to detect active phenomena on the Sun during the observing session under these limitations. Dedicated telescope is needed to study the solar active phenomena. It is necessary to cover the whole solar disk and to observe the Sun in routine base. Japanese solar radio astronomy group discussed these possibilities for more than a decade and finally we could start to construct a new radio heliograph in 1990.

There is also another big project during this active period for the study of high energy phenomena on the Sun. That is the SOLAR-A satellite[3] dedicated for solar X-ray observation which will be launched by Institute of Space and Astronautical Science. Major instruments are a soft X-ray telescope, a hard X-ray telescope, a wide band spectrometer covering soft X-ray to gamma ray, and a Bragg crystal spectrometer. This satellite will be launched on summer 1991 and will observe the Sun during the active period. Combination of SOLAR-A and the 17GHz Radio Heliograph will give us a unique opportunity for the study of high energy phenomena on the Sun.

In the following sections, design philosophy and the actual design of the 17GHz Radio Heliograph are described.

2. SCIENTIFIC OBJECTIVES AND NECESSARY PERFORMANCES OF THE NEW RADIO HELIOGRAPH

Main objectives of the heliograph is to study solar flare processes: energy build-up, triggering, acceleration of high energy electrons and heating of plasma by energy release. It is widely accepted that the flare energy is stored in sheared magnetic fields as electric currents. Magnetic loops can trap the high density plasma in corona and these plasma can emit microwave through f-f emission mechanism. So that the magnetic field structure in corona where the energy is stored, can be mapped by microwave and the study of the energy build-up process and the triggering process of flares can be done using these maps. Accelerated electrons produce strong microwave emission when they interact with magnetic field. As the flare occurs in magnetically complex region, the position and the structure of microwave emission give us useful information about the acceleration and the triggering mechanism of flares.

Study of evolution process of active regions is another objectives of the heliograph. Quiet phenomena such as networks, bright points, dark filaments, and coronal holes will be detected by the heliograph and the evolution process of these phenomena can be routinely monitored. Activity of polar cap brightening is also an interesting subject to study.

High spatial resolution is essential for most of the study of solar phenomena to resolve each radio sources. High positional accuracy is also necessary relative to the solar disk to co-align the radio maps with other data such as Hα picture, magnetograms and X-ray images to study the relation between the radio source and the chromospheric or photospheric phenomena and to study the physical conditions of the radio sources.

For the study of dynamic phenomena related to solar flares, high time resolution mapping capability is necessary. Field of view should cover the whole Sun to catch flares which cannot be known exactly in advance, where and when they will occur.

High energy electrons accelerated in flares radiate short centimeter to millimeter emission. So that the operating frequency of the heliograph should be at these frequencies and preferably multi frequency to determine the spectrum which gives us information about the energy spectrum of the accelerated electrons.

Circular polarization is another important information about the magnetic field with which high energy electron interact.

For the study of quiet phenomena, high quality map (or high dynamic range map) is necessary because they are weak compared to the solar disk component.

3. ARRAY DESIGN AND THE PERFORMANCE OF THE 17GHZ RADIO HELIOGRAPH

To realize high spatial resolution, high temporal resolution and wide field of view mapping capabilities, we use two dimensional interferometer. The diameter of each antenna element should be small to cover the whole solar disk. Operating frequency is 17GHz to detect microwave emission from high energy electrons and the electronic components at this frequency are commercially available at reasonable price. High spatial resolution require long baselines. Short baselines also needed to cover the whole solar disk. As the diameter of element antenna is small, cosmic point sources cannot be used as calibrators. We use the source itself as a calibrator while observing the Sun by redundant antenna spacings.

Taking the above requirements into account, we have chosen a regularly spaced T shape array. The total number of antennas are determined by dividing the total length of the baseline by the shortest spacing which is determined by the field of view (about 40 arc min. to cover the whole solar disk). We have to reduce the number of antennas due to budgetary limitations without cutting the major requirements: field of view, spatial and temporal resolution. The longest baseline is limited by the land available at Nobeyama. Figure 1. shows the array which we have selected. The array and antenna parameters are listed in Table 1. Figure 1 shows the array configuration.

Table 1. The array and the antenna parameters of the 17GHz Radio Heliograph.

operating frequency	: 17GHz (+34GHz in near future)
antenna size	: 80 cm diameter
number of antennas	: 84
longest baseline	: 488.96 m for EW array
	220.03 m for NS array
shortest baseline	: 1.528 m (86.647λ)

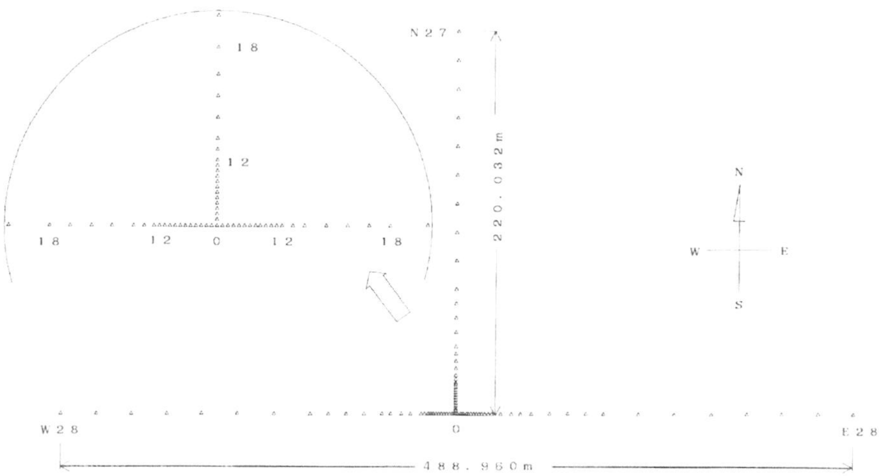

Fig. 1. The array configuration of the 17GHz Radio Heliograph.

High temporal resolution is realized by a multi correlator system. The number of complex correlators are ${}_{84}C_2=3486$ for both right hand circular and left hand circular polarization. It is necessary to resolve the motion of electron beam in the solar atmosphere. It takes about 50 millisecond for the electron beam with the velocity of light to cross the projected area of the half power beam width of the array.

The UV coverage of the array is shown in Figure 2a. Each dot corresponds to the baseline vector of the antenna pair. The visibility functions of the radio sources are sampled on these points. The point spread function or the beam pattern of the array is shown in Figure 2b. It is the Fourier Transform of the UV coverage. Due to the reduction of number of antennas, sidelobe is rather high. Image processing such as CLEAN is necessary to suppress these sidelobe effects.

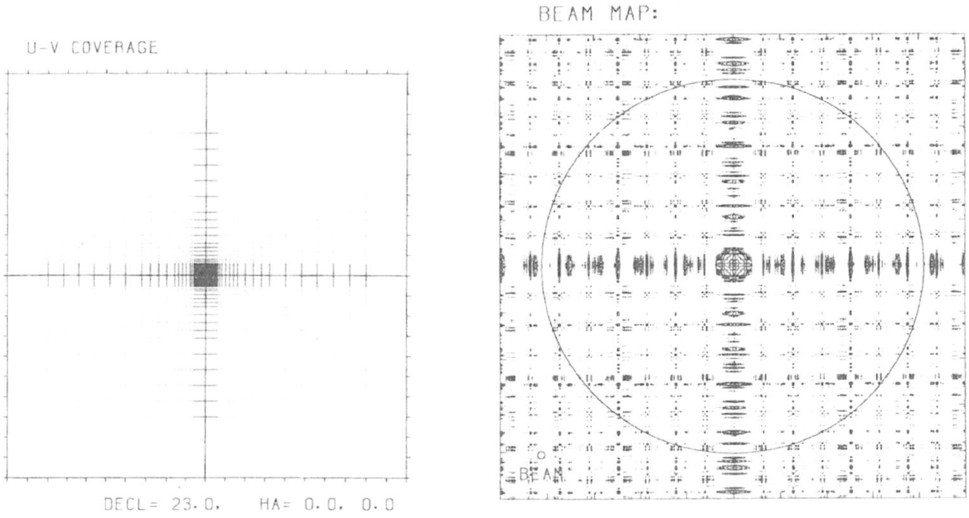

Fig. 2. a)UV coverage of the antenna array (left). b)The point spread function (right).

The dynamic range of the restored image depends on the errors which relate to the pairs of antennas. They are correlator errors, delay tracking errors, band pass characteristics of the receivers, and so on. Errors which relate to each antenna are removed when we calibrate the data using the redundant measurements. The expected dynamic range of the restored image is 50:1. The performance characteristics are summarized in Table 2.

Table 2. The performance of the 17GHz Radio Heliograph.

Field of view	: 40 arc min.
Spatial resolution	: 10 arc sec. (5 arc sec. in E-W one-dimension)
Temporal resolution	: 50 msec
Polarization	: right hand circular and left hand circular (time sharing: 25msec(R) + 25msec(L))
Dynamic range of images	: 50 : 1 (500 : 1 for rotational synthesis)

4. ROTATIONAL SYNTHESIS

Weak and stable structures on the Sun can be mapped using the rotational synthesis technique. We can increase the dynamic range of the image by the rotational synthesis.

Figure 3a shows the UV coverage during the one hour synthesis near local noon on the day of summer solstice. Trajectories of the long baseline fill the outer gaps in the UV plane. This causes low sidelobe in Fig. 3b. We expect to increase the dynamic range of the synthesized maps by factor of ten.

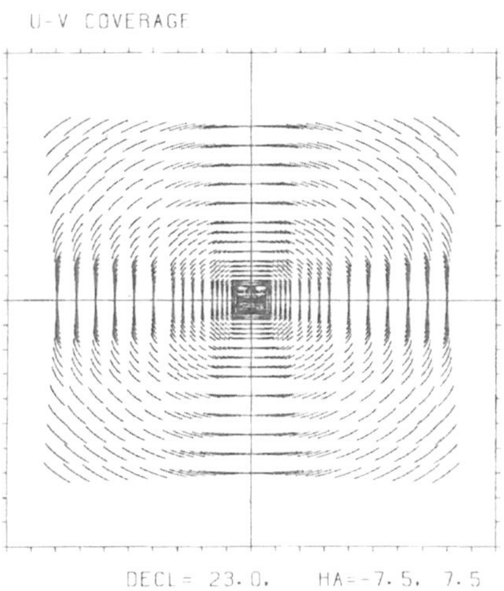

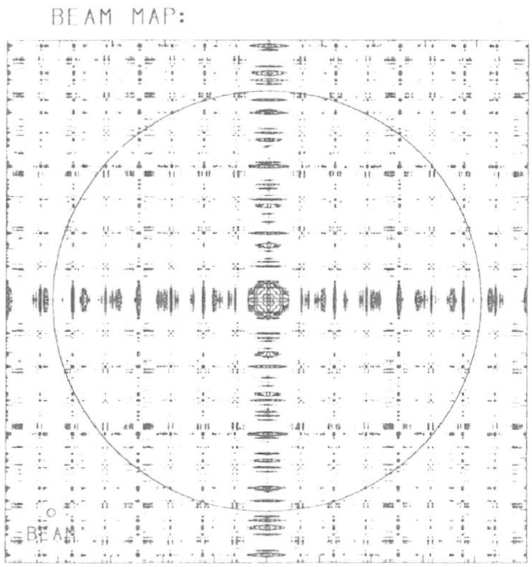

Fig. 3. a)UV coverage during the one hour synthesis near local noon on the day of summer solstice (left). b)The point spread function (right).

There are limitations for the rotational synthesis of the Sun because the Sun is extended, rotating, and active. We need modification in the rotational synthesis of the Sun compared to that of cosmic radio sources which are stable over several months.

5. SCHEDULE OF THE CONSTRUCTION

We start construction of the antennas, the receivers, and the ground base for the antenna setting on the summer 1990. It takes about two years to complete the whole system and to be ready for observation. According to the statistics of solar activity, energetic events tend to occur after the period of sunspot maximum. The sunspot maximum of the present solar cycle is expected to be in 1990. There will be many chance to catch energetic events by the 17GHz Radio Heliograph during the present solar activity period. Cooperative observations of solar flares together with the Solar-A satellite and with ground based optical telescopes will give us new insight into the energetic phenomena on the Sun.

Fig. 4. An artist's impression of the Nobeyama 17GHz Radio Heliograph.

REFERENCES

1. Napier, P.J., Thompson, A.R., and Ekers, R.D. (1983) The Very Large Array: Design and Performance of a Modern Synthesis Radio Telescope. Proc. IEEE, 71: 1295-1320.
2. Hogbom, J.A., Brouw, W.N. (1974) The Synthesis Radio Telescope at Westerbork. Principles of Operation, Performance and Data Reduction. Astron. Astrophys., 33:, 289-301.
3. Ogawara, Y. (1989) The Solar-A Mission. Adv. Space Res., 8: (11)87-(11)92.

7.5 X-Ray Astronomy Satellite Ginga and Observations

F. MAKINO
Institute of Space and Astronautical Science, Kanagawa, 229 Japan

ABSTRACT

The x-ray astronomy satellite Ginga carries three scientific instruments, the Large Area proportional Counters (LAC), All Sky x-ray Monitor (ASM) and Gamma-ray Burst Detector (GBD). The LAC is the main instrument with an effective area of 4000 cm^2 giving it the highest sensitivity to hard x-rays so far achieved. Ginga observed about 250 targets up to the end of 1989. Important results are the discovery of cyclotron absorption features in x-ray spectra from several x-ray pulsars, the discovery of transient x-ray pulsars in the region of 5 kpc ring of the Galaxy, the discovery of diffuse iron line emission peaked at the Galactic center, the detection of x-ray scattering by cosmic dust grains, the discovery of red-shifted iron line emission from quasars, the detection of spectral structures and variations of Seyfert galaxies and the detection of x-rays from SN 1987A.

1. INTRODUCTION

Ginga is the third Japanese x-ray astronomy satellite following Hakucho (Kondo et al. 1981) and Tenma (Tanaka et al. 1984). The fabrication, launching and operation of Ginga were carried out entirely by the Institute of Space and Astronautical Science (ISAS). General properties of Ginga has been described in the literature (Makino et al. 1987) and are summarized in Table 1. Ginga was launched on February 5, 1987 from Kagoshima Space Center of ISAS by M-3S-II launch vehicle. The initial orbital parameters were a perigee height of 505.5 km, an apogee height of 673.5 km, an inclination of 31.1 degree and a period of 96.5 min. The altitude of Ginga has decreased rapidly since

Launch	February 5, 1987	
Launch Vehicle	M-3S-II-3	
Initial Orbit	Perigee : 505.5 km	Apogee : 673.5 km
	Inclination : 31.1 degree	Period : 96.5 min
Mass	Total : 420 kg	Experiments : 105 kg
Experiments	Large Area Proportional Counters (LAC)	
	All Sky X-ray Monitor (ASM)	
	Gamma-ray BUrst Detector (GBD)	
Stabilization	Biased momentum three-axis control	
Attitude Sensors	Two CCD Star Trackers, Sun Sensor, Gyroscopes and Geomagnetic Aspectmeters	
Telemetry Bit Rate	High:16384 bps, Medium:2048 bps, Low:512 bps	
Data Recorder	41.9 Mbit Bubble Memory	
	Recording Time : 42m40s(H), 5h41m(M), 22h44m(L)	
Power consumption	150 W	

Table 1. General characteristics of Ginga.

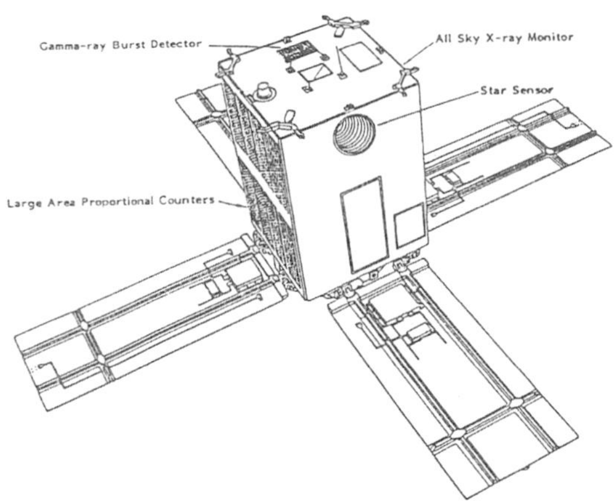

Fig. 1. The outside view of Ginga

early 1989, because of high solar activity. The perigee and apogee height are 575.1 km and 471.0 km respectively in April 1990.

The total mass of the satellite is 420 kg of which 105 kg is for scientific instruments, and the rectangular main body measures 1mx1mx1.5m with four deployable solar panels 1.7m long and 0.75m wide. A maximum power of 500W can be generated from the solar panels and normal power consumption is about 150W.

The scientific instruments are the Large Area proportional Counters (LAC), All Sky x-ray Monitor (ASM) and Gamma-ray Burst Detector (GBD). The LAC and GBD are prepared in collaboration with University of Leicester and Rutherford/Appelton Laboratory in the U. K. and Los Alamos National Laboratory in the U. S. A. respectively.

All the instruments have functioned normally to date, April 1990. Total number of targets observed to the end of 1989 is about 250, 51 % are Galactic sources and 49 % extra-Galactic sources.

2. SCIENTIFIC INSTRUMENTS

2.1 Large Area proportional Counters (LAC)

The LAC is the primary instrument of Ginga and consists of eight identical counters (Turner et al., 1989). The total effective area is 4000 cm^2. Each counter is a multi-cell proportional counter which comprises 13 cell counters of four layers. The cell counters sided by counter wall are connected together and used for anti-coincidence to lessen background, as well as the signals from the end plates. Because most of the background counts are produced by Compton scattering of hard x-rays and gamma-rays in the counter body. The anodes of the central two layers are connected together and those of the cell counters sided by the entrance window are connected alternately. The signals from these three anode groups are analyzed by the on-board data processor and are transmitted. Mutual anti-coincidence among these three anode groups is also employed for background reduction.

The entrance window is beryllium foil of 62 micron thick which is supported by honeycomb type collimators made with stainless steel plates. The counters are filled with a gas mixture of argon (75%), xenon (20%) and carbon dioxide (5%) at 2 atm at 293 K. The sensitive energy range is from 1.5 keV to 35 keV. The collimator field of view is elliptical of 1 degree by two degree (FWHM).

The counting rate of x-rays from Crab Nebula is about 10000 cps, while the background rate is 70 cps of which 18 cps is due to cosmic diffuse x-ray background (CDXB). The detection limit of the LAC is confusion limit which is defined as sky to sky fluctuation of CDXB due to intensity-source number relation. The 3 sigma limit is 2.1 cps, roughly equal to 0.2 mCrab (Hayashida et al. 1989). The sensitivity is the highest so far achieved in this energy region.

The time resolution depends on the number of energy channel for pulse

height analysis and the telemetry bit rate. The highest is 1 ms in the case of 2 energy channel and 16 kbps, and the lowest is 16 s in the case of 48 energy channel and 500 bps. However, a temperature dependence of the clock frequency has appeared after launch, but can be compensated for using temperature data for high resolution timing analysis (Deeter and Inoue, 1990).
No degradation of the LAC has been observed to the end of April 1990.

2.2 All Sky x-ray Monitor (ASM)

The ASM consists of two identical proportional counters, each containing three independent counters with veto cell counters at the back (Tsumeni et al. 1989). These six counters are attached to a collimator of 1 degree by 45 degree (FWHM) of various slant angles. The monitoring of a wide sky region is conducted by slow rotation of the satellite around the z-axis. The position of the source can be determined by the time deferences of peak position of the source for each counter. The normal frequency of scanning is once per day. The sky region which can be monitored is constrained to the equatorial region of satellite coordinate system determined by the LAC target.
The effective area of each counter is 70 cm^2. The counters are filled with xenon (96.7%) and carbon dioxide (3.3%) at 1 atm. The thickness of beryllium windows are 50 micron. The detectors are sensitive to x-rays from 1 kev to 30 keV. The detection threshold is about 50 mCrab but becomes higher with increasing elevation angle from x-y plane of the satellite.

2.3 Gamma-ray Burst Detector (GBD)

Two detectors, a NaI(Tl) scintillation counter and a Xe-filled proportional counter have been used for gamma-ray burst observation (Murakami et al. 1989). The NaI(Tl) is 8.8 cm in diameter and 1 cm in thickness, and measures hard x-rays from 13 keV to 400 keV. The proportional counter has an effective area of 63 cm^2, covering energy region from 2 keV to 30 keV. The x-ray observation of gamma-ray bursts has been conducted for the first time.
The gamma-ray bursts are unpredictable transient events shorter than 1 min. Two memory systems are employed to record bursts. The one is a fast memory triggered by the rise of the burst and is frozen until read out by command from the ground station. The sampling time is 31 ms for pulse counts and 0.5 s for pulse height distributions. The data in the time interval 16 s before and 48 s after the onset of the burst are recorded. The "time-to-spill" mode of data acquisition was used to cover wide dynamic range. The onset time are measured with accuracy of 244 micro-second. Only one burst can be stored in the fast memory. The other memory is a continuously sampling slow memory which sampling rate is either 125 ms, 1 s or 4 s depending on bit rate.
The GBD includes small semiconductor detector to monitor radiation belt (RBM) and generate RBM flag to turn off high voltage supplies of LAC, ASM and GBD.

3. ATTITUDE CONTROL SYSTEM

The attitude of Ginga is stabilized by three axis control system of bias momentum (Ninomiya et al. 1984). The rotation of the satellite around z-axis is controlled by momentum wheel, while the motion perpendicular to z-axis is due to magnetic torquers which are used also for unloading and nutation dumping. The momentum wheel has an angular momentum of 41 N m s and the maximum magnetic moment of the torquer is 48 ATm^2. The attitude sensors are geomagnetic aspectmeters, sun sensor, gyroscopes and CCD star sensors. Normally, an error signal is given by the difference between integrated counts of gyroscope signal and the position of the target. The signals which drive actuators are calculated by on-board computer. The drift of the gyroscopes is compensated occasionally by the position of reference stars observed with CCD star tracker of which the detection threshold is 6th magnitude. Both manual compensation by commanding and automatic compensation by on-board cpu are available. The stability of the attitude around z-axis and perpendicular to z-axis are 1.5 arcsec and 5 arcsec respectively. The better accuracy about z-axis is due to quicker response of the momentum wheel than that of the magnetic torquer. The maneuvering that is executed mostly by use of magnetic torquers is slow and the angular velocity is roughly 0.1 degree / min.

The sun sensor consists of two linear CCD sensors, perpendicular to each other and has an accuracy of 1.5 arc min. The geomagnetic aspectmeters are used for coarse determination of the attitude and for switching of polarity of the torquers.

4. NOTABLE RESULTS OBTAINED WITH GINGA

4.1 Cyclotron features of pulsar spectra

The merit of the LAC is high sensitivity in the energy region higher than 10 keV. Spectral structures in the hard x-ray region were detected from five x-ray pulsars, Her X-1 (Mihara et al. 1990), 4U1538-52 (Clark et al. 1990), V0332+53 (Makishima et al. 1990), 1E2259+58 (Koyama et al. 1989a) and 4U0115+634 (Nagase et al. 1990). The most pronounced spectrum was obtained from V0332+53 on October 1, 1989, as shown in Fig.2. The spectrum can be expressed by power law with resonant absorption at 28.5 keV.

Possible interpretation of this structure is cyclotron absorption by

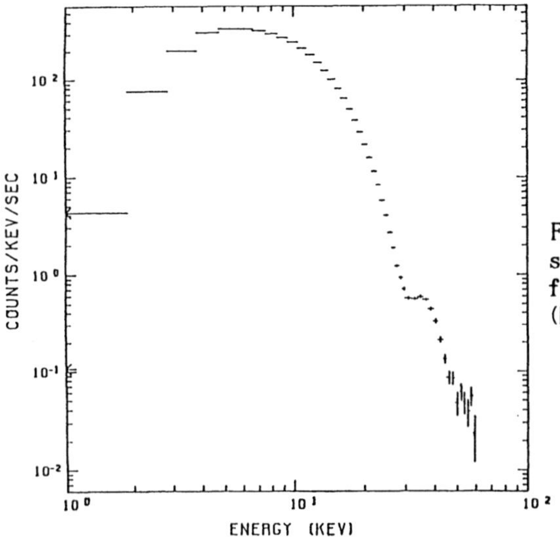

Fig.2. The spectrum of x-ray transient pulsar V0332+53. An absorption feature at 28 keV is clearly seen (Makishima et al. 1990).

electrons in highly magnetized plasma. The magnetic field can be estimated as $2.5X10^{12}$ gauss, disregarding garvitational red-shift. Further analysis, such as pulse phase dependence of absorption energy and depth will reveal x-ray transport in pulsar plasma.

4.2 X-ray scattering by interstellar dust grains

The moon occulted accidentally transient x-ray source (GS 1741-28) appeared near the Galactic center. The observation was conducted on October 26, 1987. The light curves in three energy bands are shown in Fig. 3. The gradually decreasing and increasing component around the point source, which corresponds to sharp fall and rise is clearly seen at ingress and egress respectively. The light curves show that this extended component decreases with increasing x-ray energy. One can conclude that the extended component is due to scattering by cosmic dust grains from the energy dependence of relative intensity and angular diameter.

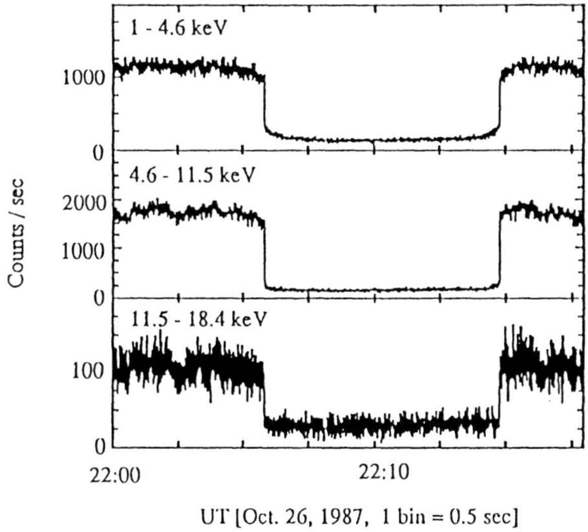

Fig.3. Light curves of transient x-ray source GS 1741-28 in a occasion of lunar occultation. The slowly varying component is clearly visible at ingress and egress (Mitsuda et al. 1990).

The size and chemical composition responsible for x-ray scattering were estimated as about 0.1 micron in diameter and iron compound.

4.3 Discovery of new transient x-ray pulsars

Five x-ray pulsars, Cep X-4, GS1843+00, GPS1722-362, GS1843-02 and Sct X-1 have been discovered with Ginga (Koyama and Takeuchi, 1989). The scanning observation of Galactic plane discovered several transient x-ray sources. Most of new sources are faint and have the hard x-ray spectra that is characteristic of x-ray pulsars. The low energy part of the spectra is heavily absorbed by inter stellar matter, and hydrogen column densities are higher than 1023. These new transients which are candidate pulsars and new pulsars are concentrated in Scutum region, possibly located in 5 kpc Galactic ring. The total number of x-ray pulsars known before Ginga is 26 and they are located within a few kpc from solar system. Ginga observations suggest existence of many faint pulsars associated with inner Galactic arms. This discovery will solve the mystery of uneven distribution of pulsars in Galactic plane.

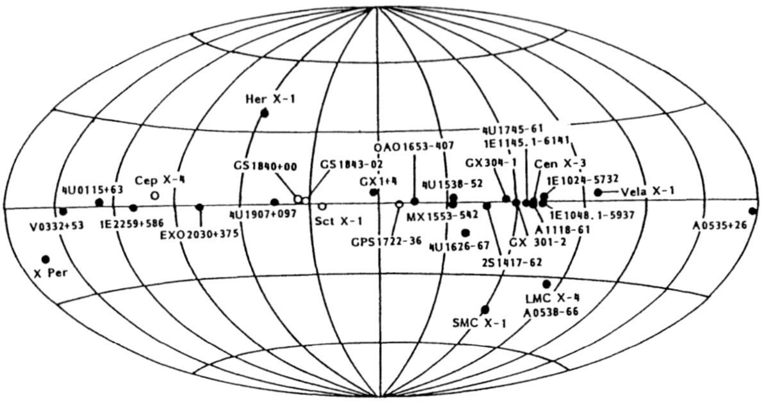

Fig. 4. Distribution of x-ray pulsars in Galactic coordinate. New pulsars discovered with Ginga are indicated by filled circles.

4.4 Intense iron line emission from Galactic center

The scanning observation along Galactic plane revealed diffuse emission of iron line peaked at Galactic center (Koyama et al. 1989b). The line energy of 6.7 keV suggests the emission is from high temperature plasma. The distribution of iron line shows sharp peak at Galactic center with angular size diameter of 1.8 degree corresponding to 300 pc at the Galactic center. The peak flux is 1.5 $X10^{-7}$ erg/s/sr/cm^2 and integrated luminosity of iron line is 2.3$X10^{36}$ erg/s. The distribution of iron line extends to Galactic ridge. A possible origin of high temperature plasma is supernova remnants. The distribution in such a narrow region suggests that successive explosion occurred at the epoch not earlier than 10^5 years.

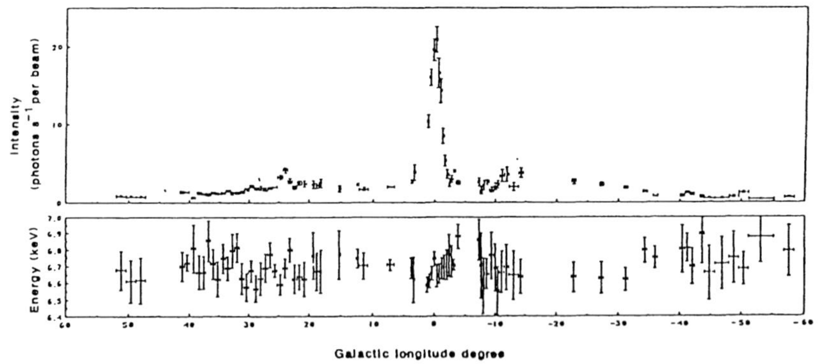

Fig. 5. Galactic longitude distribution of iron K-line. Lower panel shows line energy (Koyama et al. 1989b).

4.5 Spectral features and variability of Seyfert galaxies

Important discoveries were made on x-ray emission from Seyfert galaxies by Ginga. The iron line emission and K-absorption edge were observed from all Seyfert galaxies observed with Ginga (Pounds, 1989, Matsuoka et al., 1990,

Kunieda et al., 1990). The line energy is 6.4 keV that is fluorescent x-rays from cold matter. The equivalent widths of iron line were about 100 eV or more. While the absorbing matter density determined from spectra below 2 keV is less than 10^{22} H-atom/cm^2. Therefore, the reprocessing matter is not distributed along the line of sight but it should subtend a wide solid angle. A possible configuration is an accretion disc (George et al. 1989). The variability of iron line emission from the low luminosity Seyfert 1 galaxy, NGC 6814 was observed by Kunieda et al. (1990). They obtained a positive correlative variation between the iron line and the continuum, and time lag between them was shorter than 250 s. This is comparable to observed variability time scale. This suggests that a geometrically thin cold accretion disc extends to the vicinity of central source. On the other hand, the continuum emission from NGC 4051 and MCG-6-30-15 varied in correlated way with spectral indices (Matsuoka et al., 1990). The spectra became steeper with increasing flux. This may be a further evidence for the existence of scattered continuum which variability time scale is longer than the main component.

Ginga observed Seyfert 2 galaxies, NGC 1068 (Koyama et al., 1989c), Mkn 348 (Warwick et al., 1989) and Mkn 3 (Awaki et al.,1990). Intense iron line emission at 6.4 keV was observed from NGC 1068 and Mkn 3. The equivalent widths were as high as 1 keV. The absorbing matter densities were in the order of 10^{23} H-atom/cm^2 for Mkn 348 and Mkn 3. These facts support an idea that the Seyfert 2 galaxies are obscured Seyfert 1 galaxies. This may have an impact on the origin of the cosmic diffuse x-ray background.

4.6 X-rays from SN 1987A

The x-ray observation of SN 1987A started from February 27th, 1987, 3 days after the explosion. Positive detection was obtained in August 1987 (Dotani et al. 1987). The monitoring has been continued in every three weeks since then. Soft x-rays below 15 keV exhibited a flare peaked in January 1988 and lasted for about one month. The x-ray flux has decayed below detection limit 2 years after explosion. Pulsar search has been conducted from early stage of the observation in the frequency range of 0.03-512 Hz, taking into account frequency change. The upper limit of the pulse amplitude of 10-40 % was obtained so far (Mitsuda et al., 1988). Hard x-rays can be interpreted by the Compton scattered photons from ^{56}Co which was instantaneously synthesized at the moment of the core collapse. Early emergence of gamma-rays suggested mixing of heavy elements into outer layer (e.g. Kumagai et al., 1989). Possible explanation of the soft component is the radiation from high temperature

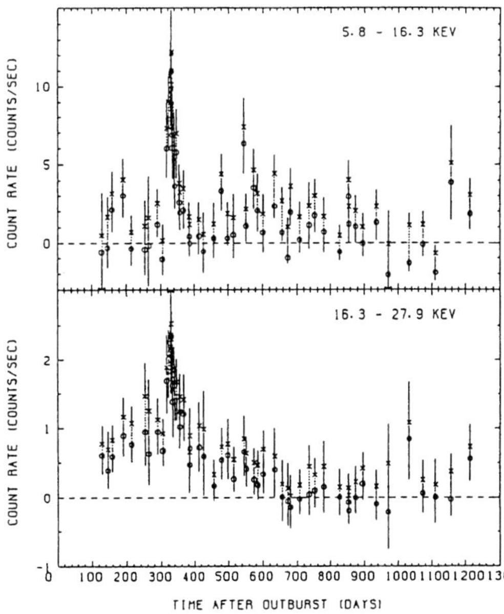

Fig. 6. X-ray light curve of SN1987A.

plasma produced by an interaction of expanding ejecta with circum-stellar matter which was lost from the progenitor.

5. CONCLUDING REMARK

Ginga has achieved highest sensitivity in the energy range from 1.5 keV to 40 keV and has revealed new aspects of x-ray astronomy.

ACKNOWLEDGMENT

Author would like to thank all the members of Ginga team for their preparation and operation of Ginga and for providing him with results of the observations. He acknowledge Dr. C. S. R. Day for careful reading of the manuscript and for comments.

REFERENCES

Awaki, H., et al. 1990, Publ. Astron. Soc. Japan, submitted.
Clark, G. W., et al. 1990, Astrophys. J., 353, 274.
Deeter, J. E., and Inoue, H., 1990, ISAS Research Note 430.
Dotani, T., et al., 1987, Nature 330, 230.
George, I. M., et al., 1989, ESA SP-296, 2, 945.
Hayashida, K., et al., 1989, Publ. Astron. Soc. Japan, 41, 373.
Kondo, I., et al., 1981, Space Sci. Instr., 5, 211.
Koyama, K., and Takeuchi, Y., 1989, ESA SP-296, 1, 483.
Koyama, K., et al., 1989a, Publ. Astron. Soc. Japan, 41, 461.
Koyama, K., et al., 1989b, Nature, 339, 603.
Koyama, K., et al., 1989c, Publ. Astron. Soc. Japan, 41, 731.
Kumagai, S., et al., 1989, Astrophys. J. 345, 412.
Kunieda., H., et al., 1990, Nature, 345, 786.
Makino, F., et al., 1987, Astrophys. Lett. and Commun. 25, 223.
Makishima, K., et al., 1990, in preparation.
Matsuoka, M., et al., 1990, Astrophys. J. in press.
Mihara, T., et al., 1990, Nature, in press.
Mitsuda, K., et al., 1988, in Physics of Neutron Stars and Black Holes, ed. Y. Tanaka, Universal Academy Press, Inc., Tokyo.
Mitsuda, K., et al., 1990, Astrophys. J., 353, 480.
Murakami, T., et al., 1989, Publ. Astron. Soc. Japan, 41, 405.
Nagase, F., et al., 1990, in preparation.
Ninomiya, K., et al., 1984, Proc. of IFAC meeting, 2915.
Pounds, K. A., 1989, ESA SP-296, 2, 753.
Tanaka, Y., et al., 1984, Publ. Astron. Soc. Japan, 36, 641.
Tsunemi, H., et al., 1989, Publ. Astron. Soc. Japan, 41, 391.
Turner, M. J. L., et al., 1989, Publ. Astron. Soc. Japan, 41, 345.
Warwick, R. S., et al.. 1989, Publ. Astron. Soc. Japan, 41, 739.

8 Man-Made Noise Environment and EMC

8.1 Man-Made Noise Sources and Environments Interference

O.V. Betsky
Institute of Radio Engineering and Electronics, Academy of Science, Moscow, USSR

Life on the Earth originated under conditions of continuos electromagnetic interference, the sources of natural interference including thunderstorm discharges, tectonic activity of the Earth's crust and outter space. Live organisms, therefore, had to develop mechanism for adaptation to various electromagnetic fields of low intensity.

The electromagnetic background experienced a drastic jump at the end of the XIXth century, when information transmission systems using electromagnetic waves to span great distances were invented. From this time onwards, radio and electric engineering inslallations radiating both useful and spurious signals experienced a rapid growth, following an exponential law.

The appearance of each new radiating device leads to an increase in the power of the electromagnetic field, this posing two major problems: 1) that of electromagnetic compatibility of radio and electric engineering equipment; and 2) that of the effect of electromagnetic interference of diverse intencity in various frequency bands of living organisms.

A. Consider some of the problems of electromagnetic compatibility.

The source of artificial electromagnetic interference is various man-made devices - from computers, power distribution systems, and to microwave devices using in radar, telecommunications, etc.

Specially designed signal generators are prone to generate spurious oscillations as well, due to nonlinear nature of interaction between electric charges and electromagnetic field.

The amplitude spectrum of spurious oscillations is a highly individual characteristic of the electron devise used. The magnitude of spurious oscillations in specific microwave electron devices are illustrated in the following Table:

Device type	Spurious oscillation level, dB
Magnetrons	-5 to -40
Travelling-wave tubes	-10 to -60
Multicavity clystrons	-40 to -50
Amplitrons	-10 to -40
Microwave solid state devices	-5 to -30

Special measures allow substantial suppression of the level of spurious oscillatios, but even so, taking into account the power capacity of modern electron devices, interference from spurious oscillations remains a significant factor.

The major sources of spurious oscillations are:

1. Nonlinear interactions are the main source of high-intencity components of the spectrum of spurious oscillations. Harmonics of the fundamental oscillation in the output signal spectrum are inherent to electron devices.
2. All electrodynamic (resonant) structures are multimode and are therefore prone to spurious resonances.
3. Self-sustained spurious oscillations in the electron devices with

resonant properties.

4. The electron beam itself is a source of discrete and noise oscillations.

5. Interference effects, if the electron devices receives external signals.

A number of new high-power sources of electromagnetic radiation, both c.w. and pulse, have appeared recently, such as various modifications of relativistic devices: gyrotrons, high-power nanosecond pulse devices. Pulse powers up to 10 GW have been attained with this latter.

Wideband noise generators of high power have been developed for countermeasure systems.

The main methods of controlling spurious oscillations are well known:

1. Filtering out spurious oscillations.

2. Optimisation of the interaction between the electron beam and electromagnetic wave in the interaction space.

3. Regulation of the device's supply voltage and magnetic field.

4. Using internal microwave absorbers.

B. Despite all these measures to suppress spurious oscillations, electromagnetic pollution of the environment is continuously growing, and this called forth the notion of *electromagnetic ecology*. Consequently, the problem of electromagnetic compatibility is gradually evolving into the urgent problem of *electromagnetic safety*.

The problem of how electromagnetic artificial interference affects live organisms, and primarilly - humans, is a most urgent one for all of us at present.

A general pattern may be discerned in the effects of

electromagnetic waves on live organisms: at small doses (low intensity) of irradiation it can be harmless, and in certain cases even medically beneficial, while at large dosage (thermal) as a rule it proves harmful, except in special cases, where high-power radiation, with certain precautions, is used for therapy (as in hyperthermy).

As is well known, such patterns are common to the effects on live organisms of small doses of alcohol, narcotic drugs, ionisation radiation. Of especial importance is the effect of pulse electromagnetic irradiation on live organisms.

Experimental studies, carried out at the Institute of Radio Engineering and Electronics of the USSR Academy of Sciences, allow some preliminary conclusions to be made, that at electric field intencities of the order of several doxen kV/cm radiation does not produce harmful effects in biological membtanes. Our experiments were carried out on frog skin, using a 3 cm relativistic generator of 15 ns pulses with output powers up to 50 MW.

It merits mention, that in a waveguide of standard cross section such a generator can provide a field intencity of the order of 10^5 V/cm, this being comparable to the intensity of the natural electric field of live biological membranes.

Harmful effects of electromagnetic waves in live organisms are widely addressed in literature, and the main problem at present is that of maximally allowable power density at workplaces.

I would like to draw your attention to the unusual properties of millimetre electromagnetic waves of low intencity, up to 10 mW/cm^2, when the integral heating of the irradiated area is, as a rule, below 0.1^oC and does not usually cause any physiological effects. I would also point out, that these wavelengths are greatly absorbed by the Earth's atmosphere, except some transparency bands. Futhermore, as is well known, artificial sources of emission of these wavelength appeared only in the middle 1960s and, naturally, radiation of natural origin could hardly significantly affect evolution of life on the Earth.

Numerous experiments, carried out for about the last two decades in the USSR and in other countries, indicate that at power intencities ranging from 0.5 mW/cm^2 to 10 mW/cm^2 millimetre waves have a therapeutic effect on the human body. At present, medical applications of millimetre wave for treatment of various deseages of the stomach, heart, circulatory system are being actively explored.

This radiation can also be used in biotechnology. Jointly with the Moscow State University, we have carried out experiments with blue-green algae during the last three years. The results obtained indicate a faster growth of the biomass under irradiation of millimetre waves, especially by greater secretion of biologically active substances into their environment.

These examples are used here to stress the main ideas of this report, which I would like to present as a conclusion.

1. The problem of artificial electromagnetic interference is largely an ecology problem, rather than the technical problem of electromagnetic compatibility of radio engineering devices.

2. We are evidencing the appearance of the problem of electromagnetic ecology and electromagnetobiology.

3. In the area of electromagnetobiology, determination of medically useful frequency and power ranges of electromagnetic waves is an urgent task. As a joke, but a joke with a great deal of sense, it may

be stated that we should seek such a "pollution" of the electromagnetic environment, at which the electromagnetic background will be not only harmless or neutral, but even would produce medicinal effects in human beings.

4. This latter statement carries the main idea of this report. A positive example of this approach has been arrived at with electromagnetic waves og the millimetre waveband.

8.2 Electromagnetic Ecology and Waves in Nonlinear Electromagnetics

V.V. KISLOV and I.V. TARANOV
USSR Academy of Sciences, Institute of Radioengineering & Electronics, Moscow, USSR

In recent years the activity of man-made sources of electromagnetic radiation (EMR) increases. So one may pose question whether this increasing activity affects living systems and surrounding these sources. The answer is of course yes. Since a huge part of industrial applications of EMR lays in microwave region, consider electromagnetic waves in this range. There was written a lot about possible biological action of microwaves, both in the sense of "electromagnetic ecology" (due to growing applications of microwaves and creation of appropriate "electromagnetic smog"), and medical cure effects of low-intensity microwave radiation [1-3].

Here we discuss, however, another aspect of the problem, concerning growing application of pulses of EMR [4]. These can be both man-made sources, in digital electronics, and rather strong (though short) pulses of EMR used in radars and other systems.

We will concentrate on two problems.

A) Whether these pulses (initially of high intensity) should be rapidly diminishing in environment, or may travel (say, like solitons) even in dissipative media and can propagate to long distances?

B) If they are propagating still with a significant amplitude, whether they could affect living systems and biologically essential processes there?

Firstly, we show that under certain conditions, in nonlinear electromagnetics, propagating EM waves may form special stable formations - solitons, travelling without essential diminishing of their amplitude on long distances. Some evaluations of the length of transparency are made for concrete media.

The important result of analytical investigation is that exact solution was found for the propagating EM solitonic formation even in a medium with dissipation with a help of Inverse Scattering Transform (IST) Method. This result is important even by itself, but also showing the possibility of analyzing length of transparency in nonlinear electromagnetics and again the possibility of propagation for pulses of EMR on substantial distances, preserving there possibilities of affecting organized electronic systems (being man-made, bioelectronics, etc.).

Secondly, we illustrate some possibilities of a short external EM pulse action on organized systems, including computer analysis of such an action on collective phenomena in a quasi-one-dimensional biomacromolecules or inducing changes of conformation in macromolecules, which could affect biologically essential processes in above mentioned systems. Also experimental approaches for investigating effects of pulse EMR action on biosystems, developed in our laboratory, are discussed, including generation, propagation of pulses and control of state of systems, exposed to EMR.

Considering above mentioned problem A, we should point out work of McCall & Hahn [5], who discovered the possibility of Self-Induced Transparency (SIT) phenomena for the intensive and short pulse, interacting with dielectric medium at a frequency, close to the resonant frequency of this medium. In the simplest case of SIT, a medium consists of two-level atoms, being not degenerated in dipole moment of transition. SIT arises, when EM pulse is formed in time in such a way, that the front of a pulse gives energy (coherently) to the medium, which stores it, returning this energy (coherently) back to the pulse. Resonant atoms remain in a ground state, effective energy transfer from pulse to medium equals zero, and pulse propagates in the medium, which is as if transparent, with a velocity less than light velocity. Main equations for this process (namely Maxwell-Bloch equations), are integrable in the sense of IST and allow solitonic solution [6], which describes propagation of EM pulse through the medium.

We considered the medium with dissipation, i.e. taking into account the processes of relaxation for the medium, which means, that now energy transfer from pulse to medium is nonzero, because of spurious processes (say, heating) in the medium. Mathematically, this means appearance of dissipative terms in Maxwell - Bloch equations:

$$
\begin{aligned}
& i P_T + i \Delta P - E \eta = 0 \\
& i \eta_T + 2 i \gamma \eta + \frac{1}{2} \left(\overset{*}{E} P - E \overset{*}{P} \right) = 0 \qquad (1) \\
& i E_X + P = 0,
\end{aligned}
$$

where $P = P(T,X)$ is polarization of individual atoms of resonant medium in the point (T, X), where

$$
\begin{aligned}
& T = \tau - x, \quad = \delta \omega_0 t, \quad X = \delta k_o x, \\
& \delta^2 = \frac{N_o \mu c^2}{2 \hbar \omega_o} \ll 1 \qquad (2)
\end{aligned}
$$

(the condition of low-density distribution of resonant dipoles), N_o being concentration of atoms, μ - dipole moment of transition, c - light velocity, $\hbar$ - Plank's constant, ω_o - frequency of transition in two-level medium (close to the frequency of EM-field in a pulse), $k_o = \omega_o / c$ is a wave-vector of EM field, and x, t - space and time coordinates, $\eta = \eta(T, X)$ is difference populations of ground and exited states, E = E (T, X) - complex amplitude of EM field, slowly changing in space and time. Coefficients are: $\Delta = (\delta\omega_0 T_2)^{-1}$, T_2 being time of relaxation for difference of populations, and

$\gamma = (2\delta\omega_0 T_1)^{-1}$, T_1 being time of relaxation for polarization.

System (1) appeared to be integrable by IST in case of

$\gamma = \Delta$ (i.e. $T_2 = 2T_1$) with (v, V) pair [7] found to be:

$$U = i(\lambda G_3 + \hat{\mathcal{E}}),$$
$$V = -\frac{i\hat{\rho}}{\lambda - i\Delta}, \tag{3}$$

where λ - spectral parameter, G_3 - Pauli matrix,

$$\hat{\epsilon} = \begin{pmatrix} 0 & \mathcal{E} \\ \bar{\mathcal{E}}, & 0 \end{pmatrix}, \quad \hat{\rho} = \begin{pmatrix} \rho_3, & \rho \\ \bar{\rho}, & -\rho_3 \end{pmatrix}, \tag{4}$$

$$\rho = P, \quad \bar{\rho} = \rho^* e^{2T} \quad \epsilon = E, \quad \bar{\epsilon} = E^* e^{2T}, \quad \rho_3 = \eta e^{2T\Delta}$$

This gives N-solitonic solution for Maxwell-Bloch equations in disipative medium. Thus we get for EM field (see (4)):

$$\hat{\epsilon} = -2\sum_{k=1}^{N} [n_k \otimes \bar{m}_k, G_3], \tag{5}$$

where bracket [....] means commutation, $\otimes$ - direct multiplication of matrices, and:

$$\bar{m}_k = \bar{A}_k \exp[-(U_o(\lambda_k)T + V_o(\lambda_k)X],$$

$\bar{A}_k$- two-component line (A_k - corresponding column) of constants of integration, determined by initial conditions,

$$U_o = U_o(\lambda) = i\lambda G_3,$$

$$V_o = V_o(\lambda) = \frac{-i G_3 \rho_3(0)}{\lambda - i\Delta},$$

$\rho_3^{(0)}$ being initial difference of populations (before the action of EM pulse), λ_k are complex constants, determined by initial conditions, connected with parameters of soliton; h_k is determined uniquely from linear algebraic system:

$$\bar{m}_k + \sum_{\ell=1}^{N} \frac{(\bar{m}_k\, m_\ell)}{\lambda_k - \bar{\lambda}_\ell}\, \bar{n}_\ell = 0,$$

and $\bar{\lambda}_\ell = \lambda_\ell^* + 2i\Delta$, λ_ℓ^* - complex conjugate of λ_ℓ, $\bar{m}_k$ means two-component line connected with a column m_k as:

$$\bar{m}_k\,(\lambda_\ell) = m_k^*\,(\bar{\lambda}_\ell),$$

where m_k^* means hermitian conjugate of m_k.

In case of 1-solitonic solution (being named sometimes as "simulton") we may write directly for EM field:

$$G = 2i\,(\eta_0 - \Delta)\, e^{2(T-\varkappa)\Delta} \,*$$

$$* \frac{\exp\left\{ -2i\eta_0 \left[\tau + \left(\frac{\rho_3^{(0)}}{\xi_0^2 + (\eta_0 + \Delta)^2} - 1 \right) x \right] + i\varphi_0 \right\}}{\operatorname{ch}\left\{ 2(\eta_0 - \Delta) \left[\tau - \left(\frac{\rho_3^{(0)}}{\xi_0^2 + (\eta_0 + \Delta)^2} - 1 \right) x \right] + \varkappa_0 \right\}}, \tag{6}$$

where spectral parameter λ_0 (λ_k from (5) for 1-solitonic case) is: $\lambda_0 = \xi_0 + i\eta_0$, and other constants, determined by initial conditions, are: $\varkappa_0$ - initial coordinate of the center of soliton, and φ_0 - initial phase of EM field in a pulse.

Thus, the solitonic solution (6) for system (1) with dissipation describes also diminishing amplitude of EM field in a pulse with a "length of transparency":

$$L_0 = (2\,\Delta\, k_0\, \delta)^{-1}. \tag{7}$$

This solitonic formation is propagating with a velocity:

$$v = c\left(1 + \frac{\rho_3^{(0)}}{\xi_0^2 + (\eta_0 + \Delta)^2} \right)^{-1} \tag{8}$$

and has spatial "length" $L^* = v\, t_0$, where "pulse duration" t_0 is:

$$t_0 = \operatorname{mod}\,(2\,(\eta_0 - \Delta)\,\omega_0\,\delta)^{-1}. \tag{9}$$

mod means here absolute value.

When dissipation is finite, but not very huge, there may occur localized EM pulse, propagating with a velocity v, having "length" L* and still capable for essential EM effects on distances about L_0 from the source. Thus the solution (6) with (7-9) allows direct analysis of possible effects for EM pulses in a given medium.

Therefore, under certain conditions even in media with dissipation, the analogue of SIT phenomenon is possible, with propagating EM pulse in a resonant (or close to resonant), i.e. nonlinear electromagnetics, preserving their form and loosing rather slowly their amplitude. Since the possibility of this phenonemon is shown in principle, we may consider problem B of our report, concerning possible effects of EMR action on organized systems (including electronics, bioelectronics, etc.).

In these aspects one should first of all evaluate possible effects of direct action of EM field on electrons, ions, structural groups of macromolecules, etc. When considering these, effects on subunits of electronics (bioelectronics) systems, also excitation of EM field due to interaction with dipole structures and connected acousto-electronic effects were taken into account [8]. In that work the intensity of EM pulse was evaluated, necessary to produce biologically essential effect on macromolecules, leading to change in their conformation. In particular, for real macromolecules with dissipation processes taken into account, effects of microwave pulse action on conformation were considered. For molecular length of about $L = 10^{-6}$cm and frequencies $\omega_0 \sim 2\pi\delta_{sound}/L$ (where $\delta_{sound} \sim 10^5$cm/sec, $\omega_0 \sim 10^{11}$ sec^{-1}) the relaxation time is $\tau_h \sim 10^{-10}$sec for hydrodynamics relaxation and $\tau_p \sim 10^{-9}$ sec for phonon relaxation if external solvate surrounding of macromolecules is almost at rest (oscillating internal parts of macromolecules). That means qualities of oscillations on a longitudinal mode of macromolecules of about $Q \simeq 10 - 100$. So during the action of a macrowave pulse only about $N_{tot} \simeq \tau_{pulse} \nu_0 \simeq 150$ periods of field are acting effectively, storing in a resonant case. Longer duration of pulses lead to heating, but not to coherent storage of energy on a longitudinal mode with a dipole moment. To store energy, exceeding biologically essential for fermentative reaction value $E_f \sim 0.5$ ev ($\sim$20 kT), (kT is thermal energy), one need intensities of EM pulse:

$$W_0 \geq \frac{Mc}{2\pi d^2} \cdot \frac{E_f L^2}{\tau^2}, \qquad (10)$$

where M molecular weight (let it be $\simeq 10^4$ Dalton), L - length of molecule ($\simeq 10^{-6}$cm), light velocity c = 3.10 ^{10}cm/sec, dipole moment $d \simeq 100$ Debay, and with relaxation time $\tau \simeq 10^{-9}$sec, one get

$$W_0 \geq 3 \text{ Mega Watt/cm}^2 = 3.10^6 \text{ w/cm}^2.$$

We investigate, in our laboratory experimentally, action of EM pulses with intensities up to $W \simeq< 10^8$ w/cm^2 and pulse duration of about $\tau \simeq 10^8$ sec. Due to short pulses used in experiments, we avoid direct heating effects, but initiating conformational changes directly by EM field of a pulse. In a special experimental set up, experimenters are analyzing by optical control the sequence of conformational changes in films and suspensia of macromolecules. This work should be enlarged to investigate also other frequency bands and intensities action, which are of common use in radars, military and industrial applications, etc. Some evidence of short pulse external action on collective phenomena (i.e. propagating solitons etc.) has been obtained during computers simulations for model system [9], where a possibility of some resonant sensitivity to such action was discovered. In particular, some existing regimes in a system were completely destroyed by a short pulse, field with a resonant EMR, depending on concrete parameters of given systems.

The examples considered above are only a small illustration of a problem, concerning investigation of EM pulses propagation and action in real (and very often - nonlinear) electromagnetic environment, including bioelectronic applications.

REFERENCES

[1] N.D. Devyatkov et. al.//Sov. Phys.(Uspekhy), Engl. Transl., 1974, 16, 568.
[2] H. Frohlich, F. Kremer//ed. Coherent excitations in biological systems, Springer-Verlag, 1983.
[3] N.D. Devyatkov//ed. Low-intensity microwaves applications in medicine and biology. Moscow, USSR, Ac. of Sci., Inst. of Radieng. & Electronics, 1989 (in Russian).
[4] H. Kikuchi//ed. Nonlinear and Environmental Electromagnetics, Elsevier, 1985.
[5] S.L. McCall, E.L. Hahn//cited in Bull. Am. Phys. Soc. 10, 1189, (1965).
[6] M.J. Ablowitz, H. Segur//Solitons and the Inverse Scattering Transform, SIAM, Philadelphia, 1981.
[7] V.E. Zakharov, S.A. Manakov, S.P. Novikov, L.P. Pitaevsky//Theory of Solitons, Moscow, 1980 (in Russian).
[8] N.D. Devyatkov, V.V. Kislov, Z.S. Chernov//Prep. N6 (481), 1988, USSR, Ac. of Sci., Inst. of Radioeng. & Electr., (in Russian).
[9] V.V. Kislov, Yu.A. Kriksin//Mathematical Simulation, 1989, 1, N4, 87.

8.3 Determination of the Mutual Inductance Between Parallel Lines by Making Use of Struve's Function

Y-G. GAO and Y-H. LU

EMC Laboratory, Beijing University of Posts and Telecoms., Beijing, PR China

Abstract: This paper shows an exact formula of mutual inductance (MI) emploied special functions, and points out that the result from Carson's formula is better than that from Pollaczek's formula under the conditions of small distance between wires and small earth conductivity. Finally by using special functions, the method to determine the MI between three phase power lines and a communication line is given.

1. An Exact MI Formula of Infinite Length Parallel Wires

The solutions of the mutual inductance of infinite parallel wires with earth return were given by Pollaczek [1] and Carson [2]. But these solutions are approximate forms which should be used in different forms for different earth conductivity. In 1959, the author gave an exact formula of determining mutual inductance by making use of Struve's function [3]. Here is the main point.

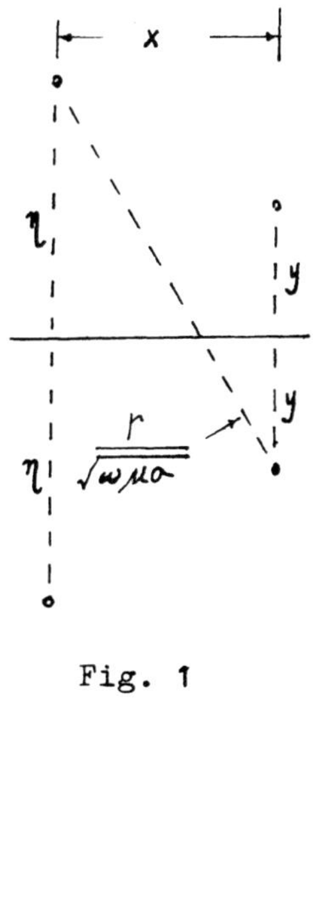

Fig. 1

The mutual inductance between wires (shown in Fig. 1) is

$$M = M_0 - j\,4J(p,q) \qquad 10^{-7} \quad h/m$$

where

$$M_0 = 2ln\sqrt{\frac{x^2 + (y+\eta)^2}{x^2 + (y-\eta)^2}}$$

$$J(p,q) = P + j\,Q = \int_0^{\infty} (\sqrt{u^2 + j} - u)e^{-pu}\cos(qu)du$$

Another form of J(p,q) is

$$J(p,q) = \int_0^{\infty} (\sqrt{u^2 + j} - u)e^{-pu}\cos(qu)du$$

$$= \int_0^{\infty} \sqrt{u^2 + j}e^{-pu}\cos(qu)du - \int_0^{\infty} ue^{-pu}\cos(qu)du -$$

$$= \frac{1}{2}\int_0^{\infty} \sqrt{u^2 + j}e^{-(p-jq)u}du + \frac{1}{2}\int_0^{\infty} \sqrt{u^2 + j}e^{-(p+jq)u}du -$$

$$-\frac{1}{2}\int_0^{\infty} ue^{-(p-jq)u}du - \frac{1}{2}\int_0^{\infty} ue^{-(p+jq)u}du$$

By using the relation [3], [6]

$$\int_0^\infty e^{-\beta\tau}(\tau^2+\alpha^2)^{n-\frac{1}{2}}d\tau = (\frac{\alpha}{\beta})^n(\frac{\pi}{2})(2n-1)!![H_n(\alpha\beta) - Y_n(\alpha\beta)] \quad (1)$$

where

$H_n(\alpha\beta)$——Struve's function of order n;

$Y_n(\alpha\beta)$——second kind Bessel function of order n.

We get

$$J(p,q) = \frac{\pi}{4r}\left\{e^{j\varphi_1}[H_1(re^{j\varphi_2}) - Y_1(re^{j\varphi_2})] + e^{j\varphi_2}[H_1(re^{j\varphi_1}) - Y_1(re^{j\varphi_1})]\right\} - \frac{cos(2\theta)}{r^2}$$

where

$p = (y+\eta)\sqrt{\omega\mu\sigma}$;

$q = x\sqrt{\omega\mu\sigma}$;

$r = \sqrt{p^2+q^2}$; $\varphi_1 = 45° + \theta$; $\varphi_2 = 45° - \theta$;

$\theta = arctg(\frac{q}{p}) = arctg(\frac{x}{y+\eta})$;

x, y, η measured in meter;

σ——measured in siemens / meter;

μ——measured in henrys / meter.

and

$$M = \left\{2ln\sqrt{\frac{x^2+(y+\eta)^2}{x^2+(y-\eta)^2}} + j\frac{4cos(2\theta)}{r^2} + \right.$$
$$\left. + \frac{2}{r}[e^{-j\varphi_1}F_1(re^{j\varphi_1}) + e^{-j\varphi_2}F_1(re^{j\varphi_2})]\right\} \quad 10^{-7} \quad h/m \quad (2)$$

where

$$F_1(re^{-j\varphi}) = \frac{\pi}{2}[H_1(re^{j\varphi}) - Y_1(re^{j\varphi})]$$

The formula of mutual inductance with special function expression was introduced in 1963 by CCITT [see p.133, (4)].

2. Comparing Carson's Formula with Pollaczek's Formula

With small distance from one wire to other wire and small earth conductivity, more exact mutual inductance formula will be required for determining whether induction voltage exceed a criterion, and which protection measure should be taken. So it arises a question which formula (Carson's and Pollaczek's) is better. Here the author analyses these formulae. The conclusion drawn from the exact formula is that Carson's formula is better than Pollaczek's. The proof is given as following.

When s is very small, we have

$$\frac{\pi}{2}[H_1(s) - Y_1(s)] \simeq \frac{s^2}{3} - [J_1(s)ln(\frac{1.78s}{2}) - \frac{1}{s} - \frac{s}{4} + \frac{5s^3}{64}$$

$$\simeq -\frac{s}{2}ln(\frac{1.78s}{2}) + \frac{1}{s} + \frac{s}{4} \qquad (J_1(s) \simeq \frac{s}{2})$$

thus

$$M = 2ln\sqrt{\frac{x^2 + (y+\eta)^2}{x^2 + (y-\eta)^2}} - 2ln\frac{1.78r}{2} - j\varphi_1 - j\varphi_2 + 1$$

$$= 2ln\frac{2}{1.78\sqrt{\omega\mu\sigma}\sqrt{x^2 + (y-\eta)^2}} + 1 - j\frac{\pi}{2} \qquad 10^{-7}\ h/m \tag{3}$$

This is Pollaczek's approximate formula under the condition of small distance from one wire to others and small earth conductivity.

As s is very small if taking

$$\frac{\pi}{2}[H_1(s) - Y_1(s)] \simeq \frac{s^2}{3} - [J_1(s)ln(\frac{1.78s}{2}) - \frac{1}{s} - \frac{s}{4} + \frac{5s^3}{64}]$$

and

$$J_1(s) \simeq \frac{s}{2} - \frac{S^3}{16}$$

we get Carson's formula under the same conditions [3]

$$M = M_0 - j\,4P + 4Q \qquad 10^{-7}\ h/m \tag{4}$$

where

$$M_0 = 2ln\sqrt{\frac{x^2 + (y+\eta)^2}{x^2 + (y-\eta)^2}}$$

$$P \simeq \frac{\pi}{8} - \frac{r\cos\theta}{3\sqrt{2}} + \frac{r^2\cos(2\theta)}{16}(0.6728 + ln\frac{2}{r}) + \frac{\theta r^2\sin(2\theta)}{16}$$

$$Q \simeq -0.386 + \frac{1}{2}ln\frac{2}{r} + \frac{r\cos\theta}{3\sqrt{2}}$$

It is obvious that Carson's formula has better accuracy, because of using special function for the approximate formula with better accuracy.

3. The Mutual Inductance between Three Phase Power Lines and a Communication Line

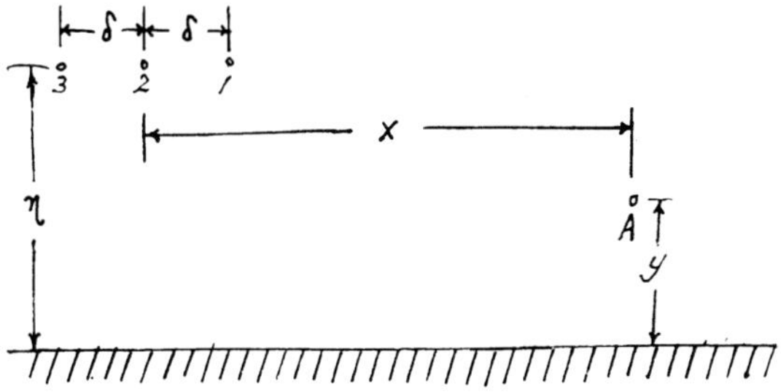

Fig. 2

The case is shown in Fig. 2. Assume the currents in the three phase power lines to be

$$I_1 = Ie^{j\frac{2\pi}{3}}, \qquad I_2 = I, \qquad I_3 = Ie^{-j\frac{2\pi}{3}}$$

then, the induced emf. per unit length along the communication line is

$$E = -j\,\omega M_{1A} - I_1 - j\,\omega M_{2A} I_2 - j\,\omega M_{3A} I_3$$

the mutual inductance between the power lines and the communication line is

$$\begin{aligned} M_{123A} &= M_{1A} e^{j\frac{2\pi}{3}} + M_{2A} + M_{3A} e^{-j\frac{2\pi}{3}} \\ &= \frac{1}{2}(M_{2A} - M_{1A}) + \frac{1}{2}(M_{2A} - M_{3A}) + j\frac{\sqrt{3}}{2}(M_{1A} - M_{3A}) \end{aligned}$$

where

$$M_{2A} - M_{1A} = M(x) - M(x-\delta) \simeq \frac{dM}{dx}\delta$$

$$M_{2A} - M_{3A} = M(x) - M(x+\delta) \simeq \frac{dM}{dx}(-\delta)$$

$$M_{1A} - M_{3A} = M(x-\delta) - M(x+\delta) \simeq \frac{dM}{dx}(-2\delta)$$

then

$$\begin{aligned} M_{123A} &\simeq \frac{1}{2}\frac{dM}{dx}\delta + \frac{1}{2}\frac{dM}{dx}(-\delta) + j\frac{\sqrt{3}}{2}\frac{dM}{dx}(-2\delta) \\ &\simeq -j\sqrt{3}\frac{dM}{dx}\delta \end{aligned} \tag{5}$$

Now we determine the mutual inductance M [5]. Differentiating the both sides of equation (1) with respect to β, we

$$\int_0^{\infty} \tau e^{-\beta\tau}(\tau^2+\alpha^2)^{\frac{1}{2}} d\tau = \frac{\alpha}{\beta^2}(\frac{\pi}{2})[H_1(\alpha\beta)-Y_1(\alpha\beta)] - \frac{\alpha^2}{\beta}(\frac{\pi}{2})[H_1'(\alpha\beta)-Y_1'(\alpha\beta)]$$

from reference [6]

$$\frac{dH_1(x)}{dx} = H_0(x) - \frac{1}{x}H_1(x)$$

$$\frac{dY_1(x)}{dx} = Y_0(x) - \frac{1}{x}Y_1(x)$$

inserting above equations into the differential equation, we get

$$\begin{aligned} \frac{dM}{dx} = & \frac{8xy\eta}{x^2+(y-\eta)^2}\frac{\omega\mu\sigma}{r^2} - 2\sqrt{\omega\mu\sigma}\, e^{j\frac{3\pi}{4}} \left\{ \frac{2}{r^3}[e^{-\beta\varphi_2} - e^{-\beta\varphi_1}] + \right. \\ & + \frac{1}{r}[e^{-j\varphi_2}F_0(re^{j\varphi_2}) - e^{-j\varphi_1}F_0(re^{j\varphi_1})] - \\ & \left. - \frac{2}{r^2}[e^{-j2\varphi_2}F_1(re^{j\varphi_2}) - e^{-j2\varphi_1}F_1(re^{j\varphi_1})] \right\} \end{aligned} \quad (6)$$

where

$$F_0(s) = \frac{\pi}{2}[H_0(s) - Y_0(s)]$$

$$F_1(s) = \frac{\pi}{2}[H_1(s) - Y_1(s)]$$

then, inserting (6) into (5), we get

$$\begin{aligned} M_{123A} = & \left\{ -j\frac{8\sqrt{3}xy\eta\delta}{x^2+(y-\eta)^2}\frac{\omega\mu\sigma}{r^2} - \frac{8\sqrt{3\omega\mu\sigma}}{r^3}\delta sin(3\theta) - \right. \\ & - \frac{2\sqrt{3\omega\mu\sigma}}{r}\delta e^{j\frac{\pi}{4}}[e^{-j\varphi_2}F_0(re^{j\varphi_2}) - e^{-j\varphi_1}F_0(re^{j\varphi_1})] \\ & \left. + \frac{4\sqrt{3\omega\mu\sigma}}{r^2}\delta e^{j\frac{\pi}{4}}[e^{-j2\varphi_2}F_1(re^{j\varphi_2}) - e^{-j2\varphi_1}F_1(re^{j\varphi_1})] \right\} 10^{-7} h/m \end{aligned} \quad (7)$$

When s is a complex number, $F_0(s)$ and $F_1(s)$can be found in the table of reference [7].

REFERENCES

〔1〕 Pollaczek: Uber das feld einez unendlich langen wechselstromdurch flossen einfachleitung, ENT, t.3, 1926.

〔2〕 Carson: Wave propagation in overhead wires with ground return, BSTJ, Oct., 1926.

〔3〕 Gao Yougang: Some applications of Bessel function in protection technique of telecommunication lines, Telecommunication USSR, No.11,1959.

〔4〕 CCITT Directives concerning the protection of telecommunication lines, p.133, 1963.

〔5〕 Lu Yinghua, Gao Yougang: The mutual coupling inductance between three phase power lines and communication lines, Journal of Beijing Univ. of Posts and Telecoms., No.1, 1983.

〔6〕 Watson: A treaties of the theory of Bessel functions, 1945.

〔7〕 Gao Yougang: Inductive coupling and resistive coupling, People's Pub. House of Posts and Telecoms., Beijing, China, 1979.

Biographical Note

Prof. Gao Yougang is a tutor and advisor of doctor aspirants of Beijing Univ. of Posts and Telecoms. He is now Chinese official member of E–Commision, URSI; Fellow of Chinese Institute of Electronics; Vice Chairman of EMC–Committee of Chinese Electrotechnical Society.

9 Noise and Communication Statistics

9.1 From Radio Interference to Electromagnetic Compatibility – Classification of the Electromagnetic Environment

F.L.H.M. Stumpers
Elzentlaan 11, Eindhoven 5611 LG, The Netherlands

In 1933 several international organisations met in Paris, to see, whether measures could be taken, to improve the quality of the reception of audio broadcasting. During the last ten years before that date, radio broadcasting had grown from reception by a few amateurs to very widespread reception. The quality of the transmitted programmes had improved and famous orchestras like the Amsterdam Concertgebouw Orchestra contributed. Under these circumstances radio interference became a real problem.

The preparatory reunion led in the following year to the foundation of the International Special Committee on Radio Interference (C.I.S.P.R.) by the International Electrotechnical Commission and the International Union of Radio Broadcasting. Representatives of C.C.I.R., C.C.I.F., (the Consultative Committees of the International Telecommunication Union), UNIPEDE (producers and distributors of electric power), UITP (electric tramways) were also present. Mr. C.C. Paterson (U.K.) presided. Two subcommittees were formed, one on Limits (chairman Mr. Braillard, UIR) and one on Measurements (chairman Dr. Balth van der Pol, Netherlands).

It was soon decided to choose a quasi-peak measuring instrument, with the parameters chosen, in accordance with auditory perception. Professor Morren (Univ. Louvain) and his collaborators constructed the first measuring apparatus, and made several copies available to different National Committees.

This was the situation around 1939, when the war intercepted the work for several years. In 1946 Dr. S. Whitehead presided over a meeting in London, and he did so too in Luzern, 1947, Paris, 1950 and London, 1953. In 1953 he retired.

His successor as president of CISPR was Mr. W.O. Humphreys, an excellent organizer, who soon decided, that biannual or triannual meetings were not enough to make real progress. He formed a Steering Committee of international experts, and soon organized several working groups on special subjects. I was one of the representatives of the Netherlands at the 1953 meeting, became a member of the Steering Committee soon afterwards, and chairman of the working group on radio and television receivers.

The original French title of CISPR referred to "perturbations radiophoniques", but with the growth of television, not only sound was important, and it was decided to change to "perturbations radioelectriques". The first CISPR receivers covered the domain from 150 KHz to 1605 KHz, but the upper limit was soon raised to 30 MHz, later 300 MHz, followed by 1 GHz and 18 GHz.

The idea had been to have one receiver made, and to deliver copies to all interested countries, but this was rather impractical, especially in view of the introduction of the transistor. It was therefore decided, to define all important parameters in Specifications. During the meetings of 1956, 1958, it was decided to follow the model of CCIR: Recommendations, Reports, Study Questions (a contraction of Questions, Study Programmes). Usually Study Questions start from the fact that there are typical sources of radio noise, then asks, how to measure them, and finally, what limits are recommended. Ignition systems, gaseous discharge devices, I.S.M. equipment were treated in this way. The requirement of a CISPR limit was clarified by the demand, that 80% of the production should be within the limit with a confidence radio of 80%.

A difficult question was S.Q.77. It is in line with recent developments in other Technical Commission of IEC, but to answer it is still difficult: The C.I.S.P.R., considering

a) that man-made noise and its effect on radio reception are generally random phenomena subjected to numerous factors, which it is difficult not only to analyse but even merely to take into account;

b) that for further development and specification of methods, it is necessary to combine various interrelated statistical parameters in appropriate ways;

decides that the following questions should be studied:

1) What are the appropriate distribution functions for describing boundaries with which the following values comply:

   1.1 The values of radio interference voltages, radio interference power or radio interference field intensity, produced by interference sources
      a. of the same type and
      b. by all interference sources relating to a given group of interference suppression limits/appliances for domestic purposes, industrial equipment, transport, etc.

   1.2 The values of decoupling factors between various interference sources and vairous types of receiver.

2) What analytic or other techniques should be used to relate the functions described in sub-clause 1.1 and 1.2 in order to establish limits that would provide a prescribed quality with a prescribed confidence in realization of the prescribed reception quality, when one or other of the limits is applied?

The limits originally followed the German legislation, but now it was found necessary to review the methods of specifying limits. This led to CISPR Report 23: Determination of Limits for Industrial, Scientific and Medical equipment.

Ideally one would have to take into account no less than 10 probability distributions and even a computer model constructed in the USA, only can give some guidance.

I.S.M. equpment is divided in groups and classes. This year the revised Publication 11 was adopted, with agreement on all major points, though not yet on limits for specially allocated I.T.U. frequencies. Most likely the revised Publication 11 will soon be adopted by the European Community and lead to legislation in the member countries within a few years.
There are already quite a number of such European standards:

Limits and methods of measurement of radio interference characteristics of HOUSEHOLD ELECTRICAL APPLIANCES, PORTABLE TOOLS AND SIMILAR ELECTRICAL APPARATUS.
Limits and methods of measurement of radio interference characteristics of FLUORESCENT LAMPS AND LUMINAIRES.
Limits and methods of measurement of radio interference characteristics of INFORMATION TECHNOLOGY EQUIPMENT.
Immunity to radiofrequency interferences of BROADCAST RECEIVERS AND ASSOCIATED EQUIPMENT.

As yet only a draft European Standard is available for:

Limits and methods of measurement of radio disturbance characteristics of BROADCAST RECEIVERS AND ASSOCIATED EQUIPMENT.

In its EMC Directive of 3 May 1989 the Council said: This Directive applies to apparatus liable to cause electromagnetic disturbance or the performance of which is liable to be affected by such disturbance. It defines the protection requirements and inspection procedures relating thereto. The apparatus shall be so constructed that:

a) the electromagnetic disturbance it generates does not exceed a level allowing radio and telecommunications equipment and other apparatus to operate as intended.

b) the apparatus has an adequate level of intrinsic immunity of electromagnetic disturbance to enable it to operate as intended.

In the Annex an EC (European Community) declaration of conformity is described as well as a list of 12 types of apparatus to be protected. There are perhaps few other laws in which electromagnetic compatibility is described in such general terms. The term "electromagnetic compatibility" dates from about 1958.

In 1961 the Plenary Assembly of C.I.S.P.R. was held in Philadelphia and during my flight Mr. Stanley, secretary from 1947 to 1961, told me, that Mr. Humphreys planned, to ask me to take charge of the CISPR Subcommittee on Limits, as Mr. Thomas, who had been chairman for Limits since 1953 had withdrawn from this position. At the time, it was known that Mr. Humphreys would retire at the end of this meeting, and that Prof. Morren, since 1953 chairman for measurements would probably succeed him. In my preparation for the Assembly I had therefore given more attention to the papers on Measurements. Quite a large number of Limits, e.g. for domestic appliances, radio and television receivers, I.S.M. apparatus, ignition noise were accepted.

I remained chairman for Limits during the Assemblies of Stockholm, 1964, and Stresa, 1967. During this latter meeting I was elected president of C.I.S.P.R. Until that time the UK Committee of I.E.C. had taken care of nearly all secretariat duties, but with the increasing number of documents, this became to great a burden for one national committee. After much discussion it was decided to prmote the working groups to Sub-Committee and to spread the secretarial work over several National Committees. There were after the 1973 Plenary Assembly of West Long Branch (U.S.A.) six Subcommittees:

- A - Measurements and statistical methods.
- B - Industrial, scientific and medical apparatus.
- C - High voltage lines and equipment. Electric traction.
- D - Ignition noise.
- E - Radio and television receivers. Immunity.
- F - Domestic appliances (including gas discharge lamps).

In 1985 another Subcommittee was formed:

- G - Information Technology.

My successor as president was Mr. Meyer de Stadelhofen, wellknown for his invention of the absorbing clamp, as a measuring instrument for the radiation of domestic appliances between 40 and 300 MHz (maybe even up to 1000 MHz). Due to his retirement in 1977, between two Plenary Assemblies, Mr. Akerlind (Sweden) as vice-president Limits filled the interval between 1977 and 1979, in which year he was succeeded by Professor Showers (Penn. Univ.) formerly chairman of sub-Committee B (Measurement) 1961-1973, and of the new sub-Committee A 1973-1979. He retired in 1985 and was succeeded by Mr. Jackson of the Electrical Research Association (UK), earlier vice-president for Limits, and chairman of the new Sub-Committee B.

In 1970 it was decided to combine all the Recommendations, Reports and Methods of Measurement, applying to a special subject in a Manual. The first publication of this type to be approved was the Manual on I.S.M. apparatus (1973). However, when it became clear, that the European Community intended to take over the rules for interference from I.S.M. equipment in its legislation, there was opposition from two sides; from airlines, who thought the limits were not strict enough in connection with their safety, and from manufacturers of I.S.M. equipment, who thought the limits were unnecessarily strict.

In 1974 we organized the second Eurocon in Amsterdam, an international technical conference of IEEE Region 8. I gave there a survey paper on electromagnetic compatibility. Professor Borgnis of the Eidgenossische Technische Hochschule in Zurich was in the audience. After the lecture he told me, that he planned to have a conference on this subject in Switzerland. Could he have the full text of my lecture? At the time the only printed text was an abstract in the Symposium Digest, but when I gave my own text to Professor Borgnis, he had it published in the Bulletin of the Swiss Engineering Society, and he distributed 5000 copies among possibly interested people. He asked me to act as program chairman of the Symposium, that took place in Montreux 1975 and again in 1977. In 1979 the Symposium was held in Rotterdam, and in 1981 in Zurich. From 1983 onwards I was vice-president of this Symposium. Our Polish colleagues, who had already organized such a Symposium in 1972, asked me to be their program chairman too, so that from 1976 on, I was every second year in Wroclaw.

For a long time the CISPR had a subcommission on terminology, that helped the IEC with the terms on radio interference in their vocabulary. Recently IEC TC 77 reviewed the terms in this vocabulary. We will give some example from Chapter 161 of the International Electrotechnical Vocabulary:

Electromagnetic Interference is the degradation of the performance of a device, equipment or system, caused by an electromagnetic disturbance.

An electromagnetic disturbance is any electromagnetic phenomenon, which may degrade the performance of a device, equipment or system, or adversely affect living or inert matter.

The electromagnetic environment is the totality of electromagnetic phenomena existing at a given location.

Electromagnetic Compatibility is the ability of a device, equipment or system, to function satisfactorily in its electromagnetic environment, without introducing intolerable disturbance to anything in its environment.

Emission is the phenomenon, by which electromagnetic energy emanates from a source.

The emission level is the level of a given electromagnetic disturbance, emitted from a particular device, equipment or system, measured in a specified way.

The emission limit is the specified maximum emission level of a source of electromagnetic disturbance.

The disturbance level is the level of a given electromagnetic disturbance, measured in a specific way.

The disturbance limit is the maximum permissible electromagnetic distirbance level.

Electromagnetic susceptibility is the inability of a device, equipment or system, to perform without degradation in the presence of an electromagnetic disturbance.

Immunity (to an electromagnetic disturbance) is the ability of a device, equipment or system, to perform without degradation in the presence of an e.m.d.

Degradation is an undesired departure in the operational performance of any device, equipment or system from its intended performance.

The immunity level is the maximum level of a given disturbance incident on a particular device, equipment or system, for which it remains capable of operating at a required degree of performance.

The immunity limit is the minimum required immunity level.

The electromagnetic compatibility level is the specified disturbance level, for which an acceptable probability of electromagnetic compatibility should exist.

Apart from CISPR several Technical Committees consider E.M.C. in relation with their own specific subject. In order to coordinate these efforts I.E.C. installed (1981) an Advisory Committee on Electromagnetic Interference (ACEC). The E.C. has set up CENELEC TC 110 for characterization of the electromagnetic environment. TC65: Industrial Process Measurement and Control has a Working Group 4, Electromagnetic Interference. It considers: The characterization of the EMC-Immunity of Industrial Process Measurement and Control equipment by EMC phenomena, and considers: a) Impulsive Interference like ESD, Electrical Fast Transients, Bursts and Surges, mainly conducted. b) Continuous wave interferences, caused by licensed radio transmission, radiated and conducted, 150 kHz-1 GHz. c) Low frequency (50/60 Hz-500 kHz) magnetic field coupling and conducted interferences.

Additionally it was felt necessary to have a Recommendation like "EMC-Environmental Classification and Installation Guideline".

TC75: "Classification of environmental conditions" set up a Preparatory Working Group: "Electrical and electromagnetic parameters and their severities".

TC77: Electromagnetic Compatibility between electrical equipment including networks" has several Working Groups. Working Group 6, Electromagnetic interference other than network frequency related, and Working Group 8, Electromagnetic interference related to the network frequency, are preparing a description of their electromagnetic environment. These Working Groups have to establish environmental levels (compatibility levels), to which susceptibility levels can be related.

The Electromagnetic Environment

1) Typical locations: Residential, Commercial, Light Industrial, Heavy Industrial, High Voltage Sub Station, Telecom Centre.

2) Interfaces: Power supply, Telecommunications, Data, Control, Screen.

Working Group 1 of TC110 decided that one environmental class could be established for the residential, commercial and light industrial locations. The vast majority of products operate in such an environmental class. It was assumed that a reasonable separation exists between any source of disturbance and potential victims. The distance to any high-voltage installation was assumed to be at least 30 m. Mains lead of equipment at least 1.5 m length. At leat 1 m separation between different pieces of equipment. Distance to any particularly strong source (e.g. amateur radio transmitter) at least 3 m.

IEC TC 77 gives the following list of main electromagnetic disturbances:
Conducted Low Frequency Phenomena: Signalling Systems, Voltage Fluctuations, Voltage dips and interruptions, Unbalance, Power Frequency variations, DC in AC networks, Harmonics and Interharmonics.

Radiated Low Frequency Phenomena: Magnetic fields, Electric Fields.
Conducted High Frequency Phenomena: Unidirectional surges*, Oscillatory transients*, Induced CW voltages or currents. (*= bursts included).

Radiated High Frequency Field Phenomena: Magnetic fields, Electric fields, Electromagnetic fields. (continuous waves and transients).
Electrostatic Discharge Phenomena.
Nuclear Electromagnetic Pulse Phenomena.

For most of these phenomena in the residential class, WG1 of TC110 has given a maximum level, more as a representation of what occurs in the environment, than as a description.

IEC TC 65 (Secr. 99) gives the following classes for Environment for Conducted and Magnetic Influences:

Level 1. Well Protected Environment. The installation is characterized by: suppression in the switched control circuits; separation between power supply lines (AC and DC), control and measurement circuits coming from other environments belonging to higher severity levels; shielded power supply cables with the screens earthed at both ends on the reference ground of the installation, with power supply protection by filtering. (e.g. computer room).

Level 2. Protected Environment. Partial suppression in the control circuits switched by relays (not contactors); separation of all the circuits from other circuits belonging to environments of higher severity levels; physical separation of unshielded power supply and control cables from signal and communication cables. (e.g. control rooms industrial plants).

Level 3. Typical Industrial Environment. No suppression in the control circuits switched by relays (not contactors); poor separation of the circuits from other circuits belonging to environments of higher levels; dedicated power supply, control, signal and communications cables; poor separation between power supply, control, signal and communications cables; availability of earthing systems represented by pipes, ground conductors in the cable trays (connected to the protection earth system) and a ground mesh. (e.g. power plants).

Level 4. Severe industrial environment. No suppression of control and power circuits switched by relays and contractors; poor separation of circuits from environments of higher severity levels; no separation between power supply, control, signal and communication cables; use of multicore cables common for control and signal lines. (e.g. power station open air H.V. switchyards).

Level 5. Special situation to be analyzed. Dubious separation of interference sources from equipment, circuits, cables, etc.; Connected lines may originate from environments of higher severity.

ACEC asked TC65, WG4, during its Paris meeting 1987 to write a document on "basic immunity requirements" applicable to TC77 and CISPR. We have already mentioned the European Standard EN 55 020: Immunity to radio frequency interferences of BROADCAST RECEIVERS AND ASSOCITED EQUIPMENT, of February 1988. It covers the frequency range 0.15 MHz to 1000 MHz. Analogous to the requirement for CISPR radiation limits, there must be 80% confidence tht at least 80% of the series produced appliances comply with these immunity limits. Repeated discussions in CISPR Plenary Assemblies have not yet led to a CISPR document for this case.

TC 65 already prepared several documents on Electromagnetic Compatibility for industrial-process measurement and control equipment.

Publication 801-1 gives the General Introduction.
The systems boundaries for penetration of the interference may be: 1) Power feed lines. 2) Input signal lines. 3) Output signal lines. 4) Equipment enclosure.

The coupling mechanisms by which interference can be injected are: 1) Common impedance (resistive). 2) Inductive coupling. 3) Capacitive coupling. 4) Electromagnetic radiation.

Sources of interference can be grouped into three main types: 1) Magnetic. 2) Electrical (broadband, narrowband). 3) Electromagnetic. Conspicuous sources are: Switchgear, contractors, relays, welders, radio and television transmitters, walkie-talkies, vehicle radio transmitters and electrostatically charged operators.

Susceptibility depends on the arrangement of the circuit (i.e. the manner of earthing the circuit and shields), the quality of shielding applied and the environment in which the system is required to work.

Installation design is intended to describe in detail various methods which can be utilized to reduce or eliminate interferences.

Publication 801-2 (revised 1989) treats: Electrostatic discharge requirements.
The contact discharge method is the preferred test method: The electrode of the test generator is in contact with the EUT, and the discharge actuated by the discharge switch within the generator. The four severity levels are 2 kV, 4 kV, 6 kV, 8 kV. In the air discharge method, the charged electrode of the test generator is approached to the EUT, and the discharge actuated by a spark to the EUT. The severity levels are 2 kV, 4 kV, 8 kV, 15 kV. The characteristics of the ESD generator are given, as well as requirements for the climatic conditions.

Publication 801-3 treats: Radiated electromagnetic field requirements.
The primary concern of this part is the radiation of the hand-held transceiver. Other sources are: fixed radio and television transmitters, vehicle radio transmitters, industrial sources (welders, thyristors, fluorescent lights, switches operating inductive loads.
Test set-up may require a shielded room, an anechoic chamber, a stripline circuit, Transverse electromagnetic cells (T.E.M.), Mode-stirred chambers.

Severity levels. (general guidelines):
Class 1: Low level electromagnetic radiation environment, typical of local radio/television stations, located at more than 1 km, or low power transceivers.
Class 2: Moderate electromagnetic radiation environments, such as portable transceivers, not closer than 1 m.
Class 3: Severe electromagnetic radiation environments, such as high power transmitters in close proximity.
Class 4: Very severe e.m. radiation. (subject to negotiation between user and manufacturer).

Publication 801-4: Electrical fast transient/burst requirements.
The repetitive fast transient test is a test with bursts consisting of a number of fast transients, coupled into power supply, control and signal inputs of electronic equipment. Significant for the test are the short rise time, the repetition rate and the low energy of the transients.
Four test levels are given. On power supply: 0.5 kV, 1 kV, 2 kV, 4 kV. On I/O signal, data and control lines use half these levels. Rise time of the pulses 5 ns, impulse duration 50 ns, repetition rate 5 kHz (at 2.0 kV: 2.5 kHZ).
Climatic conditions: ambient temp. 15°C to 35°C, relative humidity 45% to 75%, atmospheric pressure 68 kPa to 106 kPa.

Test results may be:

1) Normal performance within specification limits,
2) Temporary degradation, self recoverable.
3) Temporary degradation, needing operator intervention.
4) Not recoverable degradation.

Publication 801-5: Surge Immunity Requirements. (Draft, September 1988). The purpose of this draft standard is to verify the immunity performance of equipment under operational conditions. This part of the standard defines immunity requirements and test methods for equipment, which must withstand the surge, caused by overvoltages/currents from switching and lightning transients. The object is to establish a common refence for evaluating the performance of industrial process measurement and control instrumentation, when subjected to high energy disturbances on the power, signal and telecommunication lines.

Publication 801-6: Conducted Immunity Requirements. (to be published). This part defines the immunity requirements and test methods against modulated radio-frequency signals coupled into power mains and grounding networks. The latter will result in currents, which will run through the coverings of equipment and through the screens of interconnecting cables. These requirements are developed for and are applicable to Industrial Process Measurement and Control Equipment (IPMCE), Information Technology Equipment (ITE) and Telecommunication Equipment. The object of this standard is to establish a common and reproducible basis for evaluating the performance of electronic instrumentation when it is subjected to conducted interference sources on supply, signal and/or control lines.

Publication 801-7: EMC Environmental Classification and Installation Guidelines. (postponed).

Proposal to deal in TC77 with immunity to High Altitude Nuclear Electromagnetic Pulse.

A new Sub-Committee of TC77 will deal with the protection against NEMP in the civilian area.

It is clear, that the documents described refer directly to the expected environment.

TC77 has apart from the above mentioned list of main electromagnetic disturbances, also given attention to the parameters to be considered, and the levels to be expected.

For telecommunications they take the following environmental classes:
Class 1.1 Telecommunication centres (urban).
Class 1.3 Telecommunication centre (rural).
Class 2.1 Customers premises (urban).
Class 2.3 Customers premises (rural).
Class 3.1 Outdoor locations (urban).
Class 3.3 Outdoor locations (rural).
Class 4.1 Industrial areas.
Class 5.1 Ground vehicle installations.
These classes are discussed with reference to the following parameters:
Signal lines (many-few, long-short, buried-overhead, shielded-unshielded)
AC power lines (many-few, long-short, number of switches)

telecommunication and power installations". For tests according to the K-series of Recommendations of CCITT suitable generators are available, to give surge voltage wave shapes and surge current wave shapes, as desired. (e.g. by the GDR Institute of Posts and Telecommunications). A CCIR Study Group V sponsored session on EMC in wire communications, organized by Dr. H. Lorke, was published in the EMC book of the 1988 Wroclaw Symposium. (pages 77-138).

9.2 Generation Mechanism of R.F. Noises on Power Lines

S. NITTA[1], A. MUTOH[1], N. YAMAKOSHI[1], and T. SHIMAYAMA[2]
Tokyo University of Agriculture and Technology, Tokyo, 184 Japan[1] and Fuji Electronic Co. Ltd., Tokyo, 176 Japan[2]

[Abstract]

This paper describes that main R.F. noises induced on power supply lines are damped oscillation waveforms, clarifies their generation mechanism from the relationship between the resonance points existing on frequency-impedance characteristics measured about power supply lines and the "S" defined there and by applying the theory of resonance circuits made up of lumped constant L-C-R to the distributed constant circuits (that is, power supply lines), and insists that their suppression technique is to get rid of resonance points.

1. Introduction

Most of the R.F. noises disturbing electronic devices are coming from their power supply lines. So far, it has been thought that noises are pulsive ones, the inductive kick generating in the case that inductive loads connected to power supply lines are switched off. However, noises on power supply line generate in the case that lumped constant inductive load is not connected to it, and is actually observed as damped oscillation waveform.

It is well known that damped oscillation generates in the case that pulsive current is given to series and /or parallel resonance circuits made up of L-C-R (L: inductance, C: capacitor, R: resister).

First, this paper reviews transient phenomena (generation mechanism on damped oscillation) on series resonance circuit and parallel resonance circuit made up of lumped constant L-C-R, and, in order to apply the above theory to transient phenomena on power supply line (which is a distributed constant circuit), transient phenomena on lumped constant circuit are expressed by resonance frequency and "S" (defined at resonance point, refer to Equation (6), (25)) without using quantitative value of L-C-R.

Next, by applying the above proposed expression to the frequency-impedance characteristics curve measured of power supply line, generation mechanism of R.F. noise induced on power supply lines is clarified theoretically and experimentally.

2. Damped Oscillation on Lumped Constant Circuits.

This paragraph reviews generation mechanism of damped oscillation wave induced on series resonance circuit and parallel resonance circuit made of lumped constant L, C and R, and shows the method by which damped oscillation waveform can be got graphically from frequency-impedance characteristics representing resonance circuits characteristics, without knowing the quantitative value of L, C and R.

2.1 Series resonance circuit.

Figure 1 shows series resonance circuit diagram made up of L_2, C_2 and R_2. Capacitor C_2 is charged to voltage E_0. When switch SW is turned on, transient phenomenon will occur in the circuit, and in the case that the condition shown in Equation (1) is satisfied, damped oscillation will occur as well known. [1],[2]

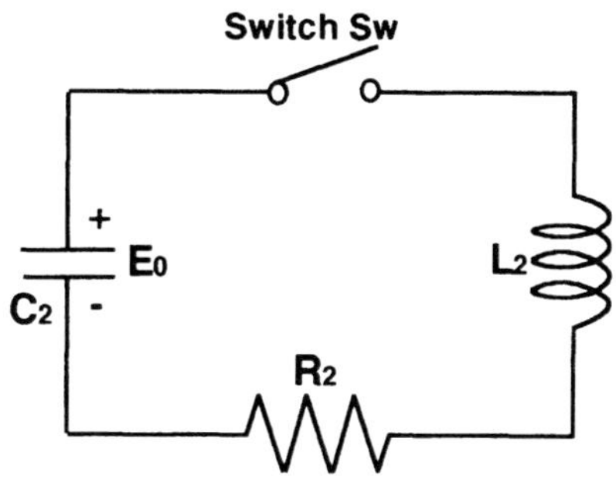

Figure 1 Series resonance circuit

$$\alpha_2 < \frac{1}{\sqrt{L_2C_2}} \qquad \text{-----(1)}$$

where,

$$\alpha_2 = \frac{R_2}{2L_2} \qquad \text{-----(1)'}$$

Angular frequency ω_2 of damped oscillation is represented by Equation (2). [3]

$$\omega_2 = \sqrt{\frac{1}{L_2C_2} - \alpha_2^2} \qquad \text{-----(2)}$$

One of purpose of this study is to analyze the damped oscillation wave occurring in the power supply lines whose resistance is very small. Therefore, resistance may be neglected and Equation (3) is used as angular frequency of damped oscillation.

$$\omega_2 \fallingdotseq \frac{1}{\sqrt{L_2 C_2}} \qquad \text{-----(3)}$$

Current i_2 flowing through the circuit oscillates as shown in Equation (4)

$$i_2 = \frac{E_0}{\omega_2 L_2} e^{-\alpha_2 t} \sin \omega_2 t \qquad \text{-----(4)}$$

In the case that resonance condition is satisfied, Equation (5) is got.

$$X_2 = \omega_2 L_2 = \frac{1}{\omega_2 C_2} = \sqrt{\frac{L_2}{C_2}} \qquad \text{-----(5)}$$

(where X_2: reactance at resonance point)

Next, S_2 which represents sharpness of resonance is defined as shown in Equation (6)

$$S_2 = \frac{X_2}{R_2} \qquad \text{-----(6)}$$

From Eq.(1)', (5) and (6), Equation (7) is drawn

$$\alpha_2 = \frac{\omega_2}{2S_2} = \frac{\pi f_2}{S_2} \qquad \text{-----(7)}$$

where, f_2: series resonance frequency

From Eq.(1) and (7), the oscillation condition is drawn as shown in Equation (8)

$$S_2 > \frac{1}{2} \qquad \text{-----(8)}$$

In the case that S_2 is larger than 1/2, damped oscillation will occur in the circuit shown in Fig.1 when switch SW is turned on.

If the circuit has no loss, the first peak current I_0 must have the value shown in

Equation (9).

$$I_o = \frac{E_o}{X_2} \qquad \text{-----(9)}$$

However, as the circuit has loss, the following attenuation shown in Equation (11) will result at the time t_2 (shown in Equation (10)) going up to maximum amplitude of current shown in Equation (11).

$$t_2 = \frac{1}{4f_2} \qquad \text{-----(10)}$$

$$e^{-\alpha_2 t_2} = e^{\frac{-\pi}{4S_2}} \qquad \text{-----(11)}$$

Therefore, the real first peak current I_1 has the value shown in Equation (12)

$$I_1 = I_o e^{\frac{-\pi}{4S_2}} \qquad \text{-----(12)}$$

The time duration t_o between first peak current I_1 and nth peak current I_n of oscillation wave train is shown in Equation (13) and I_n has the value shown in Equation (14)

$$t_o = \frac{n-1}{f_2} \qquad \text{-----(13)}$$

$$I_n = I_1 e^{-\alpha t_o} \qquad \text{-----(14)}$$

Therefore, Equation (15) is drawn,

$$\frac{I_1}{I_n} = e^{\frac{\pi(n-1)}{S_2}} \qquad \text{-----(15)}$$

In the case that I_1/I_n is equal to 23, Equation (16) is got

$$n \fallingdotseq S_2 + 1 \qquad \text{-----(16)}$$

Amplitude of 1/23 of first peak amplitude means nearly zero, and, that is to say, Eq.(16) expresses the number of peak current of damped oscillation.

Time duration t_0 is transformed to Equation (17), substituting Eq.(16) for Eq.(13).

$$t_0 = \frac{S_2}{f_2} \qquad \text{-----(17)}$$

Energy Wc stored in C_2 charged to E_0 is expressed by Equation (18)

$$W_c = \frac{1}{2} C_2 E_0^2 \qquad \text{-----(18)}$$

This is initial energy of oscillation. If energy W_c is transferred to L_2 without loss energy W_e stored in L_2 is expressed by Equation (19).

$$W_e = \frac{1}{2} L_2 I_0^2 \qquad \text{-----(19)}$$

As W_e is equal to W_c, from Eq.(18), (19) and (5), Eq.(9) is drawn, too. If we can get S_2, f_2 and X_2, damped oscillation waveform can be produced in the followings, even if the values of L_2 and C_2 can't be got.

From frequency-impedance characteristics (abbr.: f-Z CH.) of series resonance circuit shown in Figure 2, we can get f_2 and R_2 and calculate S_2 and X_2.[1][2]

Minimum value in the curve shown in Fig.2 is R_2 and frequency corresponding to R2 is series resonance frequency f_2.

Getting the difference $2\Delta f_2$ between two frequencies(f_{a2},f_{b2}) having 3dB high impedance in comparison with R_2 (refer to Fig.2), S_2 can be

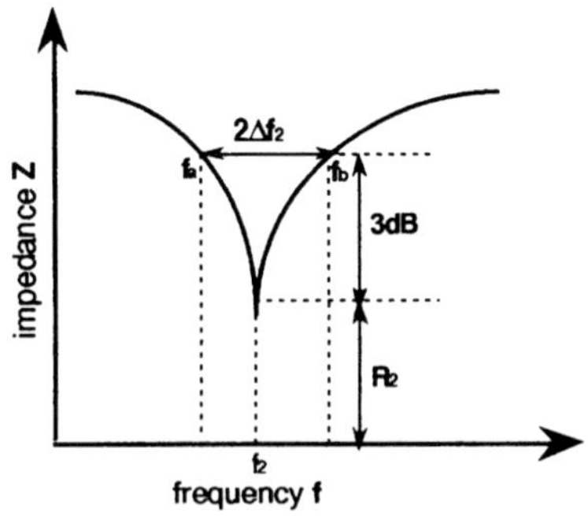

Figure 2 Frequency - impedance characteristic of series resonance circuit

calculated by Equation (20).[1I2]

$$S_2 = \frac{f_2}{2\Delta f_2} \qquad \text{-----(20)}$$

R_2 and S_2 are already known.

Therefore, X_2 can be got from Eq.(6),as shown in Eq.(21)

$$X_2 = R_2 S \qquad \text{-----(21)}$$

From the three parameters of S_2, f_2 and X_2, we can produce the damped oscillation

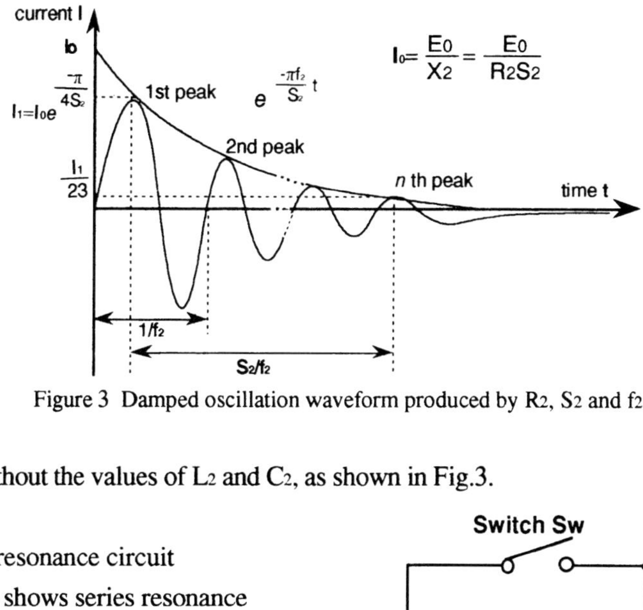

Figure 3 Damped oscillation waveform produced by R_2, S_2 and f_2

waveform without the values of L_2 and C_2, as shown in Fig.3.

2.2. Parallel resonance circuit

Figure.4 shows series resonance circuit diagram made up of L_1, C_1 and R_1. Capacitor C_1 is charged to voltage E_0. When switch SW is turned on, damped oscillation will occur in the case that the condition shown in Equation (22) is satisfied.

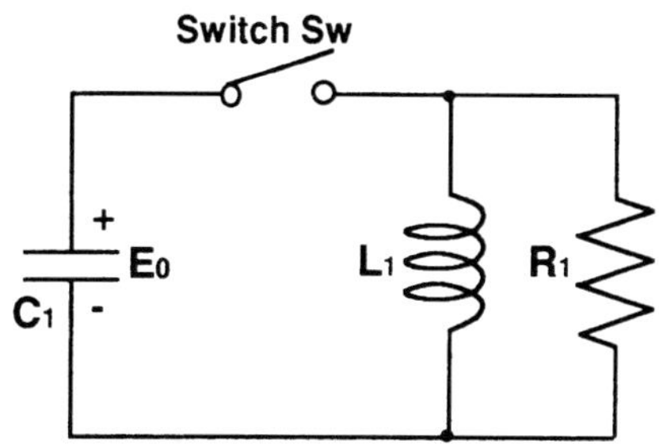

Figure 4 Parallel resonance circuit

$$\alpha_1 < \frac{1}{\sqrt{L_1 C_1}} \qquad \text{-----(22)}$$

where,

$$\alpha_1 = \frac{1}{2C_1R_1} \qquad \text{-----(22)'}$$

In this case, frequency f_1 ($\omega_1 = 2\pi f_1$: angular frequency) of damped oscillation is nearly equal to the value shown in Equation (23).

$$\omega_1 = \frac{1}{\sqrt{L_1C_1}} \qquad \text{-----(23)}$$

As the resonance condition is satisfied at ω_1, Equation (24) are got.

$$X_1 = \omega_1L_1 = \frac{1}{\omega_1C_1} = \sqrt{\frac{L_1}{C_1}} \qquad \text{-----(24)}$$

In parallel resonance circuit, S_1 is defied as shown in Equation (25)

$$S_1 = \frac{R_1}{X_1} \qquad \text{-----(25)}$$

From Eq.(22), (23), (24) and (25), the oscillation condition is drawn as shown in Equation (26).

$$S_1 > \frac{1}{2} \qquad \text{-----(26)}$$

In the same way as the way used in series resonance circuit, the number n of the oscillation whose peak value becomes 1/23 is got as shown in Equation (27).

$$n = S_1 + 1 \qquad \text{-----(27)}$$

And time duration t_0 from first peak to nth peak is got as shown in Equation (28), similarly to the case in series resonance circuit.

$$t_0 = \frac{S_1}{f_1} \qquad \text{-----(28)}$$

It is necessary to got the value of S_1 in order to produce the damped oscillation waveform. In the same way as the way used in series resonance circuit, we can get the parameters in parallel resonance circuit from f-Z CH shown in Figure 5.

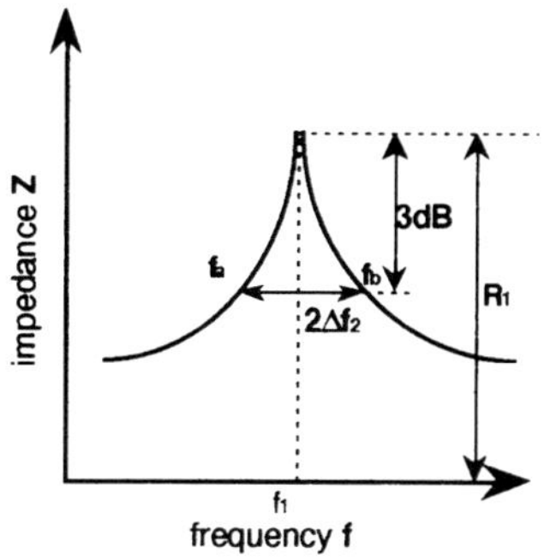

Figure 5 Frequency - impedance characteristics of parallel resonance circuit

Maximum value of impedance is R_1 and frequency f_1 at maximum value is parallel resonance frequency. Getting the difference $2\Delta f_1$ between two frequencies(f_{a1},f_{b1}) having 3dB low impedance in comparison with R_1, S_1 can be calculated by Equation (29).

$$S_1 = \frac{f_1}{2\Delta f_1} \qquad \text{-----(29)}$$

As the about described, all constants in parallel resonance circuit can be drawn in the same formats as they in series resonance circuit by utilizing the different definition of S (refer to Eq.(6) and (25)).

3. Power Supply Lines

Long power supply line is a distributed constant circuit and its f-Z CH has maximum and minimum points alternately, as shown in Figure.6. [4] In consideration of Fig.5, maximum points like Z_1, Z_3 and Z_5 mean the existence of the parallel resonance points, whose frequencies like f_1, f_3 and

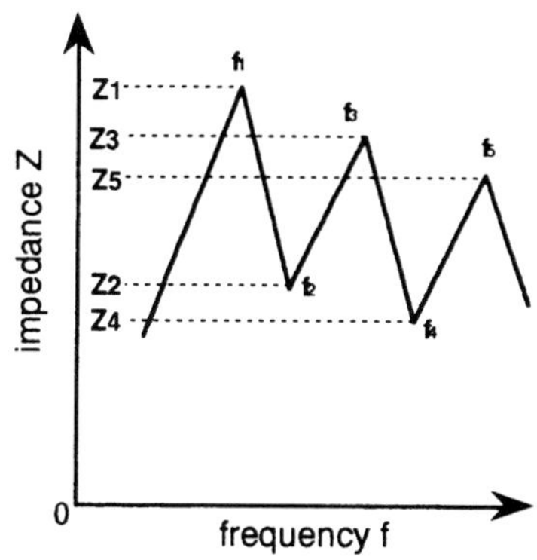

Figure 6 Typical frequency-impedance characteristics of long cable

f_5 show parallel resonance frequencies. Similarly, minimum points like Z_2 and Z_4 mean the existence of the series resonance points, whose frequencies like f_2 and f_4 show series resonance frequencies.

In this paragraph, the theory described in paragraph 2 is applied to individual resonance points observed in test circuit (long cable) simulating power supply lines, and damped oscillation waveforms at each resonance points are calculated.

3.1. Test circuit (simulation of power supply lines), and its characteristics

Figure 7 shows test circuit made up of 42 meters length cable which is a pair of parallel stranded/vinyl coated cable and is placed on a floor. And capacitor C_5 in parallel with resistor R_5 (C_5 = 22pF, R_5 = 22kΩ) are added to lowerside wire at a middle point of cable in order to make the impedance unbalance between a pair of cable and ground.

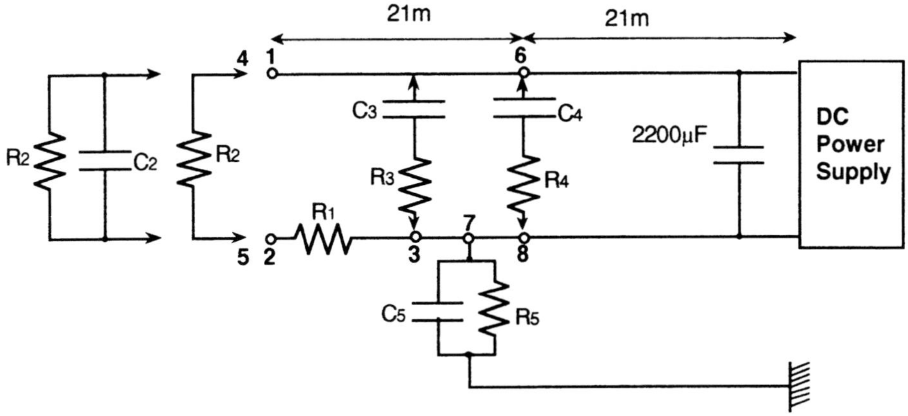

Figure 7 Network of test circuit

Curve I in Figure 8 shows f-Z CH. of each arrangement of circuit in Fig.7. As seen in Fig.8, f_1, f_3 and f_5 show parallel resonance frequencies at parallel resonance points,and f_0, f_2 and f_4 show series resonance frequencies at series resonance points.

Each parallel resonance point has the constants shown in Table 1, and each series resonance point has the constants shown in Table 2.

Humps can be seen at two frequencies of 2MHz and 4MHz. These are caused by R_3-C_3 and are neglected in this case.

In the next subparagraph, we try to calculate the damped oscillation waveforms, making use of the constants shown in Table 1 and 2.

Table 1 Constants of parallel resonance points.

f_1 ($\omega_1 = 2\pi f_1$)	1.03 MHz (6.45×10^6)
R_1	1,421 Ω
S_1	9
$X_1 = R_1 / S_1$	158 Ω

(a) Constants of first parallel resonance point

f_3 ($\omega_3 = 2\pi f_3$)	3.16 MHz (19.85×10^6)
R_1	490 Ω
S_1	11.7
$X_1 = R_1 / S_1$	41.9 Ω

(b) Constants of second parallel resonance point

Table 2 Constants of series resonance points.

f_2 ($\omega_2 = 2\pi f_2$)	2.30MHz (14.45×10^6)
R_2	29.0 Ω
S_2	9
$X_2 = R_2 / S_2$	261 Ω

(a) Constants of first series resonance point

f_4 ($\omega_4 = 2\pi f_4$)	4.68MHz (29.40×10^6)
R_2	23.6 Ω
S_2	15.4
$X_2 = R_2 / S_2$	363 Ω

(b) Constants of second parallel point

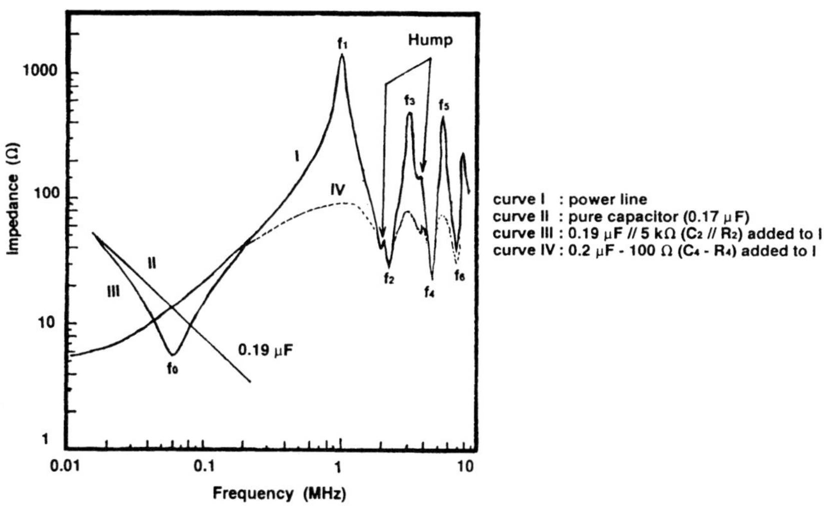

Figure 8 Frequency-impedance characteristics of test circuit

3.2. Calculation of damped oscillation waveforms (: R.F. noises)

(a) Generation of noise in parallel resonance circuit.

R.F. noises occur according to rules about parallel resonance circuits when circuit is open, that is, R_2 or $R_2//C_2$ is cut off from power supply line. In the case that voltage E_0 is given to circuit and current I_0 flows in circuit, energy stored in distributed capacity C_1 is $C_1E_0^2/2$ and energy stored in distributed inductance L_1 is $L_1I_0^2/2$. When circuit is open, the energy in C_1 doesn't charge, but voltage of L_1 di/dt is induced across terminals by the energy in L_1 as current in L_1 becomes zero. This voltage charges up C_1, and if all of the energy in L_1 is transferred to C_1, voltage E_m across C_1 has the value shown in Equation (30)

$$E_m = I_0 \sqrt{\frac{L_1}{C_1}} = I_0X_1 \qquad \text{-----(30)}$$

However, as it takes time of 1/4 f_1 until voltage E_1 reaches maximum (as shown in Eq.(10)), attenuation occurs according to Eq.(11). Therefore, first peak voltage E_1 is expressed by Equation (31).

$$E_1 = E_m\, e^{\frac{-\pi}{4S_1}} = I_0X_1\, e^{\frac{-\pi}{4S_2}} \qquad \text{-----(31)}$$

As S_1 at each parallel resonance point is larger than 1/2 as shown in Fig.8, damped oscillation will occur when circuit is open. In the case that I_0 is 35mA, damped oscillation shown in Table 3 will be resulted.

Table 3 Specification of damped oscillation

f_1	1.03MHz
$n = S_1 + 1$	10
$t_0 = S_1 / f_1$	8.7 μs
E_1	5.1 V

(a) Damped oscillation due to in first parallel resonance point

f_3	3.16MHz
$n = S_1 + 1$	13
$t_0 = S_1 / f_1$	3.7 μs
$E_1 = \frac{1}{\sqrt{3}} I_0X_1 e^{\frac{-\pi}{4S_1}}$	0.8 V

(b) Damped oscillation

(b) Generation of noise in series resonance circuit

R.F. noises occur according to rules about series resonance circuit when circuit when circuit is closed, that is, R_2 or $R_2//C_2$ is connected to power supply line. When load (connected to **1** and **2** in Fig.7) is made up of only resistor, S of the circuit is smaller than 1/2 and hence, oscillation doesn't occur. When load includes large inductance, it is hard to series-resonate with line. When load includes any capacitance, it will be series-resonated with all kinds of line and damped oscillation will occur.

In consideration of the about described facts, we study the phenomena in the case that capacitance load of C_2 (0.19μF) and R_2 (5kΩ) (; leakage resistance is not concerned in noise) is connected to power supply line, as shown in Fig.7.

Curve II and curve III in Fig.8 show f-Z CH of C_2 and it of C_2-R_2 in series with line, respectively, and cross point of curve I and II represents zeroth series resonance frequency f_0. In the case that D.C. power supply voltage E_0 is 2V, damped oscillation shown in Table 4 will be resulted.

Table 4 Specification of damped oscillation

f_0	63 kHz
R_2	5.74 Ω
S_2	2.3
$X_2 = R_2S_2$	13.2 Ω
$n = S_2 + 1$	4
t_0	36 μs
$I_2 = \frac{E_0}{X_2} e^{\frac{-\pi}{4S_2}}$	107 mA

(a) Damped oscillation in series resonance with C_2R_2

f_2	2.3 kHz
$n = S_2 + 1$	10
$t_0 = S_2 / f_2$	3.9 μs
$E_1 = E_0 / 2$	1 V
$I_2 = \frac{E_1}{X_2} e^{\frac{-\pi}{4S_2}}$	3.51and mA

(b) Damped oscillation in first series

3.3. Noise suppression technique

(a) Noise suppression in parallel resonance circuit

In order to suppression in parallel resonance circuit, S_1 must be reduced to smaller value than about 1/2. In order to fulfill these, it is required to insert the resister across lines, whose value must be equal to the characteristic impedance (about 100Ω) of line. Curve IV in Fig.8

shows f-Z CH. in the case that C_3 (0.2μF)-R_3 (100Ω) element is connected across **1** and **3** in Fig.7 and indicates flat characteristics in comparison with curve I. To made f-Z CH. flat is to suppress noise, as proved experimentally in next paragraph.

(b) Noise suppression in series resonance circuit

Similarly to parallel resonance circuit, in order to suppress noises in parallel resonance circuit, S_2 must be smaller than 1/2. For the purpose of these, it is required to insert R shunted by L serially in line. This idea is effective for high frequency noise, but not for lower one. Most effective method is to insert C_4-R_4 element (R_4 : characteristic impedance of line, C_4: stop the d.c. current) on the zone where the damped oscillation voltage is high. But, as this generates many standing wave in line, many CR elements are needed, and therefore, this is not practical. However, generally speaking, damped oscillation in series resonance points is small and doesn't disturb the electronic devices so much.

4. Experiment

Experiments are carried out on test circuit shown in Fig.6 in order to confirm the calculated result got in paragraph 3. Voltage between **1** and **2** and current flowing to **2** from **3** are observed in the case of turning on and then cutting off the load connected to **1** and **2** (Refer to Fig.7).

4.1. The case of cutting off the load(Opening circuit)

Experiment 1

The phenomena observed under the following conditions are shown in Figure 9.

Source voltage E_0: 4V

Load resistance R_0:100Ω

Current in power supply line I_0: 35mA

Results agree with data shown in Table 3(a). And oscillation with f_3(3.16MHz) shown in Table 3(b) is superposed on main oscillation (f_1=1.03MHz) and disappears at time t_0=3.7μs. Although E_0=4V is different from E_0=5V used for calculation of Table 3, first peak voltage is the same in both cases. This fact shows that damped oscillation is dominated by current change.

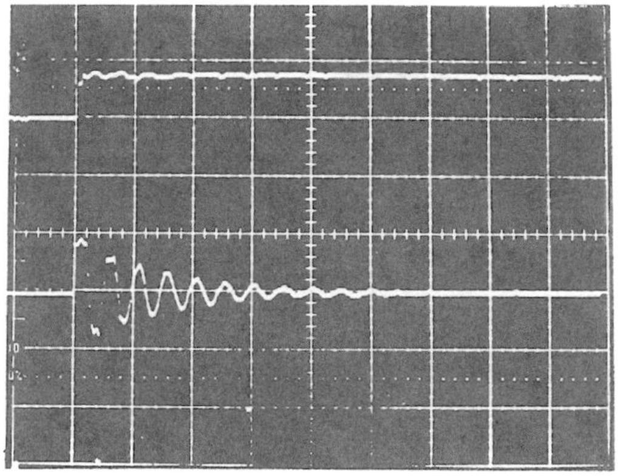

V axis
Top : Current ; **2 - 3**
50 mA / div.

Bottom : Voltage ; **1 - 3**
5 V / div.

H axis
2 μs / div.

Figure 9 Experimental result in the case of I_0 =35 mA

Experiment 2

In order to prove that first peak voltage (amplitude of oscillation) is proportional to current change I_0, experiment is carried out under the following conditions:

Source voltage E_0: 4V

Load resistance R_2: 100Ω

Current in power supply line I_0: 70mA

As shown in Figure 10, results show that amplitude is doubled when I_0 is doubled.

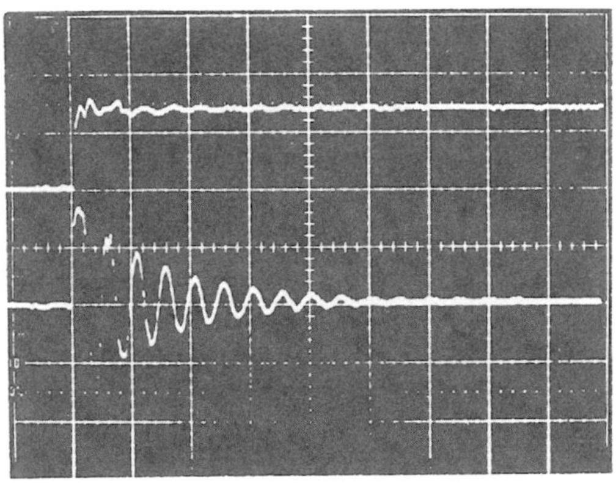

V axis
Top : Current ; **2 - 3**
50 mA / div.

Bottom : Voltage ; **1 - 3**
5 V / div.

H axis
2 μs / div.

Figure 10 Experimental result in the case of I_0 = 70 mA

Experiment 3

In order to confirm the noise suppression technique described in 3.3, the phenomena are observed in the case that C_3 (0.2μF)-R_3 (100Ω) is connected between **1** and **2**. As shown in Figure 11, damped oscillation doesn't occur. This fact proves that to suppress noise is to make S small as shown in curve IV in Fig.8.

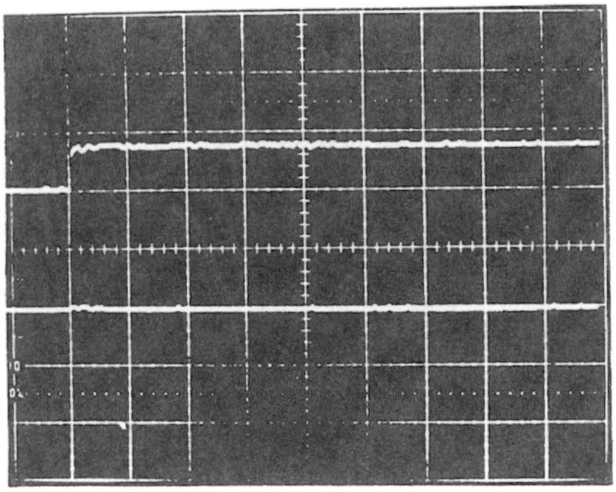

Figure 11 Current and voltage in the case that noise suppressor is applied.

4.2. The case of turning on the load (closing circuit)

As described in 3.2(b), the phenomena in the case that C_2 (0.19μF) and R_2 (5kΩ) used as the capacitive load is turned on are observed under the following condition.
Source voltage E_0: 2V

Experiment 4

The phenomena shown in Figure 12 are ones measured at the same points as the points in 4.1. Voltage of 2V (: refer to bottom in Fig.12) appears immediately after the load becomes on, and then damped oscillation voltage with frequency f_0 (63kHz) shown in Table 4(a) starts. Current goes up 90° ahead of voltage and its first peak is nearly equal to 107mA shown in Table 4(a).

As the influence of distributed capacity in line is little on 63kHz current, the line acts as lumped inductance.

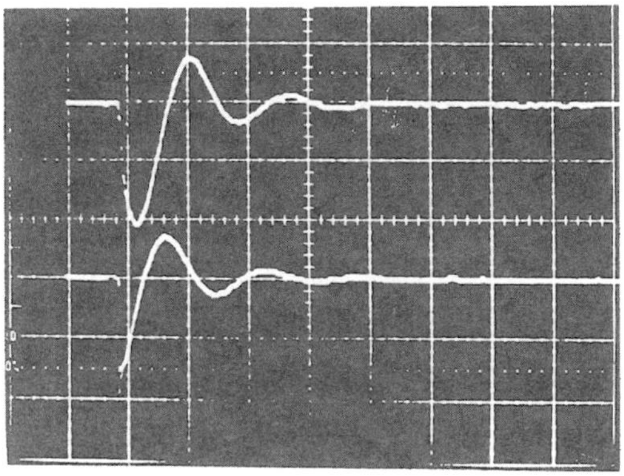

V axis
Top : Current ; **2 - 3**
50 mA / div.

Bottom : Voltage ; **1 - 2**
1 V / div.

H axis
10 μs / div.

Figure 12 Current and voltage in line when capacitive load (C_2 - R_2) is connected to line.

Experiment 5

From Table 4(b) calculated on the basis of in Fig.8, it is supposed that damped oscillation curve I with frequency f_0 (63kHz) will occur at closing circuit. As the line is distributed constant circuit, oscillation voltage has maximum and oscillation current has minimum at the middle point of line. Figure 13 shows current and voltage measured at middle point of line.

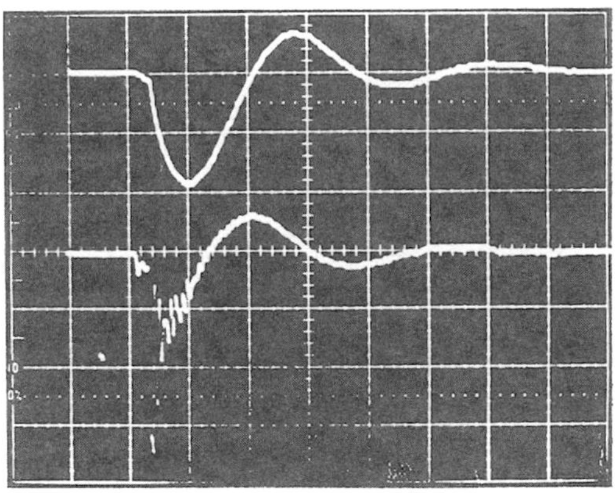

V axis
Top : Current ; **7 - 8**
50 mA / div.

Bottom : Voltage ; **6 - 7**
0.5 V / div.

H axis
5 μs / div.

Figure 13 Current and voltage at the middle point of line

Oscillation voltage with frequency f_0 (63kHz) appears strongly, and first peak voltage is 1V, about half of source voltage E_0, as it is measured at middle point of line. And damped oscillation voltage with frequency f_2 (2.3MHz) is superposed on it with frequency f_0 (63kHz). (Refer to bottom of Fig.13)

Current with frequency f_0 (63kHz) shown in top of Fig.12 is nearly equal to it measured at 2-3, and current with frequency f_2 (2.3MHz) superposed on it with f_0 is very small and therefore, can't be seen in Fig.13.

Experiment 6

This experiment is carried out to confirm noise suppression technique described in 3.3 (b).

As shown in Fig.7, C_4 (0.2μF)-R_4 (100Ω) is connected across terminal **6-8** in order to suppress noise with frequency f_2 (2.3MHz) curve IV in Fig.8 shows f-Z CH. corresponding to this case.

As shown in Figure 14, dumped oscillation with frequency f_0 is not suppressed since curve IV shows the same series resonant point as it in case of no suppression technique, but damped oscillation with frequency f_2 is suppressed.

As the above described, experimental results show good agreement with calculated results shown in 3.2.

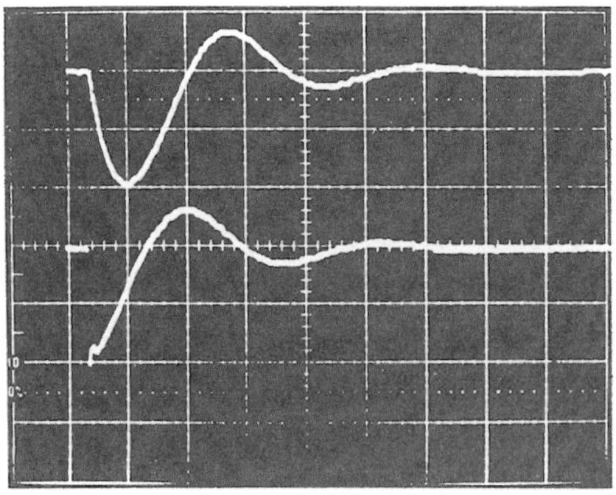

V axis
Top : Current ; **7 - 8**
50 mA / div.

Bottom : Voltage ; **6 - 7**
0.5 V / div.

H axis
5 μs / div.

Figure 14 Current and voltage at the middle point of line, in the case that noise suppression technique is applied.

5. Conclusion

R.F noises are dominated by many factors such as lines constants, constructions of machines, kinds of loads, current switching voltage, switching and so on.

This paper proved that generation mechanism of R.F. noises on power supply line is clarified by paying attention to frequency-impedance characteristics of power supply line, the existence of resonance points on f-Z CH. and several constants got from resonance points, and to suppress noises is to make S defined at resonance point to be small.

The authors think that the idea described in this paper may be applied to general generation mechanism and suppression technique of R.F. noises.

6. Reference

[1] S. NITTA, T. SHIMAYAMA: Showering Noise -Electrical Noise Generated from Inductive Circuits, Proc. of Int'l Conf. on EMI/EMC, Sept. '87, Bangalore, India, pp337-340

[2] S. NITTA, E. TAKECHI: The Noise Radiated from Showering Noise, Proc. of 9th Int'l Wroclaw Symp. on EMC, Jun '88, Poland, pp695-700

[3] J. H. MORECRAFT: Principles of Radio Communication, 1927, John-Wiley & Sons, pp247

[4] S. NITTA, O. NAKATA, T.SHIMAYAMA: A Consideration on Power-Line-Noise Estimated from Line-Impedance-Stabilization-Network-Measured Noise, 8a-C1. Proc. of 1989 Int'l symp. on EMC, Nagoya Japan, Sep. '89, pp69-74

9.3 A Rejection Method for Interference Waves Caused by Sporadic-E Ionospheric Propagation

H. ARATA and S. YAMAZAKI

NHK (Japan Broadcasting Corporation) Science and Technical Research Laboratories, Radio Engineering Division, Tokyo, 157 Japan

ABSTRACT

In Japan, VHF low channel television services are sometimes disturbed by interference from distant foreign FM transmitters. It is caused by signal which propagate via the sporadic-E layer of the ionosphere. A method is proposed for eliminating narrow band FM interference waves in television signal. The method uses quadrature synchronous detection of the television signal and Hilbert transformation. The analysis was obtained, and an experimental system was made. The system reduced interference wave for 40 dB.

1. INTRODUCTION

The sporadic-E layer (Es layer) reflects FM and TV broadcasting waves in neighboring countries. Those waves propagate abnormally to a long distance and in most cases interfere the video signal bands of Japanese television waves. This causes luminance and/or color beat disturbances and a mixing of foreign languages in the sound of TV. In a worse case, the interference drives video signals out of the synchronizing state, making TV reception totally impossible.
The Es layer usually appears from May to August. It reflects VHF low channel waves, which abnormally travel to reach as far as 1,000 to 2,000 km.[1] Es layer-caused disturbance comes in two forms: one directly affecting home TV sets and the other interfering with TV signals coming into a television translator of a rebroadcast station from its television key station. To improve these, we have tried to change the key station of the disturbed station to built a new rebroadcast station for the above-mentioned latter case. And, some researchers have carried out to find methods to eliminate Es interference by using notch filters.[2],[3]
In this paper, we discuss about a new method to cancel interference waves by introducing the quadrature synchronous detection technique. This method is effective for eliminating interference waves at any frequency within TV signal bands. Interference waves are eliminated at the IF stage. The system we have developed automatically eliminates a maximum of five interference waves and has an interference suppression capability of more than 40 dB.

2. PRINCIPLE

The Japanese TV signal is transmitted in the form of VSB-AM (Vestigial SideBand-Amplitude Modulation). (See Fig. 1a.) To demodulate this TV signal, recent TV sets employ a synchronous detection method. This method, so as to obtain video signal, reproduces the video carrier, and multiplies it by the original signal, and then yields the video signal low-pass filtering.

Through this process, mixed interference signals are demodulated together. (See Fig. 1b.)

If the interference waves are in the DSB(Double side band)-transmitted region, and these interference waves are demodulated with the quadrature synchronous detection method, in the other words, the TV signal and interference waves are multiplied by the video carrier with phase shifted by 90 degrees and then are put into low-pass filter, only interference waves are demodulated, with no video signal output in the low frequency region.

This means we can extract interference waves from the interfered TV signal. (See Fig. 1c.) These interference waves are in the relationship of Hilbert transform with the interference waves obtained by the in-phase synchronous detection method normally employed in TV set. Therefore, for eliminating interference waves, the extracted interference waves must be Hilbert-transformed and then subtracted from the synchronously detected output signals containing interference waves.

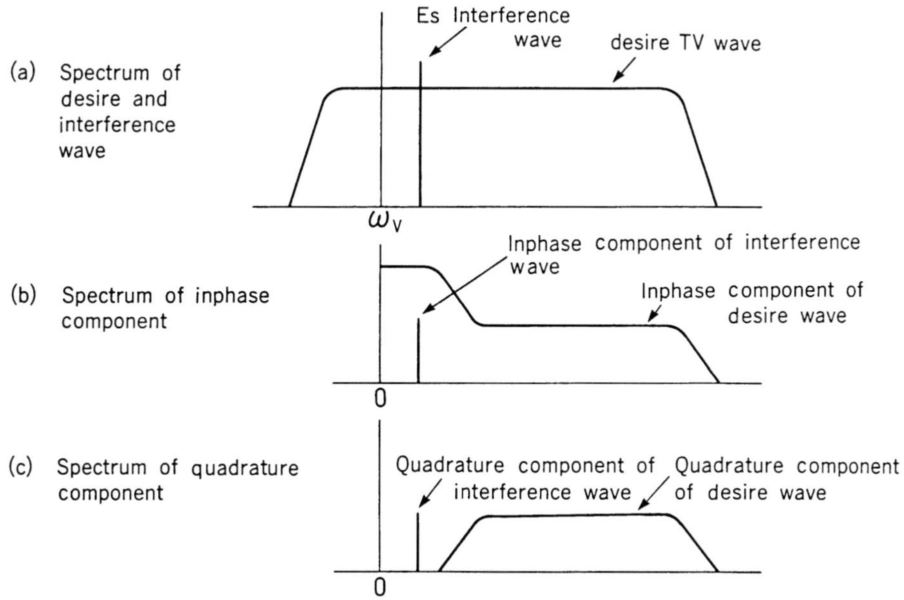

FIGURE 1 Frequency spectrum of television signal with an interference wave.

2.1 Hilbert-transform

Hilbert-transform is obtained by the following process. A baseband interference signal $S_u(t)$ is amplitude modulated by the carrier $\cos(w_c t)$. An amplitude modulated (AM) signal $f_{am}(t)$ is described as Equation (1).

$$f_{am}(t)=S_u(t)\cos(W_c t) \quad (1)$$

The lower sideband of $f_{am}(t)$ is removed by a filter and converted to a SSB-AM(Single Side Band-AM) signal $f_{ssb}(t)$.

$$f_{ssb}(t)=S_u(t)\cos(W_c t)-R_u(t)\sin(W_c t) \quad (2)$$

Between the signals $S_u(t)$ and $R_u(t)$ here is the Hilbert transform relation.[4] The SSB signal is synchronous-detected by the carrier $\sin(w_c t)$. A demodulated signal becomes

$$-R_u(t) \quad (3)$$

We get a Hilbert-transformed signal $R_u(t)$ of the $S_u(t)$.

2.2 Analysis

In the following chapters, we explain the interference elimination system. Supposing a interference wave is mixed in a TV signal band with a video carrier angular frequency of Wv, the sum of the TV signal and the interference wave will be expressed by Equation (4). The first and second terms of Equation (4) represent the in-phase component Rd and quadrature component Sd of a TV signal, respectively. Then the in-phase component Ru and quadrature component Su of a interference signal, respectively.

$$\begin{aligned} X(t) &= R_d(t)\cos(W_v t)-S_d(t)\sin(W_v t) \\ &+R_u(t)\cos(W_v t)-S_u(t)\sin(W_v t) \end{aligned} \quad (4)$$

The TV signal X(t) expressed by Equation (4) is put into a synchronous detection circuit, together with the quadrature video carrier $\sin(W_v t)$ which is obtained by reproducing the video carrier from the TV signal. Then, the TV signal is filtered with a 4.2 MHz-cutoff low-pass filter to get the baseband signal expressed by expression (5)

$$-1/2\{S_d(t)+S_u(t)\} \quad (5)$$

Signals expressed in expression (5) are multiplied by the quadrature carrier $\sin(W_v t)$ again. Expression (5) becomes

$$-1/2\{S_d(t) + S_u(t)\}\sin(W_v t) \quad (6)$$

The band of Signal is limited by the BPF to remove the lower sideband.

$$\begin{aligned} -1/4\{&-R_d(t)\cos(W_v t)+S_d(t)\sin(W_v t) \\ &-R_u(t)\cos(W_v t)+S_u(t)\sin(W_v t)\} \end{aligned} \quad (7)$$

When the interference wave is in the DSB transmission region of the TV signal, the components $S_d(t)$ and $R_d(t)$ in Expression (7) become zero. So, extract the third and fourth term of Expression (7) with a filter, and then multiply its amplitude by 4 to obtain the same interference wave component as the one represented by the third and fourth term of Equation (4).
This is the method to extract the interference wave and makes it Hilbert transformed at one time. Subtract this component from the disturbed TV signal expressed Equation (4), to eliminate the interference.
Then, what should we do if interference waves come in the region where a TV signal is SSB-transmitted? The interference waves can be removed with quite the same procedure as stated above. However, in this region, quadrature synchronous detection components contain the TV signal in the region where it is SSB-transmitted. This makes the interference wave at an extraction filter output include not only interference waves but the TV signal. But TV signal's bandwidth is restricted to the bandwidth of the extraction filter.
So the band-limited TV signal is subtract from TV signal, as well as the interference wave.
This is equivalent to removing a part of the TV signal with the filter, which causes waveform distortion. This method, nevertheless, will contribute to improving overall picture quality, if this secondary picture degradation due to the waveform distortion is less than that caused by the Es interference.
This method secures stable removal of interference, since the amplitude and phase of extracted interference waves change in exactly the same manner as those of the original interference waves even when they rapidly change due to fading, etc. Thus no adjustment for amplitude and phase changes is required.

3. PHASE ERROR RESPONSE AND AMPLITUDE ERROR RESPONSE

This system subtracts the regenerated interference waves from the original interference waves. So, it is necessary to get regenerated interference waves which have exact the same phase and amplitude as those of original interference waves.
If the phase or amplitude of the system are changed by temperature variations or other reasons, the suppression of the system will decrease. We analyze the suppression, the phase error and the amplitude error of the system.
Suppose the original interference wave is $e_u(t)$, and the regenerated interference wave is $e_u'(t)$.

$$e_u(t)=E_u\cos(W_u t) \qquad (8)$$

$$e_u'(t)=E_u'\cos(W_u t+P) \qquad (9)$$

where the P is the phase error, and the R is the amplitude error, which is given by;

$$R=E_u/E_u' \qquad (10)$$

The suppression S is expressed by Equation (11).

$$S=-10\log_{10}|1+R^2-2R\cos(P)| \qquad (11)$$

Figure 2 shows dependence of the suppression S on the phase error P and the amplitude error R. From figure 2, to get the suppression of 40 dB, the phase error should be less than 0.1 degree and the amplitude error should be less than 0.1 dB.

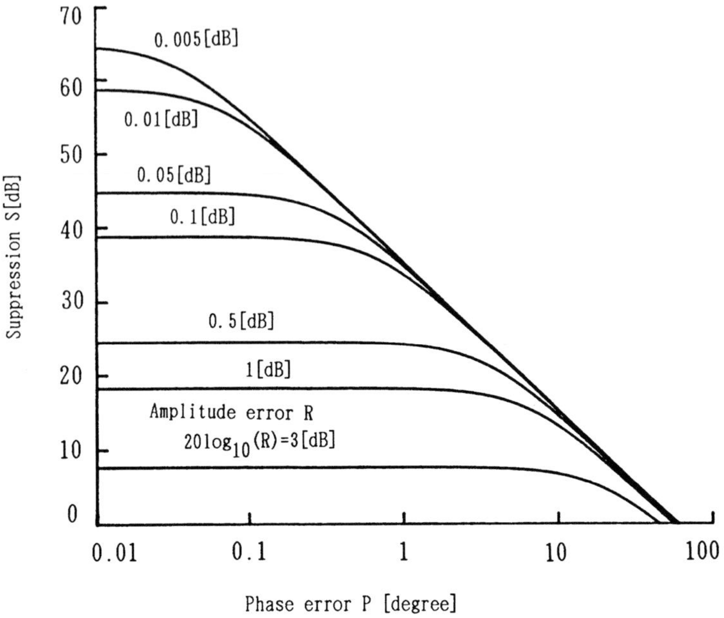

FIGURE 2 Dependence of suppression S on phase error P and amplitude error R.

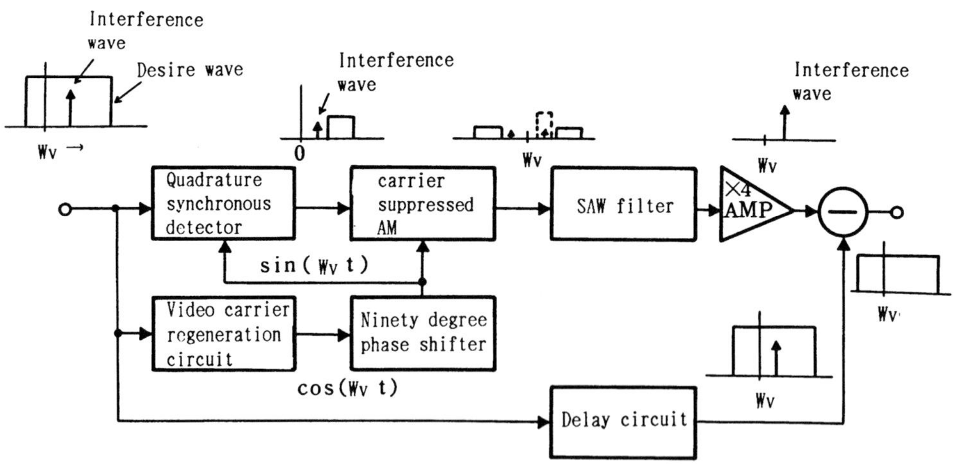

FIGURE 3 System diagram of the interference eliminating system.

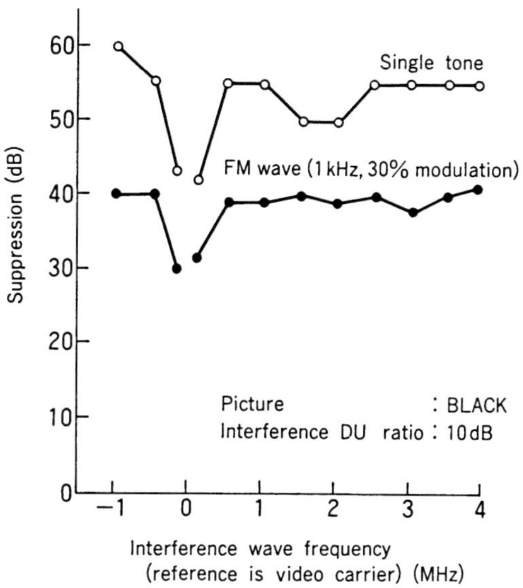

FIGURE 4 Suppression for interference as a function of frequency.

4. SYSTEM CONFIGURATION

Figure 3 shows a system diagram of the Es interference eliminating system. The specifications and performance of the system are shown in Table 1.

Table 1

The system's specifications and performance

Input/Output Frequencies:	19.5MHz$\pm$3MHz
Input/output Impedance:	50 Ohm
Input Level:	60 dBuV to 80 dBuV
Output Level:	80 dBuV
Removable Frequencies:	fv - 1.25 to fv - 0.15 MHz fv + 0.15 to fv + 4.0 MHz
Capability: (One FM interfering at DU ratio of 10 dB)	40 dB
Number of Removable Interference Waves:	5

5. SYSTEM PERFORMANCE

We conducted interference elimination tests on the experimental system by changing the frequency of interference wave.

Fig. 4 shows that an improvement of 50 dB or more is obtained for the case of a single CW interference wave. An improvement of 40 dB is also possible in the case of a frequency-modulated wave which is modulated by 1kHz sine wave with the frequency deviation of ± 22.5kHz.
According to the reference [5], the system can improve their beat disturbance to maintain the protection ratio required for viewer at DU ratios of greater than 10 dB. This means that our system has an enough practical improvement capability.
The amount of suppression is 30 dB for the disturbance within 150 kHz from the video carrier, because the bandwidth of the interference wave extraction filter is not narrow enough to discriminate between the interference wave and the video carrier.
To improve the filter characteristics, we have developed a surface acoustic wave (SAW) filter which has excellent amplitude and phase characteristics to extract interference waves. This filter improved the system performance, especially, for removing interference waves which have adjacent frequencies to the video carrier.

6. CONCLUSION

A method is proposed for eliminating narrow band FM interference waves in television signal, and an analysis is made. An experimental system is developed which automatically eliminates Es interference waves. This system suppresses by more than 50 dB a disturbing signal of sine wave and more than 40 dB for an FM interference wave, in a TV signal band. As for interference waves at DU ratios of greater than 10 dB, the system can improve their beat disturbance to maintain the protection ratio required for viewer.
The system is quite adaptable, since it is effective not only for foreign wave interferences but also for general interference waves with narrow-band spectra.
The authors are thanks to the personnel of the NEC Corporation for their collaboration in making the system.

7. REFERENCES

1. Edwards K.J. and L.F.Shrubsole (1984) Sporadic-E Propagation at Frequencies around 70 MHz, The Radio and Electronic Engineer, Vol.54, No.5, pp.231-237, May
2. J.L. Eaton and H.M.Price (1980) Co-channel interference reduction in television rebroadcast links by means of video notch filters, E.B.U. Review technical part, pp.20-23, No.179, February.
3. S.Yamazaki and H.Arata (1984) Rejection of sporadic-E interference in TV reception using notch filters",IEEE EMC Symposium, TOKYO, JAPAN, pp.2-806, October.
4. M.Schwartz (1990) Information transmission, modulation and noise, fourth edition, McGraw-Hill,chapter 4-7
5. CCIR (1963-1982) Ratio of wanted-to unwanted signal for AM vestigial sideband colour television system, Question 4/11, Study Programme 4A/11, report 306-4, p.232

9.4 The Multimode Measurements of EMC-Parameters and Statistical Theory of Multimode Systems

L.N. Geivandov

Vinnitse Polytechnic Institute, Eme Laboratory, Vinnitse, USSR

Nowadays in the force of a number of reasons known to radio-electronic aids (REA) electromagnetic compatibility (EMC) specialists much attention is paid to the investigation of multimode waveguide devices (MWD), namely multimode antennas on collateral radiation frequencies (CRF).

Wave processes in them differ in specific features, substantially complicated solution both theoretical and practical problems. Today the probabilistic approach to the solving of such problems is a common knowledge.To them we may refer the problems of complex prediction on CRF inner and outer parameters of MWD MA and REA on the whole. Such approach for some years is being developed by the author together with his colleagues. Our developed method is based on RF-section measurements which are metrologically fully supplied by the existing park of special multimode measurement devices (MWD). It allows to predict practically all inner and outer primary and secondary parameters of MWD and MA, having direct correlation to the problem of REA EMC on CRF.

For the probabilistic evaluation of the outer primary and secondary MA parameters which are of great interest from the point of view of REA EMC on CRF on the results of RF-section measurements with the help of MSDC and DCMP the possibilities of different methodics of probabilistic recalculation of measurement results in the field of radiation of MA are in the form of differentials and integral functions of probability of current density of the power radiation of MA are proposed and analysed and also in the form of their probabilistic cartographic diagrams of directivity (PCDD) which are the most important characteristics of territorial groupings of REA from the point of view of their EMC on CRF. It is justified the necéssity of the using of such methodics and on the basis of the theoretical and experimental aprobation their correctness is proved. It is pointed out that all the methodics are realized in the form of the convenient for a user the FORTRAN-program complexes for unified system computers (USC). It is pointed out that the methodics are already used and their application gives the substantial economic effect.

It is noted that the developed methodics in their main points are the basis of the statistical theory of MA (STRA) which later is necessary to apply together with already existed statistical theory of antennas (STA).

In conclusion of the report is put forward a number of perspective problems subjected to the solution: the necessity of working out of the theory of synthesis of MSDC and DCMP with large coupling; the necessity of the working out of the methodics of prediction of MA REA parameters in nearer zone and Frenel's zone; the necessity of working out the theory based on the use of MSDC and DCMP experimental analytical methods of solving inner RF-section problems of MWD electrodynamics; the necessity of the search of new constructions of MSDC for F-waves; the necessity of the search of the principally new transducers reacting on Umov-Poynting vector which having substantially simplified MSDC taking off the majority of difficulties both theoretical and constructive technological types; the necessity of the working out of automated measurement computing complexes with "built in" which is common knowledge in modern calculating technique , microprocessors and so on.

10 Space Communication and Measurement

10.1 Satellite Communications of the Future and Their EMC Problems

F.L.H.M. Stumpers
Elzentlaan 11, Eindhoven 5611 LG, The Netherlands

The problem of the possible saturation of the radio spectrum has been with us for a long time. In 1972 at the proposal of Professor Barlow, chairman of Commission VI, URSI accepted a Recommendation, that, where possible guided waves should be used instead of radiated waves. At the time he thought more of guided microwaves. In 1988 Carassa wrote: "There is no doubt that the exceptional characteristics of optical fibers and their consequent fast penetration in the terrestrial telecommunication network erode some domains which were considered available for satellite communications". Integrated Service Digital Networks are expected to be in the domain of optical fibres, but Bartholome and Catania have stressed, that, if we want to have ISDN within ten years, the help of radio (especially satellite waves), may well be necessary.

During its first two decades, the Intelsat satellite communications network has come to span almost the entire globe with over 1600 connectivity paths through nearly 800 antennas operated by more than 150 user administrations/organisations. The use of digital techniques in the INTELSAT network is old, but recently INTELSAT stressed, that the Intermediate Data Rate system, derived from the FDMA access technique, will play a major role in providing cost-effective satellite paths on a global basis (Int. J. Sat. Comm. vol 6, 365 etc. special issue, Robusto, ed.). Low rate encoding with digital speech interpolation has sufficiently matured to enable increasing the throughput of the space segment. The INTELSAT VA satellite had a capacity of 15000,64 kbps channels. The INTELSAT VI satellite has a capacity of 24000,64 kbps channels, or 120000,32 kbps/DIS channels. The TAT-8 transatlantic optical fibre cable has 7500,64 kbps channels or 37800,32 kbps/DSI channels, but this is certainly not the limit for what optical fibres can do. (Sachdev, World Telecom Forum 1987,II, 179-183.)

The provision of communication links to land mobile users is one of the most challenging tasks of satellite communications. Low gain mobile gain can be overcome by high gain in the satellite. One can use multiple overlapping spot beams to cover the desired area, but the efficient use of such beams requires a foreknowledge of the traffic distribution. Frequecny addressable fan beams give a more adaptable arrangement and provide more communication capacity for a given satellite size. In order to cover the continental United States, this region is divided into three zones, in each of which the entire allocated spectrum is used. In each zone the low end of the band is directed to the west edge of the zone, and the high end is directed to the east edge. In the mobile service one or more base stations are used to complete the communication links between the mobile users and the stationary parties. The links between the satellite and the base stations are assigned frequencies in Ku band. Three separate bands at Ku band must be used. They are converted to the single assigned band at L band by using different local oscillator frequencies. For a private network, Ku band satellite frequency addressable beams are used in connection with very small aperture terminals. (Rosen, World Telecom Forum 1987, II, 185-189).

An intelligent satellite communication system, the dynamic and adaptive demand assignment multiple access system, can realize an economical office-oriented, multi-media satellite communication system with small capacity VSAT's (Suzuki a.o., World Telecom Forum 1987, II, 191-195).

There are frequency allocations for the mobile satellite service in the range around 1.6 GHz, but here satellites will remain very far from the spectrum efficiency of cellular systems, as planned in the European Community, as even a 30 m antenna diameter gives a footprint of 300 km diameter. Frequencies are allowed for ships, aircraft, and land mobile service. A small band was allowed near 1.6 GHz for the transmission of air passengers' in flight telephone calls to terrestrial stations. Various systems have been designed for determining the location of ships, aircraft or land vehicles by radio from a central station. Some of these systems use satellites, like Geostar in U.S.A. and Locstar in Europe. Geostar is allocated 1610-1626.5 MHz for up-links and 2483.5-2500 HMz for down links. The U.S.A. has established a constraint on the use of Geostar to prevent interference from mobile station transmitters to radio astronomy observatories, receiving the hydroxyl line in the band 1610.6-1613.8 MHz. Under this constraint, within 25 km of an observatory receiving in that band, a mobile station may transmit only in the first 200 ms of each second of Coordinated Universal Time, leaving the remaining 800 ms per sec for radioastronomy. For mobile stations on aircraft the protected area has a radius of 150 km. It was suggested that the constraint be used internationally. In general it was found impossible to increase the total bandwidth allocated to the land-mobile satellite services, as much as desired by several countries and the Conference resolved that another conference should be convened, not later than 1992, with more general authority to revise frequency allocations between 1 and 3 GHz. (Space radio services and WARC-MOB-87, Withers, Int. J. Sat. Comm. 25-28, 1988).

Frequencies above 10 GHz have advantages, possibilities for smaller antennas, no terrestrial interference. Rain attenuation and depolarization are disadvantages. See Stutzman: "Prediction of rain effects on earth-space comm. links operating in the 10 to 35 GHz frequency range". (Int. J. Sat. Comm., vol 7, 37-45, 1989.) Fade countermeasures at Ka band of Olympus were disucssed by Willis and Evans (Int. J. Sat. Comm. 301-311, 1988). Rainstorms passing through satellite links cause a high degree of fading, but usually a rain cell covers only a few square kilometres. Therefore, spatial diversity with two earth stations at 10 km distance is a possible solution, giving 10 dB gain at 30 GHz. Adaptive forward error correction, or variable rate coding schemes are another possibility. Up-link power control may be used, giving 10 dB extra power if a fade of 10 dB occurs.

For intersatellite links micro/mm wave implementation and optical implementation are evaluated. (Int. J. Sat. Comm. 107, 117, 1988).
VSAT's can be defined as a class of very small aperture, intelligent satellite earth stations suitable for easy on-premise installation, usually operating in conjunction with a large size hub earth station and capable of supporting a wide range of two-way, integrated telecommunication and information services. IEEE Communications Magazine had a series of articles on this subject in 1988 and 1989.

Space in the geostationary orbit is limited, the distance between satellites in orbit was already changed from 4 to 0. In C-band the satellite size was about 30 m. At higher frequencies the design of satellite clusters, operating as many as 10 to 12 satellites in a single module point of the geostationary orbit, or the use of a large platform with the effective performance of 50 or even 100 satellites are ways of dealing with the limitations of the geostationary orbit. There has been some research on the possibility of geosynchronous orbits below the Clarke orbit. A high tensile strength suspension cable might allow a suspended communications satellite to be maintained in orbit thousands of kilometers below the "anchor" satellite. NASA may try this approach in the 1990'S. Another approach is SPIDERSAT, a very low mass, gossimer-like communications satellite, the power for the satellite's operation and for its stabilization in geosynchronous orbit some 800 km above the earth's surface, being provided by ground-based energy transmission.

Direct Broadcast Satellites are already functioning in several countries. In most cases they are received via cable (CATV). Nevertheless it is unlikely, that the broadcasting authorities would be willing to give up frequencies, e.g. because they will need more bandwidth for high definition television. In Japan it is expected, that HDTV broadcasts may be put into regular service in 1990 by use of their next operational broadcasting satellite BS3. (Nakamura "Possibilities of high definition television and its future". World Telecom Forum. II, 129-131, 1987.)

INMARSAT takes care of communication with ships. In coastal areas radio transmission via microwave transmitters is also possible.

There is a potential demand of large European road freighters for a private wide area network. Geostationary and non-geostationary options are studied. Originally USSR was using the Molnya and Tundra satellites in elliptic orbit, but later they reverted to the geostationary orbit. Recently the U.K. has also suggested a highly elliptic orbit satellite option for land mobiles. Probably three satellites would be used in such a service, so that one satellite would always be available to the mobile station.

10.2 Efficient Use of the Geostationary Satellite Orbit: Allotment Planning and Its Practical Use

Y. Ito and M. Yamada
KDD Meguro Research and Development Laboratories, Tokyo, 153 Japan

Abstract

The Allotment Plan was created and adopted in WARC-ORB2 held in 1989. For the practical use of the plan, especially when accommodating modifications or new requirements, it is still necessary to establish a workable algorithm to efficiently preserve margins for the unforeseen use. This paper proposes a concept of c/e ratio which is useful to find locations of additional requirements. It is also shown that the c/e ratio when combined with the concept of the predetermined arc can improve the use of the allotment plan.

1. INTRODUCTION

Communication networks based on geostationary satellites have become one of the most important infrastructure for the growing needs of telecommunication. However, the geostationary satellite orbit (GSO), the circular orbit 36000 Km above the equator, is considered a finite resource in which the maximum number of deployable satellites is limited by the mutual interference between nearby systems.

As for the use of the GSO, despite the finite nature of the capacity, a first-come-first-served rule has been long used. Along with the rapid development of the satellite communication systems, however, requirements of equitable access to the GSO has been raised at the World Administrative Radio Conference - General (WARC-G) held in 1979. In order to achieve this objective two series of sessions, WARC-ORB 1 and 2, were held in 1985 and in 1988 where the principle of the planning was adopted in the first session and the actual plans were approved in the second session. The allotment plan, which is the subject of this paper, was developed for the 800 MHz out of 6/4 GHz band and 13/11 GHz band in which at least one satellite per ITU member country has been allocated. [1]

Based on the above described background, this paper focuses on the allotment plan and first overviews the result of the plan. The allotment plan obtained at WARC-ORB2 is known to include 152 dual-band satellites each of which assumed to utilize both 6/4 and 13/11 GHz bands. In the practical use of the allotment plan, however, a systematic method seems still necessary to make the plan workable.

In this paper a concept called c/e ratio is first introduced. This concept is considered useful to preserve some margin for unforeseen requirements. Secondly, the other key element of the planning, use of the predetermined arcs, will be demonstrated. The predetermined arc is defined as the range of the service arc in which an allotment is obliged to move so as to absorb the impact of modifications such as addition of sub-regional systems. This concept is considered to enhance flexibility in the planning. An investigation will be made to evaluate this flexibility by introducing a hypothetical satellite in Region 1. The result shows how an additional system can be introduced in the allotment plan.

2. THE ALLOTMENT PLAN

In the WARC-ORB2 Conference held in 1988 the Allotment Plan which was adopted in the previous session was put into development. The plan was made based on the criteria summarized below. [1]

Frequency Bands	4,500-4,800 MHz (Down-link)
	6,725-7,025 MHz (Up-link)
	10.70-10.95 GHz (Down-link)
	11.20-11.45 GHz (Down-link)
	12.75-13.25 GHz (Up-link)
Up-link C/N	23 dB
Down-link C/N	17 dB (faded condition)
Total C/N	16 dB
Aggregate Interference	26 dB
Minimum Elevation Angle	10° for Climatic Zone A to G
	20° for Climatic Zone H to L
	30° for Climatic Zone M, N
	40° for Climatic Zone P
Rain Margin	8 dB
Earth Station Parameters	
Antenna Diameter	7 m (6/4 GHz)
	3 m (13/11 GHz)
Noise Temperature	140K (6/4 GHz)
	200K (13/11 GHz)
Antenna Efficiency	70%

Antenna Pattern	32-25logϕ side-lobe R.R.-AP29 main beam
<u>Space Station Parameters</u>	
Beam Shape	Circular or Ellipse
Antenna Pattern	SAT-83 Pattern
Noise Temperature	1,000K (6/4 GHz) 1,500K (13/11 GHz)
Antenna Beam Width	1.6° minimum (6/4 GHz) 0.8° minimum (13/11 GHz)
Antenna Efficiency	55%
Antenna Pointing Error	0.1°

Based on the system parameters described above, a dual band (6/4 GHz and 13/11 GHz) satellite is allotted for each ITU member country. However, because of the limited computer resource used in the Conference and because of the stringent requirements such as the high elevation angle required for climatic zone P, relaxed side-lobe of 32-25logϕ, and most significantly the 36 already registered satellites called existing systems, the synthesized plan was not compliant with the above conditions. 16 systems in 6/4 GHz band and 21 systems in 13/11 GHz band still receives C/I of less than 26 dB.

Presently, mutual agreements were made among those administrations concerned. However, future improvement is expected along with the development of the actual use of the plan. The key point to improve the plan is considered a use of the concept of predetermined arcs.

3. PREDETERMINED ARC

In the allotment plan, in addition to the nominal locations of satellites a predetermined arc is attached to each satellite. The predetermined arc is defined as a range of orbital arc where a satellite is obliged to move when neighboring systems asked for modifications or to accommodate a new allotment. The length of the predetermined arc is defined depending on the status of each satellite. The status and the corresponding length of the predetermined arc is as follows.

planning stage	-----	nominal position ± 10°
design stage	-----	nominal position ± 5°
operational stage	-----	nominal position ± 1°

The key of realizing flexible operation of the allotment plan is depending on the use of the concept of predetermined arc. However, this concept is <u>frequently misunderstood</u> in such a way that the predetermined arc is the range of arc segment where an allotted satellite can be freely moved and registered at any position within this range.

4. HOW TO IMPROVE THE PRESENT SITUATION

In order to improve and enhance the practical use of the allotment plan, two methods might be seriously considered. One is the relaxation of the planning parameters and the other is the positive use of the predetermined arc concept.

4.1 Relaxation of C/I

First approach is to simply accept reduced C/I ratio. Presently, minimum C/I of 26 dB is required for each allotment so that minimum C/N of 16 dB may be obtained. The important characteristics of satellite communication is the overall C/N value rather than the C/I value. Therefore, if degradation of C/I does not significantly affect on the value of overall C/N, slight reduction of C/I must be accepted to help improving the tight situation. The simple relation of overall C/N versus C/I is shown in Figure 1.

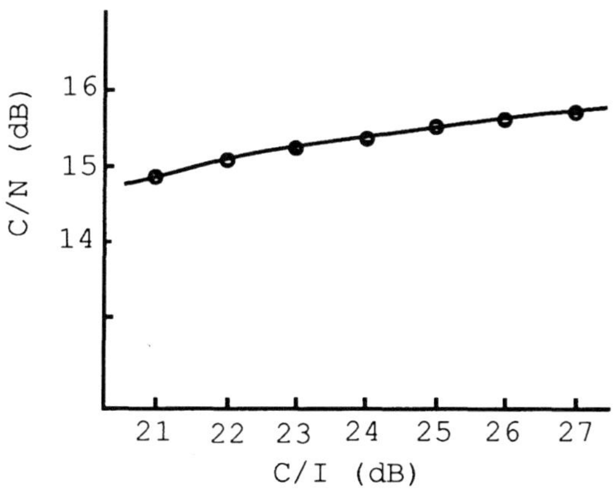

Fig. 1 Effect of Relaxed Interference Criterion

It is understood from Fig.1 that degradation of overall C/N value is only 0.4 dB even if the required C/I is relaxed from 26 dB to 23 dB. This relaxation will help most of the situation.

4.2 c/e Ratio

Second approach is based on the positive use of the predetermined arcs. The idea is to cooperatively adjust the location of neighboring satellites so as to accommodate a regional system or to accept a modification of parameters. In order to carry out this type of operation, a useful concept of "c/e ratio" is proposed hereafter.

The impact of a new entrant is apparently a function of the part of the orbital arc. If a specific portion of the orbital arc is less congesting, even a satellite with extreme characteristics may be accommodated with no coordination. On the other hand, even a moderate system may not be accommodated if congestion of that portion of the arc is severer.

The objective of c/e ratio is to evaluate a location where the smallest impact is expected. The expression "c/e" corresponds to the abbreviation of "contraction/expansion" and is computed by the procedure described below.

Procedure of Expansion

(1) Denote the service arc of a new entrant as $\alpha \leq S_0 \leq \beta$ and take a range L_0 such that $(\alpha - p°) \leq L_0 \leq (\beta + p°)$. A protection range of p° is selected so that the edge effect or the existence of satellites outside the finite arc may be minimized.

(2) Consider all satellites in the range L_0 and consider each possible inserting position as shown in Fig.2. For each inserting position, compute the expansion ratio following the steps described below.

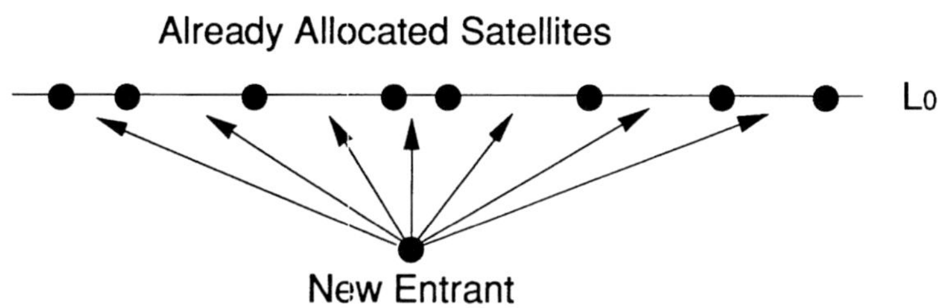

Fig.2 Search of Inserting Position

(3) Compute the carrier to interference ratio for each satellite and adjust the locations of all satellites within their service arcs. The order of satellites shall not be interchanged in this operation. In most cases, spacings near the new entrant will be increased and sometimes the interference criterion may not be satisfied because the location of neighboring satellites are kept within their service arcs.

(4) When the interference criterion is not satisfied, the entire arc segment L_0 is expanded. The service arcs are all proportionally expanded and the adjustment of the positioning is made for each small increment of the arcs. The adjustment of the positioning is made as described in the Annex.

(5) After the expansion and the adjustment, if the interference criterion is satisfied for all the satellites involved, denote the expanded arc as L. This procedure is shown in the upper half of Fig.3.

(6) Take the numerical ratios, L/Lo. This value is defined as the expansion ratio.

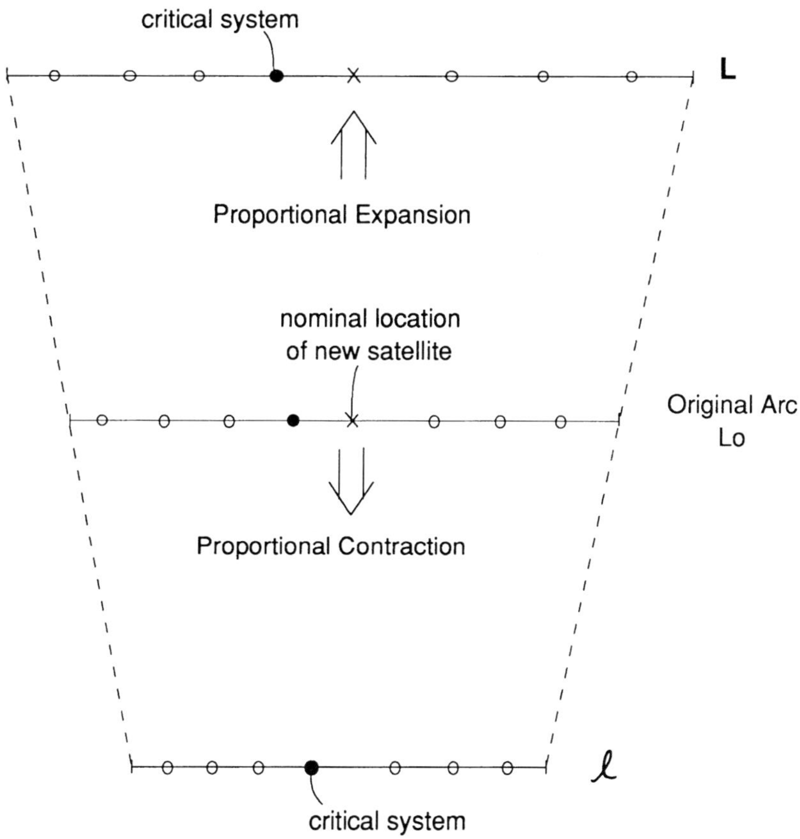

Fig. 3 Expansion (or Contraction) of Spacings

Procedure of Contraction

Procedure of contraction is performed not including the new entrant. This procedure is to identify the minimum length of the orbital arc before introduction of the new entrant.

(1) Without taking into account the new system, compute the carrier to interference ratio for each satellite and adjust the location of all satellites within their service arcs so that the length of the arc segment L_0 be minimized. In this operation, C/I for each satellite is constrained within the criterion value and the order of satellites shall not be interchanged. The minimization can be carried out using the mathematical formula given in the Annex. Denote the resultant length of the total arc as ℓ.

(2) Take the numerical ratios ℓ/Lo. This value is defined as the contraction ratio. The procedure is shown in lower half of Fig.3.

c/e ratio

The c/e ratio is merely the numerical ratio of the contraction ratio and the expansion ratio. It is considered that the larger the contraction ratio the higher the influence of the new entrant on that portion of the orbital arc.

The c/e ratio defined above is interpreted as follows.

(a) If the expansion ratio is greater than 1, the new system has an immediate influence and it must go through the coordination procedure between the critical system.

(b) If the c/e ratio is greater than 1 but if the expansion ratio is less than 1, the new system has no immediate influence.

(c) If the c/e ratio is equal to 1, it is concluded that the system has no influence presently. Other systems are mainly contributing to the capacity of that part of the orbital arc.

(d) The less the c/e ratio, it is concluded that the less the influence of the new system might be.

5. EXAMPLE USE OF c/e RATIO AND PREDETERMINED ARCS

Two examples are shown in this section. First example shows a simple illustration of the typical use of c/e ratio. Second example shows a practical application of c/e ratio combined with the concept of the predetermined arcs. In this example an additional sub-regional system is inserted in the allotment plan.

5.1 Simple Use of c/e Ratio

For the purpose of simplification, system parameters used in the allotment plan are assumed. In this basic configuration with 10 hypothetical systems shown by dotted lines in Fig.4, a new system is introduced in the mid-point of the arc. Three types of new entrants are also shown in Fig.4 by solid lines. A satellite Ex.1 covering a wide area is first introduced (coverage is shown as Ex.1 in Figure 4). A satellite Ex.2 covering a small area is then introduced in place of Ex.1. It is intuitively known that the satellite Ex.2 has less impact than Ex.1 because of its smaller coverage. Finally, a satellite Ex.3 covering an isolated area is introduced. The impact of this satellite is expected to be negligible.

Table 1 shows the initial status of 10 satellites covering the regions specified in Fig.4. The satellites are placed every 4 degrees except between #5 and #6 where 8 degree spacing is given. A new satellite is now introduced between #5 and #6. For these conditions, the aggregate interference criterion of 24 dB and the single entry interference criterion of 28 dB are applied to determine the expansion ratio and the contraction ratio.

Table 2 shows the obtained values of the c/e ratio, the expansion ratio and the contraction ratio. It is observed in Table 2 that the expansion ratio for satellite Ex.1 is 0.86. This value indicates that the satellite can be presently accommodated in the arc with no difficulty. However, the c/e ratio of 1.2 shows that the impact of this satellite is rather high which requires 20% of the arc space even in the optimum spacing is achieved. The effect might appear if additional satellites are launched in this part of arc. On the other hand, satellite Ex.2 has less impact as expected which shows the c/e ratio of 1.07, the expansion ratio of 0.76 and the contraction ratio of 0.72. The result for satellite Ex.3 shows that it is harmless for this part of arc.

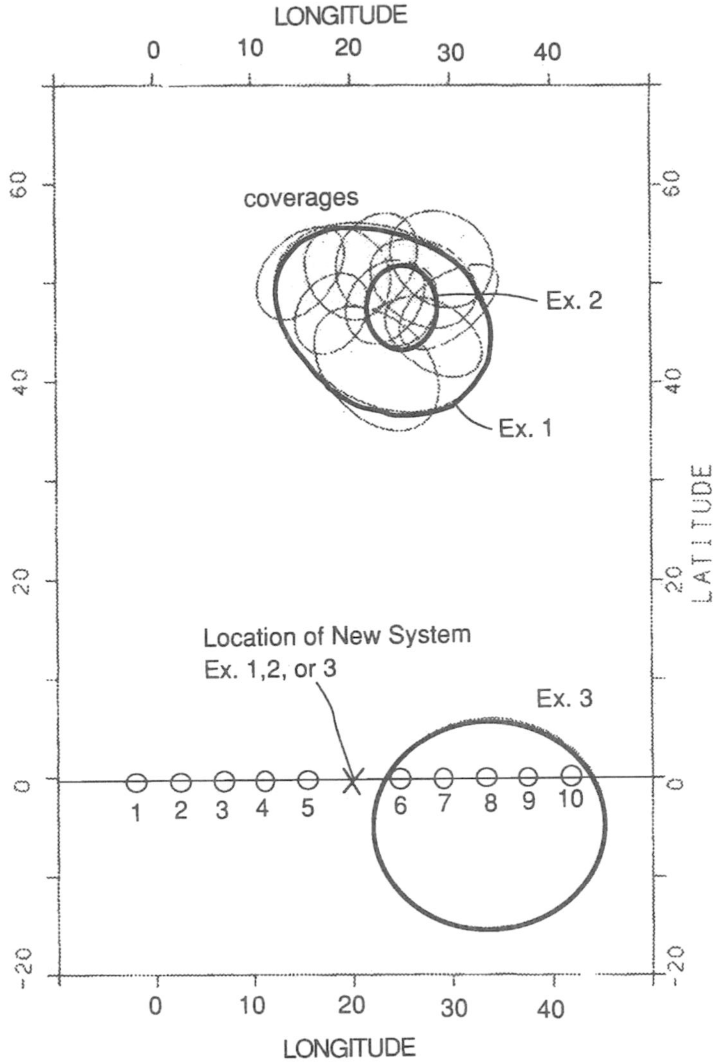

Fig.4 Coverages and Locations of 10 Satellites and Three Types of New Entrants

Table 1. Initial Status of 10 Satellites

Satellite #	Location	C/I (dB)
1	0°	38.9
2	4°	35.3
3	8°	28.0
4	12°	25.8
5	16°	27.3
6	24°	26.5
7	28°	29.4
8	32°	26.5
9	36°	26.3
10	40°	29.1

Table 2. Example Results

	various ratios		
	c/e ratio	Expansion	Contraction
Ex. 1	1.20	0.86	0.72
Ex. 2	1.07	0.76	0.72
Ex. 3	1.00	0.72	0.72

5.2 Application to the Allotment Plan

Use of the c/e ratio to the allotment plan is an important objective. In this example, an allotment is randomly selected and a location of its additional requirement is searched based on the c/e ratio. The procedure proposed here is as follows.

(1) For every possible inserting position within the service arc of the additional requirement, apply the concept of c/e ratio and find a position which produces the minimum c/e ratio. For the existing allotments, their service arcs are considered the same as their corresponding predetermined arcs.

(2) Assume that the positions of all allotments are fixed as given in the Plan and place the new entrant at the position obtained in (1). If C/I's of already allotted systems are greater than 26 dB (or at least same value given in the Plan for the satellites with less than 26 dB), the additional requirement can be accommodated with no difficulty.

(3) If C/I's of some of the neighboring systems become lower than 26 dB (or lower than the value specified if C/I's are already less than 26 dB in the Plan), the procedure of adjusting the mutual spacings of satellites will be needed. The procedure starts from (4).

(4) For the above identified satellites, assume their service arcs equivalent to their corresponding predetermined arcs. For the new entrant assume the original service arc. For the rest of the satellites, fix their location so as not to involve these systems in the coordination. The interference criterion for the allotments are same as in the Plan but the one for new entrant is based on its requirement which may be assumed to be less than 26 dB.

(5) Under the parameters specified above, optimize the locations of satellites. The mathematical formula of optimization is shown in the Annex. If C/I of all satellites are within the target value, the procedure stops. However, if the C/I's of some systems are not satisfactory, find additional satellites which interact with the satellites not fixed. For these satellites remove the constraints of fixed positions, provide service arcs equivalent to the predetermined arcs and return to (4).

An example of the above procedure was made to demonstrate the effect of the proposed method. In the example, a hypothetical satellite covering the lower part of Europe was added. The result is shown in Table 3 where the affected satellites are listed.

It is observed in the Table that C/I of 25 dB is possible for the new entrant named New. On the other hand, among the affected systems, relocations requested are in the order of 1/10 degree (except for satellite #11) which is practically considered no impact.

The impact on the interference value is also negligible. It is observed in Table 3 that C/I value of only one satellite is degraded by 4/100 dB which is again considered negligible.

Table 3. List of Affected Satellites

Sat. Name	Old Pos.	New Pos.	Old C/I	New C/I
#1*0000	26.4	26.4	26.03	25.99
#2*0000	27.0	27.1	34.12	32.41
#3-0000	40.5	40.2	28.18	26.26
#4-0000	42.5	41.8	27.69	28.57
#5-0000	43.1	42.5	26.10	26.11
Exist-#1	45.0	44.8	-	-
#6-0000	46.8	46.7	25.01	25.32
New*	-	47.5	-	30.10
New	-	47.5	-	24.94
#7-0000	48.0	48.6	28.05	29.31
#8-0000	50.4	50.1	26.08	26.65
#9-0000	51.0	51.3	33.75	34.90
#10-000	52.0	52.7	26.28	28.90
#11-000	52.7	54.3	27.76	28.80

- * in the list indicates a satellite operating at 13/11 GHz
- Exist-#1 is an existing system
- New is the new entrant
- - shows that the value is not available

6. CONCLUSION

Two methods are proposed to practically use the allotment plan. First is to simply relax the required C/I value which only degrade the C/N by 0.4 dB. Second is to introduce the concept of "c/e ratio" which is able to measure the impact of new entering systems. Then the concept of the predetermined arc is fully utilized to flexibly adjust the locations of already allotted systems. It is envisaged that those methods introduced in this paper will be useful to improve the use of the allotment plan.

References

[1] CCIR, "Final Acts of the World Administrative Conference on the Use of the Geostationary Satellite Orbit and the Planning of Space Services Utilizing It (ORB-88) ", Geneva, 1988

[2] Y.Ito, T.Mizuno and T.Muratani, "Efficient Utilization of Geostationary Orbit Through Optimization", IEEE Trans. on Comm. vol.Com-27 No.10, October 1979

[3] T.Mizuno, Y.Ito and T.Muratani, "Computer Tools for Optimizing Orbit Use", AIAA 10th Communication Satellite Systems Conference, pp 549-557, March 1984

Annex

Mathematical Expression of the Minimization of the Total Arc

In order to obtain the expansion ratio and contraction ratio, it is necessary to know the minimum arc length to accommodate all requirements. The mathematical expression of this minimum arc length problem can be generally expressed as follows. [2],[3]

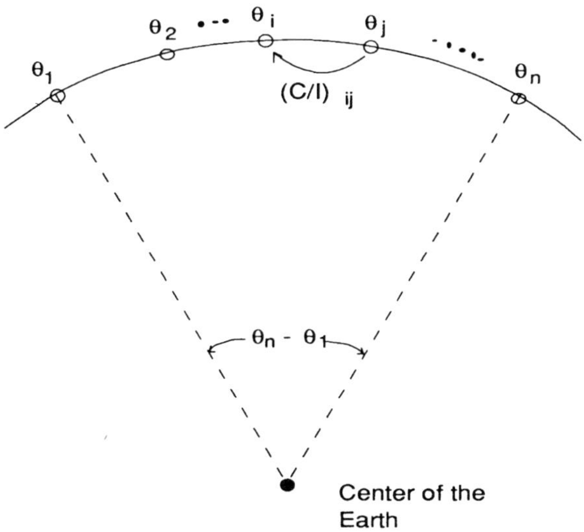

Fig. A-1 Minimization of the Total Arc

$$\text{Minimize } (\theta_n - \theta_1) \qquad \text{(A1-1)}$$

$$\text{Subject to} \quad (C/I)_i^{-1} \leq \sum_{j=1,\, j\neq i}^{n} \left((C/I)_{ijup}^{-1} + (C/I)_{ijdown}^{-1} \right) \quad \text{for all } i \qquad \text{(A1-2)}$$

$$(C/I)_{ij}^{-1} \leq (C/I)_{single}^{-1} \quad \text{for all } i \text{ and } j \qquad \text{(A1-3)}$$

$$a_i \leq \theta_i \leq b_i \quad \text{for all } i \qquad \text{(A1-4)}$$

where a_i and b_i are the left edge and the right edge of the service arc.

Referring to Figure A-1, the objective of this expression is to minimize the total arc occupied by n satellites under the condition that C/I is always greater than the criteria values.

In the equations;

$$(C/I)_{ij\ up} = \frac{P_{ei}\, G_{ei}(0)\, g^{r}_{ij}}{P_{ej}\, G_{ej}(\phi)\, g^{r}_{ij}}$$

$$(C/I)_{ij\ down} = \frac{P_{si}\, g^{t}_{it}\, G_{ri}(0)}{P_{sj}\, g^{t}_{ji}\, G_{ri}(\phi)}$$

and $(C/I)^{-1}_{ij} = (C/I)^{-1}_{ij\ up} + (C/I)^{-1}_{ij\ down}$.

where, system parameters are given by the following definition.

P_{ei}, P_{ej}	: Up-path available carrier power at earth stations of the I and J networks
$G_{ei}(\phi)$, $G_{ej}(\phi)$	: Earth station transmit antenna gain pattern
$G_{ri}(\phi)$	: Earth station receive antenna gain pattern
g^{t}_{ji} :	Space station transmit antenna gain in the j network in the direction of the i network's coverage
g^{r}_{ij} :	space station receive antenna gain in the i network toward coverage area of the j network

The mathematical problem stated in expressions (A1-1) to (A1-3) can be numerically solved by the method of non-linear programming which is well known in mathematical programming. A detail discussion is given in reference [2] and [3].

10.3 Development of AROW System for Measurement of Atmospheric Parameters Using Audio, Radio, and Optical Waves

Y. SONODA, N. KAWANO, M. FUKUMA, S. MATSUDA, and T. GONDA
Technical Center of Knowledge and Information Development, Kyushu Tokai University, Kumamoto, 862 Japan

ABSTRACT
The AROW (Audio, Radio and Optical Wave Ranging) system is proposed as a new means to measure the averaged values of the atmospheric parameters along the propagation path between two observation points, which can be used to improve the measurement accuracy of EDM (Electro Optical Distance Measurement). In this paper, the principle of this new technique is presented, theoretically indicating that the distance can be determined with an accuracy better than $1*10^{-6}$ by corrections for the atmospheric parameters, if we can determine the temperature and water vapor pressure within accuracies of about 0.3 K and 1 mb respectively by the AROW system. Further more, some results obtained from the experiments which were carried out to verify and/or realize such theoretical considerations are shown.

1.INTRODUCTION

The estimation of atmospheric parameters, such as temperature, pressure and humidity, strongly affects the accuracy of surveying by EDM. In such surveying, the distance determined with the correction using the local surface temperature measured at one spatial point shows variations with amplitude of about $8*10^{-6}$ in Japan.[1] The AROW system consists of the audio, radio and optical instruments and accurately measures the propagation times of these waves between the two points of interest. Integrated values of the atmospheric parameters along the propagating path, which should be used for the precise estimation of the distance by EDM, are derived from these mesured propagation times and their differences. The theory of the AROW and some experiments relating to the audio interferometer and DMTD (Dual Mixing Time Difference) signal processing method are presented.

2.THEORETICAL CONSIDERATIONS (PRINCIPLE OF AROW SYSTEM)

The velocities or the propagating times between two point of interest of the audio, radio and optical waves depend on the atmospheric condition along the propagating path. Inversely, we can say that the atmospheric parameters such as the water vapor pressure and temperature, especially the averaged values along the path, can be determined by exactly measuring and comparing these mean velocities or propagating times. The AROW system is composed of the audio, radio and optical sections, as shown in Fig.1, and can give the informations about the atmospheric parameters averaged through the propagating path of the EDM.

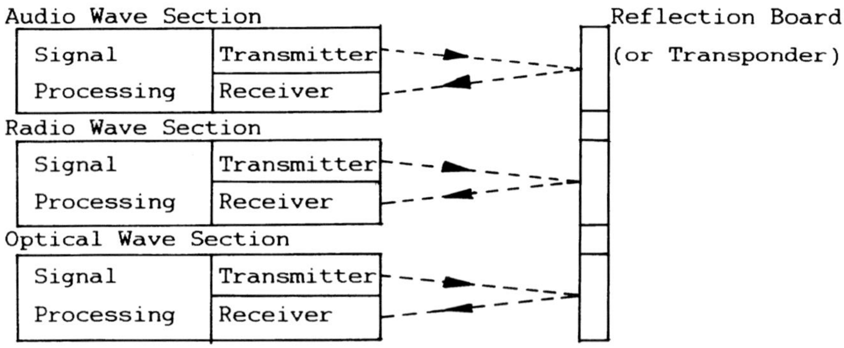

Fig.1 Block Diagram of AROW System

2-1.Comparison of Velocities in the Atmosphere of Audio, Radio and Optical Waves

The acoustic velocity in the atmosphere, c_a, is given by the following expression,

$$c_a = 20.067\{T(1 + 0.3192\ e/P)\}^{1/2} \qquad (1)$$

$$\simeq 20.067\ T^{1/2}(1 + 0.1596\ e/P) \qquad (m/s) \qquad (2)$$

where T(k), e(mb) and P(mb) are the temperature, the water vapor pressure and the total pressure respectively. Further, the water vapor pressure can be expressed by

$$e = e_s(T)(u/100)[1 - (1 - U/100)e_s(T)/P]^{-1} \qquad (3)$$

$$e_s(T) = 6.105\exp[25.22\{(T - 273.16)/T\} - 5.31\log(T/273.16)] \qquad (4)$$

where $e_S(T)$ and U(%) are the saturated pressure of water vapor depending on T and the relative humidity. From equation (2), we get the derivative of c_a as the following equation,

$$\Delta c_a \simeq 10.0\Delta T/T^{1/2} + 3.2T^{1/2}\Delta e/P - 3.2T^{1/2}e\Delta P/P^2 \qquad (5)$$

As the total pressure P is regarded as being almost constant over the wide region on the ground, the third term in equation (4) can be neglected and the local value of P, measured at the observing station by any standard device , can be used as a representative value instead of the averaged. If we assume T = 293 (K), P=1013(mb) and U=60(%) as the typical weather condition for equation (5), from a view of the effect to the acoustic velocity, the temperature change of $\Delta T=1(K)$ is equivalent to the pressure change of $\Delta e=10.8$ (mb). As the total pressure can be generally determined within $\Delta P=1$ (mb) over the wide region of a few km spread, we can say that the exact value of temperature is obtained by measuring the acoustic velocity if we can independently know the water vapor pressure by the radio and optical waves propagation, as described below.

Next, the velocity of the radio wave in the atmosphere, c_r, is given by the following formula.

$$c_r = c / n_r$$

$$\simeq c \{1 - (n_r - 1)\} \qquad (6)$$

$$(n_r - 1) \times 10^6 = 77.6P/T + 0.373 \times 10^6 e/T^2 \qquad (7)$$

where c and n_r are the light velocity in vacuum and the refractive index in the atmosphere, respectively. Equation (6) shows the change of temperature make little effect on the radio velocity, compared with the case of audio wave.
On the other hand, the refractive index of visible light, n_o, is given as follows.

$$(n_o - 1) \times 10^6 = 78.2P_d/T + 0.405 e \qquad (8)$$

where P_d is the pressure of dry air.
Comparing equation (8) with equation (7), the values of the first terms of both are nearly equal to each other. On the other hand, the second term of equation (7), which strongly depends on the water vapor pressure, is about ten times larger than that of equation (8), under the typical atmospheric condition.
Therefore, it is possible to extract the information of the water vapor pressure averaged through the propagating path by the delay-times of the radio and the optical waves.

2-2. Determination of Average Temperature and Humidity by Delay Time Measurements

From equations (8) and (7), the difference of delay-times between the radio and optical waves, t_{r-o}, is given by

$$
\begin{aligned}
t_{r-o} &= (2L/c)\{(n_r - 1) - (n_o - 1)\} \\
&\simeq (2L/c)\{0.373 \times 10^6 e/T^2 - 0.405 e\} \times 10^{-6} \\
&\simeq 0.676Le/cT^2 \qquad (9)
\end{aligned}
$$

which gives

$$\Delta t_{r-o} \simeq 0.676L\,\Delta e/cT^2 - 1.352Le\,\Delta T/cT^3$$

$$\simeq 0.676L\,\Delta e/cT^2 \qquad (10)$$

If we assume L=1000(m) and T=293(K), the measurement accuracy of the delay-time is required to be less than $2.6*10^{-11}$ (s) in order to obtain Δe=1 (mb), which corresponds to the phase difference of 10.8 degrees for 1 GHz modulation and 4.3 degrees for 400 MHz. Therefore, in order to determine the water-vapor pressure within an error of 1mb, the accuracy of the phase-measurement($\Delta\Phi$) is demanded to be better than 4.3 degrees for 400MHz , which is not easy but possible to be realized, as experimentally shown in the next chapter.

2-3. Determination of Average Temperature by Audio Wave Propagation

When the audio wave propagates the distance 2L (m), the delay-time t is given as

$$t = 2L/c_a \simeq L(1 - 0.1596e/P)/10.03T^{1/2} \qquad (11)$$

Inversely, we get

$$T = (L/10.03)^2\, t^{-2}(1-0.1596e/P)^2 \qquad (12)$$

$$\Delta T \simeq 2.0T\Delta L/L - 20.06T^{3/2}\,\Delta t/L - T\Delta e/3.13P$$
$$\simeq -20.06T^{3/2}\,\Delta t/L \qquad (13)$$

If we assume T=293(K), P=1013(mb) and L=500(m), the time resolution $\Delta t=4.97*10^{-3}$ (s) is needed to satisfy ΔT=1 (K). Therefore, we can conclude that in order to determine the temperarure with an error less than 1K, the accuracy of the delay-time measurement of the audio wave is needed to be less than about 5 ms, which value is enough to be realized, as experimentally shown in the chapter 3.

2-4. Improvement of EDM's Error Caused by Wheather Condition

Finally, we proceed to consider the improvement of EDM by applying the AROW system. The delay-time measured by EDM, t, is

$$t = (2L/c)n_o$$
$$= (2L/c)\{1 + (78.2P_d/T + 0.405\ e)\times 10^{-6}\} \qquad (14)$$

which gives

$$L = (ct/2)\{1 - (78.2P_d/T + 0.405\ e)\times 10^{-6}\} \qquad (15)$$

By differentiating equation (15), we get

$$\Delta L \simeq L\{78.2\ P_d\ \Delta T/T^2 - 0.405\Delta e\}\times 10^{-6} \qquad (16)$$

Under the typical condition (T=293(K), P=1013(mb), U=60(%), P_d=P−e=999.053(mb), ΔP_d=0.5(mb)), if we get $\Delta T=\pm 0.3$ (K) and Δe=1(mb) by the AROW system as shown in the preceding sections, the measurement accuracy of the distance of $\Delta L/L=8*10^{-7}$ can be attained.

3. EXPERIMENT

3-1. Audio Wave Experiment

The AROW system measures time difference between an audio signal reflected by a reflector(or a transponder) and transmitted one, in order to obtain mean temperature along a propagation path between the reflector and the receiver(transmitter) on the ground as shown in Fig.2. The received audio signals include not only direct waves reflected by the reflector but also multi-path waves which are reflected by the ground, buildings, trees etc. around the receiver. The multi-path waves disturb the phase of the direct waves, so that it is impossible to measure the time difference

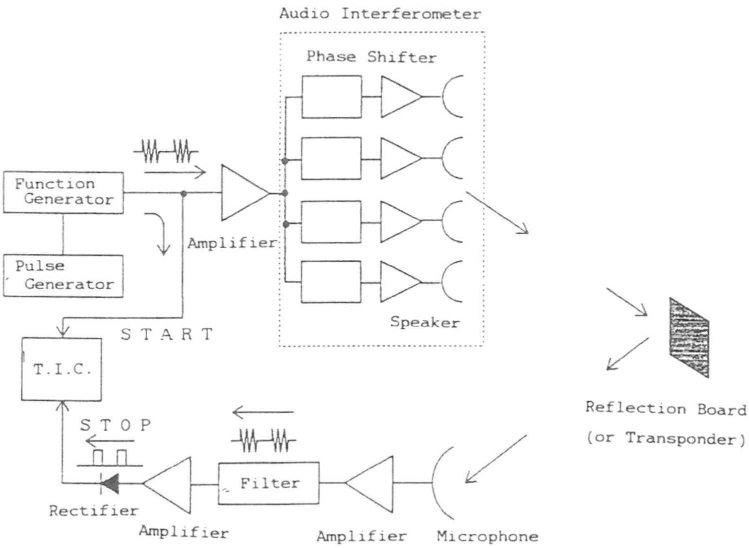

Fig.2 Block Diagram of Audio Wave Section

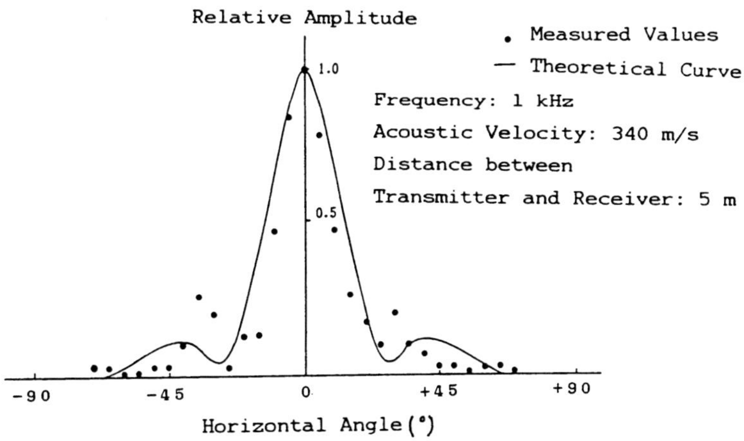

Fig.3 Field Pattern of Audio Interferometer

precicely. For this reason, it is a key technique how to eliminate the multi-path waves. One idea is to use a loud speaker and a microphone with narrow beam. A radio interferometer has acheived a narrow beam antenna and has been used in various fields such as radio astronomy and communication. We intended to make an audio interferometer with narrow beam. In this section, concept of the audio interferometer and its performance are described.

By the array theory, if each element of the array is assumed isotoropic, the normalized field pattern E_A is given by the following formula.

$$E_A = E/mE_o$$
$$= (1/m)\sin(m\psi/2)/\sin(\psi/2) \qquad (17)$$

where m:number of element, $\psi = \beta\ d\ \sin\phi + \delta$

d: distance between two elements, δ: phase difference between the signals received by neighboring elements, $\beta = 2\pi/\lambda$,

λ:wavelength of audio wave, E_0:maximum amplitude of field pattern by one element , E: field pattern of one element.

Further, the HPBW (Half Power Beam Width) and the shifted angle of the main beam ϕ are given, respectively, as follows.

$$L_\lambda = L_a/\lambda \qquad (18)$$
$$\phi = \sin^{-1}(-\delta\lambda/2\pi d) \qquad (19)$$

where L_a is the span of the array.

The audio interferometer, composed of 4 elements, was constructed and some properties were measured. Figure 3 shows an example of measured field patterns, where the theoretical curve well fits to the measured values and the high directivity is attained.

The relation between HPBW and spacing of two elements was shown in Fig.4, where the former is inversely proportional to the latter, as indicated by the theory. On the other hand, the shifted angle of the main beam is measured, controlling the phase of each element. The experimental results are shown in Fig.5.

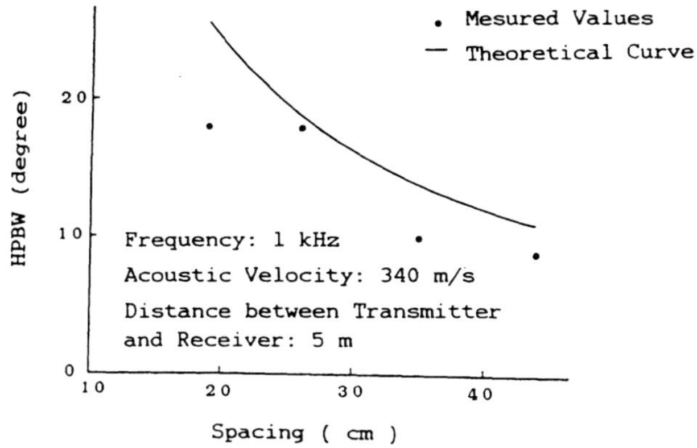

Fig.4 Relation between HPBW and Spacing

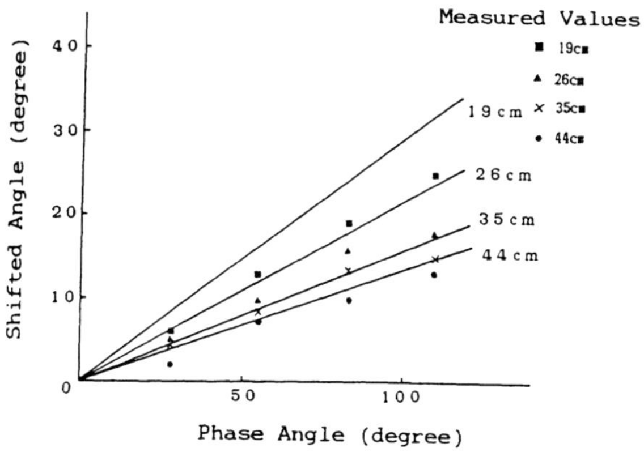

Fig.5 Relation between Shifted Angle and Phase Angle

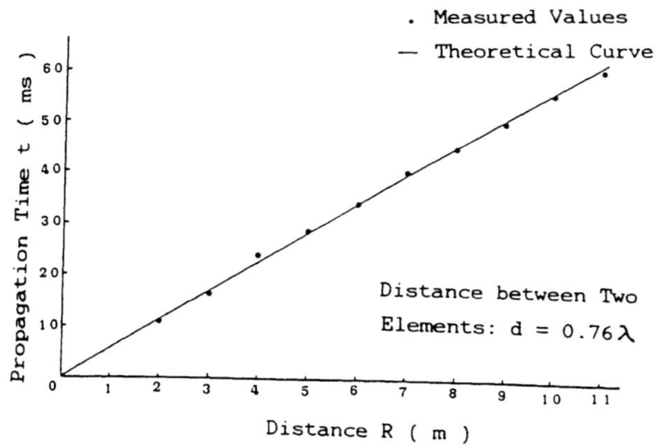

Fig.6 Result of Propagation Time Measurement

Though the shifted angle can be made large when the spacing decreases, it conflicts with the demand of the narrow HPBW. Therefore, we need to select the best condition in practically applying it to the AROW system.
Finally, the delay-time was measured by using the audio wave measurement system shown in Fig.2. In this system, the instrumental delay time was experimentally estimated to be less than 16 μs and can be neglected. The relation between the propagation time and the distance is shown in Fig.6, from which we can certify that the time resolution of a few ms demanded for the audio apparatus in AROW system can be easily attained.

3-2. Signal Processing Technique Needed for Measurement of Optical and Radio Waves

As described in section 2-2, for measurement of the water vapor pressure by using the radio and optical waves, the measurement accuracies of the phase are required to be

$\Delta\Phi$ = 30 degrees for the modulation frequency of 400 MHz,

$\Delta\Phi$ = 70 degrees for 1 GHz.

The feasibility study of the signal processing to realize such phase measurement was carried out by adopting the dual mixing time difference method (DMTD). Figure 7 shows the basic blockdiagram of the DMTD method. The optical waves are demodulated to the signal. The signal has the same frequency as that of the received radio signal. The two signals are mixed with a common local signal. The phase difference between the outputs from the mixers are measured with a TIC. In this method, the phase noise from oscillator is canceled out and the high sensitive phase mesurement can be attained. The frequency stability (Alann Variance) of DMTD was measured at the frequency of 10 MHz, and its result is shown in Fig.8. The straight line shows the frequency stability to obtain the phase error of 3 degrees. The measured values are far less than the required phase error of 3 degrees. From these results, we can conclude that it is possible to attain the phase measurement error of 30 degrees needed for the AROW system.

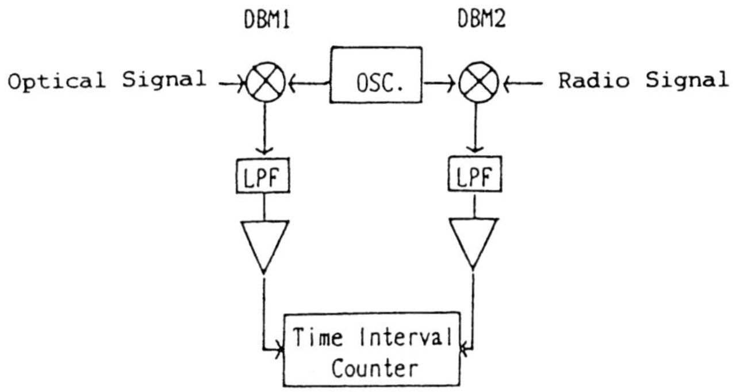

Fig.7 Block Diagram of Phase Measurement by DMTD Method

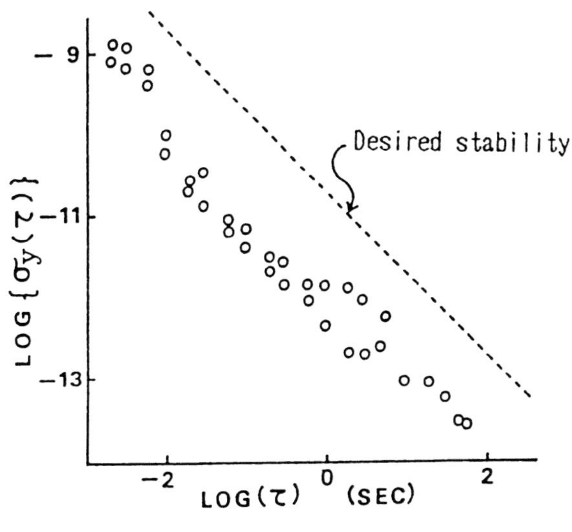

Fig.8 Alann Variance of DMTD

4.SUMMARY

We have presented the AROW system as a new means to measure the atmospheric parameters averaged along the path, which can improve the distance measurement accuracy of EDM. By the aid of this technique, the accuracy of order of $1*10^{-7}$ of the EDM is expected to be realized. It, however, demands that the phase measuremnt accuracy of a few ten degrees for 400 MHz should be achieved in the AROW system. We have experimentally shown that becomes possible by adopting the DMTD signal processing method. Further, the audio interferometer was construted to get high directivity and tested, giving good properties as an audio transmitter of the AROW system. Though we have not refered to the maximum distance so far, which the AROW system can be applied to, it must be experimentally determined because it depends on the peculiar weather condition.

The construction and basic experiments of the AROW system involving the radio and optical sections, which are not mentioned in this paper, are also in progress.

References

1.Kimata F. and Yamauchi T. (1982) A correction for temperature on electro-optical distance measurement. J. Geod. Soc. Jpn. 27 : 173-182.

11 High Power Electromagnetics

11.1 A Magnetic Field Shielding Practice Against the Heavy Lightning Electromagnetics

S. KAWAMATA[1], Y. EDANO[1], and K. WAKAMATSU[2]

Electrical Engineering, Faculty of Engineering, Utsunomiya University, Utsunomiya, 321 Japan[1]
and Tokyo Electric Power Co. Inc. Tokyo, Japan[2]

ABSTRACT

Authors represent the results of mathematical analyses to know the shield effect of a very low resistive loop in opposition to the magnetic field caused by the heavy lightning current.

1. INTRODUCTION

Until now a practice to shield the electro-static field have been used for the purpose of shield against the electro-magnetic noise. In general, the electro-magnetic energy consists of 50 percent of static energy and 50 percent of magnetic energy, and in case of the defect of one portion energy after passing through a obstacle for electro-magnetic wave, the energy is always supplied from the other portion to balance the energy between them. So the shield of one portion has somewhat effect for electro-magnetic noise.
Authors take aim the diminution in both portions of electro-magnetic energy, so the practice of magnetic shield come to be very important.
In first, a mathematical analysis to get the induced current in a rectangular loop coil caused by magnetic field change due to the lightning current is carried out by authors.
The results of numerical calculation followed by the analysis show some differnt from the wave shape of the original lightning current except for the loop condition of its value of $R/L=0$. Where R is the resistance in the loop, and L is the inductance in the same loop, respectively.
In most cases the induced currents give the refusal of an interlinkage magnetic flux in the loop due to the lightning current for corresponding to the initial stage of the lightning current.
On the contrary, most aerial loops on the ground which consist of an aerial line, two grounding wires and the earth have high R/L value larger than 1500 or more, so the induced currents in these loops show a reversal phenomena for next stage. The stage corresponds to the wave tail of the lightning impulse current. The reversal currents which

occur for next stage are smaller than the currents which flow for initial stage. But the currents come to intensify the interlinkage magnetic flux.

In second, a current in a loop which is to make the same magnetic flux due to a lightning current is supposed by authors.

In third, the supposed current in a loop is compared with a induced current which has opposite sence for initial stage to make sure of shield effect.

In fourth, a rectangular loop is made by way of trial to make sure of available and the lowest value of R/L in the loop.

2. INDUCED CURRENT IN A LOOP DUE TO THE LIGHTNING CURRENT

Authors represent the solution of a differential equation concerning to the induced current in a rectangular loop due to the magnetic field change of a lightning current. And the process to solve the equation is shown in the followings.

If we put a wave shape of the original lightning current as the form in following,

$$I(t) = \{I_0 / f(W)\} \times (e^{-\alpha_1 t} - e^{-\alpha_2 t}) \tag{1}$$

where I_0 is the peak current of the original lightning current. The function of $(e^{-\alpha_1 t} - e^{-\alpha_2 t})$ is smaller than unit. The maximum value of its function may be put as a notation of $f(W)$.

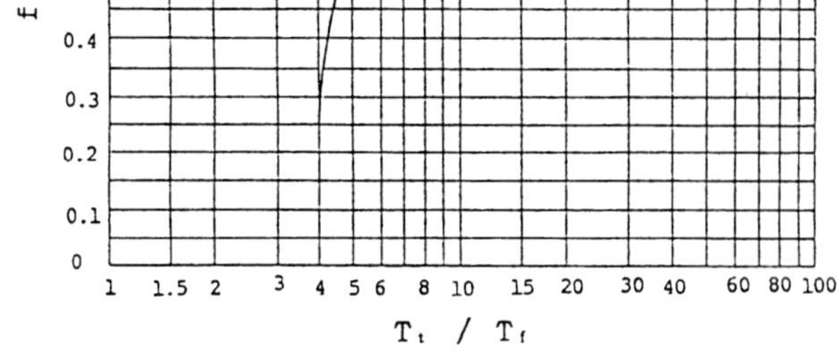

Fig.1. Correlation between f(W) and the ratio of T_t/T_f.

So, the function f(W) is shown in the following,

$$f(W) = \left(e^{-\frac{\alpha_1}{\alpha_1 - \alpha_2}\log\frac{\alpha_1}{\alpha_2}} - e^{-\frac{\alpha_2}{\alpha_1 - \alpha_2}\log\frac{\alpha_1}{\alpha_2}} \right) \tag{2}$$

and $I(t)max = I_0$.

In equation (1) and (2), α_1 and α_2 are the well known parameters correlating to the wave shape of the original lightning current, and they are also the function of T_t/T_f. Where T_t is the time duration in tail of a wave shape of the original lightning current, and T_f is also the time duration but in front of the same wave shape, respectively. Figure 1 shows the correlation between f(W) and the ratio of T_t/T_f. Maxwell's equation to be used in the paper are followings.

$$\mathrm{rot}H = I \tag{3}$$

$$\mathrm{rot}E = -\frac{\partial B}{\partial t} \tag{4}$$

$$B = \mu H \tag{5}$$

From the equation of (3), the following equation can be obtained.

$$H = \frac{I(t)}{2\pi r} = \frac{I_0}{2\pi r f(W)}\left(e^{-\alpha_1 t} - e^{-\alpha_2 t} \right) \tag{6}$$

then, $$B = \mu H = \frac{\mu I_0}{2\pi r f(W)}\left(e^{-\alpha_1 t} - e^{-\alpha_2 t} \right) \tag{7}$$

From the equation (4), the following relation can be obtained.

$$\int E \cdot dl = -\frac{\partial}{\partial t}\int B \cdot ds = -\frac{\partial}{\partial t}\phi(t) = V(t) \tag{8}$$

From these results the interlinkage flux can be obtained by the use of some assumptions. Assumptions are followings.

A loop is the rectangular shape and is situated as a side and its opposite side in parallel to the ground. Notation of h shows the height of the rectangular loop, that is a length of the other sides in vertical to the ground, and r_0 and r_1 are the nearest and the fathermost distances from the strike point of the lightning to the vertical sides, respectively.

$$\begin{aligned}\phi(t) &= \int B \cdot ds \\ &= \frac{\mu I_0 h}{2\pi f(W)}\left(e^{-\alpha_1 t} - e^{-\alpha_2 t} \right)\int_{r_0}^{r_1}\frac{dr}{r} \\ &= \frac{\mu I_0 h}{2\pi f(W)}\log\frac{r_1}{r_0}\left(e^{-\alpha_1 t} - e^{-\alpha_2 t} \right)\end{aligned} \tag{9}$$

From the equation of (9), the induced loop voltage can be obtained by next process.

$$\begin{aligned}V(t) &= -\frac{\partial}{\partial t}\phi(t) \\ &= \frac{\mu I_0 h}{2\pi f(W)}\log\frac{r_1}{r_0}\left(\alpha_1 e^{-\alpha_1} - \alpha_1 e^{-\alpha_2} \right)\end{aligned} \tag{10}$$

By the use of equation (10) on the right hand side of an equation, a following differential equation concerning with a loop current is shown in the following.

$$L\frac{di(t)}{dt} + Ri(t) = \frac{\mu I_0 h}{2\pi f(W)} \log \frac{r_1}{r_0} (\alpha_1 e^{-\alpha_1 t} - \alpha_1 e^{-\alpha_1 t}) \qquad (11)$$

To obtained an integration constant in the solution, the following condition should be used, that is the condition of $i(t)=0$, when $t=0$. From the condition the following solution can be obtained.

$$i(t) = \frac{\mu I_0 h}{2\pi f(W)} \log \frac{r_1}{r_0} \frac{1}{L}\left\{ \frac{\alpha_1}{R/L-\alpha_1} e^{-\alpha_1 t} - \frac{\alpha_2}{R/L-\alpha_2} e^{-\alpha_2 t} + \left(\frac{\alpha_2}{R/L-\alpha_2} - \frac{\alpha_1}{R/L-\alpha_1} \right) e^{-\frac{R}{L}t} \right\} \qquad (12)$$

Equation (12) is the solution of the differential equation of (11). To normalize the value of $i(t)$, we should be used the values of R/L, and α_1 and α_2 as the parameters in the equation (12). After normalization, the current wave shapes of $i(t)$ in a loop is shown in Fig.2, by the use of the resistances in Ohm as the parameters in the loop which has fixed value of the inductance in its loop of 1.82×10^{-4} in Henry, and applied wave shape of $T_f \times T_t$ of an original lightning current of 2×20 in micro-second.

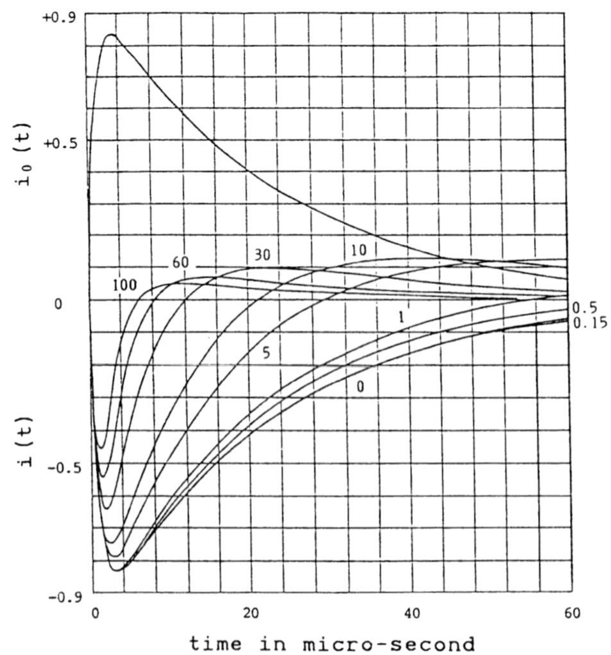

Fig.2. Wave shape of the induced currents and a supposed current.

A supposed current $i_0(t)$ is also shown in Fig.2. In the figure the induced currents i(t) are shown in opposition to the supposed current $i_0(t)$ for corresponding to the initial stage of the lightning current. But in case of the loops which have the higher resistance than the value of 0.5 in Ohm in the loop, the current reversal phenomena occur after passing through the time corresponding to the peak value of the supposed current.

Figure 3 shows the other current wave shapes of i(t) in the same loop conditions except for a different wave shape of the lightning current from Fig.2 and is used of 5X20 in micro-second.

In the Figure, the current reversal phenomena also occur under the same loop condition.

Comparing Fig.2 with Fig.3, authors may conclude that the current reversal phenomena are more dominant in case of Fig.3 than the case of Fig.2.

In these figures, a loop is assumed by authors as a span between two poles of the aerial distribution line which consist of a aerial ground wire, two grounding systems and the earth. So the inductace of the loop is asuumed as a metallic circuit which has the same dimentions by authors. The value of inductance come to be 1.82×10^{-4} in Henry.

In Fig.2 and Fig.3, the parameters of R/L in the loop may be varied directly as variation with the resistances in the loop.

For examples, the variation of the resistances may be varied in succession as 0.05, 0.1, 0.2, 0.5, and 1 in Ohm the corresponding parameters to be varied in succession as 375, 549, 1099, 2747, 5495, respectively.

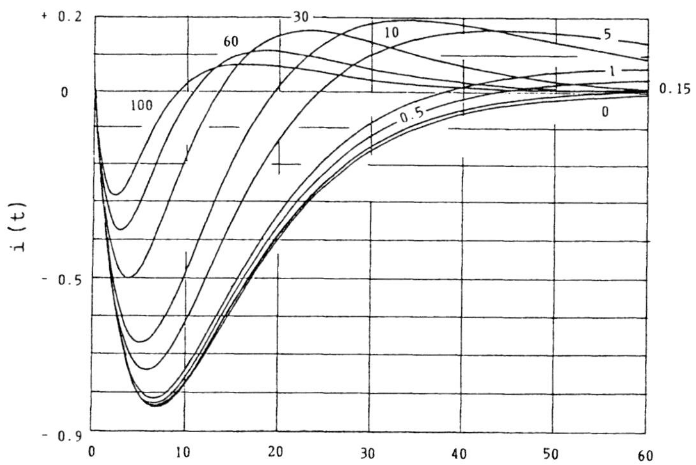

Fig.3. Wave shape of the induced currents and a supposed current.

Time variation of the induced current depends only on the terms in the brackets of the equation (12). So we can judge the appearance of the reversal phenomena by the use of numerical calculation of the terms with fixed values of α_1, α_2 and the time, and varying parameter of R/L.

As a wave shape of the original lightning current is selected of 5X20 in micro-second. So the values of $\alpha_1 = 0.0975 \times 10^6$ and $\alpha_2 = 0.2 \times 10^6$ should be used in calculation, respectively.

A time t in the brackets of 50 micro-second is also used in calculation. For the reason of the time of 50 micro-second is thought by authors the most preferable time to judge the reversal phenomena. Table 1 shows the relation between the parameter of R/L and the calculated value in the brackets of equation (12). In table 1 the calculated values are included the positive or negative notation, and the conversion of their notations show the boundary of occurring the reversal induced current.

Table 1. Calculated values in the brackets of equation (12)

R / L	
50	-0.0073346
100	-0.0070721
200	-0.0065501
500	-0.0050062
750	-0.0037445
1000	-0.0025054
1500	-0.0000931
2000	+0.0022332
3000	+0.0066379
4000	+0.0107275
5000	+0.0145198
6000	+0.0180318
7000	+0.0212795
8000	+0.0242780
9000	+0.0270419
10000	+0.0295848
15000	+0.0394122
20000	+0.0453855
50000	+0.0441315
75000	+0.0316928

The more correct value of the boundary of occurring the reversal

induced current exists between 1518 and 1519.
The reversal induced current in the loop come to intensify the interlinkage magnetic flux.

3. SUPPOSED CURRENT IN THE LOOP

Authors have to imagine a supposed current of $i_0(t)$ which is the equivalent current to make an interlinkage magnetic flux due to an original lightning current. If the interlinkage magnetic flux of $\phi(t)$ is to be represented as in equation (9), the next relation may consists.

$$\phi(t) = Li_0(t) \tag{13}$$

The equation (13) is transformed to another form, so the next relation can be obtained.

$$i_0(t) = \frac{1}{L}\phi(t)$$

$$= \frac{\mu I_0 h}{2\pi f(W)} \log \frac{r_1}{r_0} \frac{1}{L}\left(e^{-\alpha_1 t} - e^{-\alpha_2 t}\right) \tag{14}$$

If we put the ratio of R/L to zero, that is corresponding to the super conductive condition, and the next relation may be obtained.

$$i_0(t) + i(t) = 0 \tag{15}$$

To explain more precisely, the correlation between a magnetic flux of $\phi^*(t)$ and the induced current of $i(t)$ is supposed by authors, and is shown in the following.

$$\phi^*(t) = Li(t) \tag{16}$$

So the next relation may consists.

$$L\{i_0(t) + i(t)\} = \phi^*(t) + \phi(t) \tag{17}$$

The value shown in equation (17) is a residual magnetic flux in the loop. A model of the residual magnetic flux and the deformed line of magnetic force due to the induced current in the loop are shown in Fig.4.

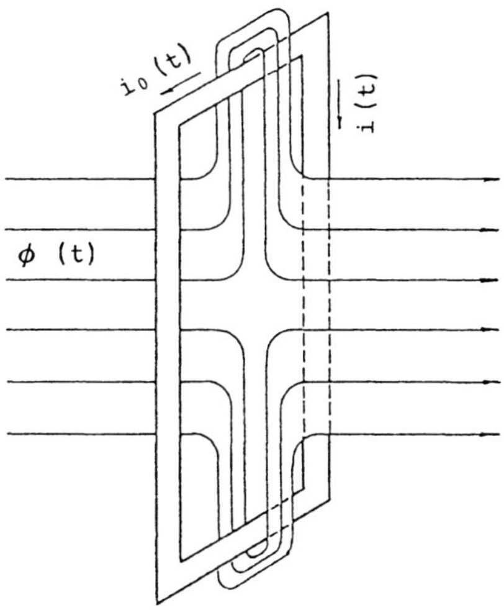

Fig.4. Diminution scheme by the use of the conductive loop.

4. SHIELDING EFFECT OF THE MAGNETIC FIELD

Shielding effect of the magnetic field should be evaluated by the use of suitable method. So the ratios of i(t) value at the instant of $i_0(t)_{max}$ to $i_0(t)_{max}$ are selected and taken by authors.
In general, the instants to attain the peak value of i(t) are some different from the instant to attain the peak value of $i_0(t)$.
The ratios are thought to represent a diminution factor or the shielding effect of the magnetic field by authors.
Figure 5 shows the correlation between the ratio of i(t) value at the instant of $i_0(t)_{max}$ to $i_0(t)_{max}$ and the variation of the resistances in the loop which has the fixed inductace of 1.82×10^{-4} in Henry, and wave shape of the lightning current of 2×20 in micro-second.
R/L parameter depends on variation with the resistances in the loop, and directry proportional to them.
Uunder the condition to make different value of the resistances as 0.15, 0.2, 0.5, 1 and 2 in Ohm, the diminution factors are changed in succession to 0.9994, 0.9981, 0.9937, 0.9875 and 0.9751, respectively.

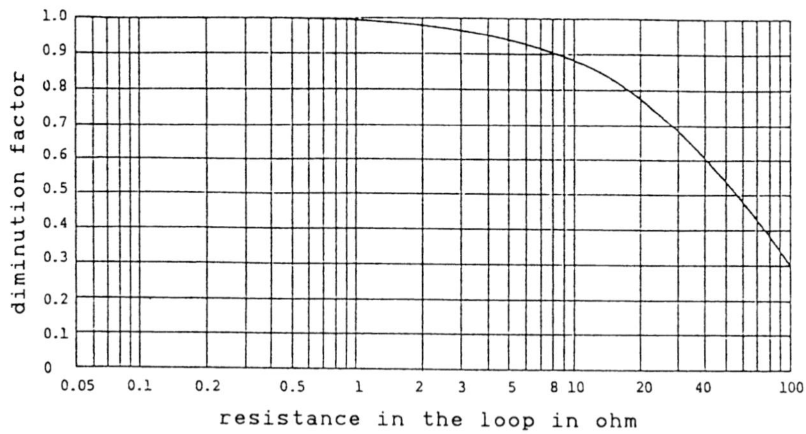

Fig.5. Correlation between the diminution factor and the resistances in the loop.

5. PRACTICE AND APPLICATION

From the result in measuring the R/L value of the rectangular loop coil which was made of copper tube shows very low value to 330. The squre coil was made of its dimentions of 1×1 in meter, and sectional view of the tube is also square of 2×2 in cm. The thickness of the

tube is 2 in mm.
But in general, the ordinary loops have the higher R/L value than the value of 1500.
The wire netting made of thin copper, the grounding system of the transmission and distribution lines which include the aerial ground wire and the steel enhanced concrete or steel structure buildings have the loops of high R/L value. The loops always show the dominant reversal phenomena of the induced current. The phenomena have been called as Lenz's low [1].
From the result of the lowest value of R/L in the loop which was made of electrolytic copper shows 330, so in the other cases which the loop were made of the other metals, that are steel, brass and alminium alloy show high value of R/L larger than 1500. For reason of the other metals have high specific resistances larger than five times that of electrolytic copper.
Figure 6 shows an application of the conductive loops to a buildings.

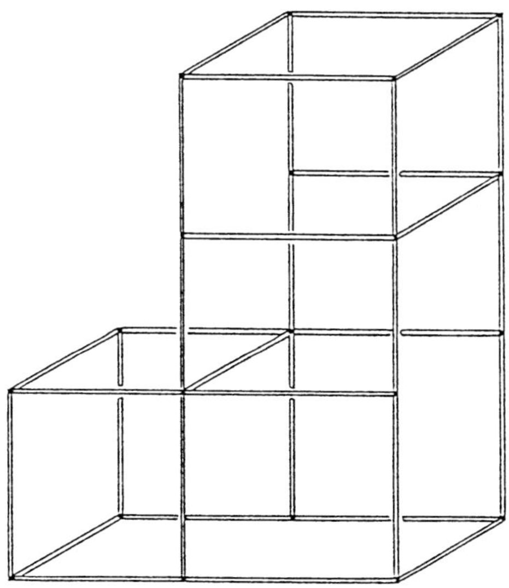

Fig.6. An application of the loop coils to the building. The coils may be used besides the structual steels.

The large number of electrical and electronic apparatus are set in the building, and they are sometimes disturbed and damaged by electromagnetic noise caused by lightning. So we hope to take the practices to diminish the noise, and to take the condition in the building with free from the noise.
The figure shows a proposal that may be applied to the building. Many

loop coils made of conductive bar or tube are assembled as the framework, and are set besides the structual steel or the steel bars to enhance the concrete of the building.

The walls of the building act as to diminish the electromagnetic noise travelling to the both directions, one is the noise from open air to the building and other is the noise from the room in the building to open air.

On the contrary, in the building the small electrical signals that are the signal of pocket bell and the signal of cordless telephone may be useful among the different floors and rooms.

6. REFERENCE

1. Lenz's law (1977) McGraw Hill Encyclopedia of Science and Technology. Vol.7. McGraw Hill. New York

11.2 Simulation of Electromagnetic Aspects of Lightning

C.E. Baum

Phillips Laboratory, Kirtland Air Force Base, NM 87117, USA

ABSTRACT

In testing complex electronic systems for their vulnerability to lightning interference and electrical damage, it is necessary that the test include a simulation. For a simulation one requires that the experimental conditions be sufficiently close to the true physical environment that valid conclusions can be drawn concerning the response in such a criterion environment. Present commonly used test procedures, particularly for the direct-strike case, are in general not simulations. For the case of distant lightning, commonly used EMP simulation techniques are applicable with some modification in the sources (pulsers).

Electromagnetic processes peculiar to the direct-strike case are reviewed with respect to their implications for lightning electromagnetic simulation. At low frequencies (quasistatic) there are important surface-charge-density and corona effects in addition to the surface-current-density effects. At resonant frequencies the frequency-spectral content of the excitation and properties of the arc (attachment, detachment, time history, spatial distribution, resistance, etc.) are significant. Of great complexity in all this are the nonlinear aspects of the arc and corona around the system of interest. The complexity of these various processes requires rigor in the simulator design. Potential simulation concepts are presented and their relative merits are discussed.

1. INTRODUCTION

In an engineering discipline concerned with the reliable performance of some complex electronic system in an intense electromagnetic environment, there comes the question of the demonstration of the performance in such an environment. Such a system is in general so complex that one cannot have a complete understanding of its response to such an environment from first-principle calculations. While one may be able to calculate whether or not the signals at some electronic elements are sufficient to cause failure, some is not good enough (even though the information can still be useful). Given some definition of the mission of the system (in the electromagnetic environment) one must be able to determine whether or not the mission will be accomplished in the presence of the environment. This requires that there be no failures (in the mission accomplishment sense) in any of a certain subset of the electronic elements; such electronic elements (black boxes, subsystems, etc.) are usually referred to as <u>mission critical</u>. More importantly it is required that one know that there be no such failures with high confidence. There are possible exceptions to this stringent a requirement if sufficient redundancy is built into the system, but this is a more complicated question. In general the system complexity and the possibility (or even probability) that there are important signal paths which are not even identified (even implicitly) in the formal statement of the system design (blueprints, etc.), makes a reliance on first-principle analysis usually untenable for system vulnerability assessment.

Since system-level testing (experimentation) is required for high confidence assessment, the question arises as to what is an adequate test. One might think of an arbitrary electromagnetic stimulus as a test, but what assurance would one have that the system response would bear any significant relationship (for assessment) to its response in the real environment of interest. For the test to be useful it should be reasonably closely related to the real environment of interest and this relationship should be quantitative. Ideally this real environment is summarized in the form of a valid criterion. Paraphrasing an earlier definition [6],

> A lightning electromagnetic criterion is:
>
> > a quantitative statement of the physical parameters of the lightning environment relevant to the electromagnetic response of a system of interest in a volume of space and a region of time and/or frequency extended to contain all physical parameters having a non-negligible influence on any of the electromagnetic response parameters.

Normally one will have to state some range or bounds of the parameters, and the time functions (waveforms) and/or frequency spectra may be specified in simplified analytic form, a form which, however, should quantitatively include all relevant environmental parameters.

To test to such a criterion in a way which is quantitatively related to it requires a special kind of test referred to as a simulation. For lightning (electromagnetic) simulation the definition of nuclear electromagnetic pulse (EMP) simulation can be adapted [7].

> Lightning (electromagnetic) simulation is an experiment in which the postulated lightning exposure situation is replaced by a physical situation in which:
>
> 1. the lightning sources are replaced by a set of equivalent sources which to a good approximation produce the same excitation (including reconstruction to the extent feasible) to the total system under test or some portion thereof as would exist in the postulated lightning environment, and
>
> 2. the system under test is configured so that it reacts to sources (has the same Green's function) in very nearly the same way and to the same degree as it would in the postulated lightning environment.
>
> A lightning (electromagnetic) simulator is a device which provides the excitation used for lightning simulation without significantly altering the response of the system under test by the simulator presence.

Lightning (electromagnetic) simulation naturally divides into types according to the types of lightning environments to be simulated. If the system of interest is sufficiently distant from the lightning stroke and its associated ionization, the simulation problem is somewhat simplified due to the ability to separate the incident and scattered fields at the system; this case is briefly discussed later. A much more difficult case is that of a direct lightning strike to the system because of the complex electromagnetic field structure, the time-varying electromagnetic properties (conductivity, etc.) of the lightning arc and corona around the system, and the nonlinear properties of the lightning arc and corona. This latter case is very important because of both the intense electromagnetic fields present and the poorly understood nonlinear and time-varying electromagnetic parameters of importance in this lightning source region. These source-region phenomena are of fundamental importance to the lightning simulation problem (for the direct strike); some of the implications of these phenomena for simulator design are discussed in this paper.

There is a fundamental limitation in how far we can go in designing a lightning (electromagnetic) simulator; namely one must know what he is to simulate. The detailed physical processes of the lightning arc and corona are poorly quantitatively understood. How then does one simulate it? One can try to have "real" arcs and corona, but how can one be sure that all the relevant physical parameters have been properly controlled. In a direct-strike situation such questions can be very important, while for distant strokes the problem can be reduced to the locally incident electromagnetic fields and electromagnetic properties of the local materials. Particularly in the direct strike situation the reader should note that the present considerations concerning a lightning simulator are based on the current limited understanding of the lightning arc and corona. As our understanding of the lightning physics becomes more detailed, more rigorous design constraints may be placed on the simulator.

2. SIMULATION OF DISTANT LIGHTNING

If, as indicated in Fig. 1, the lightning arc is distant from the system of interest, the lightning interaction is greatly simplified. Let S be a closed surface bounding an interior volume V which contains the system of interest. For simplicity let the system be in "free space" such as an in-flight missile or aircraft.

Let all the current and ionization associated with the lightning arc be outside S (i.e., not in V). Then the field equivalence principle can be invoked by noting that the incident field in V (in the absence of the system) is determined by imposing the tangential components of the original $\vec{E}$ and $\vec{H}$ on S via equivalent electric and magnetic surface current densities on S [8, 11]. These equivalent currents also give zero fields outside S. Having referred the incident fields to equivalent sources on S, note that the incident fields are somewhat decoupled from the lightning arc. We do not need to understand the details of the lightning arc if we have sufficient knowledge of the incident fields (say by measurement of such fields).

Introducing the system into V there are now scattered fields which propagate away from the system through S. These fields can scatter from the lightning arc (in general a nonlinear process) and in turn rescatter from (interact with) the system thereby changing the system response. To avoid this interaction of the system with the lightning arc one can require that the system be sufficiently far from the arc that the interaction via the multiply scattered waves be sufficiently small as to be insignificant. In this latter approximation the system can be considered as responding to the lightning incident fields in a manner decoupled from the arc physical processes.

In this case of the system distant from the lightning arc the simulation problem is greatly simplified. The problem becomes similar to that found in many EMP simulation problems in which the system is away from the source region. Referring to Fig. 1 the equivalent sources on S can be approximately synthesized with sets of electric and magnetic dipoles to correspond to any desired incident field (consistent with Maxwell's equations with no sources in V). This is the PARTES simulation concept which is quite general, but perhaps complex to implement [8].

If the system is sufficiently distant from the lightning arc that the incident fields can be considered as an approximate plane wave, then the problem further simplifies. The high-altitude EMP (below

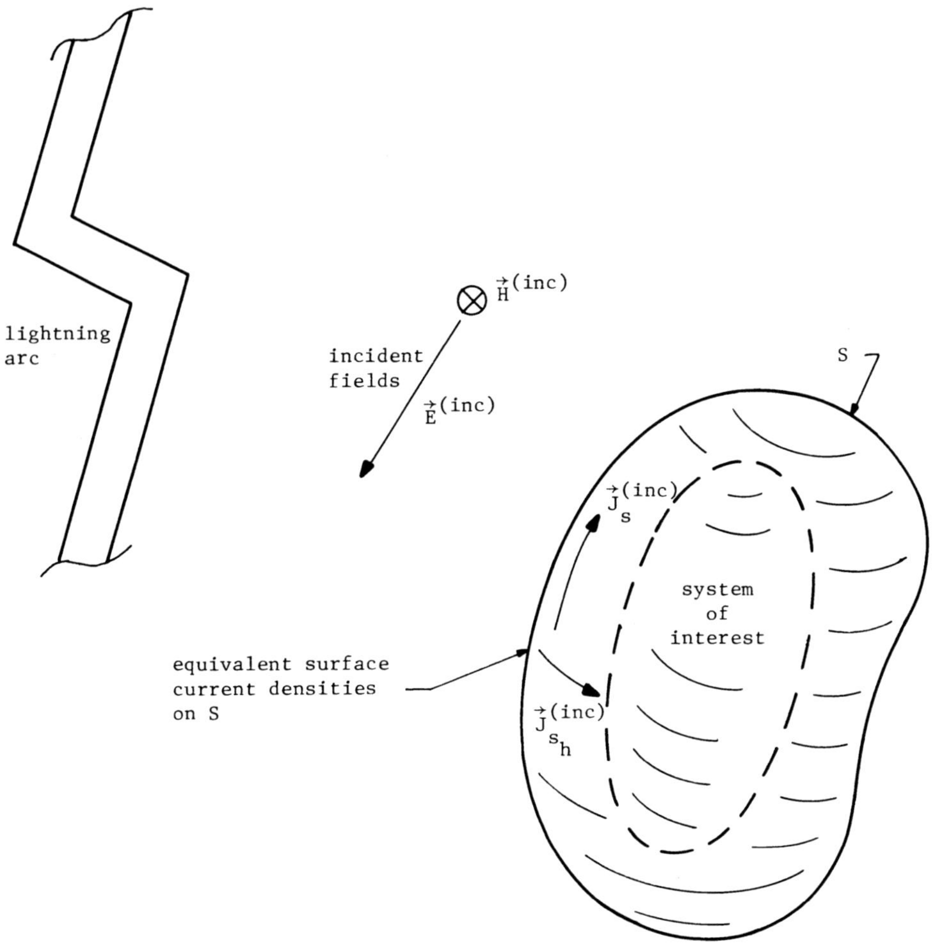

Fig. 1. Fields Incident on System Away from Lightning Arc

the source region) is also approximated as a free-space plane wave; various EMP simulator types produce an approximation of this type of field [7]. One of the most applicable types is the guided wave of TEM-transmission-line simulator constructed with parallel plates and conic sections for launch and/or termination of the wave. Note, however, that while the spatial forms (plane waves) are common between the cases of distant lightning and high-altitude EMP, the temporal forms (waveforms) are not the same. Thus while the simulator proper (electromagnetic-field-forming structure (waveguide, antenna, etc.)) can be used for both simulation problems, one needs different waveforms with different frequency spectral contents; this can be achieved by the use of different sources (pulsers, etc.) to drive the simulator proper. Hence, some EMP simulators can also be used for lightning simulation for the case of distant lightning provided there are changes in the driving sources.

The discussion here has centered on the case of an in-flight system distant from lightning because of the simplifications thereby introduced. Similar considerations apply to the case of a system on or near the earth surface. If the system is sufficiently distant from the lightning stroke a plane-wave approximation can still be made, except that the ground reflection should also be included. This problem is also encountered in the case of the high-altitude EMP incident on systems near the ground surface. A commonly used simulator for this type of the EMP is a hybrid EMP simulator shaped as an impedance loaded arch (half loop) connected to the earth with a generator in one position in the arch [7]. This simulator includes the incident and reflected waves at the ground surface, but again for application to the simulation of distant lightning the required waveforms are different requiring a modification in the simulator sources.

3. SIMULATION OF DIRECT-STRIKE LIGHTNING

A more interesting and more difficult type of lightning (electromagnetic) simulation is that concerned with a direct strike including arc attachment to and detachment from the system and corona surrounding the system. Since in general the stimulus is larger than that from distant lightning, the direct strike case is important to understand and design against.

The first and fundamental problem to observe is that the system is in contact with the nonlinear and time-varying source. As such the separation into incident fields followed by system response discussed in the previous section is no longer applicable. A surface S surrounding the system as in Fig. 1 now has current passing through S into (and out of) V. The problem no longer separates into incident and scattered fields. The system is in the source and influences the evolution of the lightning arc. The interaction of the fields with the system is in turn influenced by the nonlinear, time-varying arc which can change the electromagnetic properties of the system (e.g., natural frequencies), and by the nonlinear, time-varying corona which can influence the response of the external penetrations (apertures and antennas) on the system.

Subsequent sections of this paper consider some of the important aspects of the interaction of direct-strike lightning with the system, and the implications of the processes for lightning simulator design. Beginning with the recognized low-frequency current effects, the lightning interaction and simulator design are extended to include charge (more generally normal-current-density) effects and the associated corona. Then the arc-conductance effects are considered as well as the interaction of peripheral parts of the simulator with the test object. The inclusion of all these effects leads to a more rigorous simulator design.

4. QUASISTATIC (LOW-FREQUENCY) CONSIDERATIONS

Consider first the case that wavelengths of interest are large compared to the exterior system dimensions; this is the low-frequency of quasistatic regime. For this part of the lightning interaction problem one can think of the current flowing through and charge on the aircraft as producing responses which are separable from each other, at least as an approximation.

While the magnetic fields associated with the lightning-arc current flowing on the system surface can penetrate through a metallic surface at sufficiently low frequencies, this type of penetration is usually not of dominant significance because the shield inductance and resistance (of the basic metal) make a low-pass filter with a very low roll-off frequency. (Note that for some composite materials with conductivities much lower than those of typical metals this magnetic distributed type of penetration can be of greater significance.) For typical systems such as aircraft the important penetrations are generally more discrete (spatially localized) in nature; they include apertures, antennas, and various direct conductive penetrations such as power, signal lines, mechanical control cables, etc. For the present discussion consider the cases of small apertures and antennas as illustrated in Fig. 2.

Consider, as in Fig. 2A, the case of small apertures. In the linear approximation the fields inside an exterior conducting system surface (at positions not near the aperture) are derived from equivalent dipole moments at the aperture. By an equivalent dipole moment is meant that vector quantity which, when substituted in the formulas for fields in free space [9], approximately gives the correct fields over a volume of space of interest, at least in the low-frequency asymptotic form for positions not close to the aperture. In this sense we can write

$$\begin{aligned}
\vec{\tilde{p}}_{a_{eq}}(s) &= \overleftrightarrow{\tilde{P}}_{a_{eq}}(s) \cdot \vec{\tilde{D}}_{s.c.}(s) = \varepsilon_0 \; \overleftrightarrow{\tilde{P}}_{a_{eq}}(s) \cdot \vec{\tilde{E}}_{s.c.}(s) \\
&\equiv \text{equivalent electric dipole moment} \\
\vec{\tilde{m}}_{a_{eq}}(s) &= \overleftrightarrow{\tilde{M}}_{a_{eq.}}(s)\; H_{s.c.}(s) = \frac{1}{\mu_0} \; \overleftrightarrow{\tilde{M}}_{a_{eq}}(s) \cdot \vec{\tilde{B}}_{s.c.}(s) \\
&\equiv \text{equivalent magnetic dipole moment} \\
s &\equiv \Omega + j\omega \equiv \text{complex frequency or Laplacetransform variable}
\end{aligned} \tag{4.1}$$

$\sim$ indicates Laplace transformed quantity (for, in general, two – sided Laplace transform)

where the subscript s.c. ($\equiv$ short circuit) indicates the electromagnetic fields on the system exterior with the aperture closed (shorted). These short-circuit fields can also be related to the corresponding surface current and surface charge densities via the unit outward-pointing normal vector $\vec{1}_{S_s}$ to the system boundary surface S_s as

$$\vec{E}_{s.c.} = \frac{1}{\varepsilon_0}\, \rho_{s s.c.}\, \vec{1}_{S_s}, \quad \vec{H}_{s.c.} = -\, \vec{1}_{S_s} \times \vec{J}_{S_{s.c.}} \tag{4.2}$$

Assuming for the foregoing that the system exterior (outside S_s) is free space with

$$Z_0 = \frac{\mu_0}{\varepsilon_0} \simeq 377\ \Omega \qquad \text{(wave impedance)}$$

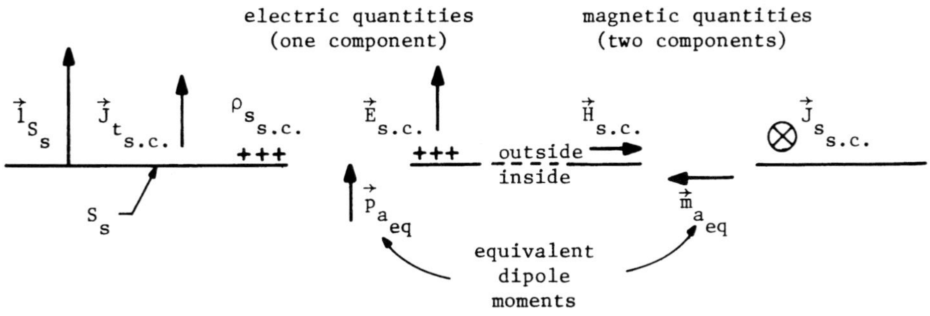

A. Apertures

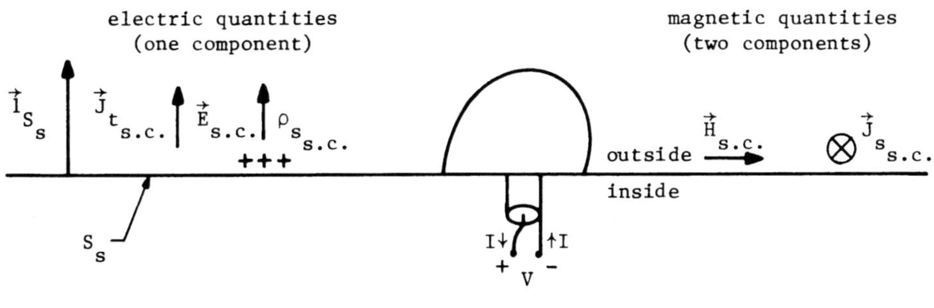

B. Antennas

Fig. 2. Quasistatic Interaction Mechanisms for Small Penetrations.

$$c = \frac{1}{\sqrt{\mu_0 \varepsilon_0}} \simeq 3 \times 10^8 \; m/s \quad \text{(speed of light)} \tag{4.3}$$

one can make some observations concerning the relative magnitudes of the electric and magnetic types of aperture penetration. Considering the case of an open aperture (say a circular hole) first note that the equivalent polarizabilities

$$\overleftrightarrow{P}_{a_{eq}}(s) \equiv \vec{1}_{S_s}\vec{1}_{S_s}\tilde{P}_{a_{eq}}(s) \qquad \text{(equivalent electric polarizability)}$$

$$\overleftrightarrow{M}_{a_{eq}}(s) \equiv \left[\overleftrightarrow{1} - \vec{1}_{S_s}\vec{1}_{S_s}\right] \cdot \overleftrightarrow{M}_{a_{eq}}(s) \quad \text{(equivalent magnetic polarizability)} \tag{4.4}$$

have units (meter)^3, making them some kind of equivalent volumes. Furthermore, an open hole (perhaps covered with an insulator) has polarizability components of order d^3 where d is a characteristic dimension of the aperture (say the radius of a circular hole). Specifically the dominant components of the electric and magnetic polarizabilities of an unloaded circular aperture are about equal, and frequency independent for wavelengths $\lambda \gg$d.

For the case that the short-circuit electric field is the same magnitude as Z_0 times the short-circuit magnetic field, i.e., for

$$\left| \overleftrightarrow{E}_{s.c.}(s) \right| \simeq Z_0 \left| \vec{H}_{s.c.}(s) \right| \tag{4.5}$$

and assuming (as above) comparable equivalent aperture polarizabilities, i.e.,

$$\left| \overleftrightarrow{P}_{a_{eq}}(s) \right| \simeq \left| \overleftrightarrow{M}_{a_{eq}}(s) \right| \tag{4.6}$$

where for dyads (or matrices) the magnitude $||$ is the 2 norm (or spectral or euclidean norm) [10], then we have from (4.1)

$$\left| \overleftrightarrow{p}_{a_{eq}}(s) \right| \simeq \frac{1}{c} \left| \overleftrightarrow{m}_{a_{eq}}(s) \right| \tag{4.7}$$

provided $\overleftrightarrow{H}_{s.c.}$ is oriented approximately in a direction corresponding to the largest components of $\overleftrightarrow{M}_{a_{eq}}$. This result (4.7) is precisely that for making the far fields (r^{-1} terms) in the dipole formulas comparable for both electric and magnetic dipoles [5]. The near fields (r^{-2} and r^{-3} terms) are dominantly electric for electric dipoles and magnetic for magnetic dipoles. Thus, for a significant class of apertures, comparable short-circuit exciting fields (related by Z_0 as in (4.5)) give comparable fields penetrating the aperture.

Now consider direct-strike lightning. Fig. 3 illustrates the case of some elongated object of radius a (with length $\gg$ a) which might represent a missile or aircraft. Let arc 1 first attach to one end followed by arc 2 detaching from the other end. For quasistatic (slow time variation) considerations with current I flowing through the system we have a typical short-circuit magnetic field on the system surface

$$H_{s.c.} \simeq \frac{I}{2\pi a} \tag{4.8}$$

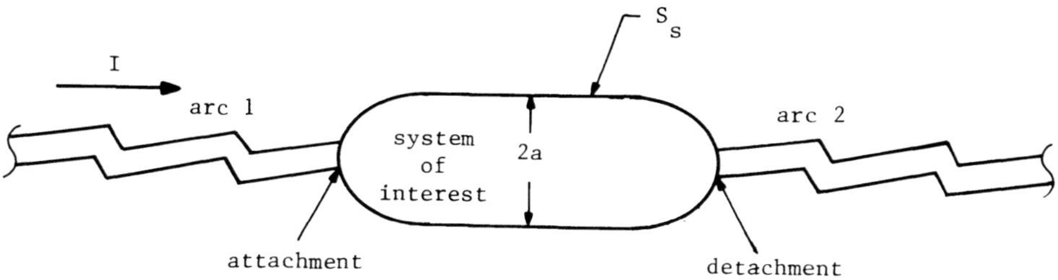

Fig. 3. Quasistatic Surface Fields for Direct-Strike Lightning

The corresponding short-circuit electric field can be estimated by noting that an arc leaving the system surface occurs when the breakdown electric field of air has been exceeded. This implies an electric field of

$$E_{s.c.} \simeq 3 \text{ MV/m} \tag{4.9}$$

or a little less because of the system's operating altitude. Letting

$$a \simeq 1\ m \tag{4.10}$$

let us consider two cases: Case 1:

$$I \simeq 200\ kA,\ H_{s.c.} \simeq 30\ kA/m,\ \frac{E_{s.c.}}{H_{s.c.}} \simeq 100\Omega \tag{4.11}$$

Case 2:

$$I \simeq 20\ kA,\ H_{s.c.} \simeq 3\ kA/m,\ \frac{E_{s.c.}}{H_{s.c.}} \simeq 1000\Omega \tag{4.12}$$

These cases of interest then give quasistatic E/H ratios on the system surface comparable to Z_0. Thus the quasistatic electric and magnetic fields are of comparable importance for penetration through a class of apertures.

For this result the effect of corona on the polarizabilities has not been included. Our estimate of the quasistatic electric field in (4.9) is based on streamers (arcs) leaving S_s. One form of such breakdown can be considered a corona around much of S_s. One can readily show that the magnetic-field change associated with bounded variations in the air conductivity around local perturbations of S_s (such as apertures) is small in the quasistatic (low-frequency) limit [2]. Basically, a bounded conductivity change over a small region does not significantly alter an external magnetic field provided the skin depth (or diffusion depth) in the region is large compared to the size of the region. However, the case for the electric field is quite different. The air conductivity σ directly combines with the electric field $\vec{E}$ producing the current density $\vec{J}$ $(= \sigma\ \vec{E})$. The low-frequency local continuity of the current density clearly indicates that changes in σ change $\vec{E}$ in a comparable way. this change in σ can then change the equivalent electric dipole moment and the corresponding electric polarizability by changing the charge distribution in the vicinity of the aperture, including in the air as well. Note that the nonlinear (and poorly known) character of the corona near the aperture significantly complicates the problem and requires the analysis to be conducted in time domain. It may be interesting to view this electric response in terms of the total current density

$$\vec{J}_{t_s.c.} \equiv (\sigma + \epsilon \frac{\partial}{\partial t}) \, \vec{E}_{s.c.} \tag{4.13}$$

which includes the local corona conductivity.

The case of electrically small antennas on S_s is indicated in Fig. 4B. Analogous to the aperture-penetration formulas (4.1) one has the linear response of such antennas as

electric:

$$\tilde{V}_{o.c.}(s) = - \, \tilde{\vec{E}}_{s.c.}(s) \cdot \, \tilde{\vec{\ell}}_{eq}(s), \; \tilde{I}_{s.c.}(s) = -s \, \tilde{\vec{D}}_{s.c.}(s) \cdot \, \tilde{\vec{A}}_{eq}(s) \tag{4.14}$$

magnetic:

$$\tilde{V}_{o.c.}(s) = s \, \tilde{\vec{B}}_{s.c.}(s) \cdot \, \tilde{\vec{A}}_{eq}(s), \; \tilde{I}_{s.c.}(s) = \tilde{\vec{H}}_{s.c.}(s) \cdot \, \tilde{\vec{\ell}}_{eq}(s) \tag{4.15}$$

for open-circuit voltage and short-circuit current. Note that the electrically small antennas are usually characterized by equivalent lengths (or heights) and equivalent areas. While these are generally functions of frequency, simple electric- and magnetic-dipole antennas have frequency-independent equivalent lengths and areas in the electrically small regime [4]. In the quasistatic regime of the system the direct-strike lightning current is important for magnetic antennas, and the lightning charge is important for electric antennas.

Note that as in the case of apertures the corona in the vicinity of the antenna needs to be considered. The effect of corona conductivity on the response of magnetic antennas is not of first-order importance at low frequencies [2]. It is, however, of fundamental importance to the response of electric (or current-density) antennas [1, 3].

Noting the importance of both current and charge at low frequencies on the system of interest, one might attempt to simulate this part of the direct strike lightning environment as illustrated in Fig. 4. The technique indicated in Fig. 4A is currently commonly employed, but note that as used it gives a large current with a (relatively) small voltage on the system. Beyond present practice, various types of impedance loading in the leads connecting to the pulse generator (and various impedances with the pulse generator) can be used to tailor the driving waveform and the response of the system by controlled loading of the system exterior.

The complementary simulation for the low-frequency charge is indicated in Fig. 4B. this charge or voltage simulation raises the potential of the system relative to some ground reference, such as a ground plane. This can be thought of as a high-impedance simulation in contrast to the low-impedance simulation for current. The capacitance of the system with respect to the ground reference is used to determine the required voltage (V) needed to produce the desired total charge (Q) on the system.

For both the current and charge quasistatic simulation one should be careful of lead placement and proximity of the system of interest to other objects (such as buildings, earth, etc.) because these

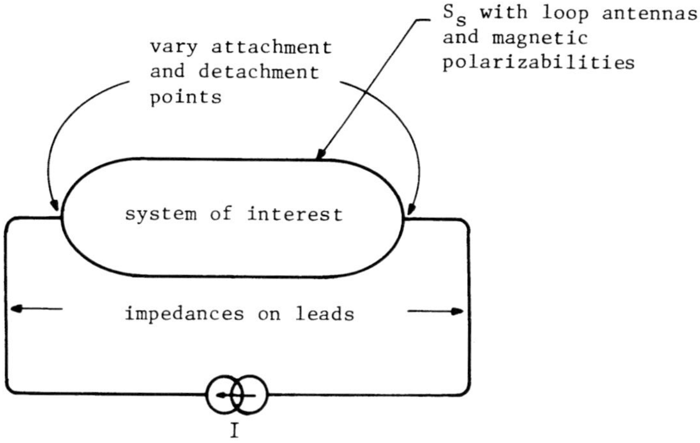

A. Current (or tangential magnetic field)

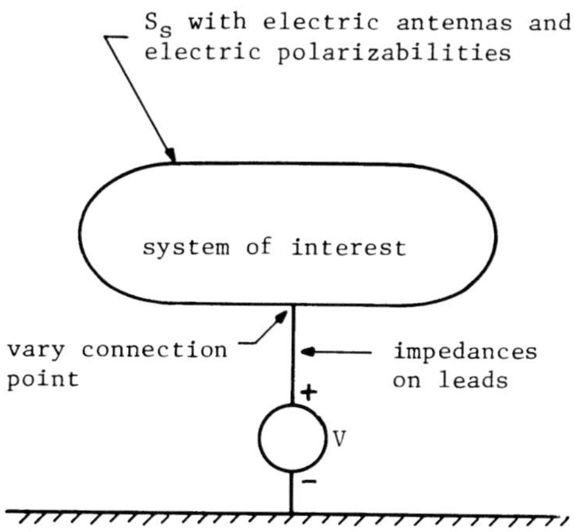

B. Charge (or normal electric field)

Fig. 4. Current and Charge Low-Frequency Simulation

all influence the distribution of the surface current and charge densities on S_s. In other words one should minimize what is referred to as the simulator/test-object interaction [7]. Combining the two quasistatic techniques one might have a somewhat more complete simulation as indicated in Fig. 5. Here one has two sources V_1 and V_2 with associated impedances $\tilde{Z}_1$ and $\tilde{Z}_2$ together with a ground reference to establish the desired V and I on the system at low frequencies.

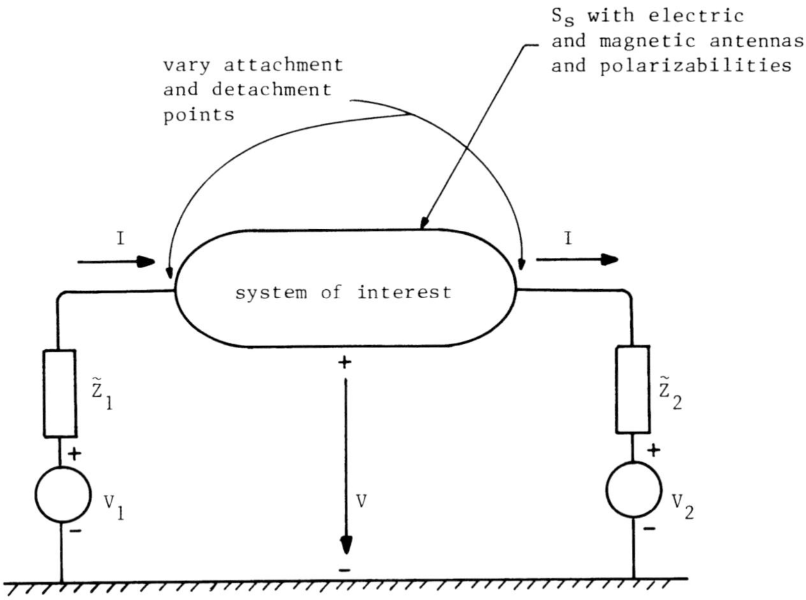

Fig. 5. Combined current and Charge Low-Frequency Simulation

Note that, while our discussion in this section has been centered on the quasistatic regime for the system, the nonlinear character of the lightning arc and corona on S_s limits one's ability to completely separate the quasistatic regime from the higher-frequency regime. The early-time and resonant response all affect the lightning arc and corona which in turn influence the quasistatic response.

5. EXTERIOR SYSTEM RESONANCES

It is well known that typical electromagnetic scatterers resonate in a manner characterized by damped sinusoids in time domain. These complex natural frequencies s_α are poles in the complex-frequency or s plane. Generalizing on this observation has led to the singularity expansion method (SEM) for the representation of the electromagnetic scattering process for linear (and to date time invariant) scatterers. There is a considerable literature now developed to which the reader may refer [13, 14]. We are not concerned here with the general theory, but are concerned with some of its implications for the lightning external interaction and corresponding lightning simulation.

Consider an integral equation of the general form [14]

$$\left\langle \overleftrightarrow{\tilde{Z}}(\vec{r},\vec{r}\,';s) \;+\; \overleftrightarrow{\tilde{Z}}_\ell(\vec{r},s)\,\delta(\vec{r}-\vec{r}\,');\; \vec{\tilde{J}}(\vec{r}\,',s)\right\rangle = \vec{\tilde{E}}_s(\vec{r},s)$$

$$\begin{aligned}
\vec{\tilde{E}}_s(\vec{r},s) &\equiv \text{source or incident electric field}\\
\overleftrightarrow{\tilde{Z}}_\ell(\vec{r},s) &\equiv \text{impedance loading (such as ligtning arc)}\\
\vec{\tilde{J}}(\vec{r},s) &\equiv \text{responsecurrentdensity}\\
\overleftrightarrow{\tilde{Z}}(\vec{r},\vec{r}\,';s) &\equiv \text{impedance kernel}
\end{aligned} \tag{5.1}$$

with <,> indicating the domain of integration (the region over which the current density (or at least its relevant portion)) exists.

Ignoring for the moment the nonlinear and time-varying character of the air conductivity in the lightning arc, let us approximate the impedance loading by the reciprocal of the air conductivity (times the dyadic identity). Without the arc one can find the exterior natural frequencies of the system from

$$\left\langle \overleftrightarrow{\tilde{Z}}(\vec{r},\vec{r}\,';s_\alpha^{(0)});\; \vec{j}_a^{\,(0)}(\vec{r}\,')\right\rangle = \vec{0}$$

$$\begin{aligned}
\vec{j}_\alpha^{\,(0)}(\vec{r}) &\equiv \text{natural mode without arc}\\
s_\alpha^{(0)} &\equiv \text{natural frequency without arc}
\end{aligned} \tag{5.2}$$

where $\vec{r},\vec{r}\,'$ is here only over the system. In the moment method (MoM) [12] the impedance kernel is converted into a matrix $(\tilde{Z}_{n,m}(\mathrm{s}))$ which allows one to find the natural frequencies from

$$\det\left(\left(\tilde{Z}_{n,m}(s_\alpha^{(0)})\right)\right) = 0 \tag{5.3}$$

Now including the arc conductivity in an approximation as linear and time-invariant, we have a new equation for natural frequencies as

$$\left\langle \overleftrightarrow{\tilde{Z}}(\vec{r},\vec{r}\,';s_\alpha^{(1)}) \;+\; \overleftrightarrow{\tilde{Z}}_\ell(\vec{r},s_\alpha^{(1)})\,\delta(\vec{r}-\vec{r}\,');\; \vec{j}^{\,(1)}(\vec{r}\,')\right\rangle = \vec{0}$$

$$\begin{aligned}
\vec{j}_\alpha^{\,(1)}(r) &\equiv \text{natural mode with arc}\\
s_\alpha^{(1)} &\equiv \text{natural frequency with arc}
\end{aligned} \tag{5.4}$$

with $\vec{r},\vec{r}\,'$ over the system plus arc. In general one does not expect the set of $s_\alpha^{(1)}$ to equal the set of $s_\alpha^{(0)}$ given their different defining equations. Furthermore, as the arc conductivity is varied the natural frequencies $s_\alpha^{(1)}$ will in general also vary. Similar observations can be made concerning the natural modes $j_\alpha^{(1)}(\vec{r})$ and the amplitudes of these resonances known as coupling coefficients.

In order to have the $s_\alpha^{(1)}$ be correctly included in the simulation it is in general necessary that the arc conductance per unit length be properly included in the simulation. At least near the system, then, the leads from the pulse generators in Figs. 4.3 and 4.4 can be impedance loaded to approximate some desired arc conductance per unit length. If the lead lengths from their connections to the system are sufficiently long one can get at least some of the natural frequencies and natural modes (on the system) to approximate those appropriate to the desired arc conductance per unit length.

One should regard this arc-conductance part of the simulation as a necessary condition in the resonance regime of the system. It is not in general a sufficient condition because of the nonlinear and time-varying character of the real lightning arcs. In addition, there is the corona surrounding the system under direct-strike lightning conditions which may also have some influence on the resonance-region response. To include these nonlinear and time-varying conductivities in the simulation may require a very realistic simulation with "real" current levels, voltages, and surrounding atmosphere with arcs and corona.

6. MAKING SIMULATOR A "PART" OF A LONG LIGHTNING ARC

Previous sections have considered some of the aspects of the interaction of direct-strike lightning with electronic systems from the viewpoint of simulating such aspects with pulsers and impedances directly connected to S_s. However, such simulation is limited by the nature of the lightning arc, especially in its nonlinear and time-varying characteristics. This is further complicated by the limited state of quantitative knowledge concerning the detailed physical processes in the arc and the resulting conductance per unit length, tortuosity, etc.

A possible approach to lightning simulation which at least partly avoids some of these difficulties consists in constructing an arc in air and letting this arc attach to and detach from S_s. This arc is generated by an appropriate high-voltage pulse generator with impedance loading as illustrated in Fig. 6. The arc might be initiated at some high-voltage electrode, propagate toward the system of interest (perhaps meeting streamers from the system), attach to the system, charge the system, detach from the system, propagate toward a return conductor (such as a ground plane), and close to the return conductor, thereby completing the current path through the pulse generator.

This type a simulation might be referred to as dual-arc lightning simulation, referring to the two arcs connected to the system in Fig. 6. How closely this simulates the direct-strike lightning phenomenon depends potentially on various physical parameters. The arcs should be sufficiently long to simulate the important aspects of the natural phenomenon. For the simulation of in-flight conditions for the system (as for an aircraft or missile) one may wish to control the local air density, water-vapor content, etc. Clearly, what is required for the environmental details can be a rather complex question.

A less complete form of the type of simulation in Fig. 6 would use only one arc in air. One might have an attachment single-arc simulation by electrically connecting the system to the return conductor to the pulser (through perhaps some distributed impedance). An alternate approach would be a detachment single-arc simulation obtained by connecting the system to the high-voltage electrode in Fig. 6 (again perhaps with special impedances). Both of these techniques allow charging of the system (with resulting large electric fields). However, the sequential charging and discharging in the nonlinear, time-varying arc manner is not fully accomplished.

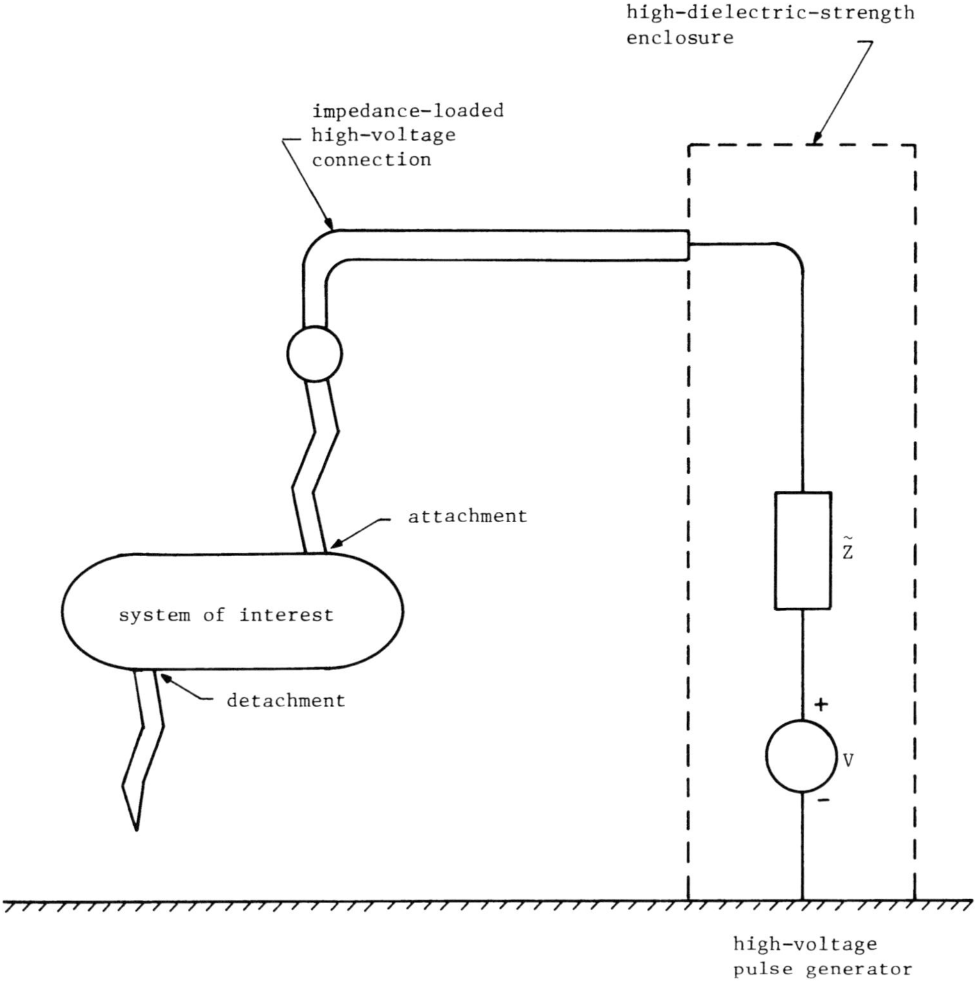

Fig. 6. Dual-Arc Lightning Simulation

7. SUMMARY

As one may now realize, there are several possible improvements that can be incorporated into lightning testing to make the test a simulation. Associated with various identifiable physical processes in the lightning interaction with electronic systems, one can formulate corresponding constraints on the simulator design. While this leads to improved simulator designs, this is not necessarily complete because of the limited understanding of the lightning physical processes by the scientific/engineering community. For distant lightning, except for some waveform questions, the simulation problem is similar to a class of EMP simulation and thereby relatively well known. For direct strike lightning the situation is quite complex and little understood by comparison.

For direct-strike lightning one can consider the physical processes involved to develop simulator designs. Quasistatic considerations lead to the importance of both current and charge on the system, thereby requiring the simulator to produce large voltage as well as large current. In the resonance region SEM considerations lead to the requirement of simulating the lightning arcs in both geometry and impedance properties, at least near the system. Both low frequencies and resonance frequencies require that the non-arc conductors and other objects be positioned away from the system under test so as to not undesirably modify the system response. At high frequencies (short wavelengths compared to system dimensions) the problem is very messy making it difficult to identify specific aspects of the simulation associated with this regime.

By successively imposing the various constraints on lightning simulation one can progressively improve the realism of the simulation. Given the state of lightning understanding at a given time (such as the present) one can design a simulator which is consistent with this understanding. Such understanding already indicates that considerable improvement in lightning simulation is needed as discussed here. However, there is still the fundamental need of obtaining an adequate understanding of the lightning electromagnetic environment.

8. REFERENCES

1. C.E. Baum, Radiation and Conductivity constraints on the Design of a Dipole Electric Field Sensor, Sensor and Simulation Note 15, February 1965.

2. C.E. Baum, The Influence of Radiation and Conductivity on B-Dot Loop Design, Sensor and Simulation Note 29, October 1966.

3. C.E. Baum, Two Types of Vertical Current Density Sensors, Sensor and Simulation Note 33, February 1967.

4. C.E. Baum, Parameters for Some Electrically-Small Electromagnetic Sensors, Sensor and Simulation Note 38, March 1967.

5. C.E. Baum, Some Characteristics of Electric and Magnetic Dipoles Antennas for Radiating Transient Pulses, Sensor and Simulation Note 125, January 1971.

6. C.E. Baum, Extrapolation Techniques for Interpreting the Results of Tests in EMP Simulators in Terms of EMP Criteria, Sensor and Simulation Note 222, March 1977.

7. C.E. Baum, EMP Simulators for Various Types of Nuclear EMP Environments: An Interim Categorization, Sensor and Simulation Note 240, January 1978, and IEEE Trans. on Antennas and Propagation, pp. 35-53, January 1978, and IEEE Trans. on Electromagnetic Compatibility, pp. 35-53, February 1978.

8. C.E. Baum, The PARTES concept in EMP Simulation, Sensor and Simulation Note 260, December 1979, and Electromagnetics, pp. 1-19, 1983.

9. C.E. Baum, Emerging Technology for Transient and Broad-Band Analysis and Synthesis of Antennas and Scatterers, Interaction Note 300, November 1976, and Proc. IEEE, pp. 1598-1616, November 1976.

10. C.E. Baum, Norms and Eigenvector Norms. Mathematics Note 63, November 1979.

11. A.E.H. Love, The Integration of the Equations of Propagation of electric Waves, Phil. Trans. Roy. Soc. London, ser. A, vol. 197, pp. 1-145, 1901.

12. R.F. Harrington, Field Computation by Moment Methods, MacMillan, 1968.

13. C.E. Baum, The Singularity Expansion Method, in L.B. Felsen (ed.), Transient Electromagnetic Fields, Springer-Verlag, 1976.

14. C.E. Baum, Toward an Engineering Theory of Electromagnetic Scattering: The Singularity and Eigenmode Methods, in P.L.E. Uslenghi (ed.), Electromagnetic Scattering, Academic Press, 1978.

11.3 Some Features of Waveguide/Horn Design

C.E. Baum

Phillips Laboratory, Kirtland Air Force Base, NM 87117, USA

ABSTRACT

In transitioning a sinusoidal electromagnetic wave from source to radiating antenna a conducting waveguide operating in the lowest order H mode is often used. This waveguide can be optimized for its power handling capabilities. As a common rectangular waveguide the $H_{1,0}$ mode has geometric properties which allow E-plane subdivision into some number of rectangular guides. This in turn allows convenient division or recombination of electromagnetic waves. This concept is extended to pyramidal horns which can launch the waves from one or more waveguides. The E-plane subdivision can be used for designing a metallic grating which allows passage of the wave while mechanically supporting a dielectric sheet separating vacuum from gas regions.

1. INTRODUCTION

This paper considers some of the aspects of waveguides for transmitting electromagnetic waves from one or more sources to one or more horns for constructing a radiating electromagnetic wave at the aperture plane or array surface (for an array of horns). Such a waveguide/horn system can be an integral part of a source/antenna system such as discussed in [1].

Beginning with a review of the modes in a perfectly conducting waveguide the power handling capability is considered. Constraining operation to the lowest order mode (an H mode) a dimensionless efficiency factor is defined. Paying special attention to the usual rectangular waveguide the various parameters are discussed for the $H_{1,0}$ mode.

Concerning the lowest order H mode of a perfectly conducting waveguide it is possible to place perfectly conducting sheets which do not interfere with this mode. For the rectangular waveguide these sheets are planes. These can be used to divide the $H_{1,0}$ mode of one waveguide into the same mode of several rectangular waveguides but with a smaller height-to-width ratio. Alternatively the same technique can combine the waves in several waveguides (as from several sources) into one waveguide provided that the amplitudes and phases of the several waves are properly matched into the single waveguide.

Expanding the waveguide into a horn (a conical waveguide) the electric field can be reduced to go from a medium of high dielectric strength (such as high vacuum) to a lower strength medium (such as air or some other gas). Noting the electromagnetic field distribution in the lowest H mode as in the waveguide similar conducting sheets can be placed in non-interfering positions in the horn. One use of these is as a metal grating to support a dielectric pressure barrier between high vacuum and some gas (such as air or a special high-dielectric-strength gas). Expanding this special grating back into the throat of the horn the conducting sheets can be modified via corrugation to give a slow-wave character to the waves passing between the sheets and thereby form a lens to modify the phase distribution on the horn aperture.

2. MODES IN CLOSED PERFECTLY CONDUCTING WAVEGUIDES

The basic theory of closed conducting waveguides is now summarized following [3]. As in Fig. 1 we have a rectangular coordinate system

$$\vec{r} = (x, y, z) \tag{2.1}$$

where the waveguide cross section is independent of z. Designate the cross section surface (on the z = 0 plane) as S with outer boundary C oriented with tangent vector $\vec{1}_C$. The unit normal to S

$$\vec{1}_S = \vec{1}_z \tag{2.2}$$

with $\vec{1}_C \times \vec{1}_z$ as the outward unit normal to C.

The modes come in precisely two kinds (no hybrid modes as on an open dielectric waveguide). For the H (or TE) modes we have a real valued scalar function $\psi_{h,\beta}$ satisfying a two-dimensional Helmoltz equation

$$\left[\nabla_t^2 + k_{h,\beta}^2\right] \psi_{h,\beta} \;(x,y) = 0 \;,\; \nabla_t^2 = \frac{\partial^2}{\partial x^2} + \frac{\partial^2}{\partial y^2} \;,\; \left[\vec{1}_C \times \vec{1}_z\right] \cdot \nabla_t \, \psi_{h,\beta} \;(x,y) = 0 \text{ on C}$$

$$\nabla_t = \vec{1}_x \frac{\partial}{\partial_x} + \vec{1}_y \frac{\partial}{\partial_y} \;,\; \beta \equiv \text{mode index set} \tag{2.3}$$

The $k_{h,\beta}$ are real-valued cutoff wave numbers [2]. The fields take the form

$$\begin{aligned} H_{z_{h,\beta}} &= k_{h,\beta}^2 \; e^{\pm\gamma_{z_{h,\beta}} z} \; \psi_{h,\beta} \;(x,y) \;\; \text{(hence an H mode)} \\ \vec{H}_{t_{h,\beta}} &= \pm\gamma_{z_{h,\beta}} \; e^{\pm\gamma_{z_{h,\beta}} z} \; \nabla_t \psi_{h,\beta} \;(x,y) \\ \vec{E}_{t_{h,\beta}} &= \pm Z_{h,\beta} \; \vec{1}_z \times \vec{H}_{t_{h,\beta}} = s \; \mu_0 \; e^{\pm\gamma_{z_{h,\beta}} z} \; \vec{1}_z \times \nabla_t \; \gamma_{h,\beta} \;(x,y) \end{aligned} \tag{2.4}$$

Here a subscript "t" indicates the transverse part. Various other parameters are

$$\begin{aligned} \gamma &= \frac{s}{c} \;\; \text{(free - space propagation constant)} \\ Z_0 &= \sqrt{\frac{\mu_0}{\varepsilon_0}} \simeq 377 \; \Omega \;\; \text{(free - space impedance)} \\ c &= \frac{1}{\sqrt{\mu_0 \varepsilon_0}} \simeq 3 \times 10^8 \; \frac{m}{s} \;\; \text{(free - space propagation speed)} \\ \gamma^2 &= \gamma_{z_{h,\beta}}^2 + \gamma_{h,\beta}^2 = \gamma_{z_{h,\beta}}^2 - k_{h,\beta}^2 \\ Z_{h,\beta} &= \frac{s \; \mu_0}{\gamma_{z_{h,\beta}}} = \frac{\gamma}{\gamma_{z_{h,\beta}}} \; Z_0 \;\; \text{(modal impedance)} \end{aligned} \tag{2.5}$$

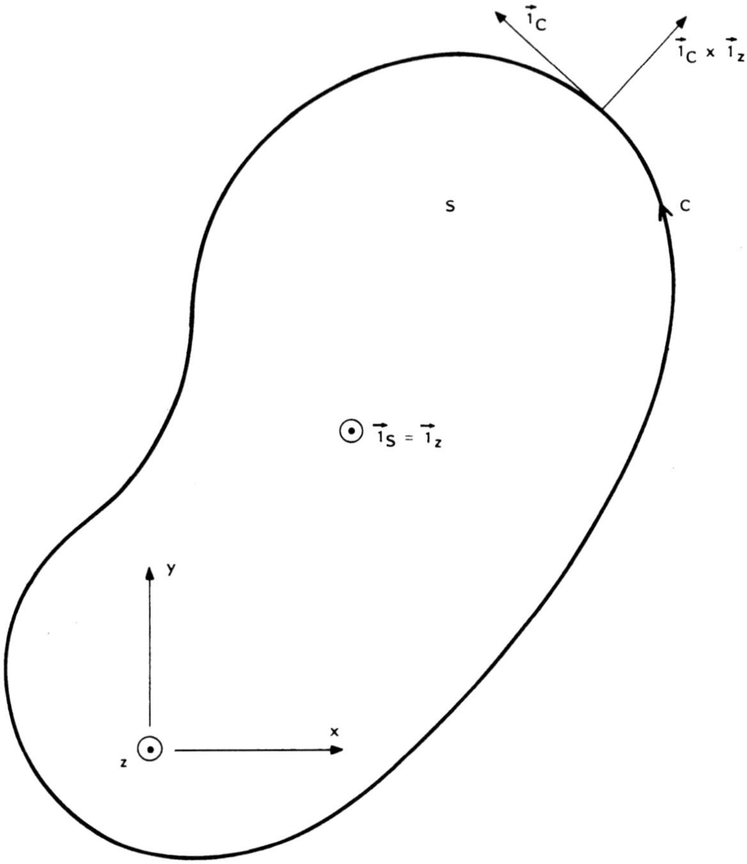

Fig. 1. Cross Section of Closed Perfectly Conducting Waveguide

In terms of the jω axis (ω real) we have

$$\gamma = jk, \; s = j\omega, \; k = \frac{\omega}{c}$$

$$\gamma_{z_{h,\beta}} = jk_{z_{h,\beta}} \text{ (guide propagation constant)}$$

$$\varsigma_{h,\beta} = \left[1 - \left(\frac{k_{h,\beta}}{k}\right)^2\right]^{\frac{1}{2}} < 1 \text{ for } 0 < \frac{k_{h,\beta}}{k} < 1$$

$$Z_{h,\beta} = \varsigma_{h,\beta}^{-1} Z_0$$

$$k_{z_{h,\beta}} = k \, \varsigma_{h,\beta} \text{ (guide wave number)}$$

$$f_{h,\beta} = \frac{1}{2\pi} \omega_{h,\beta} = \frac{c}{\gamma_{h,\beta}} \text{ (cutoff frequency)}$$

$$\lambda_{h,\beta} = \frac{2\pi}{k_{h,\beta}} \text{ (cutoff wavelength)} \tag{2.6}$$

For the E (or TM) modes we have a real valued scalar function $\psi_{e,\beta}$ satisfying a two-dimensional Helmholtz equation

$$\left[\nabla_t^2 + k_{e,\beta}^2\right] \psi_{e,\beta}(\mathrm{x,y}) \,,\; \psi_{e,\beta}(\mathrm{x,y}) = 0 \text{ on C} \tag{2.7}$$

Again the $k_{e,\beta}$ are real-valued cutoff wave numbers [2]. The fields take the form

$$E_{z_{e,\beta}} = k_{e,\beta}^2 \, e^{\pm z_{e,\beta} z} \, \psi_{e,\beta}(\mathrm{x,y}) \text{ (hence an E mode)}$$

$$\vec{E}_{t_{e,\beta}} = \pm \gamma_{z_{e,\beta}} \, e^{\pm\gamma_{z_{e,\beta}} z} \, \nabla_t \, \psi_{e,\beta}(\mathrm{x,y})$$

$$\vec{H}_{t_{e,\beta}} = \mp Y_{e,\beta} \, \vec{1}_z \times \vec{E}_{t_{e,\beta}} = -s \, \varepsilon_0 \, e^{\pm\gamma_{z_{e,\beta}} z} \, \vec{1}_z \times \nabla_t \, \psi_{e,\beta}(\mathrm{x,y}) \tag{2.8}$$

Various other parameters are

$$\gamma^2 = \gamma_{z_{e,\beta}}^2 + \gamma_{e,\beta}^2 = \gamma_{z_{e,\beta}}^2 - k_{e,\beta}^2$$

$$Y_{e,\beta} = \frac{s \, \varepsilon_0}{\gamma_{z_{e,\beta}}} = \frac{\gamma}{\gamma_{z_{e,\beta}}} Y_0 \text{ (modal admittance)}$$

$$Y_0 = Z_0^{-1} \text{ (free – space admittance)} \tag{2.9}$$

In terms of the $j\omega$ axis we have

$$\gamma_{z_{e,\beta}} = jk_{z_{e,\beta}} \text{ (guide propagation constant)}$$

$$\varsigma_{e,\beta} = \left[1 - \left(\frac{k_{e,\beta}}{k}\right)^2\right]^{\frac{1}{2}} < 1 \text{ for } 0 < \frac{k_{e,\beta}}{k} < 1$$

$$Y_{e,\beta} = \varsigma_{e,\beta}^{-1} Y_0$$

$$k_{z_{e,\beta}} = k\ \varsigma_{e,\beta} \text{ (guide wave number)}$$

$$f_{e,\beta} = \frac{1}{2\pi}\,\omega_{e,\beta} = \frac{c}{\lambda_{e,\beta}} \text{ (cutoff frequency)}$$

$$\lambda_{e,\beta} = \frac{2\pi}{k_{e,\beta}} \text{ (cutoff wavelength)} \tag{2.10}$$

As shown in [3] there are a set of mode orthogonality conditions which we summarize as

$$\int_S \psi_{\nu_1}\,\psi_{\nu_2}\,dS = 0\,,\ \int_S [\nabla_t\psi_{\nu_1}]\cdot[\nabla_t\psi_{\nu_2}]\,dS = 0$$

$$\int_S \left[\vec{1}_z\times\nabla_t\,\psi_{\nu_1}\right]\cdot\left[\vec{1}_z\times\nabla_t\,\psi_{\nu_2}\right]dS = 0\,,\ \int_S \left[\vec{1}_z\times\nabla_t\,\psi_{\nu_1}\right]\cdot[\nabla_t\,\psi_{\nu_2}]\,dS = 0$$

$$\nu = ({}^{e}_{h},\beta) \equiv \text{ index set designating any mode (E or H)} \tag{2.11}$$

These results apply for any combination of modes with the sufficient condition

$$k_{\nu_1} \neq k_{\nu_2} \text{ for } \nu_1 \neq \nu_2 \tag{2.12}$$

Now while this is a sufficient condition there are many cases of waveguides involving symmetrical cross sections where the degeneracy of equal cutoff wave numbers for different ν_1 and ν_2 does not lead to exceptions to (2.11). So it is known that (2.12) is often not necessary.

3. POWER IN WAVEGUIDE MODES

The peak power in a waveguide is

$$P = \int_S \left[\vec{E}\times\vec{H}^*\right]\cdot\vec{1}_z\ \mathrm{dS}\,,\ = 2\,P_{avg}\,,\ P_{avg} \equiv \text{average power} \tag{3.1}$$

where we have chosen the mode to be propagating in the +z direction and set z=0 for the integration. Also with s=$j\omega$ we have assumed that k is greater than the cutoff wave number. For H modes only the transverse components contribute and (2.4) gives

$$P_{h,\beta} = -k^2 Z_0\,\varsigma_{h,\beta}\int_S \left\{\left[\vec{1}_z\times\nabla_t\,\psi_{h,\beta}(x,y)\right]\times[\nabla_t\psi_{h,\beta}(x,y)]\right\}\cdot\vec{1}_z\,dS$$

$$= k^2 Z_0\,\varsigma_{h,\beta}\int_S |\nabla_t\,\psi_{h,\beta}(x,y)|^2\,dS \tag{3.2}$$

Similarly for E modes we have from (2.8)

$$P_{e,\beta} = k^2 Y_0\,\varsigma_{e,\beta}\int_S \left\{[\nabla_t\,\psi_{e,\beta}(x,y)]\times\left[\vec{1}_z\times\nabla_t\,\psi_{e,\beta}(x,y)\right]\right\}\cdot\vec{1}_z\,dS$$

$$= k^2 Y_0\,\varsigma_{e,\beta}\int_S |\nabla_t\,\psi_{e,\beta}(x,y)|^2\,dS \tag{3.3}$$

Alternate expressions can be found using the relation valid for both H and E modes [3].

$$\int_S |\nabla_t \psi_\nu (x,y)|^2 \, dS = k_\nu^2 \int_S \psi_\nu^2 (x,y) \, dS \tag{3.4}$$

4. POWER HANDLING CAPABILITY OF WAVEGUIDE IN LOWEST ORDER H MODE

Now let us consider that the waveguide is being operated in its lowest order mode, an H mode. Let the subscript "0" be used to designate parameters for this mode. Now we have $d > k_0$ for this mode to propagate. As frequency is increased other modes can also propagate. Let the subscript "1" be used to designate frequency/wavelength parameters for the next mode (or modes if more than one have the same cutoff frequency). Then to operate in the lowest order mode we have

$$k_0 < k < k_1 \, , \; f_0 < f < f_1 \, , \; \lambda_0 > \lambda > \lambda_1 \tag{4.1}$$

Let us define a relative bandwidth for the lowest order mode as

$$\eta_f \equiv \frac{f_1}{f_0} = \frac{k_1}{k_0} = \frac{\lambda_0}{\lambda_1} \tag{4.2}$$

One measure of waveguide optimization is the maximization of this relative bandwidth, such as for communications purposes. It is known that the lowest order mode is an H mode as is the next mode, so only H modes need to be considered to determine both k_0 and k_1 [2].

From (3.2) the power (peak) in the lowest order H mode is

$$P_0 = k^2 \, Z_0 \, \varsigma_0 \int_S |\nabla_t \psi_0 (x,y)|^2 \, dS \, , \; \varsigma_0 = \left[1 - \left(\frac{k_0}{k} \right)^2 \right]^{\frac{1}{2}} \tag{4.3}$$

From (2.4) we have the maximum electric field magnitude on S as

$$E_0 = \max_S \; \omega \mu_0 \, |\nabla_t \psi_0 (x,y)| \tag{4.4}$$

Defining a normalizing power density as

$$p_0 \equiv \frac{E_0^2}{Z_0} \tag{4.5}$$

The power can be normalized to give an effective guide area as

$$A_0 \equiv \frac{P_0}{p_0} = \left[\max_S \; |\nabla_t \psi_0 (x,y)| \right]^{-2} \varsigma_0 \int_S |\nabla_t \psi_0 (x,y)|^2 \, dS \tag{4.6}$$

For some specified peak electric field E_0 (related to breakdown) then $A_0 p_0$ gives the power handling capability of the guide (peak, not average). In maximizing A_0 note the geometric part involving ψ_0. There is also a frequency part contained in ς_0. For a given waveguide size and shape we can maximize ς_0 subject to (4.1). Taking the limit as $k \to k_1$ we define

$$\varsigma_1 \equiv \left[1-\left(\frac{k_0}{k_1}\right)^2\right]^{\frac{1}{2}} = \left[1-n_f^{-2}\right]^{\frac{1}{2}}$$

$$A_1 \equiv \left[\max_S |\nabla_t \psi_0(x,y)|\right]^{-2} \varsigma_1 \int_S |\nabla_t \psi_0(x,y)|^2 \, dS \tag{4.7}$$

Now A_1 is only a function of the size and shape of the waveguide via ψ_0 and ς_1.

We can define as associated effective length which we wish to maximize as

$$\ell_1 \equiv A_1^{\frac{1}{2}} \tag{4.8}$$

Noting a frequency/distance inverse scaling define a dimensionless parameter as

$$\eta_1 \equiv \frac{f_1 \ell_1}{c} = \frac{\ell_1}{\lambda_1} \tag{4.9}$$

For a given desired operating frequency f_1 we wish to maximize ℓ_1 to maximize the power handling capability of the guide in its lowest mode.

Combining the above results gives

$$\begin{aligned} \eta_1 &= \frac{1}{\lambda_1}\left[\max_S |\nabla_t \psi_0(x,y)|\right]^{-1} \left\{\varsigma_1 \int_S |\nabla_t \psi_0(x,y)|^2 \, dS\right\}^{\frac{1}{2}} \\ &= \frac{2\pi}{\lambda_1 \lambda_0}\left[\max_S |\nabla_t \psi_0(x,y)|\right]^{-1} \left\{\varsigma_1 \int_S \psi_0^2(x,y) \, dS\right\}^{\frac{1}{2}} \\ &= \frac{\eta_f}{\lambda_0}\left[\max_S |\nabla_t \psi_0(x,y)|\right]^{-1} \left\{\varsigma_1 \int_S |\nabla_t \psi_0(x,y)|^2 \, dS\right\}^{\frac{1}{2}} \\ &= \frac{2\pi\,\eta_f}{\lambda_0^2}\left[\max_S |\nabla_t \psi_0(x,y)|\right]^{-1} \left\{\varsigma_1 \int_S \psi_0^2(x,y) \, dS\right\}^{\frac{1}{2}} \end{aligned} \tag{4.10}$$

as several alternate forms of this dimensionless efficiency factor. Maximizing η_1 optimizes the shape of the waveguide cross section for maximum power handling.

In some cases the electric field E_0 for breakdown may be more important on C than throughout S. For example, under sufficiently high vacuum conditions one is concerned about field emission from the waveguide walls. Then one can replace

$$\max_S |\nabla_t \psi_0(x,y)| \to \max_C |\nabla_t \psi_0(x,y)| \tag{4.11}$$

in the previous formulae. To distinguish this case η_1 might be replaced by η_1'.

5. RECTANGULAR WAVEGUIDE

As a common example consider the widely used rectangular waveguide as exhibited in Fig. 2. Let the guide be described by

$$(x,y) \in S \quad \text{for} \begin{cases} 0 \leq x \leq a \\ 0 \leq y \leq b \end{cases}$$

$$0 < b \leq a \tag{5.1}$$

Following [3] the H modes have

$$\psi_{h,n,m} = \cos\left(\frac{n\pi x}{a}\right) \cos\left(\frac{m\pi x}{b}\right), \; n = 0,1,2,3,..., \; m = 0,1,2,3,...$$

$$\text{but not both n} = 0 \text{ and m} = 0, \; \lambda_{h,n,m} = \frac{2\pi}{k_{h,n,m}} = 2\left[\left(\frac{n}{a}\right)^2 + \left(\frac{m}{b}\right)^2\right]^{-\frac{1}{2}} \tag{5.2}$$

The lowest order H mode is the $H_{1,0}$ mode for which

$$\lambda_{h,1,0} = 2a \; , \; \psi_{h,1,0} = \cos\left(\frac{\pi x}{a}\right) \; , \; \nabla_t \, \psi_{h,1,0} = -\frac{\pi}{a} \sin\left(\frac{\pi x}{a}\right) \vec{1}_x \tag{5.3}$$

Then $\vec{H}_t$ has only an x component and $\vec{E}_t$ has only a y component, both with the sine variation as above going to zero on the side walls.

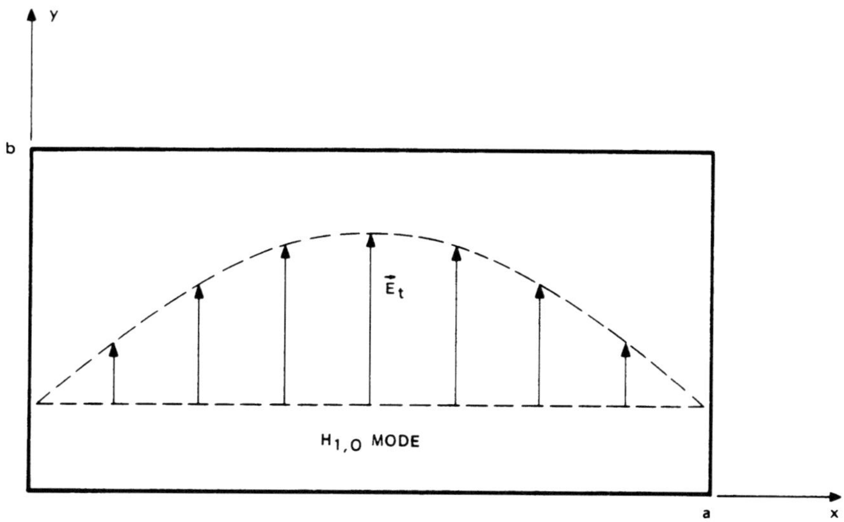

Fig. 2. Rectangular Waveguide Cross Section and Lowest Order H Mode.

Similarly the E modes have

$$\psi_{e,n,m} = \sin\left(\frac{\pi x}{a}\right) \sin\left(\frac{\pi x}{b}\right) \ , \ n = 1,2,3,\ldots \ , \ m = 1,2,3,\ldots$$

$$\lambda_{e,n,m} = \frac{2\pi}{k_{e,n,m}} = 2\left[\left(\frac{n}{a}\right)^2 + \left(\frac{m}{b}\right)^2\right]^{-\frac{1}{2}} = \lambda_{h,n,m} \tag{5.4}$$

Traditionally the rectangular guide is constructed with

$$b = \frac{a}{2} \tag{5.5}$$

This gives cutoff wavelengths for the next few modes as

$$\lambda_{h,2,0} = a \ , \ \lambda_{h,0,1} = a \ , \ \lambda_{e,1,1} = \lambda_{h,1,1} = \frac{2a}{\sqrt{5}} \simeq .89a \tag{5.6}$$

For a given width a (and hence $\lambda_{h,1,0}$) the height b has been chosen as in (5.5) so that b is maximized without making $\lambda_{h,0,1}$ (=2b) any greater than $\lambda_{h,2,0}$. For this choice we have

$$\eta_f = \frac{\lambda_{h,1,0}}{\lambda_{h,2,0}} = \frac{\lambda_{h,1,0}}{\lambda_{h,0,1}} = 2 \tag{5.7}$$

This gives the maximum relative bandwidth for the lowest order mode of a rectangular guide.

For this lowest order $H_{1,0}$ mode with (5.5) we have

$$\varsigma_1 = \frac{\sqrt{3}}{2} \simeq .866, \quad \max S \ |\nabla_t \psi_0 (x,y)| = \max C \ |\nabla_t \psi_0 (x,y)| = \frac{\pi}{a}$$

$$\int_S |\nabla_t \psi_0 (x,y)|^2 \ dS = b\left(\frac{\pi}{a}\right)^2 \int_0^a \sin^2\left(\frac{\pi x}{a}\right) dx = \left(\frac{\pi}{a}\right)^2 \frac{ba}{2} = \left(\frac{\pi}{a}\right)^2 \frac{a^2}{4} \tag{5.8}$$

$$\lambda_0 = 2a, \lambda_1 = a, \ \eta_1 = \eta_1' = \left[\frac{\sqrt{3}}{8}\right]^{\frac{1}{2}} \simeq .465$$

This efficiency can serve as a standard for comparison to other shapes of waveguide cross section.

6. PLACING PERFECTLY CONDUCTING SHEETS IN WAVEGUIDE CROSS SECTION WITHOUT INTERFERING WITH A PARTICULAR H MODE

For various purposes one may wish to place conductors in a waveguide, or its extension as a horn. This is applicable to merging and dividing waveguides for mechanical strengtheners as we shall see later.

Consider the modes as discussed in Section 2. Note that the E modes in (2.8) have both an E_z and an $\vec{E}_t$ and that these have different distributions over S. One might place a conducting wire on some cross section (at a particular z) with this wire orthogonal to $\nabla_t \ \psi_{e,\beta}$(x,y) for some βth mode of interest. However this wire cannot in the general case be extended in the z direction to form a conducting sheet without interfering with this same βth mode. To avoid this the sheet must be placed on a contour where $\psi_{e,\beta}$ is zero and this contour must also be orthogonal to $\nabla_t\ \psi_{e,\beta}$ (parallel to $\vec{1}_z \times \nabla_t\ \psi_{e,\beta}$) on S. There can be special cases of this (such as the $E_{0,m}$ rotationally symmetric modes of a circular waveguide) but this is not the usual case.

Now the H modes are a different matter. As in (2.4) there is no E_z. Thus we only need to consider $\vec{E}_t$. Then conducting sheets can be extended arbitrarily in the z direction. In the transverse sense these conductors can follow contours on S that are perpendicular to $\vec{1}_z \times \nabla_t\ \psi_{h,\beta}$, or equivalently parallel to $\nabla_t\ \psi_{h,\beta}$. Note that this direction can be expressed as a real valued vector since the $\psi_{h,\beta}$ can be expressed as real valued functions on S. To see this note that with real-valued $k^2_{h,\beta}$ then (2.3) is satisfied by both $\psi_{h,\beta}$ and $\psi^*_{h,\beta}$. Adding the two Helmoltz equations shows that the Helmholtz equation is also satisfied by Re$[\psi_{h,\beta}]$ (and Im$[\psi_{h,\beta}]$ as well). So Re$[\psi_{h,\beta}]$ can be used as the mode (or any complex constant times $\psi_{h,\beta}$ for that matter).

Note also for the $\psi_{h,\beta}$ mode then $\vec{H}_t$ is parallel to the conducting sheet as is H_z. This is another way of looking at the boundary conditions at the conducting sheet inserted in the waveguide. There can be no normal magnetic field at a perfect conductor.

As an example consider the rectangular guide as in Fig. 3. In the lowest order $H_{1,0}$ mode the electric field is purely vertical and the magnetic field is purely horizontal (including the longitudinal part) as indicated in Fig. 3A. The variations of the fields is independent of y with sine variation with respect to x for the transverse fields as in Fig. 2. The longitudinal H_z has a cosine variation with respect to x (see (5.3)).

In Fig. 3A N-1 conducting sheets are added on planes of constant y within the waveguide. This gives N waveguides, the heights of which sum to b, and the widths of which are all a. Now as in Fig. 3B let these sheets begin on some cross section to divide the one waveguide on the left into N waveguides on the right. Then one can make each dividing sheet become two sheets which are gradually separated so that N separated waveguides are formed.

If the $H_{1,0}$ mode is propagating from the left, it divides into the $H_{1,0}$ modes of the N waveguides, the power in each guide being proportional to its height. If desired each of these N guides can be gradually expanded to the usual height of a/2 for lower losses. This type of matched power divider might have various applications, such as to feed an array of N radiating horns.

Conversely, one can have N waves propagating from the right in the $H_{1,0}$ modes of the N waveguides. Provided the fields arrive at the cross section where the N guides become one and the fields match between adjacent guides then one $H_{1,0}$ mode propagates out to the left. This requires that not only the amplitudes, but phases as well match at this cross section. For simplicity one might have the heights of each guide on the right as b/N with equal powers in each of these guides. Then the important issue is to have the phases of the N waves all match at the interface to the guide to the left. If the N sources driving the waves in the N guides are all of the same frequency then phases need to be adjusted in the sources and/or the waveguide delays to establish the matching conditions.

Note that the N guides on the right are in general curved different amounts so that the delays in the different guides are perturbed differently (as compared to a straight waveguide). Symmetry can be used to provide some exception to this. For example for N=2 one can have a symmetry plane (one guide being the mirror image of the other) as the two guides are bent away from each other (E plane bends). The right guides in Fig. 3B can have one bend into and the other bend out of the page (H plane bends). This is an example of rotation symmetry about the axis (x = a/2, y = b/2) of the left guide. Combinations of various bends and twists of the two guides

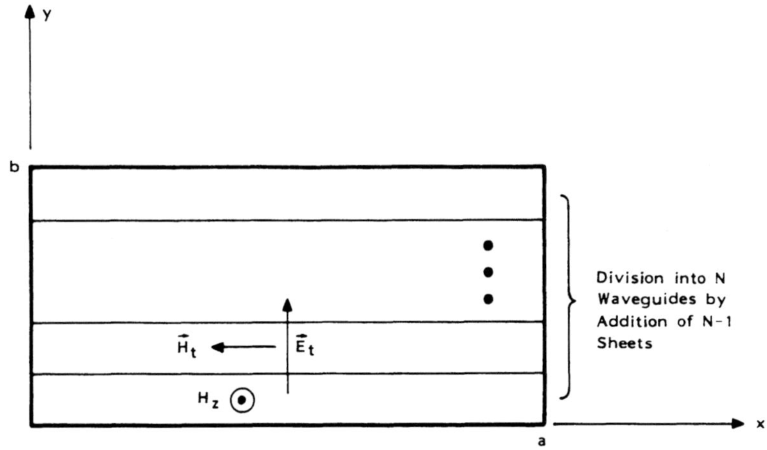

A. Addition of Conducting Sheets Orthogonal to Electric Field of $H_{1,0}$ Mode

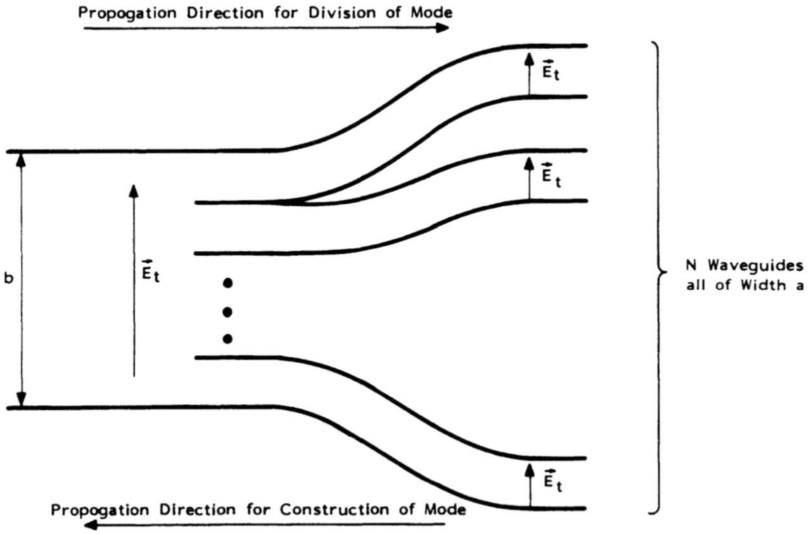

B. Division of $H_{1,0}$ Mode or Construction of $H_{1,0}$ Modes from Matched $H_{1,0}$ Modes

Fig. 3. Division of Rectangular Waveguide Operating in $H_{1,0}$ Mode.

can be made subject to the symmetry that rotation by π around the object leaves the geometry unchanged. Note that for b = a/2 the two waveguides each have width a and height a/4.

Using this 2-symmetrical a by a/4 guide system one could connect the guides to two sources with identical phases and have the two signals combine without reflection into the single a by a/2 guide. Such might be two ports on a single magnetron.

This concept can be extended by dividing a locally straight section in each a by a/4 guide into two a by a/8 guides. Then proceeding with each pair as above we have four a by a/8 guides which can be used to combine four signals without reflection into the single a by a/2 guide. Such might be four ports on a single magnetron.

7. VACUUM INTERFACE AT HORN EXIT

Another application of these non-interfering sheets in a waveguide cross section is as a grating to support a dielectric sheet serving as a gas barrier. Suppose for simplicity we consider a rectangular or pyramidal horn as in Fig. 4. The horn might be considered as a conical waveguide or a gradually expanding rectangular waveguide. Care should be taken that at the mouth (aperture or exit) of the horn the fields are like the $H_{1,0}$ mode, specifically that the electric field is at least approximately perpendicular to the metal slats (grating) placed across the mouth of the horn.

One purpose of the horn is to expand the power in the waveguide so that the electric field is considerably reduced at the aperture. If the waveguide is operated under high-vacuum conditions it can propagate very large electric fields. In order to bring the power out of the waveguide the fields need to be reduced if the dielectric medium is air or some high dielectric strength gas (such as SF_6 at one atmosphere). Choosing the aperture dimensions to avoid breakdown one transitions these back to the waveguide via the evacuated horn.

Note that at the aperture the wave is ideally to pass through the grating with no disturbance. Of course there will in general be some perturbation due to the grating. It may be desirable to avoid resonant currents on the slats. This can be done by making the slat length an odd number of quarter wavelengths, or by addition of appropriate impedance loading.

Variations on the design of the horn aperture are possible as indicated in Fig. 5. Begin with a flat dielectric sheet in Fig. 5A. Note the metal slats supporting the gas pressure through the dielectric sheet. In Fig. 5B one might replace the flat dielectric sheet by some sort of ridged sheet, both for pressure reasons and dielectric-strength reasons. Figure 5C shows an example in which the grating is not against a single flat surface, but pushes the center of the aperture out so that the electric field is not parallel to the dielectric sheet.

8. ARRAYS OF HORNS

For various reasons one may have the electromagnetic wave to be released into the air (or other gas medium) present on more than one waveguide. Then one may transition the M waveguides via M horns with the aperture surfaces all on a common desired surface such as a plane or sphere. The M signals are assumed to be all in phase with respect to this surface.

There are various ways that one may arrange horns to form a two-dimensional array. For simplicity approximating the array surface as a plane, the array may be considered as a periodic structure,

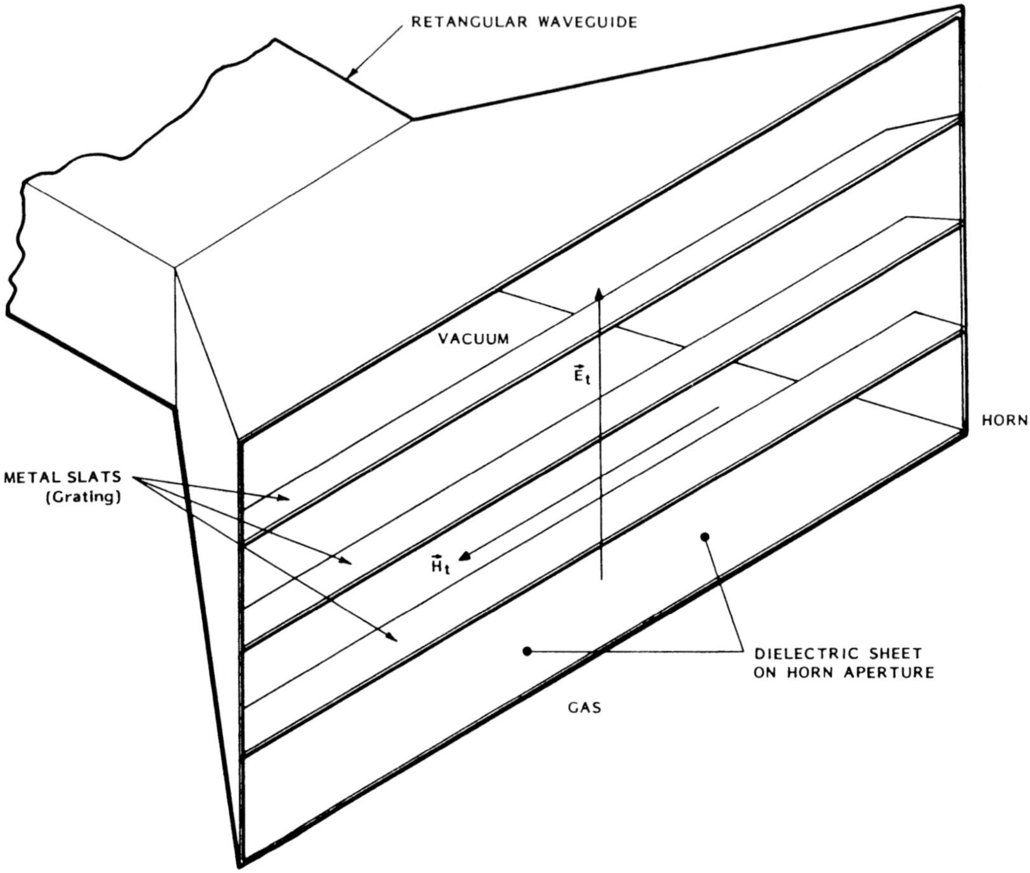

Fig. 4. Dielectric Sheet and Metal Grating at Horn Aperture.

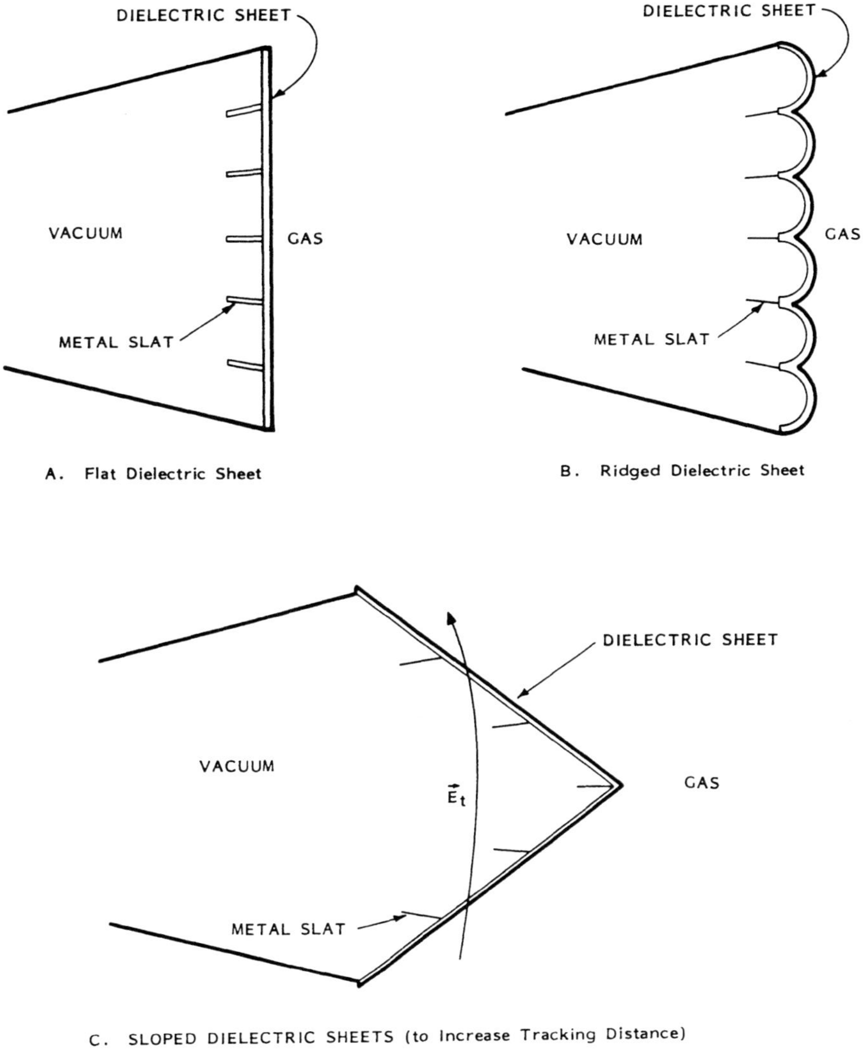

Fig. 5. Vacuum/Gas Interfaces at Horn Aperture.

a few examples of which are illustrated in Fig. 6. Using rectangular horns (a by a/2) we have the in-line and staggered arrays as in Fig. 6A and Fig. 6B. Note the presence of the gratings with parallel slats. A limitation of such an array is that the $H_{1,0}$ mode is not uniform across each horn, but has $\vec{E}_t$ going to zero at the "side" walls (walls parallel to the E field direction). This means that the electric field distribution over the array surface is undesirable) nonuniform.

An alternate approach is illustrated in Fig. 6C. The rectangular waveguides are transitioned to rhombic (diamond shaped) waveguides with about a 2 to 1 width to height ratio so that the corresponding dominant H mode is effectively excited. These waveguides are then expanded as conical horns with rhombic cross sections. These horns fill the array plane as in Fig. 6C. Noting the symmetry of these diamonds at least one conducting slat can be placed across the center of each and be perpendicular to $\vec{E}_t$. However, off the symmetry plane the slats are in general not flat for the lowest order H mode. In this geometry the distribution of $\vec{E}_t$ is such that it is perpendicular to the diamond boundaries. This gives a smooth transition from one horn to the next. While the rectangular $H_{1,0}$ distribution has zero electric field along each "side" boundary, the corresponding diamond distribution crowds this null into the "corners".

In general there are various details for optimizing horn arrays. The horn shape and arrangement as above is but one question. Tapering the distribution across the array (say by variation in the horn design, especially near the edges of the array) is another interesting question. Perhaps other horn shapes for arrays need to be considered.

9. LENSES IN HORNS

In designing the horn or array of horns that launches a wave one may be concerned with controlling the phase distribution on the horn aperture(s). This depends on the type of wave (e.g., planar or spherical) that one wishes to synthesize, and on the uncorrected phase distributions in the horn aperture(s). One can insert some kind of lens in the horn to adjust the phase distribution in the horn aperture toward some more desirable distribution. Reflections from this lens should also be minimized since these would reduce the power radiated.

One interesting kind of lens involves modifications to the grating design in Section 7. As indicated in Fig. 7A extends the grating slats back into the throat of a pyramidal horn. Note that these sheets, if extended, would have a common apex, the same apex as the pyramidal horn. As discussed before in the context of the grating slats these sheets (assumed perfectly conducting) do not interfere with the field distribution of the $H_{1,0}$ mode.

Now modify these conducting sheets so as to create a slow wave structure [4] as in Fig. 7B. The magnetic field H_t can penetrate into the grooves (including on both sides if desired). As the horn expands so that there are many wavelengths in the H direction (between the side walls), one can think of the wave propagating between two sheets as locally a TEM plane wave. Corrugating the sheets then can be approximately analyzed as a two-dimensional problem [4]. The top and bottom surfaces of the horn (E plane walls) can also be corrugated as in a typical corrugated pyramidal horn [5].

By varying the corrugations (depth, width, and/or spacing) along the length and width of the sheets various propagation speeds (phase speeds) can be achieved. Thereby, the phase at the horn aperture can be controlled to some extent. Note that one may wish to vary the corrugation parameters slowly so as to not introduce significant reflections. Furthermore, the conducting sheets need to be close enough together (say less than a half wavelength at the horn aperture) to

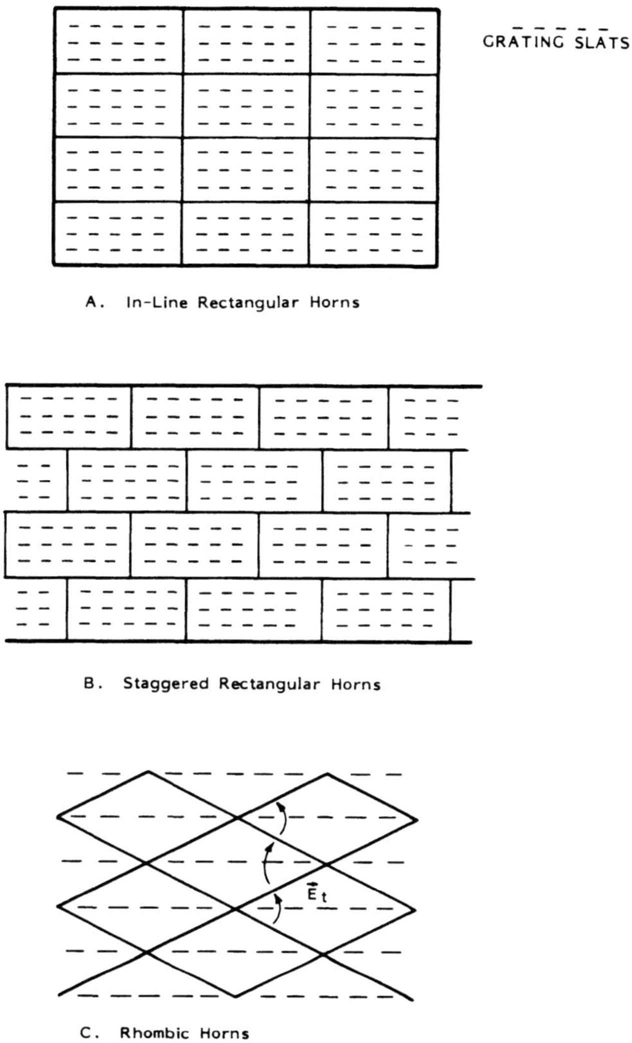

Fig. 6. Horn Arrays with Aperture Gratings.

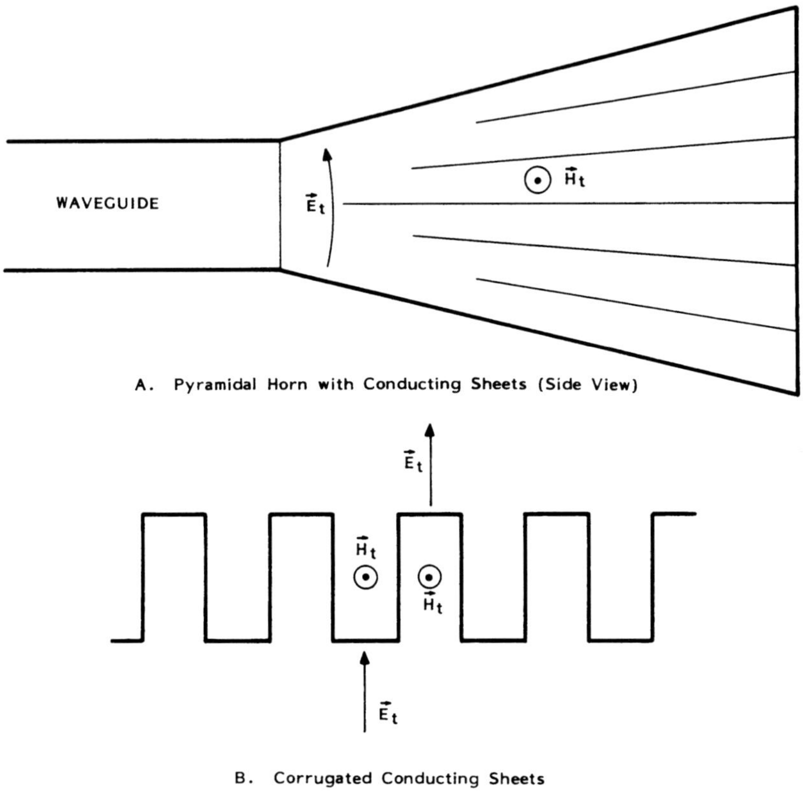

Fig. 7. Corrugated Conducting Sheets in Horn Making Slow-Wave Structure for Lens.

avoid introduction of higher order modes between the sheets by the effective impedance boundary condition of the corrugations.

10. REFERENCES

1. C.E. Baum, Maximization of Electromagnetic Response at a Distance, Sensor and Simulation Note 312, October 1988.

2. F.E. Borgnis and C.H. Papas, Electromagnetic Waveguides and Resonators, in S. Flugge/Marburg (ed.), Encyclopedia of Physics, vol. 16, Springer-Verlag, 1958.

3. R.E. Collin, Field Theory of Guided Waves, McGraw Hill, 1960.

4. R.E. Collin, Foundations for Microwave Engineering, McGraw Hill, 1966.

5. C.A. Balanis, Horn Antennas, in Y.T. Lo and S.W. Lee (eds.), Antenna Handbook, Van Nostrand Reinhold, 1988.

11.4 Energy Norms and 2-Norms

C.E. Baum
Phillips Laboratory, Kirtland Air Force Base, NM 87117, USA

ABSTRACT

This paper defines an energy norm or e-norm based on the energy delivered to some port in a system. It is expressible as a weighted 2-norm. Criteria are developed for the frequency spectrum of time-domain excitation waveforms to bound the 2-norms and e-norm at the port. Canonical incident waveforms are introduced to give convenient bounds.

1. INTRODUCTION

In characterizing system response to electromagnetic excitation it is useful to use norms because of the inequalities (bounds) which are an integral part of their definition. In addition, some norms can be related to damage/upset mechanisms such as energy, peak signal, etc. Various previous papers [1 - 5] deal with norms of time-domain waveforms. This paper extends our consideration of the 2-norms of such signals (which might be called the action-integral norm) to what can be called the energy norm or e-norm. This can be expressed as a weighted 2-norm and represents the energy delivered to some port of interest.

Fundamental to the norms in this paper is the relation between norms in frequency and time domains which are extended here. This also shows how the frequency spectrum of excitation waveforms must be controlled to achieve the correct norm values. Canonical incident waveforms for plane-wave excitation are discussed from the point of view of bounding the response norms for a variety of a possible excitations. Step-function excitation is particularly interesting leading to the concept of an effective volume to characterize the energy absorption at the port.

2. GENERALIZED PARSEVAL THEOREM

The well-known Parseval theorem states [2]

$$\int_{-\infty}^{\infty} f^2(t)dt = \frac{1}{2\pi}\int_{-\infty}^{\infty} \tilde{f}(j\omega)\tilde{f}(-j\omega)d\omega$$

$$s = \Omega + j\omega \equiv \text{complex frequency or Laplace transform variable} \qquad (2.1)$$

$$\sim \; \equiv \; \text{Laplace transform (two sided) designator}$$

In this form f(t) can even be complex, but normally we are concerned with real-valued time functions. Stated in the form of a 2-norm we have

$$\|f(t)\|_2 \equiv \left\{\int_{-\infty}^{\infty} f^2(t)dt\right\}^{\frac{1}{2}} = \frac{1}{\sqrt{2\pi}}\left\{\int_{-\infty}^{\infty} \tilde{f}(j\omega)\tilde{f}(-j\omega)d\omega\right\}^{\frac{1}{2}}$$

$$\equiv \frac{1}{\sqrt{2\pi}}\left\|\tilde{f}(j\omega)\right\|_2 \tag{2.2}$$

where now f(t) is assumed real so that (conjugate symmetry)

$$\tilde{f}(-j\omega) = \tilde{f}^*(j\omega) \equiv \text{complex conjugate} \tag{2.3}$$

This result expresses a concept that the energy in a time-domain signal is also computable in frequency domain.

As shown in [2] the Parseval theorem is a special case of

$$\int_{-\infty}^{\infty} f_1(t)f_2(t)dt = \frac{1}{2\pi j}\int_{Br} \tilde{f}_1(s)\ \tilde{f}_2(-s)ds$$

$$\text{Br} \equiv \text{Bromwich contour in strip of convergence in s plane} \tag{2.4}$$

The strip of convergence is to the right of singularities of $\tilde{f}_1(s)$ and to the left of singularities of $\tilde{f}_2(-s)$ provided $\tilde{f}_1(t)$ and $\tilde{f}_2(t)$ are zero before some t (a real variable, taken here as time). Setting $\tilde{f}_1(t)$ and $\tilde{f}_2(t)$ equal to $f(t)$ gives

$$\int_{-\infty}^{\infty} f^2(t)dt = \frac{1}{2\pi j}\int_{Br} \tilde{f}(s)\tilde{f}(-s)ds \tag{2.5}$$

which is equivalent to (2.1) if the Bromwich contour is taken as the $j\omega$ axis. As used in [4] this form involving the integral of an analytic Kernel (specifically no conjugation) along a contour in the complex frequency plane allows one to deform the contour for convenience of evaluation of the integral.

3. ENERGY NORM

Suppose now that we interpret $f_1(t)$ as voltage and $f_2(t)$ as current at some port in a system of interest. With sign convention giving positive energy into the port and assuming a passive system with no initially stored energy we have

$$\int_{-\infty}^{t} V(t')I(t')dt' \geq 0 \text{ for all t} \tag{3.1}$$

as discussed in [2]. Applying (2.4) we have

$$\int_{-\infty}^{t} V(t)I(t)dt = \frac{1}{2\pi j}\int_{Br} \tilde{V}(s)\,\tilde{I}(-s) = \frac{1}{2\pi j}\int_{Br} \tilde{V}(-s)\,\tilde{I}(s)ds$$

$$= \frac{1}{2\pi}\int_{-\infty}^{\infty} \tilde{V}(j\omega)\,\tilde{I}(-j\omega)d\omega = \frac{1}{2\pi j}\int_{-\infty}^{\infty} \mathrm{Re}\left[\tilde{V}(j\omega)\,\tilde{I}(-j\omega)\right]d\omega$$

$$= \frac{1}{2\pi}\int_{-\infty}^{\infty} \tilde{V}(-j\omega)\,\tilde{I}(j\omega)d\omega \qquad (3.2)$$

giving several alternate forms for evaluation in frequency domain.

Assuming that the port of interest is linear and time invariant, we can characterize it by an impedance or admittance

$$\tilde{V}(s) = \tilde{Z}(s)\,\tilde{I}(s),\ \tilde{I}(s) = \tilde{Y}(s)\,\tilde{V}(s),\ \tilde{Z}(s)\tilde{Y}(s) = 1 \qquad (3.3)$$

which in time domain is

$$V(t) = Z(t)\ o\ I(t),\ I(t) = Y(t)\ o\ V(t),\ Z(t)\ o = [Y(t)\ o]^{-1},\ [Y(t)\ o]^{-1} = Z(t)\ o$$

$$Z(t)\ o\ Y(t)\ o = Y(t)\ o\ Z(t)\ o\ = \delta(t)\ o,\ o \equiv \text{convolution} \qquad (3.4)$$

The requirement that the port be passive for every frequency leads to the concept of a p.r. function [7] which has properties

$$\tilde{Z}(s^*) = \tilde{Z}^*(s),\quad \tilde{Y}(s^*) = \tilde{Y}^*(s)$$

(conjugate symmetry implying real values for real s) (3.5)

$$\mathrm{Re}[\tilde{Z}(s)] \geq 0,\ \mathrm{Re}[\tilde{Y}(s)] \geq 0 \text{ for } \mathrm{Re}[s] \geq 0$$

Such functions have the properties that there are no singularities in the right half plane (RHP), poles on the $j\omega$ axis must be simple with positive real residues, and that as s $\rightarrow\ \infty$ in the RHP they behave as s^{ν} times a positive real constant with $-1 \leq \nu \leq 1$ (and for lumped element networks ν = -1, 0, 1).

So the integrals in (3.2) can be restated in terms of either voltage or current with admittance or impedance as a kind of weighting function. Stating this in terms of a norm (the energy norm, or e-norm) we have

$$\|V(t), Y(t)\ o\ \|_e \equiv \left\{\int_{-\infty}^{\infty} V(t)[Y(t)\ o\ V(t)]dt\right\}^{\frac{1}{2}}$$

$$= \|I(t), Z(t)\ o\ \|_e \equiv \left\{\int_{-\infty}^{\infty} I(t)\ [Z(t)\ o\ I(t)]dt\right\}^{\frac{1}{2}}$$

$$= \left\{\int_{-\infty}^{\infty} V(t)\ I(t)dt\right\}^{\frac{1}{2}} \qquad (3.6)$$

Notationally we can think of this as a weighted norm with the convolution operator after the comma. Note that it is like a 2-norm in that V(t) and I(t) appear quadratically in the integrand and the square root is taken of the integral. Note that if the operator were $\delta(t)o$ (the identity) this would be precisely a 2-norm.

To see that this is a norm go to frequency domain as in (3.2) giving

$$\begin{aligned}\int_{-\infty}^{\infty} V(t)I(t)dt &= \frac{1}{2\pi j}\int_{Br} \tilde{V}(s)\ \tilde{I}(-s)ds = \frac{1}{2\pi j}\int_{Br} \tilde{V}(-s)\ \tilde{I}(s)\\ &= \frac{1}{2\pi j}\int_{Br} \tilde{V}(s)\ \tilde{Y}(-s)\ \tilde{V}(-s)\ ds = \frac{1}{2\pi j}\int_{Br} \tilde{V}(s)\ \tilde{Y}(s)\ \tilde{V}(-s)\ ds\\ &= \frac{1}{2\pi j}\int_{Br} \tilde{I}(s)\ \tilde{Z}(-s)\ \tilde{I}(-s)\ ds = \frac{1}{2\pi j}\int_{Br} \tilde{I}(s)\ \tilde{Z}(s)\ \tilde{I}(-s)\ ds \qquad (3.7)\end{aligned}$$

Note now that the convolution in time domain has been replaced by multiplication in frequency domain. Further specializing to the $j\omega$ axis we have

$$\begin{aligned}\int_{-\infty}^{\infty} V(t)I(t)dt &= \frac{1}{2\pi}\int_{-\infty}^{\infty} |\tilde{V}(j\omega)|^2\ \mathrm{Re}[\tilde{Y}(j\omega)]d\omega\\ &= \frac{1}{2\pi}\int_{-\infty}^{\infty} |\tilde{I}(j\omega)|^2\ \mathrm{Re}[\tilde{Z}(j\omega)]d\omega \qquad (3.8)\end{aligned}$$

Note the use of the real part of the admittance and impedance since the integral must be real valued (as on the left). Now return to (3.5) showing that these real parts must be non-negative. Restricting these functions to have only isolated zeros of the real parts on the $j\omega$ axis then we can say that

$$\begin{aligned}\left\|\tilde{V}(j\omega),\tilde{Y}(j\omega)\right\|_e &\equiv \left\{\int_{-\infty}^{\infty} |\tilde{V}(j\omega)|^2\ \mathrm{Re}[\tilde{Y}(j\omega)]d\omega\right\}^{\frac{1}{2}}\\ &= 0 \text{ iff } \tilde{V}(j\omega)\equiv 0\end{aligned}$$

(except at isolated ωs) provided Re $[\tilde{Y}(j\omega)] > 0$ (except at isolated ωs)

$$\begin{aligned}\left\|\tilde{I}(j\omega),\tilde{Z}(j\omega)\right\|_e &\equiv \left\{\int_{-\infty}^{\infty} |\tilde{I}(j\omega)|^2\ \mathrm{Re}[\tilde{Z}(j\omega)]d\omega\right\}^{\frac{1}{2}}\\ &= 0 \text{ iff } V(j\omega) = 0\end{aligned}$$

(except at isolated ωs) provided Re $[\tilde{Z}(j\omega)] > 0$ (except at isolated ωs) (3.9)

Thus provided the weight is positive (except at isolated ωs) we can take this as a norm, specifically a weighted 2-norm. In fact, we have the alternate representations

$$\begin{aligned}\left\|\tilde{V}(j\omega),\ \tilde{Y}(j\omega)\right\|_e &= \left\|\tilde{V}(j\omega)\left[\mathrm{Re}[\tilde{Y}(j\omega)]^{\frac{1}{2}}\right]\right\|_2\\ \left\|\tilde{I}(j\omega),\ \tilde{Z}(j\omega)\right\|_e &= \left\|\tilde{I}(j\omega)\left[\mathrm{Re}[\tilde{Z}(j\omega)]^{\frac{1}{2}}\right]\right\|_2 \qquad (3.10)\end{aligned}$$

Since clearly scalars come out of a weighted 2-norm as magnitude, this leaves only the triangle inequality to establish the norm property. This follows in voltage form as

$$\begin{aligned}\left\|\tilde{V}_2(j\omega)+\tilde{V}_2(j\omega),\tilde{Y}(j\omega)\right\|_e &= \left\|\tilde{V}_1(j\omega)[\mathrm{Re}\,[\tilde{Y}(j\omega)]]^{\frac{1}{2}}+\tilde{V}_2(j\omega)[\mathrm{Re}\,[\tilde{Y}(j\omega)]]^{\frac{1}{2}}\right\|_2\\ &\le \left\|\tilde{V}_1(j\omega)[\mathrm{Re}\,[\tilde{Y}(j\omega)]]^{\frac{1}{2}}\right\|_2+\left\|\tilde{V}_2(j\omega)[\mathrm{Re}\,[\tilde{Y}(j\omega)]]^{\frac{1}{2}}\right\|_2,\\ &= \left\|\tilde{V}_1(j\omega),\ \tilde{Y}(j\omega)\right\|_e+\left\|\tilde{V}_2(j\omega),\tilde{Y}(j\omega)\right\|_e \end{aligned} \quad (3.11)$$

and similarly for the current form. Thus the e-norm satisfies all the properties of a norm with suitable restriction on $\tilde{Y}(s)$ and $\tilde{Z}(s)$. While our discussion of the norm properties has been in the frequency domain form, since by (3.6) and (3.8) there is a strict equality with the time domain form as

$$\begin{aligned}\|V(t),Y(t)\ o\|_e &= \frac{1}{\sqrt{2\pi}}\left\|\tilde{V}(j\omega),\tilde{Y}(j\omega)\right\|_e\\ \|I(t),Z(t)\ o\|_e &= \frac{1}{\sqrt{2\pi}}\left\|\tilde{I}(j\omega),\tilde{Z}(j\omega)\right\|_e\end{aligned} \quad (3.12)$$

then all the norm properties apply to the time-domain form as well.

4. BOUNDS ON E-NORM IN TERMS OF 2-NORM

As in [5] we have the Hölder inequality now applied to functions of ω over the range $-\infty$ to ∞ as

$$\begin{aligned}&\left|\int_{-\infty}^{\infty} g_1(\omega)\ g_2(\omega)d\omega\right| \le \|g_1(\omega)\|_{p_1}\ \|g_2(\omega)\|_{p_2}\\ &1=\frac{1}{p_1}+\frac{1}{p_2},\ p_1\ge 1, p_2\ge 1\end{aligned} \quad (4.1)$$

where limiting cases of 1 and ∞ for p are permissible. For the above formula ω can just as well be t.

Consider first the integrals involving V and I as in (3.2). Interpreting p_1 and p_2 as both 2 we have

$$\begin{aligned}0 &\le \int_{-\infty}^{\infty} V(t)\ I(t)dt \le \|V(t)\|_2\ \|I(t)\|_2\\ 0 &\le \int_{-\infty}^{\infty} \tilde{V}(j\omega)\ \tilde{I}(-j\omega)d\omega \le \|\tilde{V}(j\omega)\|_2\ \|\tilde{I}(j\omega)\|_2\end{aligned} \quad (4.2)$$

Stated in norm form

$$\begin{aligned}\|V(t),\ Y(t)\ o\|_e &= \|I(t),Z(t)\ o\|_e \le \{\|V(t)\|_2\ \|I(t)\|_2\}^{\frac{1}{2}}\\ \|\tilde{V}(j\omega),\ \tilde{Y}(j\omega)\|_e &= \|\tilde{I}(j\omega),\tilde{Z}(j\omega)\|_e \le \left\{\|\tilde{V}(j\omega)\|_2\|\tilde{I}(j\omega)\|_2\right\}^{\frac{1}{2}}\end{aligned} \quad (4.3)$$

indicating that the e-norm is bounded by the geometric mean of the 2-norms of the voltage and current. As in (3.12) the frequency and time forms above are all simply related.

Now look at the integrals as in (3.9). We have the simple bounds

$$\|\tilde{V}(j\omega),\ \tilde{Y}(j\omega)\|_e \ \leq\ \|\tilde{V}(j\omega)\|_2 \left\{ \sup_{\omega} \ \text{Re}\,[\tilde{Y}(j\omega)] \right\}^{\frac{1}{2}}$$

$$\|\tilde{I}(j\omega),\ \tilde{Y}(j\omega)\|_e \ \leq\ \|\tilde{I}(j\omega)\|_2 \left\{ \sup_{\omega} \ \text{Re}\,[\tilde{Z}(j\omega)] \right\}^{\frac{1}{2}} \tag{4.4}$$

provided the real parts of the admittance and/or impedance are bounded on the jω axis.

5. CHOOSING INCIDENT SPECTRUM TO BOUND e-NORM AND 2-NORM OF RESPONSE

Let there be transfer functions $\tilde{T}_V(s)$ for voltage and $\tilde{T}_I(s)$ for current from some incident or source waveform $F^{(inc)}(t)$. this incident waveform may represent an incident field (electric or magnetic), a voltage or current source in the system, etc. Then we have norms

$$\|V(t)\|_2 \ =\ \frac{1}{\sqrt{2\pi}}\|\ \tilde{V}(j\omega)\|_2 \ =\ \frac{1}{\sqrt{2\pi}}\|\ \tilde{T}_V(j\omega)\ \tilde{F}^{(inc)}(j\omega)\|_2$$

$$\|I(t)\|_2 \ =\ \frac{1}{\sqrt{2\pi}}\|\ \tilde{I}(j\omega)\|_2 \ =\ \frac{1}{\sqrt{2\pi}}\|\ \tilde{T}_I(j\omega)\ \tilde{F}^{(inc)}(j\omega)\|_2$$

$$\begin{aligned}
\|V(t),\ Y(t)\ o\|_e \ &=\ \|I(t),\ Z(t)\ o\|_e \ =\ \left\{\frac{1}{2\pi}\int_{-\infty}^{\infty} \tilde{V}(j\omega)\ \tilde{I}(-j\omega)d\omega\right\}^{\frac{1}{2}} \\
&=\ \left\{\frac{1}{2\pi}\int_{-\infty}^{\infty} \tilde{T}_V(j\omega)\ \tilde{T}_I(-j\omega)|\ \tilde{F}^{(inc)}(j\omega)|^2\ d\omega\right\}^{\frac{1}{2}} \\
&=\ \left\{\frac{1}{2\pi}\int_{-\infty}^{\infty} |\tilde{T}_V(j\omega)\ \tilde{F}^{(inc)}(j\omega)|^2\ \text{Re}\,[\tilde{Y}(j\omega)]\ d\omega\right\}^{\frac{1}{2}} \\
&=\ \left\{\frac{1}{2\pi}\int_{-\infty}^{\infty} |\tilde{T}_I(j\omega)\ \tilde{F}^{(inc)}(j\omega)|^2\ \text{Re}\,[\tilde{Z}(j\omega)]\ d\omega\right\}^{\frac{1}{2}}
\end{aligned} \tag{5.1}$$

Next consider some other waveform F(t) which is directly compared to $F^{(inc)}(t)$ in that it represents the same physical parameter with the only difference being the temporal or frequency dependence. Specifically, we require that the transfer functions remain unchanged. Suppose we constrain

$$|\tilde{F}(j\omega)| \geq \left|\tilde{F}^{(inc)}(j\omega)\right| \ \text{for all}\ \omega \tag{5.2}$$

Then we have the norm inequalities

$$\left\|\tilde{T}_V(j\omega)\ \tilde{F}(j\omega)\right\|_2 \ \geq\ \left\|\tilde{T}_V(j\omega)\ \tilde{F}^{(inc)}(j\omega)\right\|_2,$$

$$\left\|\tilde{T}_I(j\omega)\ \tilde{F}(j\omega)\right\|_2 \ \geq\ \left\|\tilde{T}_I(j\omega)\ \tilde{F}^{(inc)}(j\omega)\right\|_2$$

$$\left\|\tilde{T}_V(j\omega)\ \tilde{F}(j\omega),\ \tilde{Y}(j\omega)\right\|_e \ \geq\ \left\|\tilde{T}_V(j\omega)\ \tilde{F}^{(inc)}(j\omega),\ \tilde{Y}(j\omega)\right\|_e$$

$$\left\|\tilde{T}_I(j\omega)\ \tilde{F}(j\omega),\ \tilde{Z}(j\omega)\right\|_e \ \geq\ \left\|\tilde{T}_I(j\omega)\ \tilde{F}^{(inc)}(j\omega),\ \tilde{Z}(j\omega)\right\|_e \tag{5.3}$$

These are tight bounds in that equality is achieved in some cases. Specifically if equality holds for some ω in (5.2), then if the transfer function is zero for frequencies except very near this ω the inequalities in (5.3) become equalities.

This concept can be extended to a set of incident waveforms $F_n^{(inc)}$ for n=1,2,...,N. Require

$$\left|\tilde{F}(j\omega)\right| \geq \max_{1\leq n\leq N} \left|\tilde{F}_n^{(inc)}(j\omega)\right| \ \text{for all}\ \omega \tag{5.4}$$

Then the inequalities in (5.3) apply for all $F_n^{(inc)}$. At least in 2-norm and e-norm senses, then this provides a way for establishing some criteria waveform F(t) which covers all of some set of environmental waveforms $F^{(inc)}(t)$. The bound must hold for all frequencies if it is to apply to a generally unknown transfer function which can emphasize any frequencies al all.

6. NORMS FOR RESPONSE TO CANONICAL INCIDENT WAVES

Consider that our system of interest is excited by an incident plane wave of the form

$$\vec{E}^{(inc)}(\vec{r},t) = E_0\ \vec{1}_e\ f^{(inc)}\left(t - \frac{\vec{1}_1\cdot\vec{r}}{c}\right)$$

$$\vec{H}^{(inc)}(\vec{r},t) = H_0\ \vec{1}_h\ f^{(inc)}\left(t - \frac{\vec{1}_1\cdot\vec{r}}{c}\right)$$

$$E_0 = Z_0 H_0$$

$$Z_0\sqrt{\frac{\mu_0}{\varepsilon_0}} \equiv \text{characteristic impedance of free space}$$

$$c = \frac{1}{\sqrt{\mu_0\varepsilon_0}} \equiv \text{speed of light}$$

$$\vec{1}_1 \equiv \text{direction of incidence}$$

$$\vec{1}_e \equiv \text{polarization (assumed constant)}$$

$$\vec{1}_1\times\vec{1}_e = \vec{1}_h,\ \vec{1}_e\times\vec{1}_h = \vec{1}_1,\ \vec{1}_h\times\vec{1}_1 = \vec{1}_e \tag{6.1}$$

the transfer functions discussed in the previous section are now taken with respect to the above excitation such that we have

$$\tilde{V}(s) \equiv \tilde{T}_V(s)\ E_0\ \tilde{f}^{(inc)}(s),\ \ \tilde{I}(s) \equiv \tilde{T}_I V(s)\ H_0\ \tilde{f}^{(inc)}(s) \tag{6.2}$$

In this form the transfer functions both have dimensions of meters and are functions of not only s, but $\vec{1}_1$ and $\vec{1}_e$ (or equivalently $\vec{1}_1$ and $\vec{1}_h$) as well. If desired, one can think of these transfer functions as the response to an incident field of 1 V/m or 1 A/m.

Looking at the energy norm we can define

$$\text{Re}[\tilde{A}_a(j\omega)] \equiv \text{Re}[\tilde{T}_V(j\omega)\tilde{T}_I(-j\omega)] = \text{Re}[\tilde{T}_V(-j\omega)\tilde{T}_I(j\omega)],\ \equiv \text{absorption area or cross section} \tag{6.3}$$

Note that one can also define a complex version of this if one wishes as

$$\tilde{A}_a(s) \equiv \tilde{T}_V(s)\ \tilde{T}_I(-s)\ = Z_0\ \tilde{Y}(-s)\ \tilde{T}_V(s)\tilde{T}_V(-s)\ = \frac{\tilde{Z}(s)}{Z_0}\ \tilde{T}_I(s)\ \tilde{T}_I(-s) \tag{6.4}$$

with the real part on the $j\omega$ axis corresponding to (6.3). Note that one could equally define $\tilde{A}_a$(s) with all signs on s reversed in (6.4). Now since our port is assumed passive we have

$$\text{Re}\ [\tilde{A}_a(j\omega)]\ \geq\ 0 \text{ for all } \omega \tag{6.5}$$

and unless the port is purely reactive this will, in general, be positive. Applying this to the energy norm we have

$$\begin{aligned} \|V(t), Y(t)\ o\|_e\ &=\ \|I(t),\ Z(t)\ 0\|_e \\ &= \left\{\frac{E_0 H_0}{2\pi}\int_{-\infty}^{\infty} \tilde{A}_a(j\omega)\left|\tilde{f}^{(inc)}(j\omega)\right|^2 d\omega\right\}^{\frac{1}{2}} \\ &= \left\{\frac{E_0 H_0}{2\pi}\int_{-\infty}^{\infty} \text{Re}[\tilde{A}_a(j\omega)]\left|\tilde{f}^{(inc)}(j\omega)\right|^2 d\omega\right\}^{\frac{1}{2}} \end{aligned} \tag{6.6}$$

Let our incident wave be a step function as

$$\vec{E}_s^{(inc)}(\vec{r},t) = E_0\vec{1}_e\ u\left(t - \frac{\vec{1}_1\cdot\vec{r}}{c}\right),\ \vec{H}_s^{(inc)}(\vec{r},t) = H_0\vec{1}_h\ u\left(t - \frac{\vec{1}_1\cdot\vec{r}}{c}\right) \tag{6.7}$$

Note that this wave has a constant energy density after the wavefront as

$$U = \frac{1}{2}\ \varepsilon_0\ E_0^2\ +\ \frac{1}{2}\ \mu_0\ H_0^2\ =\ \frac{E_0 H_0}{c} \tag{6.8}$$

For the step response (6.6) becomes

$$\|V_s(t),\ Y(t)\ o\|_e = \|I_s(t),\ Z(t)\ o\|_e\ = \left\{\frac{E_0 H_0}{2\pi}\int_{-\infty}^{\infty} \tilde{A}_a(j\omega)\omega^{-2}d\omega\right\}^{\frac{1}{2}}\ = U^{\frac{1}{2}}\ V_e^{\frac{1}{2}}$$

$$V_e = \frac{c}{2\pi}\int_{-\infty}^{\infty}\tilde{A}_a(j\omega)\omega^{-2}d\omega = \frac{c}{\pi}\int_0^{\infty}\text{Re}[A_a(j\omega)]\omega^{-2}d\omega \equiv \text{effective volume} \tag{6.9}$$

This effective volume is a function of $\vec{1}_1$ and $\vec{1}_e$ and relates how much of the electromagnetic-field energy is delivered to the port. This is similar to the equivalent volume of a sensor [9] except that this latter parameter is based on the quasi-static characteristics of an optimally designed sensor. Here V_e is somewhat more general. Note that by changing ω to wavelength the integral in (6.9) is an integral over wavelength similar to the formula for the total cross section $\mathrm{Re}[\tilde{A}_t(j\omega)]$ in [8]. Of course, since

$$\mathrm{Re}\,[\tilde{A}_t(j\omega)] \geq \mathrm{Re}[\tilde{A}_a(j\omega)] \geq 0 \tag{6.10}$$

then such an integral can bound V_e. However, note that here V_e is but one port of perhaps many in a system and A_a then may only account for a fraction of the absorbed power.

For step-function excitation we also have the 2-norms

$$\|V_s(t)\|_2 = \frac{E_0}{\sqrt{2\pi}} \left\| \frac{\tilde{T}_V(j\omega)}{j\omega} \right\|_2 , \quad \|I_s(t)\|_2 = \frac{H_0}{\sqrt{2\pi}} \left\| \frac{\tilde{T}_V(j\omega)}{j\omega} \right\|_2 \tag{6.11}$$

Using the bound in (4.3) we have

$$0 \leq V_e \leq \frac{c}{2\pi} \left\| \frac{\tilde{T}_V(j\omega)}{j\omega} \right\|_2 \left\| \frac{\tilde{T}_I(j\omega)}{j\omega} \right\|_2 \tag{6.12}$$

Note that if the port has a constant resistance, say R, for $\tilde{Z}(s)$, then $\tilde{T}_V$ and $\tilde{T}_I$ are the same except for a real constant multiplier simplifying the above results and making (6.12) an equality.

One reason for considering a step-function response is its utility for bounding some other common responses. One often considers canonical waveforms such as [8]

$$f_1(t) = \left[e^{-\alpha t} + e^{-\beta t}\right] u(t), \; f_2(t) = \left\{e^{-\alpha t} + e^{\beta t}\right\}^{-1}, \; \alpha > 0, \; \beta > 0 \tag{6.13}$$

with two-sided Laplace transforms

$$\begin{aligned} \tilde{f}_1(s) &= \frac{1}{s+\alpha} - \frac{1}{s+\beta} = \frac{\beta - \alpha}{(s+\alpha)(s+\beta)} \\ \tilde{f}_s(s) &= \frac{\pi}{\alpha+\beta} \csc\left[\frac{\pi}{\alpha+\beta}(s+\beta)\right] \end{aligned} \tag{6.14}$$

Here α^{-1} is a characteristic time for the rise and β^{-1} is a characteristic time for the fall. We have in time domain

$$0 \leq f_1(t) < 1, \; 0 < f_2(t) < 1 \text{ for all } t \tag{6.15}$$

Compared to the unit step the peak is less, but this is not the important point. On the $j\omega$ axis we have (without going into details)

$$\left|\tilde{f}_1(j\omega)\right| < \left|\frac{1}{j\omega}\right|, \; \left|\tilde{f}_2(j\omega)\right| < \left|\frac{1}{j\omega}\right| \text{ for all } \omega \tag{6.16}$$

Applying (5.2) we then have that the e-norms and 2-norms for response to f_1 and f_2 are bounded by the corresponding step response norms. Thus the results of (6.9) in terms of effective volume bound the response to waveforms in (6.13) for all positive α and β.

Besides a step response which essentially weights the transfer functions by $1/(j\omega)$ one might define other canonical examples which weight the frequencies differently. For example, we might take an impulsive wave as

$$\vec{E}_\delta^{(inc)}(\vec{r},t) = E_0\ \tau_1\ \vec{1}_e\ \delta\left(t - \frac{\vec{1}_1\cdot\vec{r}}{c}\right),\ \vec{H}_\delta^{(inc)}(\vec{r},t) = H_0\ \tau_1\ \vec{1}_h\ \delta\left(t - \frac{\vec{1}_1\cdot\vec{r}}{c}\right) \tag{6.17}$$

where τ_1 is a time to account for the area of the pulse. Then we have the norms of the δ response as

$$\begin{aligned}
\|V_\delta(t), Y(t)\ o\ \|_e &= \|I_\delta(t), Y(t)\ o\ \|_e \\
&= \tau_1\left\{\frac{E_0H_0}{2\pi}\int_{-\infty}^{\infty}\tilde{T}_V(j\omega)\ T_I(-j\omega)\ d\omega\right\}^{\frac{1}{2}} \\
\|V_\delta(t)\|_2 &= \frac{E_0\tau_1}{\sqrt{2\pi}}\left\|\tilde{T}_V(j\omega)\right\|_2,\ \|I_\delta(t)\|_2 = \frac{H_0\tau_1}{\sqrt{2\pi}}\left\|\tilde{T}_I(j\omega)\right\|_2
\end{aligned} \tag{6.18}$$

Compared to the step response this emphasizes the high frequencies. One will need the transfer functions to be small enough at high frequencies that the above integrals converge.

Another choice is a ramp wave as

$$\begin{aligned}
\vec{E}_r^{(inc)}(\vec{r},t) &= \frac{E_0}{\tau_2}\ \vec{1}_e\left[t - \frac{\vec{1}_1\cdot\vec{r}}{c}\right]\ u\left(t - \frac{\vec{1}_1\cdot\vec{r}}{c}\right) \\
\vec{H}_r^{(inc)}(\vec{r},t) &= \frac{H_0}{\tau_2}\ \vec{1}_h\left[t - \frac{\vec{1}_1\cdot\vec{r}}{c}\right]\ u\left(t - \frac{\vec{1}_1\cdot\vec{r}}{c}\right)
\end{aligned} \tag{6.19}$$

where τ_2 is a time to account for the slope of the waveform. The norms of the ramp response are then

$$\begin{aligned}
\|V_r(t),\ Y(t)\ o\|_e &= \|I_r(t),\ Z(t)\ o\|_e \\
&= \frac{1}{\tau_2}\left\{\frac{E_0H_0}{2\pi}\int_{-\infty}^{\infty}\tilde{T}_V(j\omega)\ \tilde{T}_I(-j\omega)d\omega\right\}^{\frac{1}{2}} \\
\|V_r(t)\|_2 &= \frac{E_0}{\tau_2\sqrt{2\pi}}\left\|\frac{\tilde{T}_V(j\omega)}{(j\omega)^2}\right\|_2,\ \|I_r(t)\|_2 = \frac{H_0}{\tau_2\sqrt{2\pi}}\left\|\frac{\tilde{T}_I(j\omega)}{(j\omega)^2}\right\|_2
\end{aligned} \tag{6.20}$$

Compared to the step response this emphasizes the low frequencies. One will need the transfer functions at low frequencies to be sufficiently small that the above integrals converge.

7. REFERENCES

1. C.E. Baum, Black Box Bounds, Interaction Note 429, May 1983, and Proc. EMC Symposium, Zurich, 1985, pp. 381-386.

2. C.E. Baum, Some Bounds Concerning the Response of Linear Systems with a Nonlinear Element, Interaction Note 438, June 1984.

3. C.E. Baum, Norm Limiters Combined with Filters, Interaction Note 456, August 1986.

4. C.E. Baum, Transfer of Norms Through Black Boxes, Interaction Note 462, October 1987, and Proc. EMC Symposium, Zurich, March 1989, pp. 157-162.

5. C.E. Baum, Norms of Time-Domain Functions and Convolution Operators, Mathematics Note 86, December 1985, and chapter 2, pp. 31-55, in H.N. Kritikos and D.L. Jaggard (eds.), Recent Advances in Electromagnetic Theory, Springer-Verlag, 1990.

6. K.S.H. Lee, Relations Between Electric and Magnetic Polarizabilities and Other Related Quantities, Radio Science, 1987, pp. 1235-1238.

7. E.A. Guillemin, The Mathematics of Circuit Analysis, M.I.T. Press, 1949.

8. K.S.H. Lee (ed.), EMP Interaction: Principles, Techniques, and Reference Dqata, Hemisphere Publishing Corp., 1986.

9. C.E. Baum, Electromagnetic Sensors and Measurement Techniques, in J.E. Thompson and L.H. Luessen (eds.), Fast Electrical and Optical Measurements, Martinus Nijhoff, Dordrecht, 1986, Vol. 1, pp. 73-144.

12 Seismoelectric Emissions

12.1 On Seismoelectromagnetic Phenomena

N.I. GERSHENZON and M.B. GOKHBERG
Institute of Physics of the Earth, USSR Academy of Science, Moscow, USSR

1. INTRODUCTION

During last dozens of years a lot of original papers and reviews on the electromagnetic phenomena associated with earthquake preparation processes and with earthquakes themselves have appeared. These phenomena can be divided in the following way:

- atmospheric luminosity in epicentral areas before, during, and after earthquakes,
- appearance of electrotelluric field anomalies,
- local changes of geomagnetic field,
- generation (or abnormal change) of the radio frequency electromagnetic emission,
- changes of upper and lower ionosphere parameters,
- generation of low frequency oscillations in the Earth magnitosphere.

Note that in spite of long-continued instrumental observations of different geoelectromagnetic phenomena the study is only starting and there are even more questions then positive statements in this field. For instance what is the origin of atmospheric luminosity? Is it electric discharge in thunderstorms or chemoluminescence of air? If it is the first one then what is the reason of the great magnitude electric field generation on the large area (from hundreds m^2 to dozens km^2). If it is the measurable gas luminosity then how does this gas appear before earthquakes?

There is another trouble. It is usually difficult to exclude the possibility of the relation between these anomalies and another phenomena (changes of hello-geophysical conditions,industry noises etc.). Essentially up to now one can not say for sure that the excitation of the earthquake preparation processes influence on the upper atmospheric parameters, though there are a lot of publications in this field.

The physical-mathematical model of the lithospheric geodynamic electromagnetic effects has been weakly developed, although the laboratory study of mechanoelectromagnetic phenomena (piezoelectric and piezomagnetic effects, electrokinetic effect, the Stepson effect, etc.) in crustal bodies have been sufficiently developed.

All this, as far as it seems to us, make the attempt to analyze this problem useful.

In present paper we are confined to brief discussion of electrotelluric field (EFT) anomalies (section 2), radio frequency electromagnetic emission (EME) (section 3), and seismoionospheric effects (section 4).

2. SEISMOELECTROTELLURIC ANOMALIES.

According to publications there are several hundreds of cases of the EFT anomalies appearance before earthquakes. Amplitudes of disturbances are usually of a few mV to a few tens mV, the duration is of few minutes to few weeks, the period of disturbances appearance before earthquake is from an hour to some weeks. The typical features: amplitude is practically independent from the measuring line length, the crustal inhomogeneities essentially influence on the effect that became apparent in "mosaic" structure of anomalies field.

We do not take into consideration the analysis of all possible physical-mathematical models (the brief analyze contains in the paper [Gershenzon & Gokhberg, 1989]) let's consider a model described more complete, to our opinion, morphological features of the EFT anomalies.

For the last stage of the earthquake preparation the crust strain-stress changes in the so-called preparation zone. It may result from the effective stress modulus change inside the focus, that should lead to stress changes in the surrounding space focus in the field of the tectonic tangential stresses. The abnormal changes of deformation and land tilts near future epicenters from some hours to weeks before the earthquakes have been sometimes registered [Rikitake, 1976]. If to assume that it is really true it is necessary to find a suitable mechanism, that transforms the mechanic energy into the the energy of the electric field. Well known mechanoelectric effects in dielectric crystals, for example piezoelectric effect, Stepson effect, and other effects observed during the deformation of dry rocks can not explain characteristic features of the EFT anomalies. One of the obstacle for it is the anomalies duration. The crust has a large conductivity due to the water intrusions inside the crust rocks. So the electric field that has the origin mentioned above, will be neutralized with conduction currents during less then 1 s while the anomalies duration is more for 2 to 6 orders. The existence of the field during long time is possible only in the presence of the continuous mechanoelectric mechanism. This mechanism may be the electrokinetic effect.

The crust contains water filling pore space, cracks and cavities. A change in cubic strain within the crust causes changes in fluid pore pressure, fluid current arises and electric current and field appear due to the electrokinetic effect. Fluid current in pore space is described by an equation of the parabolic type, i.e., relaxation time is defined by the fluid "diffusion" time and can be of some hours to some months depending on the typical dimensions and permeability in the medium.

It can be shown that electric and magnetic fields at the surface in a homogeneous half-space do not generate, regardless of the fluid pressure distribution. Non-zero field can only appear near inhomogeneities of the crust electrophysical properties. This feature of the electrokinetic ETF origin allows to explain "mosaic" anomalies field and weak dependence of signal magnitude on like length. In reality the electric potential jump appears on the inhomogeneity boundary so it does not important how far from the boundary the measuring electrodes locate (more detail about it see in papers [Gershenzon & Gokhberg, 1989; Dobrovolsky et al., 1989]. Finally, the estimations of field magnitude that depend on the earthquake magnitude and epicenter distance show that electrokinetic effect can explain observational data.

3. ELECTROMAGNETIC EMISSION

The EKE of the radio frequency sources are quite various. One of the main natural source is lighting discharges. It is difficult to avoid man-made noise: a number of radio stations, power lines, transport, industry. Nevertheless a number of publications on the field observations of EME anomalies have increased.

Disturbances have been registered within a frequency range from some kHz to MHz. Earthquakes with magnitude 5 and more have been "felt" for distance of about some hundreds of km. The signal intensity has been more than the background value from 2 to 10 times. Statistical analysis is difficult due to a small amount of cases in the vicinity of one of a measuring point.

However there are additional indications for the possibility of the EME existence of the lithosphere origin. For instance, it is generation of electromagnetic impulses during rock destruction that were registered in laboratory conditions and before rock strikes.

How it has been mentioned above we assume that the abnormal crust deformation exists in the zone of earthquake preparation. Rapid changes in the state of crustal stress and strain, assuming a hierarchical block structure in the crust, will necessarily lead to the local concentration of stresses and micro crack density increases in this area. It is known that crack appearance in practically all the dielectric materials are accompanied by impulses of the electromagnetic emission. Typical frequencies f are easy to estimate using the relation $f=v/l$, where l - is a typical crack size, v - is the velocity of the crack opening (about 10^3m/s). With $l=10^{-3}$- 1 m we obtain f=1 - 10^3kHz. It is difficult to find other processes in the crust that have the same frequencies.

The hypothesis about influence of earthquake preparation processes on the conditions of radio waves propagation in the Earth-ionosphere waveguide is less used. However indirect data indicate the possibility of this influence.

4. IONOSPHERE AND MAGNITOSPHERE DISTURBANCES

Some effects indicate the possible disturbance of the upper atmosphere parameters during strong earthquakes preparation:

- changes of conditions of the propagation radio waves propagation,
- increase of the night air glow intensity,

abnormal behavior of sporadic E_s layer ionosphere parameters,

- change of the electron density on the F layer level and on upper layers,
- appearance of low frequency oscillations in the magnitosphere.

In spite of the increase in publications understanding of

the relation between earthquakes preparation processes and ionosphere disturbances has not obtained.

In connection with this the following question is necessary. Did the
local disturbances observed before earthquakes connect with earthquake? May be the time connection of this events is casual or (what still less probable) they are not connected with themselves, but each of them is connected with the third event. As we have no possibility to give a full answer to this question, let's only note that there is a number of events (that are not connected with earthquakes) of the lithospheric influence on the ionosphere. This influence becomes apparent in those exotic forms as geographic alignment of intense auroral emissions along the coast, lithospheric inhomogeneity influence on the ionosphere current density, connection of the electron density at the level of 110 km and the depth of Moho zone etc. You can see in more detail about terrogenic effects in the ionosphere in paper [Popov et al., 1989].

The explanation of these effects as well as seismoionospheric effects has not been yet found. The attempts to explain the lithospheric-ionosphere interaction by the crust seismoelectromagnetic fields meet sufficient difficulties. It is known that the ionosphere parameters can be changed if the field of the amplitude not less 1 - 10 tv/m is "pushed through". For this it is necessary to have a large-scale (typical size about 100 km) field on 7 - 10 orders more. None of the known transformation of mechanoelectric mechanisms can create large-scale fields of such a large magnitude in the crust. Really observed fields are usually 7 and more orders less. The influence of nonstationary electromagnetic fields of the deformation origin on the ionosphere is restricted by their insignificant amplitude.

So, the direct influence of the electromagnetic fields of the lithospheric origin is excluded. However the possibility of the atmospheric electromagnetic field influence is not excluded. It is known that the electric field magnitude near the thunderstorm clouds can be essentially more than the usual atmospheric field (about 100 V/m) near the surface of the Earth. There are experimental data that confirm the influence of thunderstorm process on the ionosphere parameters. So it is necessary only to find a mechanism of earthquake preparation processes influence on the atmospheric electricity dynamics.

Thunderstorm clouds are charged by air convection streams that carry out pre-surface ions. The ions composition and concentrations are formed by lithospheric apportionments. As the change of the chemical composition and the concentration of underground water elements and gas apportionment intensity (radon, for example) occur before earthquakes then it is natural to suppose that this may influence on processes of the atmospheric electricity formation in the area of the earthquake preparation. However it is now impossible to obtain quantitative estimates of the given model.

REFERENCE

Dobrovolsky, I.P., Gershenzon, N.I., Gokhberg, M.B., 1989. Theory of electrokinetic effects occurring at the final stage in the preparation of a tectonic earthquake. Physics of the Earth and Planetary Interiors, 57: 144-156.

Gershenzon, N.I. and Gokhberg, M.B., 1989. On the origin of electrotelluric disturbances prior to earthquakes. EMC'89, Nagoya, Japan, September 8-10.

Rikitake, T., 1976. Earthquake Prediction. Elsevier, Amsterdam, 387 pp.

Popov,L.N., Krakovetzkiy, Yu.K., Gokhberg, M.B. and Pilipenko, V.A., 1989. Terrogenic effects in the ionosphere: a review. Physics of the Earth and Planetary Interiors, 57: 115-128.

12.2 Radio Emissions from Rock, Iron, and Wood and an Example of Radio Noises Associated with Earthquake

T. OGAWA[1], S. YANO[2], K. OIKE[3], and A.C. FRASER-SMITH[4]

Department of Physics, Kochi University, Kochi 780, Japan (Now at Science Laboratory International, Kochi, 780 Japan)[1], Department of Electrical Engineering, Kochi National College of Technology, Kochi, 783 Japan[2], Geophysical Institute, Kyoto University, Kyoto, 606 Japan[3], and STARLAB, Department of Electrical Engineering/SEL, Stanford University, Stanford, CA 94305-4055, USA[4]

ABSTRACT

The experiment was made with rock, iron, and wood by hitting those samples with either an iron hammer or the same kind of matter as the sample. Electric (E) and magnetic (H) fields of ELF/VLF radio noises were measured in the experiment. It was observed that the E field was emitted rather than the H field from rock and wood, while the H field was emitted rather than the E field from iron. As a related subject to this experiment an example of anomalous radio noises observed associated with earthquake is shown.

1. INTRODUCTION

This work is a continued work from the previous experiment by Ogawa et al.[1]. The experiment has been made in the electro-magnetically shielded room to demonstrate possibility of radio emission from rock and other material which were hit by either an iron hammer or the same kind of matter as the sample tested. Here the hitting means to give an instant press to the sample matter and relax instantly. This experiment has no direct connection with earthquake but this simple experiment is a clear demonstration of possibility of radio emission from various material, and may have a certain connection with earthquake-associated radio noise emission. Along this line ELF/VLF radio noises have been continuously observed with the Stanford University ELF/VLF radiometer since 1985 at the Earth Observatory, Kochi University (33°31'N, 133°24'E) together with the other seven stations over the world [2]. An example of detected anomalous radio noises on the occasion of an earthquake is shown.

2. EXPERIMENTS

The experiment has been made in the electro-magnetically shielded room of the size 3.5(W)×2.5(D)×2.5(H) m constructed in the Geophysics Laboratory, Department of Physics, Kochi University. A ball antenna [3] 11 cm in diameter and a solenoid 9 cm in diameter and 25 cm in length with 25,000 turns of Cu wire 0.3 mm in diameter were set 20 to 50 cm apart from

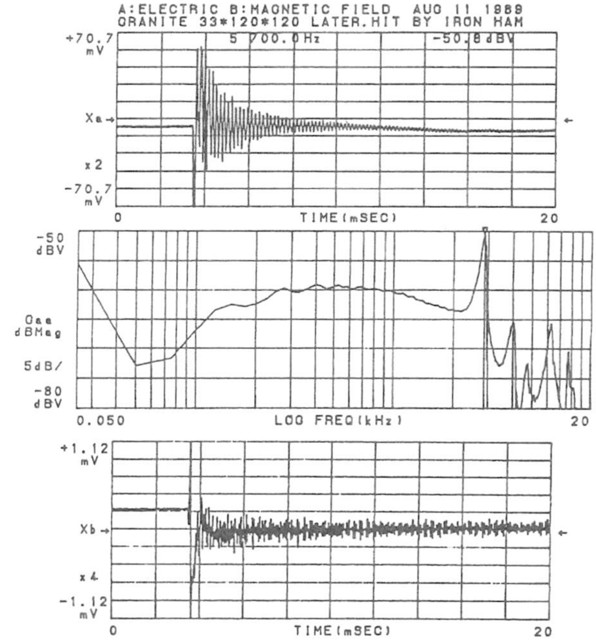

Fig. 1. First run of the experiment. The granite sample of the size of 33X120X120 mm was hung by hand and hit horizontally (latarally) by the iron hammer. E field (Xa), E field spectrum (Gaa), and H field (Xb) are shown from the top to the bottom, respectively.

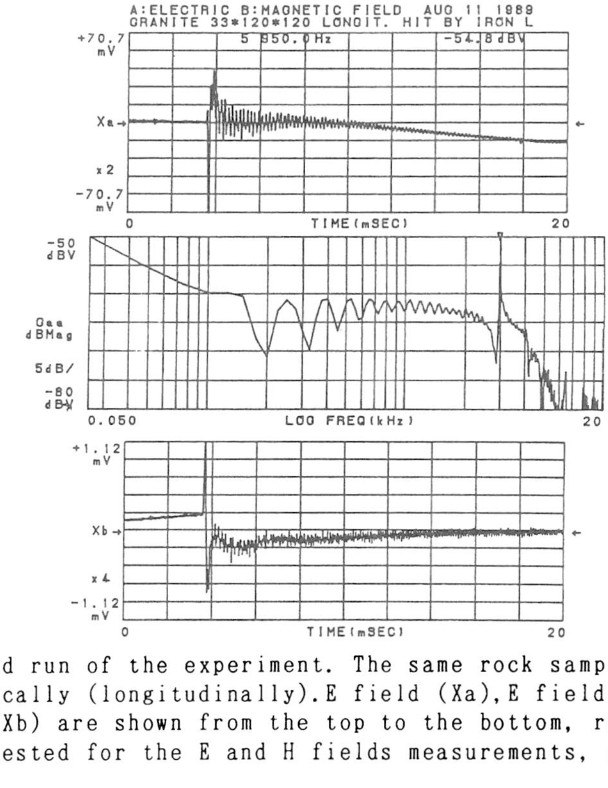

Fig. 2. Second run of the experiment. The same rock sample as in Fig. 1 was hit vertically (longitudinally). E field (Xa), E field spectrum (Gaa), and H field (Xb) are shown from the top to the bottom, respectively.
the samples tested for the E and H fields measurements, respectively.

Frequency responces of each system were 5 Hz - 20 kHz and 500 Hz - 5 kHz for the E field and H field, respectively.

The samples were hit by either an iron hammer or the same kind of matter as the samples. Emitted signals were recorded by a digital memory capable of spectral analysis. The sample material used for the experiment were rock (mainly granite), iron, and wood.

The first run of the experiment was made with an granite rock sample of the size of 33×120×120 mm. The largest plane was pollished. The sample was hung by hand and the largest plane was horizontally hit by an iron hammer. We call this situation "hit laterally". Near fields of E and H field components of the emitted noises were measured with the ball antenna and the solenoid, respectively. Figure 1 shows the E field (Xa, top panel) and H field (Xb, bottom panel) waveforms. It is noted that the H field voltage was about two orders of magnitude smaller than the E field: The recorded E field strength ranged within ±50 mV, while the H field ranged within ±0.2 mV. The E field waveform was preceeded by a large negative initial pulse followed by damping oscillations. The corresponding H field pulse was of differential form of the E pulse. See Xb upside down for the polarity identification with the E pulse. The damping oscillations of the H field was of a modulated type. The E field was sometimes followed in the later part by a large drift. The total duration of the recorded fields was as long as 40 msec, but the final portion was not shown in Fig. 1. The E field spectrum (Gaa) is also shown in the center panel in Fig. 1. The spectrum has several maxima with the largest of -50.8 dBV was observed at 5.7 kHz.

The second run of the experiment was made by hitting one edge of the same rock sample with the same iron hammer as used in the first run. We call this situation "hit longitudinally". The all experimental procedure is the same as in the first run of the experiment. The recorded E (Xa, top panel) and H (Xb, bottom panel) fields are shown together with the E field spectrum (Gaa, center panel) in Fig. 2. The H field waveform (Xb) is better to be seen upside down. The gross feature of the field pattern in the second run is the same as in the first run except that the maximum of the spectrum appeared at 5.95 kHz instead of 5.7 kHz in the first run.

The third run of the experiment was made with the iron sample of the cylindrical bar 20mm in diameter and 280 mm in length. The top of the sample was hung by hand and was hit at a middle point horizontally (laterally) by the iron hammer. The result of the experiment is shown in Fig. 3. Contrary to the granite experiments in Figs. 1 and 2, the H field is relatively larger than the E field. The H field ranged ±0.5 mV, while the E field ranged ±1 mV. The -78.5 dBV maximum of the H field spectrum appeared at 1.15 kHz. Two more peaks were seen in the spectrum. They looks the higher harmonics of the fundamental frequency.

The fourth run of the experiment was made with the same iron sample as in the third run, but the sample was hit vertically (longitudinally) from the top. The result of the experiment is shown in Fig. 4. The order of arrangement of three separate figures is the same as in Fig. 3. The -68.7 dBV peak at 9.55 kHz is prominent. This frequency was not seen in Fig.3.

The fifth and sixth run of the experiments were made with a 37 mm squared piece of timber 500 mm in length. The timber was hung by hand at its top and hit by another wood bar horizontally and vertically, respec-

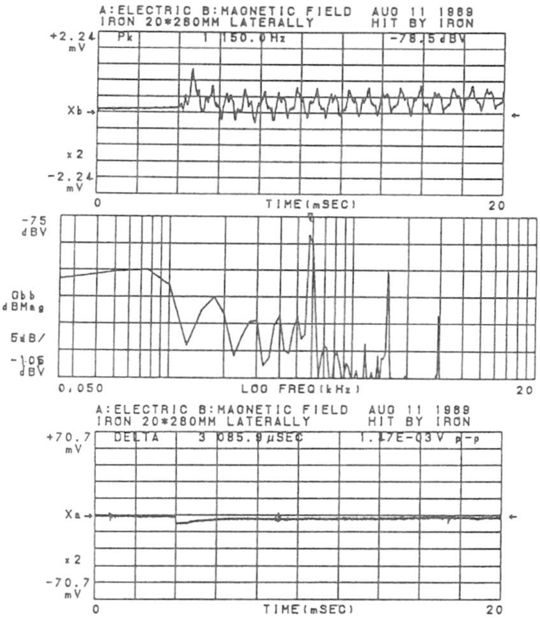

Fig. 3. Third run of the experiment. The cylindrical iron bar 20 mm in diameter and 280 mm in length was hit horizontally (laterally) by the iron hammer. H field (Xb), H field spectrum (Gbb), and E field (Xa) are shown from the top to the bottom, respectively.

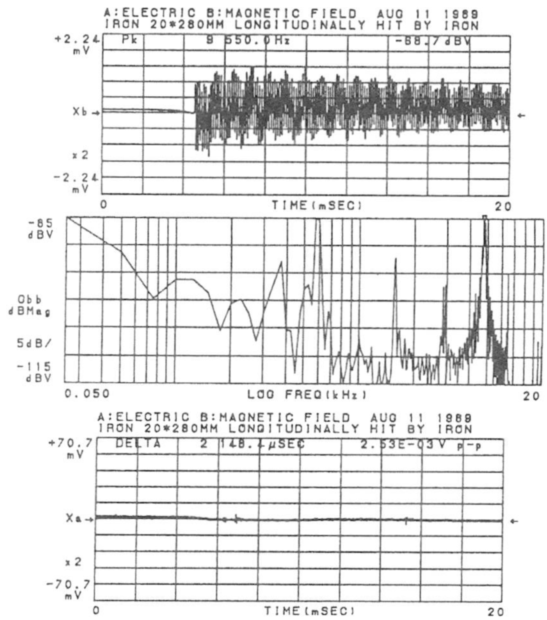

Fig. 4. Fourth run of the experiment. The same iron bar as in Fig. 1 was hit vertically (longitudinally). H field (Xb), H field spectrum (Gbb), and E field (Xa) are shown from the top to the bottom, respectively.

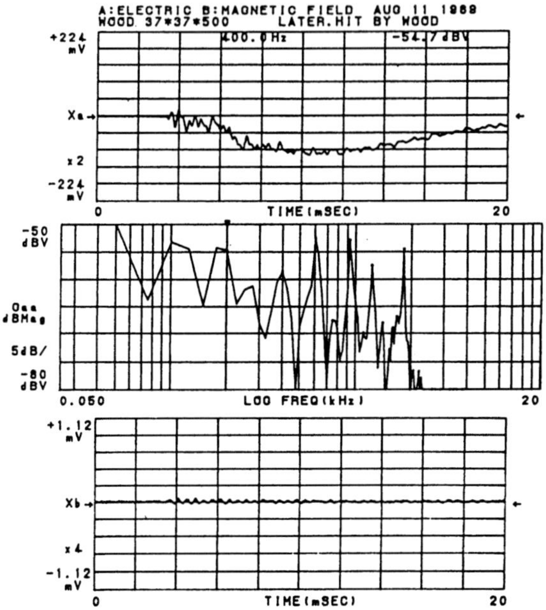

Fig. 5. Fifth run of the experiment. The 37 mm squared timber 500 mm in length was hung by hand at its top and hit horizontally (latarally) by another wood bar. E (Xa), E field spectrum (Gaa), and H field (Xb) are shown from thetop to the bottom, respectively.

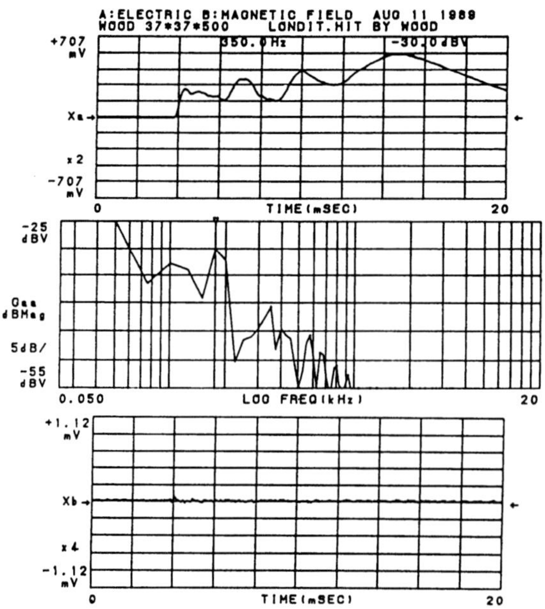

Fig. 6. Sixth run of the experiment. The same timber sample was hit vertically (longitudinally). E field (Xa), E field spectrum (Gaa), and H field (Xb) are shown from the top to the bottom, respectively.

tively. The wood experiment was done just to show possibility of emitting EM noises even from wood. The results are shown in Figs. 5 and 6. The order of arrangement of the three figures are the same as in Fig. 1 and 2 : The E field (Xa), the E field spectrum (Gaa), and the H field (Xb) are placed from the top to the bottom, respectively. We could see relatively larger E field than the H field. The common maxima of the spectra were seen at 350 ~ 400 Hz. A few other peaks of higher frequencies can also be seen.

3. RESULTS

The results of the hitting experiments with rock, iron, and wood are sammarized in the following.
(1) The E field and the H field were observed with any kinds of samples, although either the E or the H field was larger or smaller than the other.
(2) The observed waveforms from rock and iron had large positive or negative initial pulses and a succeeding damping oscillations occured.
(3) The predominant frequencies observed in the experiment depended on the kind of material tested. The frequencies of the peak field were found at 5.7 kHz ~ 5.95 kHz for granite, 1.15 kHz and 9.55 kHz for iron, and 350 ~ 400 Hz for wood.
(4) The E field dominated in rock and wood, while the H field dominated in iron.
(5) The initial pulse of H field had a differential form of the E field.
(6) The same kind of experiment produced almost the same result.

4. DISCUSSIONS AND CONCLUSION

The results of the present experiment were compared with the previous similar experiments and were discussed as follow.
(1) The EM radiations occur from shocked material.
(2) The dielectric material produces by hitting the E field rather than the H field and the magnetic material produces the H field rather than the E field.
(3) The initial pulses may be caused by contact- and tribo-electricity.
(4) The damping oscillations may be caused by piezo-electricity in the dielectric material for the E field, while they may be caused by the oscillations of magnetic dipoles in the magnetic material for the H field.

The EM radiations occur from rock samples in the frequency range, DC to some 20 MHz referring to similar experiments of fracturing various rock samples [4] [5]. Based on those experiments including the present one it is reasonable to consider that similar radiations may occur on the occasion of earthquake. Better choice of the E field or the H field for the earthquake-related measurements depends on the geological structure there. It is clear from the present experiment that if the geological structure of that earthquake area is composed of more dielectric material the E field is preferable than H field, while if it is composed of more magnetic material the H field is preferable than E field.

It is important to consider attenuation during propagation of the emitted fields for the earthquake-related problem. The DC component is

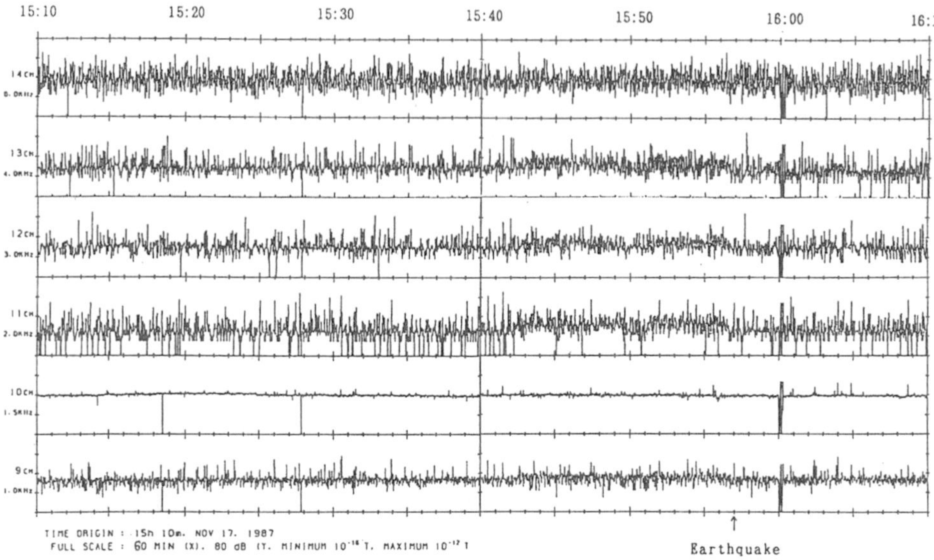

Fig. 7. An example of anomalous ELF/VLF radio noises observed with the Stanford University radiometer. The earthquake occured 200 km NW of the observatory at 15:57 on November 17, 1987. Magnitude of the earthquake was 5.4 and the focal depth was 4.0 km.

difficult to be discriminated from other natural and artificial drifting phenomena. The HF component attenuates greately during propagation through the crust. The VLF/ELF components are much covered by numerous atmospheric noises. They are, however, less attenuated during propagation. It is then appropriate to adopt ELF/VLF for the earthquake-related study. In this case the study of atmospherics waveforms is necessary.

Several examples of anomalous ELF/VLF radio noises accompanying the earthquakes have been obtained with the Stanford University radiometer. An example of these observations is shown in Fig. 7. Enhanced radio noises upto 20 dB from the normal level are seen for 15 min just before the earthquake of the magnitude (M) 5.4, and the focal depth (H) 4.0 km. It occurred about 200 km far NW from the observatory on November 17, 1987. The noises occurred in the frequency range 500 Hz to 32 kHz but a large effect was seen in the frequency range 2 kHz - 4 kHz. Such data are now accumulated and will be reported in the near future.

ACKNOWLEDGEMENT

The authors would like to thank Ken-ichi Hamamoto for his help to make Fig. 7. This work is partly supported by Hitachi Communication Systems, Inc.

REFERENCES

1. Ogawa T. Oike K. Miura T. (1985) Electromagnetic radiation from

rocks. J Geophys Res. 90:5961-5966
2. Fraser-Smith AC. Helliwell RA. (1985) The Stanford University ELF/VLF radiometer project: Measurement of the grobal distribution of ELF/VLF electro-magnetic noise. IEEE EMC-85 Int Sympo.
3. Ogawa T. Tanaka Y. Miura T. Yasuhara M. (1966) Observations of natural ELF and VLF electromagnetic noises by using ball antennas. J Geomag Geoelectr. 443-454
4. Nitsan U. (1977) Electromagnetic emission accompanying fracture of quartz-bearing rocks. Geophys Res Lett. 4:333-336
5. Warwick JW. Stoker C. Meyer TR. (1982) Radio emission associated with rock fracture: Possible application to the great Chilean earthquake of May 22, 1960. J Geophys Res. 87:2851-2859

12.3 Radio Emission from Earthquakes

J.W. WARWICK

Radiophysics, Incorporated, 5475 Western Avenue, Boulder, CO 80301, USA

ABSTRACT

There exists an extensive world literature concerned with radio emissions from earthquakes. The great majority of reports pertain to emissions at very low frequencies, in the range below 100 khz. I will describe a different kind of emission that appears to have been associated with the great Chilean earthquake of 22 May 1960. The evidence that we recorded signals related to that quake is indirect and has elicited little response from the seismological community. The principal data and interpretation were published some time ago. Now I would like to stress the opportunity this phenomenon presents for enriching the data base relevant to earthquakes, perhaps even enabling spacecraft observations of seismological events.

I became interested in radio emission from earthquakes as a result of a sequence of events involving Japan and Tokyo some 13 years ago, but extending back more than 30 years.

I was then employed at the High Altitude Observatory, at the time of the International Geophysical Year when we took on the task of implementing a world-wide network of devices that could monitor solar flares through their production of ionization in the lower edge of the D-region of Earth's ionosphere. We measured this by its associated absorptivity effects on cosmic noise at 18 MHz.

In addition, we built supporting equipment in the form of a radio interferometer, also at 18 MHz, and with a similar vertically-oriented beam as the absorption equipment.

In 1962 I summarized data observed with this and some other similar equipment that related to the question of radio phenomena produced by Earth both in its natural and non-natural (that is, man-made) modes. These were published by me at that time and included one particular event that is my subject today: a strange emission observed at no less than four stations distributed on the dimensions of a continent or larger. The network comprised Michigan, Colorado, New Mexico, and Hawaii, at each of which was recorded 18 MHz signals that were noisy in essentially identical patterns on this continental scale.

In addition, in Boulder this same pattern was recorded by the phase-switching interferometer in its average power mode, but not at all in its interferometer mode.

Finally, an identical (scaled) interferometer was operated in Boulder at 36 MHz, at which no signals were recorded either in average power or in interferometer mode.

All of this equipment used a frequency-sweeping, asymmetric time-constant device to reduce or eliminate communications (narrow-band) interference at both 18 and 36 MHz. The measure of its success is that the equipment easily detected radio star transits through the interferometer system, the galactic noise background passing through the beam, and Jupiter decametric emissions, including during the hours surrounding the event.

It also detects solar radio noise although for the times of this set of records there appeared to have been none; this is confirmed by the fact that the signal observed in Boulder's 18 MHz interferometer showed no fringes. For Sun, which of course had set in Boulder at this time, the absence of fringes is a very strong statement insofar as the fringes in the azimuth from which solar signals would have arrived as over-the-horizon events are widely-spaced and should have been recorded as strong phase-sensitive events.

In my first publication about this event I therefore indulged in some arm-waving, to the effect that our source had to hang over <u>all four stations like a cloud, fluctuating in intensity virtually simultaneously at all four</u>.

This was, and is, an interpretation presented from desperation: the event is real, it is noisy (not monochromatic), it is enormous, and it must be dealt with one way or another. That is, the event will simply not go away as a phenomenological problem.

Having published it, I forgot about it.

Until, that is, Peter Sturrock entered the scene in his inimitable way, carrying the banner of rationality toward the goal of understanding UFOs (unidentified flying objects). How better, he must have understood, to find reliable data on these enigmatic phenomena than to ask those whose careers are based on examining, and understanding, what they see in the sky. These persons (not meteorologists, since he assumed that clouds would block their view) are usually called astronomers. So, he sent out a questionnaire to the members of the American Astronomical Society asking them for events they may have seen, but not reported, in the normal course of their research activities.

My memory, thus stirred, produced the event I had studied some 15 years previously. I dutifully reported it to him and again forgot about it.

About then, in 1976 or thereabouts, an associate of mine in Boulder, Carl Kisslinger, returned from a Tokyo meeting where Chinese, Japanese, and American scientists gathered to discuss their ability to forecast the occurrence of earthquakes.

The Chinese scientists had recently successfully predicted an enormous earthquake, with the saving of many thousands of lives. But their techniques were, to me, exotic: level of water in wells and behavior of barnyard animals, for example.

If they were exotic, yet successful, I reasoned, there might be a case to be made for a seismological explanation of "my" event, so I posed the question to Kisslinger.

He was not discouraging to me, although he knew of no association between seismic disturbances and radio emissions. For example, he reminded me that piezoelectric crystals (perhaps quartz) were an abundant fraction of common rocks and that seismic strains could thereby generate electric fields.

Where, I asked, might I find data relating to earthquakes? He suggested I call the United States Geological Survey offices in Golden, Colorado.

I did so, and was honored to be helped by the keeper of the earthquake records, Waverly Person, himself (!).

He asked when my event had occurred. I said, 16 May 1960. He said he didn't need to go to the records, inasmuch as the most powerful recorded earthquake in history occurred on 22 May 1960 in southern Chile.

I was beginning to be motivated, but only became seriously turned on when my student, Carol Stoker, unearthed Nitsan's paper that had serendipitously appeared about the middle of the events I described immediately above and which discussed radio emissions from rocks fractured by hydraulic ram pressure in the laboratory.

She had read Nitsan's paper carefully and discovered its obvious weakness: he had assumed that his electric field measurements could be directly converted to the corresponding propagating field flux densities, even though they were carried out extremely close to the rock fracture.

We knew that a proper measurement would involve careful calibration and measurement with high time-resolution of both E field and B field associated with the fracture.

This we did and interpreted the data in terms of relocation of charge (electrostatic fields) and currents (magnetostatic fields). The records comprise a host of impulses on both sensors, but, aside from the storm of impulses beginning and ending simultaneously, there is little correlation between them.

Our need was for a connection between <u>these</u> data and eventual propagating waves. I assumed that acoustic velocities in rocks were kilometers per second and that these speeds measure the rate of fracturing in our granite sample.

I also assumed that the radiation mechanism was the accelerated motion of charge initially existing as polarization with the piezoelectric crystal. As the crystal fractures, a crack opens up and becomes a two-plate capacitor whose charge is measured by the electric fields and whose equivalent currents are measured by the magnetic fields. The measured time scale of the phenomenon is used to calculate the radiation per crack.

I returned to consideration of the data from the Chilean quake. I was comforted by what I considered to be a reason to think that the radio emission may have originated in phenomena preliminary to the actual rupture. That is, I am not as concerned about the few days gap between the actual full-scale rupture on 22 May 1960 and the event we observed. It seems that microfractures might be primarily involved and not necessarily full-scale breaking of rocks. The stresses within competent rocks lead to such "Griffith" cracks <u>before</u> the rock actually fractures.

But what about the absence of fringes on 16 May 1960? The interferometer fringes on the Boulder horizon can be compared with the positions of the ends of the Chilean fracture zone as would be seen along great circle paths from Chile to Boulder. One of the outstanding features of the great quake was its enormous length along the Chilean coast, approaching 800 to 1000 kilometers.

Because of its length, I believe that this quake may have "hidden" itself in our phase-sensitive records, by virtue of its having covered several positive and negative fringes.

Other questions that are relevant to this problem are: (1) will the 18 MHz radiation from Chile propagate, by ordinary ionospheric multihop paths to the four stations at which radiation was observed? The answer, supplied by the experts on this problem who are resident at government laboratories in Boulder is "yes". (2) if 18 MHz radiation from Chile, where considerable portions of the fracture zone are exposed on land, can freely escape and get to Boulder, Michigan, New Mexico, and Hawaii, will the same be possible for other major fault zones? The answer is "no" for many of the major island arcs within and around the Pacific, but it is ostensibly "yes" for the Californian faults. (3) have other major quakes occurred during the 20 (1957 - 1976) years of operation of the flare detectors and did they produce observed effects? The answers are "yes" and "no" respectively. The largest of these was the Alaskan Good Friday quake (1964), a sea floor event, but in any case not observable by us as a result of unfavorable propagation conditions.

With great trepidation, but also great awareness of the importance of the subject, I provisionally accept the validity, demonstrated in the laboratory and exhibited in Nature, of the hypothesis that radio emissions can be produced, and are likely to be observed, in earthquake zones.

Obviously, the way to enhance the credibility of the hypothesis is to achieve more and better observed events.

In the United States, the geophysical community is loathe to accept the relevance of radio physics to its researches (save perhaps VLBI!). Other nations are more enlightened on the subject, so I am not completely discouraged as to is ultimate development.

All that I have said was published long ago and is familiar to at least some of the participants in this symposium.

I would, however, like to suggest that something further might evolve. The spectrum of the fracture events we observed lies below one MHz, while we observed a putative event at 18 MHz.

We have not, therefore, "really" explained that event. Do not brush it off; insofar as terrestrial electromagnetics is concerned, this one event is the most remarkable I have seen in 34 years of low-frequency radio studies.

Therefore accept it at face value and ask what further research it might suggest. Aside from ground observations of faults, which are underway in Japan and the Soviet Union, I am encouraged to suggest that an event like this might be observed from spacecraft orbiting Earth.

The requirements of such an observation are that substantial areas of Earth be simultaneously observed in ways that respond to broadband noise, reject communications signals, and allow direction finding. I believe that the technology to do so exists. Clearly it requires high satellites, stationary above those "sides" of Earth in which events are expected and operated synoptically for extensive periods, probably measured in years.

Covering fracture zones with receivers on Earth is a useful concept, as long as noise of artificial origin is carefully controlled and if direction finding is carried out on the surviving emissions. The reason to suggest space observations instead is the much greater efficiency and global coverage possible from that perspective.

13 Ball Lightning

13.1 State Equation and Phase Diagram for Fractal Growth in Ball Lightning

G.C. DIJKHUIS

Zeldenrust College, Terneuzen, and Convectron N.V., Rotterdam, The Netherlands

ABSTRACT

A state equation for electron pressure in ball lightning plasma is derived by interpolating classical and quantum limits for pressure of free electrons with attractive exchange interaction. Vanderwaals-like isotherms with critical temperature 1844 K are shown in phase diagrams of this state equation. The locus of electron fractions coexisting at different density is obtained with Maxwell's lever rule . A subregion of the phase diagram is obtained where universal phase separation dynamics creates structures with fractal dimension.Their relevance to fractal growth in ball lightning, fireballs and laser ablation experiments is considered.

INTRODUCTION

As environmental phenomenon associated with atmospheric electricity, ball lightning is receiving increasing attention in recent years. In Hungary, Gy. Egely [1] selected 278 new cases from a polling of ca. 3 000 000 newspaper readers. Egely's collection confirms 1-5 sec. as most probable lifetime, and 20-30 cm as most probable diameter for ball lightning. 34 of his ball lightning cases existed for more than 50 sec. Egely labels 28 cases as high-energy ball lightning for doing more than 1 kilojoule material damage to the environment, ranging from splintered wood to molten metal. Biological effects including death occurred in 32 of Egely's cases.

In Austria A.G. Keul collected 150 new ball lightning cases through calls-for-reports in the media. His computer-aided analysis confirms standard ball lightning parameters, and establishes ball lightning observers as a sociologically random sample with normal eye-sight. In Japan, Y.H. Ohtsuki and H. Ofuruton obtained 2000 ball lightning reports, including photographs and videotapes and animation films. Keul and Ohtsuki reported their surveys on the International Ball Lightning Symposium in 1988 in Tokyo[2].

In a comprehensive review article B.M. Smirnov [3] derives a standard ball lightning from statistical analysis of surveys made by Rayle and McNally (both US), Charman (England) and Stakhanov (SU). Smirnov's standard ball lightning averages 28 cm diameter, 9 sec lifetime, 2 kJ energy content and 100 W light output. Smirnov divides ball lightning theories into five categories with different hypotheses for energy storage. All plasma models for ball lightning are rejected by Smirnov because room-temperature recombination rates keep a uniform sphere of ionized air much below the required energy-lifetime-product. Likewise Smirnov rejects gas models with excited atoms or molecules because quenching rates in air make relevant metastable states too short-lived. Furthermore, Smirnov rules out electric energy storage in ball lightning because dielectric materials in up-to-date capacitors allow insufficient energy density , and because the required energy density on a spherical capacitor with vacuum permittivity would cause breakdown in the surrounding air. Considering that the average energy corresponds to combustion of a mere 0.08 gram of coal dust in air , Smirnov decides on chemical reactions as primary energy source for ball lightning. Explosive reaction rates for relevant simple, single-step chemical processes in hot air lead Smirnov to an intensive and slow multi-step combustion process of ozone adsorbed on a porous framework of coal dust particles. Turning to electric discharges that form thread-like aerosols in relaxing metal vapours, Smirnov hardens his ball lightning model with a rigid skeleton that evolves within one second from nanometer-size solid particles sticking together in a cluster structure with fractal dimension 1.77-1.85 depending on cluster aggregation by random walk or gravity. Incorporating dynamic and thermal properties of air, Smirnov arrives at a coal-fired ball lightning storing 400 kilojoule in its combustible framework, and staying afloat on its own thermal updraft by heating ambient air at power level 34 kilowatt. Using water vapor as reference for outward surface tension from negative charge on the cluster framework, Smirnov calculates - 34 kilovolt as the electric potential of ball lightning , a harmless 0.02 joule as its electric energy , microwave radiation at 1-10 MHz by negative charge on vibrating cluster endings, and 10 minutes as maximum lifetime due to gradual neutralization by atmospheric ions. With blackbody radiation as dominant cooling proces for chemically inactive skeleton fibres in solidifying copper or iron vapor, Smirnov obtains 10 micron as fibre radius meeting the observed ~ 1 sec fireball lifetime. Combining line radiation with pyrotechnic data as reference for energy release by chemically active skeleton fibres, Smirnov obtains mean ball lightning power in the yellow sodium doublet from ~6000 separate luminous waves consuming 3-micron fibres in air mixed with 1% sodium at 2400 K.

While setting new standards for consistency, scope and detail in ball lightning chemistry, Smirnov's study too readily dismisses alternative models with non-chemical energy sources. Smirnov's recombination argument against self-supporting plasma models fails when the plasma is non-uniform as in Koloc's compound plasma configuration [2, 4]. His vacuum permittivity argument against self-supporting electric models breaks down for strong dielectrics as in Dijkhuis' polarized superfluid [2,5]. Smirnov labels nuclear processes as in Altschuler et al.'s beta decay model [6] "exotic" and "beneath criticism", and ignores Neugebauer's exchange model [7] along with its two-phase bosonic sequel [8,9]. We also miss clear prescriptions on experimental conditions for generating a fractal cluster structure with ball lightning properties in the laboratory.

Though excluded from Smirnov's review, ball lightning models supported by an external energy source have not been abandoned. Gladyshev [2] considers ball lightning a diffusional flame fed by constant atmospheric currents such as earlier appeared in Finkelstein and Rubinstein's direct current model [10]. In Handel's atmospheric maser model [2] lightning discharges overpopulate metastable energy levels of rotating water molecules in the thunderstorm region, and above an electric field threshold, their weakly allowed transitions feed microwave power into ball lightning plasma as in resonant cavity models by Kapitza [11] , Silberg [12] and Zou [2].

Some ball lightning theories were put to the ultimate test of experimental scrutiny. A series of videotaped discharge experiments in dilute hydrocarbon gas by Ofuruton and Ohtsuki [2] did not reproduce Barry's spherical flames [13], but recorded light emission up to 2.5 sec with a cotton fibre aerosol admixture [14]. Extending Tuck's short-circuit experiments with a complete submarine propulsion battery [15] to threshold current levels [9], Dijkhuis [16] recorded on high-speed film 10 cm luminous spheres with lifetime up to 1.2 sec forming in metal vaporizing from circuit-breaker contacts. With a high-voltage facility designed for testing expected [8, 17] and observed [18] nuclear processes in air at lightning power levels , Dijkhuis and Pijpelink [2] recorded time-delayed and intermittent pulse counts on a neutron detector monitoring electric breakdown in a cold, supersonic gas flow mixed with deuterium [2].

Very recently, Corum and Corum reported successful production of ball lightning from copper or carbon electrodes in free air using high-voltage rf equipment originally developed by Tesla around 1900 [24]. Fireballs measuring 2-3 cm and lasting several seconds were recorded on photographs and videotapes. Electrical circuit parameters of the Tesla coil and auxiliary equipment are specified in detail to facilitate fireball reproduction in other laboratories. Corum and Corum view their laboratory fireballs as explicit support (if not direct verification) for Smirnov's electro-chemical fractal cluster model for ball lightning. Subsequently Smirnov presented calculations for non-equilibrium processes leading to fractal cluster structures observed by A.A. Lushnikov in expanding metal plasma evaporating from a metal surface after a laser pulse [25].

Smirnov's ball lightning calculations rely on empirical data for vapour pressures, sublimation energies, rate coefficients and ionization potentials . The following sections derive a state equation for conduction electrons in ball lightning with all reference quantities expressed in fundamental constants. The new formula extends our earlier prescriptions for generating a bosonic and nuclear ball lightning by current interruption [9] or electric breakdown [2] to the regime of intermediate and strong electron degeneracy.

STATE EQUATION

At temperature T and concentration n, non-interacting electrons exert ideal gas pressure ranging from a classical high-temperature limit $p_\infty = n*k_B*T$ down to $p_o = (h^2/20m)*(3/\pi)^{2/3} * n^{5/3}$ in the zero-temperature quantum limit for fermions with mass m and spin 1/2 [19]. From $p_o = p_\infty$ follows that classical and quantum pressure are comparable on a curve in the T,n-plane given by:

$$n^{-1/3} = (h^2/2\pi m k_B T)^{1/2} \qquad (1)$$

Above this curve the classical ideal gas pressure p_o from thermal motion dominates because the mean distance $n^{-1/3}$ between particles is larger than their thermal de Broglie wavelength $(h^2/2\pi m k_B T)^{1/2}$, making the available energy levels plentyfull and sparsely populated. For ionized gases in the laboratory and in space the classical limit p_∞ is usually adequate. Electron temperature T and concentration n in lightning plasma are known from spectroscopic measurements by Orville [20]. From his data the ideal gas pressure $p = n*k_B*T$ of electrons in the lightning channel peaks at:

$$T = 29000\ \text{K}, \quad n = 6.0*10^{23}\ \text{m}^{-3}, \quad p = 240\ \text{kPa} = 2.4\ \text{bar} \qquad (2)$$

Ions and remaining neutrals also add to the total pressure driving rapid expansion of the lightning channel heard as thunder. Perturbation theory for non-ideal electrons gives an attractive exchange interaction diminishing ideal gas pressure by $p_{e\infty} = (-e^2h^2/32\pi^2\varepsilon_o m)*(n^2/k_B T)$ to leading order ($p_{e\infty} \ll p_\infty$) in the classical limit [18 ,19]. This gives no more than -4.0 Pa for the lightning plasma conditions of Eq'n (2).

In metals at room temperature, free electron pressure approaches the quantum limit $p_o = (h^2/20m)*(3/\pi)^{2/3} * n^{5/3}$ giving a huge outward electron pressure checked by attractive Coulomb interaction in the ion lattice , or by gravity in dwarf stars [19]. Non-ideal electrons have an attractive exchange interaction diminishing their ideal gas pressure by $p_{eo} = (-e^2/16\pi\varepsilon_o)*(3/\pi)^{1/3}*n^{4/3}$ in the quantum limit [18, 19]. Conduction electrons with concentration n give as quantum limits for ideal gas pressure p_o, exchange correction p_{eo} and real electron pressure $p = p_o + p_{eo}$ for copper respectively:

$$n = 8.45*10^{28}\ \text{m}^{-3}, \quad p_o = 38\ \text{GPa}, \quad p_{eo} = -21\ \text{GPa}, \quad p = 17\ \text{GPa} \qquad (3)$$

Iron has slightly higher quantum limits on its electron pressure because it contains $8.50*10^{28}$ free electrons/m³ .

To account for higher-order exchange terms we combine the four pressure terms in

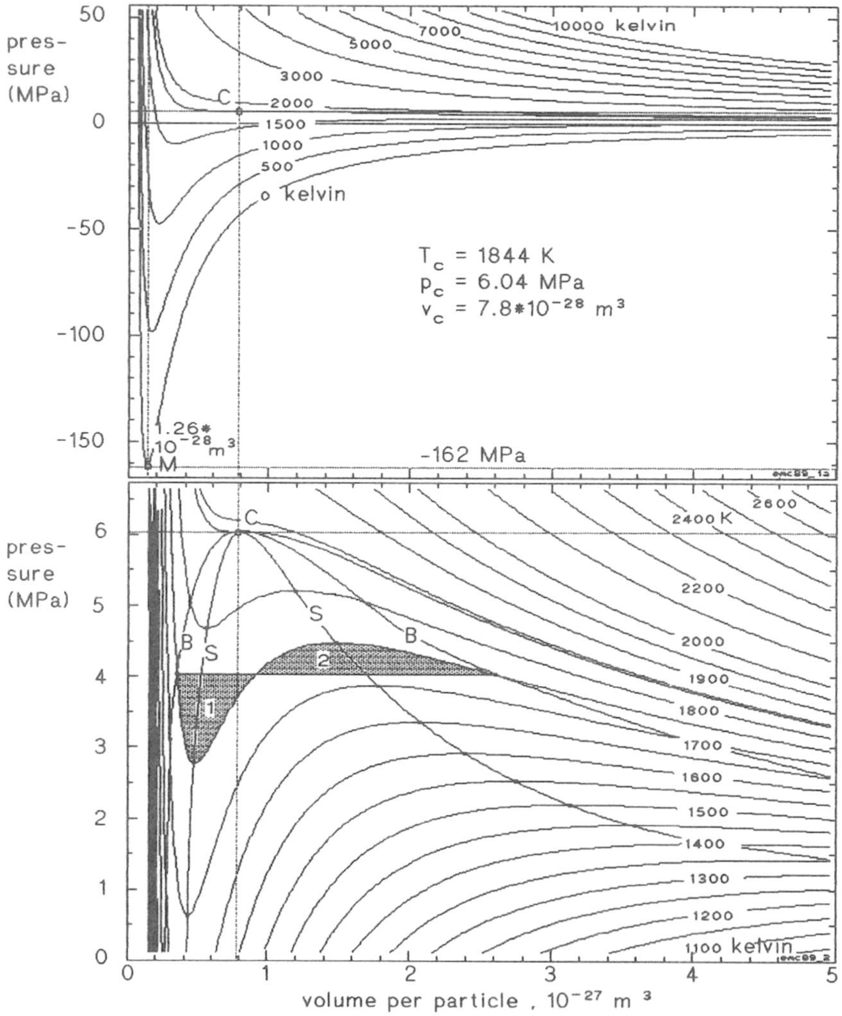

Fig. 1A, 1B Phase diagram calculated from state equation (4) showing binodal curve B obtained from Maxwell's lever rule as locus of coexisting phases in electron gas with exchange attraction on isotherms below critical point C; spinodal curve S encloses rising sections of subcritical isotherms where phase separation dynamics creates fractal structures at plasma temperatures and densities relevant to ball lightning.

the form $p_{el} = p_o + p_\infty + p_{eo}\, p_{e\infty}/(p_{eo} + p_{e\infty})$ preserving the correct classical and quantum limits. The ideal gas contributions p_o and p_∞ add up as in Dalton's law for partial pressures. The exchange corrections p_{eo} and $p_{e\infty}$ always combine to a finite exchange contribution smaller than each separate correction term . Relative to Bohr radius $a_o = \varepsilon_o h^2/(\pi m e^2) \approx 53$ pm, and hydrogen ground state energy $u_o = me^4/(8\varepsilon_o^2 h^2) \approx 13.61$ eV, our state equation for pressure p_{el} electrons with exchange interaction takes a non-dimensional form:

$$\frac{p_{el}}{u_o/a_o^3} = n a_o^3 * \frac{k_B T}{u_o} + \frac{2}{5}(3\pi^2)^{\frac{2}{3}} * (n\, a_o^3)^{\frac{5}{3}} - \frac{2\pi * (n\, a_o^3)^2}{k_B T/u_o + 4\pi * (\pi/3)^{1/3} * (n a_o^3)^{2/3}} \qquad (4a)$$

$$u_o/a_o^3 \approx 1.471*10^{13} \text{ Pa}, \quad u_o/k_B \approx 1.579*10^5 \text{ K}, \quad 1/a_o^3 \approx 6.748*10^{30} \text{ m}^{-3} \qquad (4b)$$

The references for pressure u_o/a_o^3 , temperature u_o/k_B and concentration $1/a_o^3$ outline the p,T,n-domain where Eq'n (4) deviates from classical and quantum limits for ideal gas. For T = 0 two remaining terms in Eq'n (4) give the correct quantum limit $p_o + p_{eo}$ on electron pressure. The opposite limit $T \rightarrow \infty$ preserves temperature dependence in Eq. (4A), reproducing the classical electron pressure limit $p_\infty + p_{e\infty}$. Our inductive argument underlying the analytic form of Eq'n (4) does not introduce any phenomenological constants such as appear in the van der Waals or Berthelot equations for phase transitions of atomic and molecular gases.

PHASE DIAGRAMS

Figure 1A shows relevant isotherms of state equation (4) in a p,v-diagram with volume per particle v=1/n replacing concentration n . At large volumes all isotherms properly tend asymptotically to zero pressure. A closer inspection shows that zero kelvin gives the only isotherm approaching zero pressure from below. In the opposite limit of zero volume, pressure on all isotherms rises without limit. Unlike the van der Waals- and similar equations, state equation (4A) gives finite pressures for all finite gas volumes because its derivation avoids the "hard sphere"-approximation for particle volume.

Isotherms below a critical temperature T_c in Fig. (1A) have ascending sections where negative compressibility $-v(\partial p/\partial v)$ makes the electron gas thermodynamically unstable to local density perturbations. At a critical volume v_c the critical isotherm has an inflexion point C tangent to the critical isobar at height p_c. The three algebraic equations for the critical point C are too complicated for deriving the critical parameters explicitly in fundamental constants from state equation (4). Numerically, state equation (4A) gives critical parameter values :

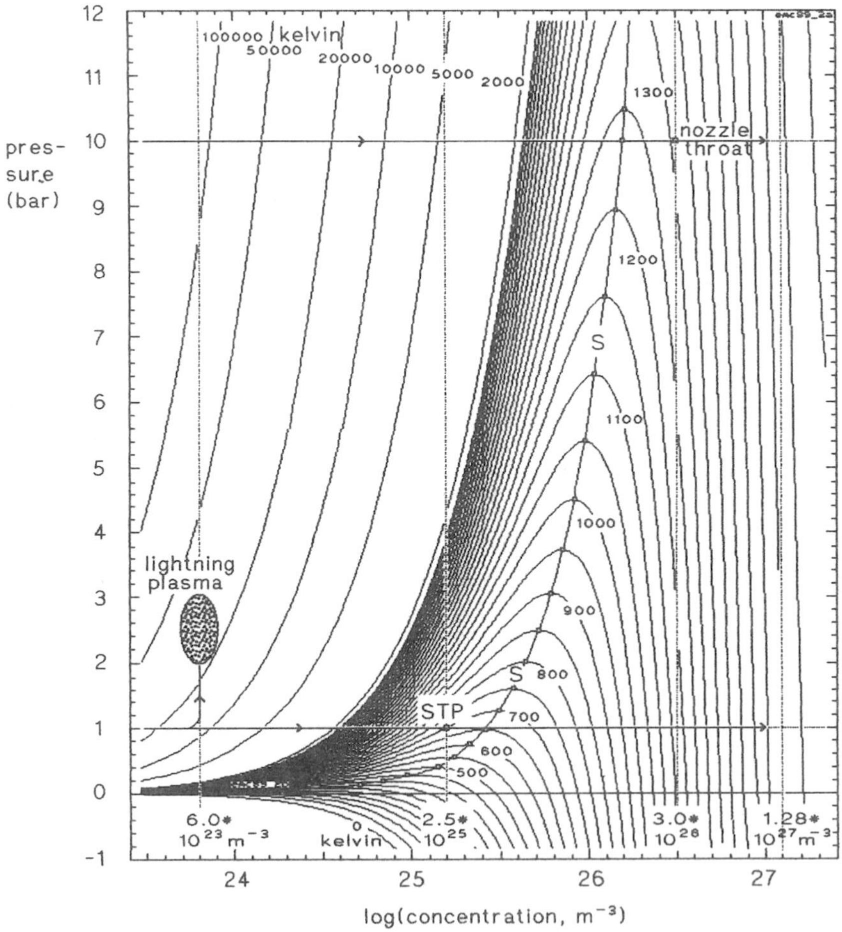

Fig. 2 Phase diagram for ball lightning formation by electric breakdown in neutral gas showing low-density side of fractal growth region with descending isotherms from Eq'n (4) relative to lightning plasma and discharge nozzle conditions; ref. 18 derives electric field threshold for reaching boundary curve S by impact ionization.

$T_c \approx 1844$ K, $\quad p_c \approx 6.04$ MPa, $\quad n_c \approx 1.28*10$ m (5)

showing that the critical electron temperature T_c is about six times higher than the temperature 288.15 K of the U.S. Standard Atmosphere at sea level, and the critical electron pressure p_c is about sixty times higher than standard atmospheric pressure at sea level, 101.325 kPa [21]. The critical electron concentration $n_c = 1/v_c$ is about fifty times higher than standard atmospheric number density $2.5470*10^{25}$ [21], but

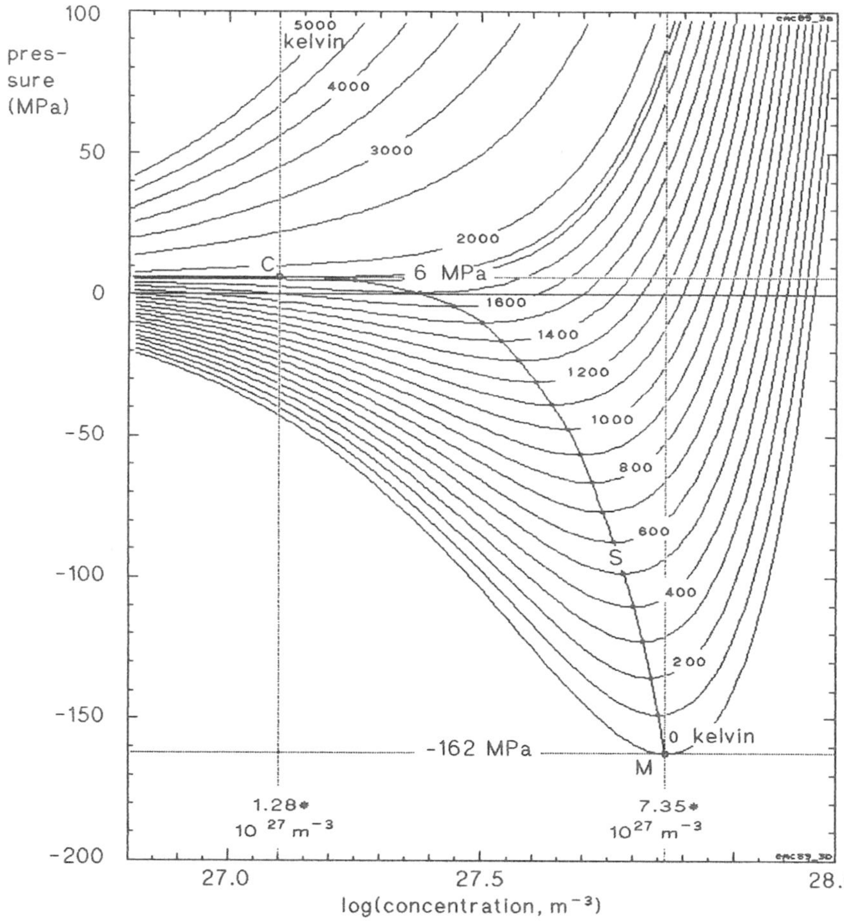

Fig. 3 State diagram for fireball generation in relaxing metal vapour showing high-density side of fractal growth region with descending isotherms from Eq'n (4); ref. 9 derives discharge current threshold for reaching boundary curve S with unit flux quantum on conduction electrons starting from metals at room temperature.

sixty six times lower than free electron concentration in copper at room temperature. Figure 1B expands the p,v-diagram of Eq'n (4) in vertical direction to show the region of coexisting phases and its subregion for fractal growth. By Maxwell's "lever rule" equal areas 1 and 2 define horizontal sections of isotherms inside the region of coexistence. This ensures equal chemical potential for electrons on boundary curve B.

Figures 2, 3 show isotherms of state equation (4) and spinodal curve S in a logarithmic (p,n)-diagram relative to STP-conditions in air, lightning channels [20], and high-voltage ball lightning experiments in a nozzle throat [2]. Descending sections of subcritical isotherms define the region of absolute phase instability where local density fluctuations will grow without activation threshold, and universal phase separation dynamics creates a condensed phase with a scale-free fractal structure [22, 23]. This spinodal decomposition mechanism forms dendritic , foam-like or sponge-like [8] structures with fractal dimension determined by details of the aggregation process. The Witten-Sander-model with diffusion-limited growth of a single cluster at fractal dimension 2.52 proved realistic for silica . Cluster-cluster aggregation as in Smirnov's chemical ball lightning model adequately models silica particle structures with fractal dimension 2.1 observed by Rouw et al. [23]. In Lushnikov's laser ablation experiment copper resolidified to threads with radius 25 nm in a tissue-like structure with fractal dimension 1.81 [25]. Absolute pressure minimum M in Fig. 3 has electron concentration close to ceramic compounds in high-T_c superconductivity [26]. Critical point C nearly coincides with critical deuterium concentration $1.21*10^{27}$ m^{-3} in cold fusion experiments with palladium.

CONCLUSIONS

Many-body theory for conduction electrons with exchange interaction at intermediate degeneracy spells out a phase transition in discharge plasma and evaporating metals with critical parameters fixed by fundamental constants. Exchange instability grows fractal structures in cold, dense plasma regimes relevant to lightning discharges, high-voltage breakdown facilities, current interruption and laser ablation experiments.

Fractal condensate in ball lightning has exchange interaction parameters similar to conduction electrons in ceramic compounds with high-T_c superconductivity . Its critical state parameters match the palladium-deuterium system known for cold nuclear fusion.

ACKNOWLEDGEMENT

The author thanks prof. B.M. Smirnov (IVTAN, Moscow) for providing unpublished data on high-voltage fireballs and fractal structures from laser ablation experiments.

REFERENCES

1. Egely G (1987) Hungarian Ball Lightning Observations (Case 1 - 278). Central Research Institute for Physics, Budapest, Preprint KFKI-1987-10/D

2. Ohtsuki YH (ed) (1989) Science of ball lightning (fireball). World Scientific Publishing Co., Singapore

3. Smirnov BM (987) The properties and the nature of ball lightning. Physics Reports 152: 177-226

4. Koloc PM (1977) Method and apparatus for generating and utilizing a compound plasma configuration. United States Patent 4,023,065

5. Dijkhuis GC (1979) Thermonuclear energy from ball lightning. In: Proceedings of the 14th IECEC (Boston). American Chemical Society. Washington D.C. p1614

6. Altschuler MD. House LL. Hildner E (1970) Is ball lightning a nuclear phenomenon? Nature 228: 545 - 546.

7. Neugebauer Th. (1937) Zu dem Problem des Kugelblitzes. Z.f.Physik 106: 474-484

8. Dijkhuis GC (1980) A model for ball lightning. Nature 284: 150-151
idem (1981) Reply to E.A. Witalis. Nature 290: 166

9. Dijkhuis GC (1982) Threshold current for fireball generation. J Appl Phys 53: 516-519

10. Finkelstein D. Rubinstein J (1964) Ball lightning. Phys Rev 135A: 390-396

11. Kapitza PL (1955) The nature of ball lightning. Dokl Acad Nauk USSR: 245-248

12. Silberg PA (1981) On the formation of ball lightning. Nuovo Cimento 14C: 221-235

13. Barry JD (1980) Ball lightning and bead lightning. Plenum Press. New York London

14. Ohtsuki YH (1989) Private communication

15. Tuck JL. (1971) Ball lightning, a status summary to november 1971. Informal Report LA-4847-MS, Los Alamos

16. Dijkhuis GC (1985) Het fase-diagram van bolbliksem: model en test. Ned. Tijdschift voor Natuurkunde B51: 125, 128; (1986) Phase diagram of ball lightning: model and test. Ball Lightning Newsletter 3 : 4 - 9 (published by Mark Stenhoff UK)

17. Shah GN. Razdan H. Bhatt CI. Ali QM (1985) Neutron generation in lightning bolts. Nature 313: 773 - 775.

18. Dijkhuis GC (1988) Scaling law for fusion power from boson vortex in ball lightning Janiszewski J. Moron W. Sega W (eds) Ninth international Wroclaw symposium on electromagnetic compatibility. Wroclaw, Poland. p21

19. Fetter AL. Walecka JD (1971) Quantum theory of many-particle systems. McGraw-Hill Book Company. New York Düsseldorf Sydney

20. Orville RE (1977) Lightning spectroscopy. In: Golde RH (ed) Lightning, volume 1. Academic Press. London New York San Francisco. p281

21. US Standard Atmosphere (1987) In: Weast RC (ed) CRC Handbook of chemistry and physics. 67th edn. CRC Press. Boca Raton, Florida p. F-141

22. Mandelbrot BB (1977) The fractal geometry of nature. W.H. Freeman . New York

23. Rouw PW. Woutersen A. Eckerson B. de Kruif CG (1989) Adhesive hard sphere dispersion V: observation of spinodal decomposition in a colloidal dispersion. Physica A 125: 1 - 25

24. Corum KL. Corum JF. (1990) High-voltage rf ball lightning experiments. Soviet Physics Uspekhi (in press)

25. Smirnov BM (submitted) The evolution of non-equilibrium laser plasma of metals with the formation of structures. Chem. Phys. Lett.

26. Jansen L (1988) Model analysis of superconductivity in Tl-Ca-Ba-Cu-O and Bi-Ca-Sr-Cu-O compounds in terms of exchange-induced Cooper pair formation via oxygen anions. Physica C 156: p501

13.2 Correlative Relationship Between the Intensity of Radiation for Ball Lightning and Its Typical Linear Dimension

A.I. GRIGORYEV, I.D. GRIGORYEVA, E.N. NOSKOV, and S.O. SHIRYAEVA
Faculty of Physics, Yaroslave University, Yaroslave, USSR

Ball lightning (BL) is a rare, nearly unknown form of thunderstorm electricity and has been studied during a long time. However, because it cannot be reproduced under laboratory conditions, the only source of new information on its properties is provided by eye witnesses and statistical analyses of their descriptions [1-5]. In this connection independent surveys of witnesses of ball lightnings have been conducted in a number of countries, concepts of BL emerging from these. The overall consistency in the "portraits" of an average BL obtained by independent investigators (see, for example, the reviews in [2,3]) is proof that the approach is reasonable. However, statistical analyses of eye witness information can provide, not only averaged data on some individual properties of BL, but also relations between these. The authors of [4,5] have tried to find correlative relationships between individual BL characteristics derived by averaging over large volumes of data. The present paper sets the same goal: 3115 descriptions of BL collected at Yaroslavl University are used to investigate paired correlation between BL brightness (playing the role of BL radiation intensity in the optical range) and typical linear dimension (diameter) by utilizing the technique described in [4,5].

Obviously, information communicated by BL eye witnesses

under natural conditions is not particularly reliable. This is associated with both individual human capacities to esti- space and time intervals, colour and brightness and with the fact that the witnesses have to make these estimates long after they actually observed the phenomenon. In order to evaluate, if only very approximately, the accuracy of eye witness communications as to BL dimensions and brightness, all witnesses whose descriptions have been used in this analysis were sent questionnnaires to fill their BL descriptions again, about a year after their first responses were obtained. A comparison of the two questionnaires reveals that the observers indicate the typical linear dimension with a relative uncertainty that varies according to the range of the quantity to be estimated up to ≈ 0.18 , the mean value over all the ranges being ≈ 0.14. The mean uncertainty in brightness descriptions was ≈ 0.3. These uncertainties also characterize data obtained below by statistical procedures.

In what follows, where correlation is sought, it is assumed that brightness J_i and diameter d_j from i-th and j-th ranges are correlated if

$$P_{ij} - P_i \cdot P_j > 0$$

where P_{ij} is the probability of finding BL with J from i-th and d from j-th ranges simultaneously, P_i and P_j are the probabilities of finding BL with J-from i-th and d from j-th ranges separately. The method of least squares was used on the pairs of brightness J and diameter d thus chosen to seek (from general physical considerations) the correlative relationship

$$J = J_o \cdot d^n \tag{1}$$

It has been found that such a statistically significant relationship exists with $n \approx 2.06$ and a coefficient of paired correlation of ≈ 0.55. The expression (1) with $n \approx 2.06$ also satisfies Fisher's F-test at 10 per cent significance level [6].

It is easy to see that n=2 if BL radiates from the surface. If, on the other hand, the radiation comes from the BL volume, then $n > 2$. When the photon free path for a typical wavelength in BL substance exceeds the BL radius, we have n=3, otherwise $2 < n < 3$. The value we have found, n=2.06, corroborates the assertion about surface radiation of BL. It should be noted, however, that the accuracy of the analysis is poor; to enhance it we need much larger volumes of raw data.

REFERENCES

1. Y.H.Otsuki and H.Ofuruton. Nature of Ball Lightning in Japan. Il Nuovo Cimento, 1987, v.10C, No.5, p.577-580.

2. B.M.Smirnov. The Properties and the Nature of Ball Lightning. Physics Reports, 1987, v.152, No.4, p.177-226.

3. I.P.Stakhanov. The Physical Nature of Ball Lightning. Moscow, Nauka, 1970. 240pp.(in Russian).

4. A.I.Grigoryev. On the correlative relationship between the life time and typical linear dimension for ball lightning. Meteorologiya i gidrologiya, 1986, No.6, p.103-104.

5. A.I.Grigoryeva and I.D.Grigoryeva. Correlative relationships between properties of ball lightning. Zhurnal tekhnicheskoi fiziki, 1989, v.59, No.2(in print).

6. E.N.L'vovsky. Statistical methods for constructing empirical formulae. Moscow, Vysshaya shkola, 1988, 239pp.(in Russian).

13.3 Experimental Research on Ball Lightning

Y.H. OHTSUKI[1] and H. OFURUTON[2]

Department of Physics, Waseda University, Tokyo, 169 Japan[1]
and Tokyo Metropolitan College of Aeronautical Engineering, Tokyo, 116 Japan[2]

Abstract

Experiments on producing ball lightning (fire balls) were made with electric discharge in fuel gas and/or aerosol. A long lifetime fire balls (1 - 2 s) were observed in 2.7% ethane and cotton fibers (100 cm^3.) In other experiment, plasma fire balls were formed in the natural atmosphere by the interference of the microwave (wave length being 12.2cm , power ranging 1kw - 5.0kw). This is the first confirmation of the possibility of the ball lightning formation proposed by Kapitza. We also report other properties of these fire balls formed in our experiment.

INTRODUCTION

In spite of the extent of many examinations, information analyses, experimental and theoretical studies, a widely accepted explanation of the ball lightning has remained elusive. However it is believable that the ball lightning is created by the energy condensation of the atmospheric electricity.[1] Barry reported the fire ball production by electric discharge in dilute fuel gases in air such as methane, propane and benzene.[2] We produced fire balls of the three types by the electric discharge in changing the fuel gas density in air contained aerosol.

On the other hand, Kapitza proposed the theory predicting the possibility of formation of the fire ball (ball lightning) by interference of the microwave in natural atmosphere.[3] Powell and Finkelstein obtained the plasma formation by the interference of the electromagnetic wave in a vacuum cube, partly confirming the Kapitza's model.[4] However, their experiment was not made in the natural atmosphere.

In this paper, we reexamined Barry's experiment and showed the change of the fire ball. However, we did not use propane for safety, and used methane or ethane. We also made a experiment of discharge in the air containing cotton fibers as aerosol. Moreover, we report the results of the microwave experiment.

It is very difficult to form the plasma fire ball in the natural atmosphere by the microwave interference. We have constructed the high power microwave source (< 5kW) with

12.2 cm wave length, and the special type "semi-cavity" at the end of the wave guide connected to the micro wave source. By changing the several types of the "semi-cavity" , we could obtain the fire ball moving up and down or floating in stable. This is the first confirmation of the Kapitza's theory for formation of the fire ball in natural atmosphere.

EXPERIMENTAL

A) Electric discharge

Figure 1 shows the shape of the electrodes we used. The electrodes were made of copper rods (3 mm in diameter) and spark gap was 3 mm. The glass chamber with interior dimensions of 734×372×429 mm was used.

Figure 2 shows the electric circuit. The first experiment was made in the methane-air mixed gas (Capacitor:0.2 μF). The second was in the air containing cotton fibers (Capacitor: 0.2μF). The last was in the air containing ethane and cotton fibers (Capacitor: 7μF).

All discharges of these experiments were made in the atmospheric pressure. Before these three groupes of experiments, discharge experiments in the normal air were done, and ordinary sparks were observed. The changes of the state in the chamber were recorded with video cameras.

B) Microwave cavity

A 2.45 GHz, 5kW microwave oscillator was operated in the air and microwave was guided to a cylindrical cavity made of copper. The dimension of the cavity was shown in Fig.3. The one side of the cavity was closed by immovable aluminum foil and the other side was closed by a movable meshed aluminum plate. The pressure of the inside of the cavity was atmosphere pressure and there is only air in the cavity, not other gas. The change of the plasma in the cavity was recorded by a video camera.

RESULTS AND DISCUSSIONS

A) Electric discharge

1) Methane-air mixture and ethane mixture

At 1 % methane experiment, the spark was the same as in normal air. At 2 % methane experiment, a spherical light (3-6 cm in diameter) was observed. At first the green light was observed which was caused by copper from the electrodes. After 0.26 s this light changed into an ordinary ignition, because the concentration of methane of the upper part of the chamber was high enough. Figure 4 shows the change of this luminous state.

At 1.4% ethane experiment, a normal air spark was observed. At 2.1% ethane experiment, a fire ball (about 4 cm in diameter) rose up from near the electrodes. The color of the fire ball was red, but at the first it contained green color. The lifetime of the fire ball was about 0.3 s. Figure 5 shows the change of this luminous state.

2) Aerosol (cotton fibers) and air mixture

Cotton fibers (about 100 cm^3 in volume) were torn into small lumps and were put in the chamber. The lumps of the fibers were scattered in the chamber by compressed air. After the discharge, about 3 cm in diameter white luminous ball was first prodused, became smaller, and disappeared within 0.8 s after the discharge. Figure 6 shows the change of the luminous ball. This luminous phenomenon should be treated as aerosol plasma theory.

3) Ethane, aerosol (cotton fibers), and air mixture

At the last experiment, 2.7% ethane and 100 cm^3 cotton fibers (torn into small lumps) were used. After the discharge two white luminous phenomena were produced near the electrodes. One phenomenon (about 5 cm in diameter) disappeared after 2 s of the spark. Another one (about 1 cm in diameter) moved downwards and upwards, and disappeared. The lifetime of the latter lightning matter was about 1 s. The experiment was done under the condition of 18 C in room temperature and 100 % relative humidity.

These two luminous balls were recorded on two video tapes. Two video cameras were set at right angles. Figure 7-1 and 7-2 show the changes of these balls.

B) Microwave cavity

We observed various kind of plasma fires at the power ranging 1kW - 2.2kW. The first type of the plasma fire was the flames located at the end of the wave guide , its color changed from white red to blue and orange . (see Fig. 8-1). The sizes of the flames were changed from 1cm to 3cm , and the lifetime was a few minutes (depending on starting a dismiss of the interference).

The second type of the plasma fire we produced was the very bright homogeneous flash as a whole of the semicavity. The colour was white , and the lifetime was very short (- 0.2 sec). just after the above bright flash , there appeared a separated plasma ball, moving up and down, horizontally, and rounding along the cavity wall.(see Fig.8-2). The colour was orange and its lifetime was 2 - 3 sec.

We are now attempting to use a various kind of semicavibty, or meshed cavity, changing the air component like aerosol, vapor and fuel gases. It was very interesting to note that the plasma fire was produced even when we deleted the meshed plate of the one side of the cavity.

ACKNOWLEDGMENTS

The authors thank Prof. K.Kohra and Mr. H.Ara for their

help. We also wish to thank Tsukuba Institute of Science and Technology for convenience for our facilities.

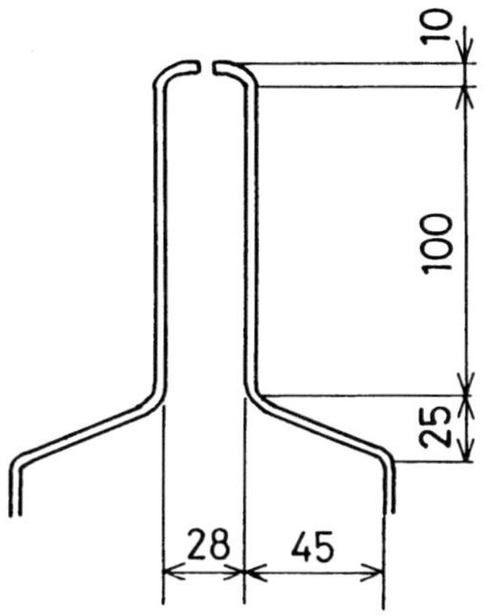

Fig.1 Electrodes (3 mm in diameter, copper rods) of discharge experiments

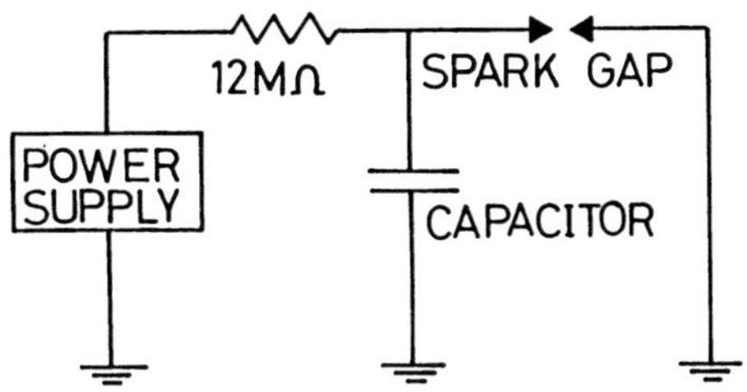

Fig.2 Electric circuit of discharge experiments

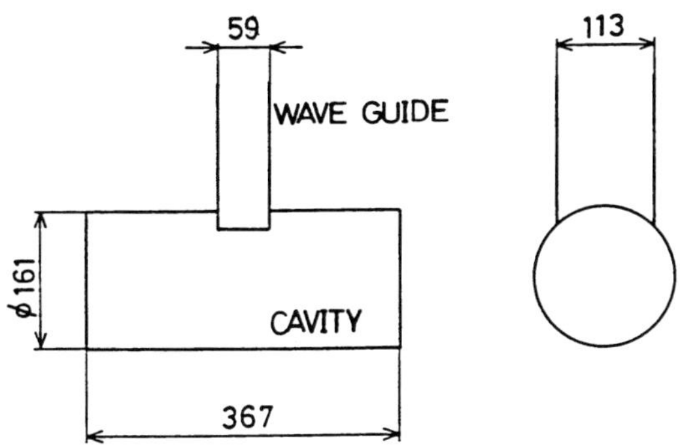

Fig.3 Microwave cavity

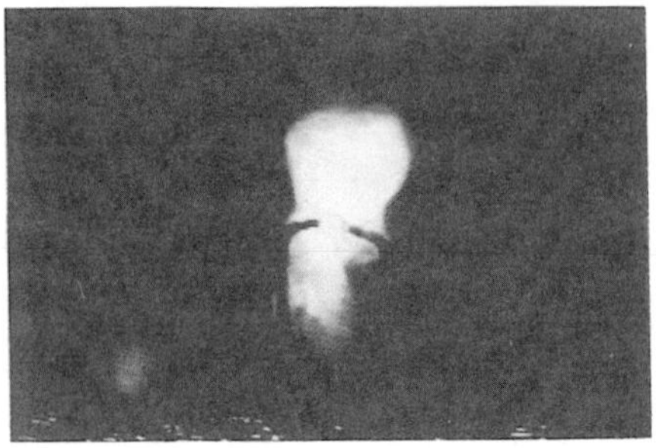

(a)

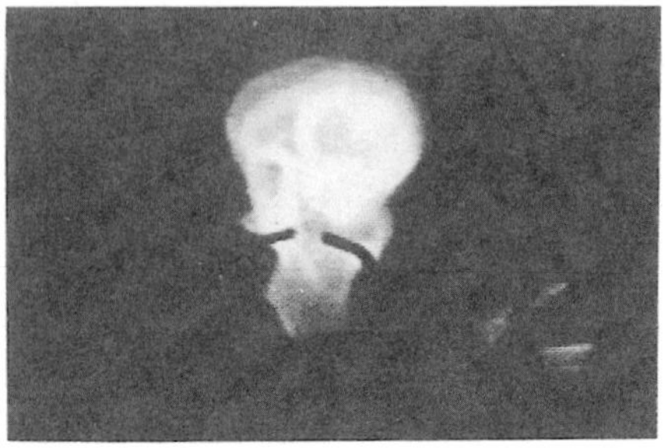

(b)

(c)

(d)

Fig.4 Methane(2%)-air mixture (discharge voltage 8 kV)
(a) 0.066 s after discharge
(b) 0.099 s
(c) 0.132 s
(d) 0.264 s

(a) (b) (c)

Fig.5 Ethane(2.1%)-air mixture (discharge voltage 10 kV)
(a) 0.033 s after discharge
(b) 0.066 s
(c) 0.132 s

(a) (b) (c)

Fig.6 Cotton fiber-aii mixture (discharge voltage 9 kV)
(a) 0.05 s after discharge
(b) 0.10 s
(c) 0.20 s

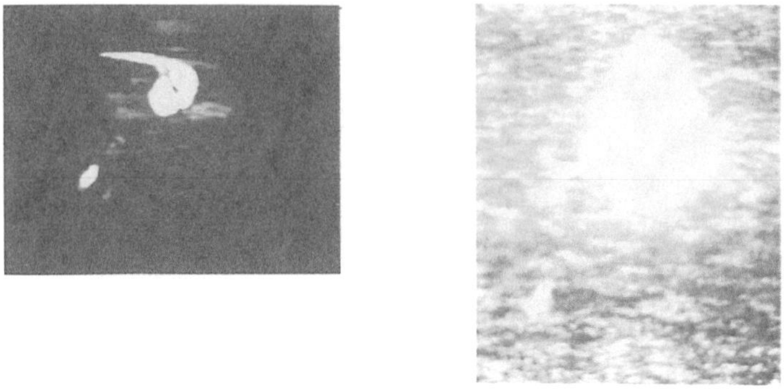

(a)

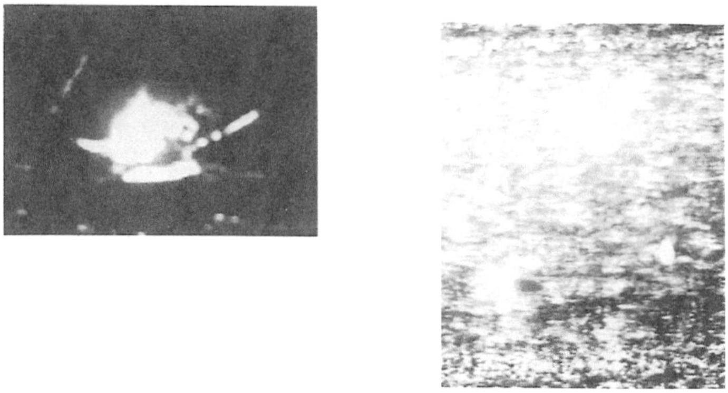

(b)

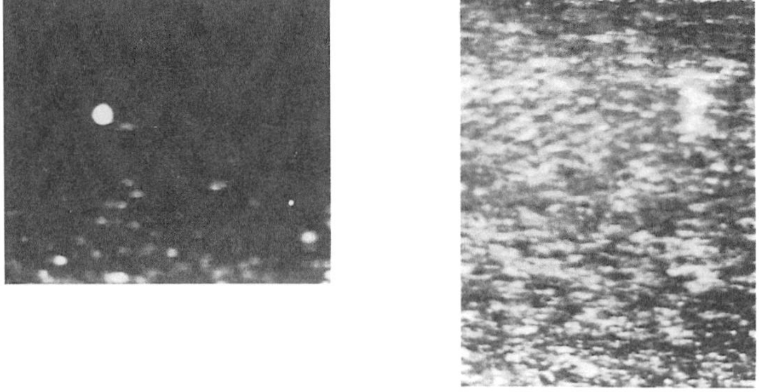

(c)

Fig.7-1 (left) and Fig.7-2 (right)
Ethane(2.7%), cotton fiber, and air mixture
(discharge voltage 8 kV)
(a) 0.264 s after discharge
(b) 0.660 s
(c) 1.254 s

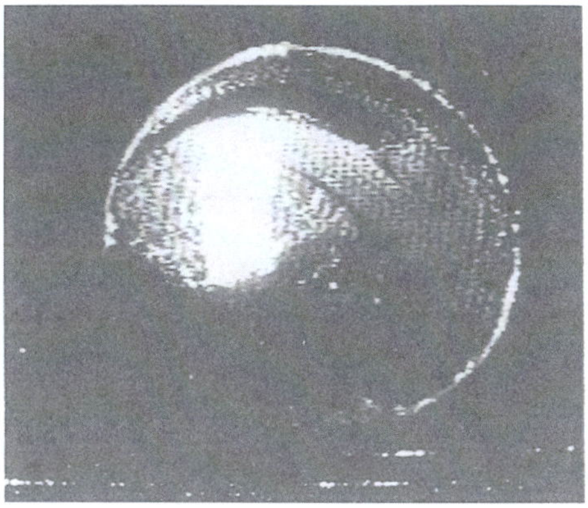

Fig.8-1 Plasma fire (Type 1)

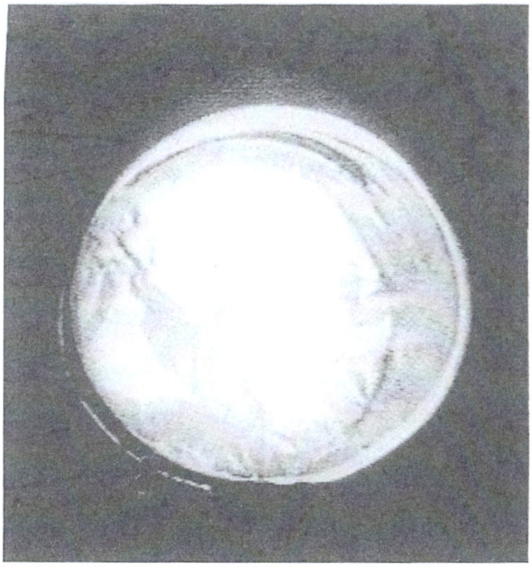

(a)

(b)

Fig.8-2 Plasma fire (Type 2)
(a) 0.12 s after the phenomenon occerred
(b) 0.33 s

REFERENCES

[1] 'Science of Ball Lightning' ed.by Y.H.Ohtsuki, World Scientific, Singapore (1989)
[2] J.D.Barry : J.Atmos.Terr.Phys. 30, 313 (1968)
[3] P.L.Kapitza : Dokl.Akad.Nauk.SSSR, 101, 245 (1955)
[4] J.R.Powel and D.Finkelstein : Am.Scientist, 58, 262 (1970)

14 Meteorologico-Electric Phenomena and EHD

14.1 Meteorologico-Electric Phenomena and Electrohydrodynamics (EHD) or Electromagnetohydrodynamics (EMHD)

H. KIKUCHI
Nihon University, College of Science and Technology, Tokyo, 101 Japan

ABSTRACT

The meteorologico-electric environment may be considered to consist of dusty plasmas with noise and chaos, containing charged dust grains, for the meteorological phenomena involving or accompanying electric discharge, ionization or electrification such as thunderstorms, planetary rings, cometary tails and dark interstellar clouds. This paper is intended, first of all, to provide several basic concepts: electric reconnection; ponderomotive force; critical ionization velocity; self-organization; dynamic chaos and so on, useful for understanding the phenomena physically, and then to furnish a new set of electrohydrodynamic (EHD) and/or electromagnetohydrodynamic (EMHD) equations with additional terms of ponderomotive forces, incorporating Maxwell's equations or Poisson's equation, relevant for the meteorologico-electric phenomena.

1. INTRODUCTION

There are a number of meteorologico-electric phenomena involving or accompanying electric discharge, ionization or electrification such as thunderstorms, cold and polar fronts, jet streams, typhoons, hurricanes, cyclones and tornadoes in the terrestrial environment, and planetary lightnings, rings, cometary tails, dark interstellar clouds in the extraterrestrial environment. For these phenomena, conventional hydrodynamics (HD) or magnetohydrodynamics (MHD) breaks down and even conventional plasma physics is also not relevant, because their environments are considered neither fully ionized gases nor collisionless plasmas as often assumed for conventional plasmas, but may be termed unconventional plasmas as a whole. They may be partially ionized and collisional gases, dusty or dirty plasmas containing charged dust grains or aerosols, which can be an environment, for example, for natural, triggered and ball lightning, sandstorms or even whirlwinds. Consequently, conventional HD or MHD should be extended to a new area of electrohydrodynamics (EHD) and/or electromagnetohydrodynamics (EMHD).

In regard to laboratory and lightning discharges, there has been considerable progress in observations for the last decades, particulary on fine structure processes by using some modern techniques of radio- and opto-electronics [1]. Surprisingly, however, there has been little progress in theory of electric discharge and ionization in partially ionized and collisional gases from the point of a unified clear view, although a variety of different approaches were devised for individual phenomena, for example, by means of a theoretical modelling and computer simulation. In this sense, it is not too much to say that the theory of electric and lightning discharges is still demanding a new development [2].

There is another class of meteorologico-electric environments as described in terms of dusty plasmas that are loosely defined as fully or partially ionized or neutral gases containing charged dust grains. Such plasmas exist in planetary rings, cometary atmospheres, and interstellar dust clouds in the extraterrestrial environment, in thundercloud, volcanic plumes, rocket exhausts, sandstorms, and whirlwinds in the terrestrial atmospheres, in nuclear fusion devices with impurity ions and atoms, and in plasma synthesis for recent industrial applications. Although term "dusty plasmas" originated among the space plasma community and has come into use over the last few years, it is not familiar to laboratory and fusion plasma physicists who use the term "impurity ions, atoms, or plasmas". Instead, the author has suggested the use of the term "dirty plasmas" in the broadest sense for laboratory and space plasmas which may partly contain aerosols, liquid drops, solid bodies or debris; and the use of the term "dusty plasmas" in a narrower sense, perhaps for small solid bodies or grains, defining their size range [2].

The dust grains are strongly coupled to the fields and particles in the plasma through their electrical charges. As a result, the motions of the grains may be dominated by electric and magnetic forces, and collective effects can play an important role. This field of study is receiving increased attention from various aspects mentioned above and a couple of workshops and organized sessions on dusty and dirty plasmas were held at the University of California, San Diego (UCSD) on 10-11 February, 1986 [3], at the Workshop on Laboratory and Space Plasmas, Tokyo on 25-26 November, 1986 [2], at Capri, Italy on 18 May - 1 June, 1989 [4] and at the URSI Symposium on Environmental and Space Electromagnetics, Tokyo on 4-6 September, 1989 [5]. Although a number of novel problems on this subject have been raised through these workshops, the studies have only just started and the whole field is still in its infancy when compared to traditional plasma physics, and correspondingly undeveloped. One of key problems on dusty plasmas concerns the effects of space charges and electric fields that become essential and significant like partially ionized and collisional gases apart from that of magnetic fields. This demands the creation of a new set of electrohydrodynamics (EHD), or electromagnetohydrodynamics (EMHD) more generally, combining plasma physics with meteorology.

To effect this, the author has introduced three basic physical concepts: (1) electric reconnection; (2) ponderomotive force; (3) critical ionization velocity, useful for understanding the meteorologico-electric phenomena in a unified manner at least qualitatively, and thus for establishing a new EHD or EMHD.

The next problem is how to treat random noise fluctuations and chaos accompanied by meteorologico-electric phenomena on the basis of a statistical approach to nonlinear and turbulent processes, incorporating two more concepts of self-organization and dynamic chaos in a new EHD or EMHD regime.

This paper, first of all, describes the former three and latter two concepts briefly, and then attempts to apply them to a couple of real phenomena, traditional and novel, such as natural, triggered, ball lightning, sandstorms and whirlwinds. Finally, a new set of EHD and/or EMHD equations with additional terms of ponderomotive forces are presented, incorporating Maxwell's equations or Poisson's equation.

2. ELECTRIC RECONNECTION AND PONDEROMOTIVE FORCE

As is well known, the concept of magnetic reconnection is so familiar to plasma physicists as well as geo-astro-physicists and has been extensively investigated for the past four decades, for example in relation to solar and auroral flares in space and fusion devices in the laboratory. The reader may remember that a Conference devoted exclusively to magnetic reconnection in space and laboratory plasmas was held at the Los Alamos National Laboratory, New Mexico on 3-7 October, 1983 and its Proceedings edited by E.W. Hones, Jr. was published by American Geophysical Union [6].

Suprisingly, however, there were no suggestions, to the author's knowledge, on the concept of electric field line reconnection or merging analogous to magnetic reconnection.

To make electric field reconnection feasible, one needs to form electric cusp or neutral point (line) analogous to magnetic cusp for magnetic reconnection.

The simplest form of an electric cusp can be obtained from a symmetrical quadrupole configuration as shown in Fig. 1 (a) and its center of a diagonal forms an electric cusp or neutral point (line) around which a low electric field region is attained as indicated by a stippled area. Although the right panel Fig. 1 (b) constitutes a dipole configuration above the ground, this is equivalent to the left panel as far as the electric field configuration of the upper half region is concerned. Because the right panel actually goes to the left panel by putting an image of the dipole instead of the ground plane. As a result, the left shows an idealized configuration of an electric cusp, perhaps only attainable in the laboratory, while the right provides a more realistic situation which could represent a horizontal configuration of charged clouds to be discussed later.

Let us consider a case when a spherical solid particle, conducting or dielectric, be inserted in the cusp region as shown in Fig. 2. Then, the positive and negative charges are induced or polarized on the surface of the particle, facing the negative and positive electrodes applied, respectively. As a result, electric field line merging takes place toward the solid particle from the four quarters and reversed field lines tend to reconnect with each other through the particle as seen in Fig. 2 (a,b), corresponding to Fig. 1 (a,b). Then, an electrical Lorentz or ponderomotive force is exerted by a localized large electric field on the particle, accelerating it to a very high speed, since the electrostatic or electric energy accumulated is released from the environment and tends to be converted into the kinetic energy of the particle. Thus, electric field line reconnection is achieved in principle in the laboratory and occurs in space and in the atmosphere in a variety of more relaxed and realistic configurations. The concept of electric reconnection analogous to magnetic reconnection was originally proposed by the author in 1985 in relation to his model of rocket-triggered lightning during the winter thunderstorms in Japan to be touched on later briefly [7, 8].

3. CRITICAL IONIZATION VELOCITY

The critical ionization effect was first proposed by Alfvén in 1954 in the context of space plasmas as the base of his model for the origin and evolution of the solar system [9, 10]. This effect occurs for a neutral gas moving through a magnetized plasma and leads to rapid ionization and braking of the relative motion due to the transfer of the kinetic energy into ionization when "the critical velocity is reached. Although its experimental evidence

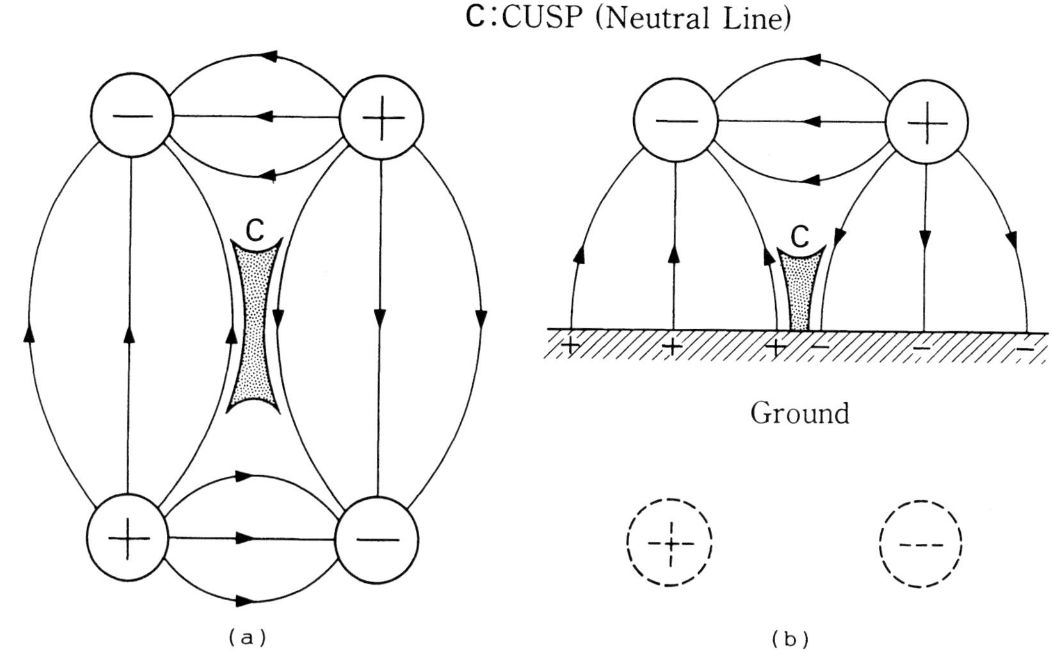

Fig. 1 Quadrupole Configuration

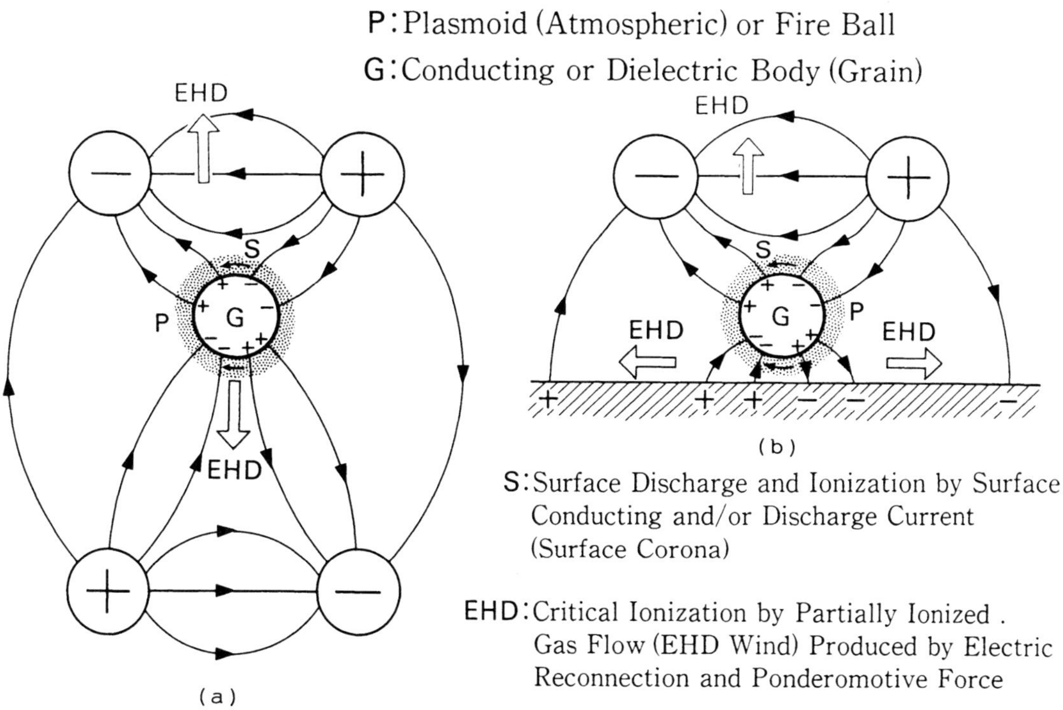

Fig. 2 Electric Reconnection (with a grain in a cusp)

has been established by a considerable number of laboratory experiments, the theoretical understanding of the process is not yet complete. Somehow, the mechanism for the transfer of the kinetic energy into ionization must rely upon a combined action of electron impact ionization and a collective plasma interaction heating electrons. Unfortunately, this subject had been left behind for a long time among the plasma community until it received a renewed interest at the Workshop on Laboratory and Space Plasmas held in Tokyo in April, 1980 [11, 12]. Since then, the critical velocity effect has occupied one of topics of plasmas, particularly from the theoretical aspect with computer simulation as the problem of a modified two-stream instability between a low-pressure neutral gas and a magnetized plasma.

The question arises what about the dependence of the critical velocity on the gas pressure, although it is known to be independent of the pressure at least in a collisionless regime. In 1986, it was found by the present author that the critical velocity concept is still valid even for a collisional regime at least with atmospheric pressure in an extended sense, based on numerous data obtained from laboratory and lightning discharge experiments [13]. There are now three (apparently but essentially two) classes of critical velocities as two threshold boundaries between three regimes, ionic (or atomic), electronic, and fully ionized. They may be defined as

$$\frac{1}{2}m_{-}v^{2}_{-c}=eV_{i} \qquad \text{for electron critical velocity} \qquad (1)$$

$$\frac{1}{2}m_{+}v^{2}_{+c} = eV_{i} \qquad \text{for ion critical velocity} \qquad (2)$$

$$\frac{1}{2}m_{a}v^{2}_{ac} = eV_{i} \qquad \text{for atom critical velocity,} \qquad (3)$$

where m, e, v_c, and V_i are the mass, electronic charge, and the ionization potential, respectively, and the subscripts −, +, and a refer to electrons, ions, and atoms, respectively.

Eq. (1) holds for a neutral gas and an electron beam in relative motion, Eq. (2) for a neutral gas and an ion beam in relative motion, and Equ. (3) holds for a neutral gas and plasma beam in relative motion, which corresponds to Alfvén's critical velocity. From Eqs. (2) and (3), the ion critical velocity agrees approximately with the atom critical velocity, since the ionic mass is nearly equal to the atomic mass. These three kinds of critical velocities have been found to be most relevant for the evaluation of both streamer and leader velocities for laboratory and lightning discharges; Eq. (1) for positive and negative streamers and for negative and dart leaders; Eq. (2) for positive leaders. Although the mechanism of the transfer of the kinetic energy into ionization is thought to rely on inelastic collisional ionization, this is a largely open problem for future investigations.

In reference to laboratory and lightning discharges, the negative and dart leader velocities or positive and negative streamer velocities are evaluated to be 2.18×10^6 m/s and 2.26×10^6 m/s from Eq. (1) for oxygen and nitrogen gases, respectively, while the positive leader velocities are 1.27×10^4 m/s and 1.41×10^4 m/s from Eq. (2) for oxygen and nitrogen, respectively in good agreement with observations [13].

4. SELF-ORGANIZATION AND DYNAMIC CHAOS

Among a variety of growing interests in nonlinear dynamics of structures for the last two or three decades, two kinds of problems are particularly intriguing, namely self-organization and dynamic chaos, and are also of great importance for the meteorologico-electric phenomena.

While a modern concept of self-organization covers the whole range of science, natural and social, the problem of self-organized formation of ordered structures or patterns of coherent motions has received growing attention among the fluid dynamics community and is considered to be most relevant to the study of the generation of larg-scale coherent structures in the meteorologico-electric environment such as typhoons, hurricanes, cyclones, tornadoes, whirlwinds, etc. In recent studies of self-organization, it has been shown that hydrodynamic helical turbulence and/or some additional symmetry-breaking factors can lead to the generation of large-scale vortex structures [14]. Essentially all of the helicity and some additional factors are of the nature of a release mechanism which makes it possible to pump some of the energy of helical turbulence into large-scale vortex structures and such a process has naturally been interpreted as a vortex dynamo. This is a typical example of self-organizational processes in nonequilibrium media and will be extended to a new electromagnetohydrodynamic regime in the last paper in the same chapter.

Chaos is a most remarkable nature as well as a ubiquitous mode of motion in nature, and its study has developed rapidly in recent years in various fields of science. Chaos is naturally generated by random flucuations that are always present in systems and the threshold for the onset of chaos is lowered with increasing noise level. It has been shown that simple scaling laws can be used to study the behavior of chaotic systems near the threshold in the presence of external noise. This kind of approach is particularly relevant for the study of noise and chaos in the meteorologico-electric environment such as the fluctuations of electrophysical parameters of the atmosphere, ionosphere, and the magnetosphere and their influence on generation and propagation of electromagnetic waves. Along this line, it has been found through a number of examples that a new kind of chaos, "fluctuational chaos", develops from weak fluctuations due to the instability of some moments of random fields [15]. Thus, the addition of external noise or weak fluctuations leads to acceleration of existing instabilities and creats new ones, including fluctuational chaos. Thus, fluctuational chaos can arise even when the amplitudes are considerably below the threshold for the onset of dynamic chaos, giving birth to new critical states and the system quikly becomes chaotic.

5. APPLICATION OF THE CONCEPTS TO METEOROLOGICO-ELECTRIC PHENOMENA

In Sections 3 and 4, it was indicated that the three concepts: electric reconnection; ponderomotive force; critical ionization velocity, are most useful for explaining natural, triggered and ball lightning phenomena, based upon a basic configuration of electric cusp in Figs. 1 and 2 which will be used all the time together with additional illustrations in more relaxed and realistic forms.

Let us proceed to explanations for real discharge phenomena. A quadrupole-like configuration of clouds and charges basically close to Fig. 1 (a) may exist in the atmosphere between four clouds in a modified form Fig. 3. If reversed field lines reconnect with each other for some reason or other near the cusp region where some dusty plasmas or another charged cloud be present, the release and transfer of the electrostatic or electric energy into the plasma flow or kinetic energy takes place in the form of leader extension between clouds, constructing discharge channels and eventually leading to cloud-to-cloud discharges.

Winter-thunderstorms and rocket-triggered lightning can also be explained on the basis of Figs. 1 and 2 (b) which indicate horizontal cloud and charge distributions favorable to triggered lightning. Consider the case when a rocket-trailing wire is impinging above on the cusp region. Then, electric field line merging to the tip of the rocket takes place, producing the ponderomotive force near the tip and forming an artificial neutral line along the wire. Leader inception and propagation occurs from the tip towards the cloud and can be explained in terms of the critical ionization velocity. In general, the drift velocity of ions or electrons in a neutral gas increases with increasing electric field below a certain critical velocity for ions or electrons where the kinetic energy of ions or electrons is thought to be transferred into ionization, causing the velocity saturation until a complete ionization of the gas is attained [7, 8, 13].

Going back to a quadrupole configuration or a dipole above the ground with a conducting or dielectric dust particle in a cusp region in Fig. 2, consider the case when the electric field is high enough to produce surface corona by surface discharge and ionization. Then, the critical velocity effect of the plasma flow outwards leads to the rapid transfer of the flow energy into ionization, producing a plasmoid or fireball which is a dust particle heavily laden with a plasma layer around a nucleus. Such a sequence of transfer processes from the electric to the flow energy and eventually into ionization makes it possible to sustain a fireball for a long time, since the electric field lines are frozen to the grain particle and are almost unchanged as long as the configuration is maintained, thus sustaining a plasmoid or fireball with continuous energy supply. Such an external source model may explain a long life time of a lightning ball [16].

Further, if the environment is considered a dusty plasma with a solid particle in a cusp, electric reconnection may produce a large-scale EHD wind flowing outwards due to the ponderomotive force in addition to a localized ionizing shock, discharge and ionization around the particle. Then, a plasmoid or fireball may have a chance to float along with the EHD wind as often observed. This is a simplest model for ball lightning and seems to provide a picture of its essential processes and features at least qualitatively [16].

Summarizing laboratory, natural, triggered and ball lightning discharges, Fig. 4 illustrates a schematic picture for all of them in a most realistic case, although their scales may be quite different. It is shown that electric reconnection or merging is the ubiquitous nature of dusty plasmas and occurs commonly when space charges are separated into a number of charge groups or clouds in the air or in the atmosphere like a system of miniature or real thunderclouds, accompanying tiny or large discharges between charge groups or parts of the clouds that may be triggered or facilitated by the EHD wind and the critical velocity effect. Thus, Fig. 4 visualizes a cascade of electric reconnection processes due to the relative movement of space charges that may be dusty charged particles, aerosol clusters or thunderclouds.

There is another class of meteorologico-electric phenomena involving or accom-

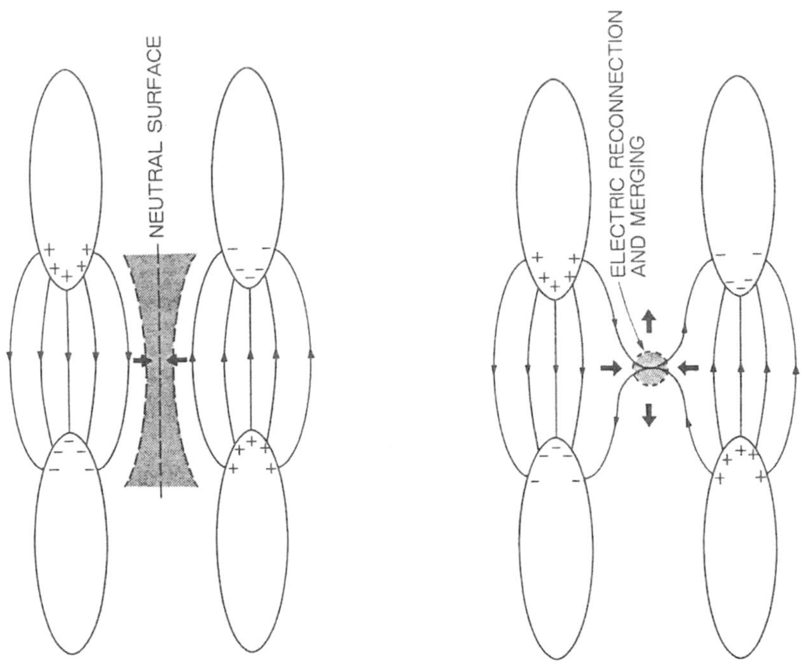

Fig. 3 Electric Reconnection (more realistic view)

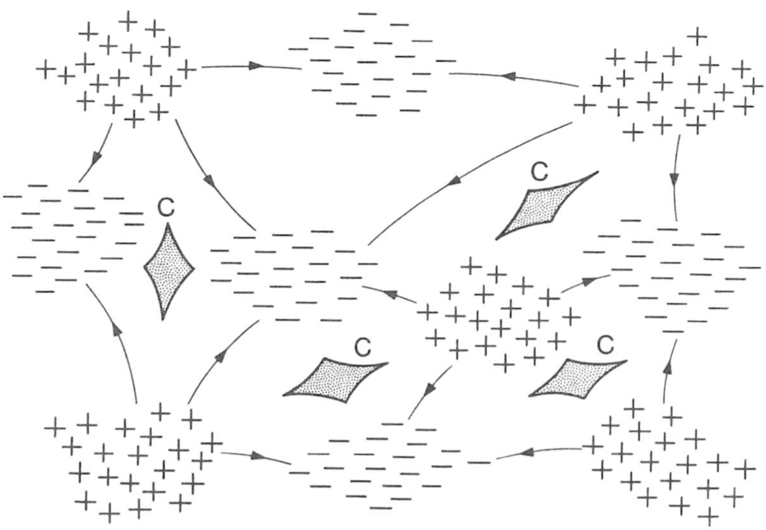

C : CUSP

Fig. 4 Electric Reconnection in Dusty Plasma or Aerosol Clusters panying some kind of electrification, not from ionization but, for instance due to frictional electricity. Possible examples are planetary rings, cometary atmospheres, intersteller dust clouds, sandstorms, whirlwinds and so on. Even for these phenomena, the concepts, electric reconnection and ponderomotive force, are still most useful, except the critical velocity, and those environments are still considered some kind of dusty plasmas. Consequently, Fig. 4 is also relevant for understanding a variety of associated phenomena. Additional two concepts, self-organization and chaos, introduced in Section 4 are becoming more and more important for possible formation of large-scale vortex structures and for noise and chaos associated with randomness of media. Phenomena of whirlwinds and sandstorms are thought to be similar to each other, although their scales may be different. The last paper in this chapter presents a new type of EMHD vortices [17, 18] derived from EMHD equations in the next section and possible evidence of it in the lower atmosphere obtained from a number of observations of patterned circle trace in a corn field in southern England [19].

6. ELECTROHYDRODYNAMIC (EHD) OR ELECTROMAGNETOHYDRODYNAMIC (EMHD) EQUATIONS

As is well known, the conventional neutral wind or flow is produced by the gradient of the mechanical gas pressure and the wind vortex by the rotation of flow, while the conducting or plasma flow and vortex are produced by the total of mechanical and magnetic forces, the latter being due to the gradient of the magnetic pressure. These HD and MHD vortices have been well investigated for large-scale structures and are presently the subject of active research for small-scale turbulence on the basis of the HD or MHD equations. As mentioned above, the meteorologico-electric phenomena, in which electric and space charge fields play a significant role, must be treated in terms of a new EHD or EMHD, taking into consideration the electric or ponderomotive force that is the gradient of electric or radiation pressure. Along this line, the present section presents the fundamental equations of EHD and/or EMHD [17].

In order to extend HD to EHD or EMHD and to establish the EHD or EMHD equations, the equation of motion of the fluid should incorporate electromagnetic forces, particularly electric or ponderomotive forces in total forces with mechanical and viscous forces. In addition, Maxwell's equations must be combined with EHD or EMHD equations and the equations of state must be accompanied for both mechanical and electromagnetic quantities. Bearing this in mind, the equation of continuity and the equation of motion of the fluid in the EHD or EMHD regime may be written, respectively, as

$$\frac{\partial \rho}{\partial t}+\nabla\cdot(\rho \underline{v})=0, \tag{1}$$

$$\rho\frac{D\underline{v}}{Dt}=\rho[\frac{\partial \underline{v}}{\partial t}+(\underline{v}\cdot\nabla)\underline{v}]=\underline{f}_T=\underline{f}_G+\underline{f}_V+\underline{f}_E+\underline{f}_M, \tag{2}$$

where ρ and $\underline{v}$ are the density and velocity of fluid, respectively, $\underline{f}_T$ the total force density,

$$\underline{f}_G = -\nabla p \tag{3}$$

the mechanical force density or gas pressure gradient,

$$\underline{f}_V=\rho[\nu\Delta\underline{v}+(\mu+\frac{1}{3}\nu)\nabla(\nabla\cdot\underline{v})] \tag{4}$$

the viscous force density, ν the kinetic viscosity, μ the second viscosity,

$$\underline{f}_E=q\underline{E}-\frac{1}{2}E^2\nabla\varepsilon+\frac{1}{2}\nabla(E^2\rho\frac{\partial\varepsilon}{\partial\rho}) \tag{5}$$

the electric force density or ponderomotive force, $\underline{E}$ the electric field, q the charge density,

$$\underline{f}_M=q\underline{v}\times\underline{B} \tag{6}$$

the magnetic force density, and $\underline{B}$ is the magnetic flux density.

The full equations of state for a perfect gas may be written

$$p=\frac{c_0^2}{\gamma}\rho_0\left(\frac{\rho}{\rho_0}\right)^{\gamma}, \tag{7}$$

$$\varepsilon_s-1=\kappa_s\rho=\rho\frac{\partial\varepsilon_s}{\partial\rho} \quad \text{or} \quad \rho\frac{\partial\varepsilon}{\partial\rho}=\varepsilon_0\kappa_s\rho=\kappa\rho, \tag{8}$$

$$\nabla\cdot\underline{D}=q=\eta\rho \quad (\eta\gtreqless 0), \tag{9}$$

$$\underline{D}=\varepsilon\underline{E}=\varepsilon_0\varepsilon_s\underline{E}=\varepsilon_0\underline{E}+\underline{P}, \tag{10}$$

$$\nabla\cdot\underline{B}=0, \tag{11}$$

$$\underline{B}=\mu_0\underline{H}, \tag{12}$$

where c_0 is the sound velocity, γ the ratio of specific heats, ρ_0 the constant equilibrium density, ε_s the specific dielectric constant, ε the dielctric constant, $\underline{D}$ the electric displacement, $\underline{P}$ the electric polarization, μ the permeability, and $\underline{H}$ is the magnetic field.

For a polytropic gas, Eqs. (3) and (5) reduce to

$$\underline{f}_G=-\nabla p=-c_0^2\left(\frac{\rho}{\rho_0}\right)\nabla\rho \qquad \text{for } \gamma = 2, \tag{13}$$

$$\underline{f}_E=\rho\left(\eta\underline{E}+\frac{\kappa}{2}\nabla E^2\right)=\rho\left(\eta\underline{E}+\kappa E\nabla E\right). \tag{14}$$

Referring to Eqs. (1) ~ (6), (9), (13), and (14) and combining Maxwell's equations, a full set of EHD or EMHD equations have the form

$$\frac{D\underline{v}}{Dt}=\frac{\partial \underline{v}}{\partial t}+(\underline{v}\cdot\nabla)\underline{v}=\nu\Delta\underline{v}-\frac{c_0^2}{\rho_0}\nabla\rho+\kappa E\nabla E+\eta(\underline{E}+\underline{v}\times\underline{B}) \tag{15}$$

$$(1+\kappa_s\rho)\nabla\cdot\underline{E}+\kappa_s\nabla\rho\cdot\underline{E}=\frac{\eta}{\varepsilon_0}\rho=\frac{q}{\varepsilon_0}, \tag{16}$$

$$\nabla\times\underline{H}=\eta\rho\underline{v}+\varepsilon_0\frac{\partial\underline{E}}{\partial t}, \tag{17}$$

$$\nabla\times\underline{E}=-\frac{\partial\underline{B}}{\partial t}. \tag{18}$$

For the quasi-electrostatic case, Maxwell's equations (17) and (18) can be replaced by Poisson's equation (extended):

$$(1+\kappa_s\rho)\nabla^2\phi+\kappa_s\nabla\rho\cdot\nabla\phi=-\frac{q}{\varepsilon_0}=-\frac{\eta}{\varepsilon_0}\rho \qquad \left(\begin{array}{l}\underline{E}=-\nabla\phi\\ \phi\text{: scalar potential}\end{array}\right). \tag{19}$$

We have thus obtained a new set of EHD and/or EMHD equations relevant for meteorologico-electric phenomena. A new finding of EMHD vortices based on the EMHD equations derived here is presented in another paper in this chapter.

7. CONCLUSION

In order to understand a variety of meteorologico-electric phenomena in the terrestrial and extraterrestrial environments physically and at least qualitatively in a unified manner, the five basic concepts: electric (field line) reconnection; ponderomotive force; critical ionization velocity; self-organization; dynamic chaos, and have proved to be useful and relevant for dusty plasmas and to real phenomena: natural lightning; triggered lightning; ball lightning; whirlwinds etc.

For these phenomena, conventional hydrodynamics (HD) or magnetohydrodynamics (MHD) breaks down and even plasma physics is also not relevant for partially ionized or electrified collisional plasmas. In fact, electrical effects such as electric and space charge fields play a significant role in those phenomena. Because of this, a new set of electrohydrodynamic (EHD) and/or electromagnetohydrodynamic (EMHD) equations have been obtained, taking into consideration the ponderomotive force and combining Maxwell's equations, and should be relevant for various meteorologico-electric phenomena.

REFERENCES

1. Degauque, P., Hamelin, J., Demoulin, B., in Nonlinear and Environmental

Electromagnetics, ed. Kikuchi, H. (Elsevier, Amsterdam, 1985) p.193.

2. Kikuchi, H., in Laboratory and Space Plasmas, ed. Kikuchi, H. (Springer-Verlag, New York, 1989) p.2, p.8.

3. Whipple, E.C., EOS, 67 (1986) 658.

4. Tsytovich, V.N., Morfill, G.E., Bingham, R., De Angelis, U., Comments Plasma Phys. Controlled Fusion, 13 (1990) 153.

5. Kikuchi, H., in Environmental and Space Electromagnetics, ed. Kikuchi, H. (Springer-Verlag, Tokyo, 1991), this issue.

6. Hones, E.W., Jr., ed. Magnetic Reconnection (AGU, Washington, D.C., 1984).

7. Kikuchi, H., in Proc. 6th Symp. Electromag. Compat. Zurich (1985) p.47.

8. Kikuchi, H., in Laboratory and Space Plasmas, ibid., p.332.

9. Alfven, H., On the Origin of the Solar System (Oxford Univ. Press, Oxford).

10. Alfvén, H., Arrhenius, G., Evolution of the Solar System (NASA, Washington, D.C., 1976),

11. Raadu, M.A., in Relation between Laboratory and Space Plasmas, ed. Kikuchi, H. (D. Reidel, Dordrecht, 1981) p.13.

12. Petelski, E.F., ibid., p.23.

13. Kikuchi, H., in Laboratory and Space Plasmas, ed. Kikuchi, H. (Springer-Verlag, New York, 1989) p.336.

14. Moiseev, S.S., Sagdeev, R.Z., Tur A.V., Khomenko, G.A., Yanovskii, V.V., Sov. Phys. JETP, 58 (6) (1983) 1149.

15. Buts, V.A., Kolykhalov, P.I., Moiseev, S.S., Pungin, V.G., in Proc. 10th Symp. Electromag. Compat. Wroclaw (1990) p.855.

16. Kikuchi, H., Annales Geophysicae, Special issue (XIV EGS General Assembly, Barcelona, 1989) p.286.

17. Kikuchi, H., in Nonlinear Dynamics of Structures, ed. Sagdeev, R.Z., Erokhin, N.S. (World Scientific, Singapore, 1991).

18. Kikuchi, H., in Environmental and Space Electromagnetics, ed. Kikuchi, H. (Springer-Verlag, Tokyo, 1991) this issue.

19. Meaden, G.T., J. Meteorology, 14 (1989) 151.

14.2 Discovery of a New Electromagnetic Phenomenon in the Atmosphere:
An Electrified Vortex and Its Physical Properties as Revealed by Patterned Ground Traces and Radio-Frequency, Electromagnetic, Acoustic and Luminous Effects

G.T. MEADEN

Circles Effect Research Group (CERES), Tornado and Storm Research Organisation (TORRO), Bradford-on-Avon, BA15 1LD, UK

An electromagnetic vortex system has been discovered which lays down spiral-centred circular patterns in crops and upon the ground. Much information is now available about the processes that create the vortices which are caused by the breakdown of a columnar eddy vortex downwind of hills. Besides eyewitness reports of the vortices in action there is proof of simultaneous acoustic, luminous and electromagnetic effects. The evidence points to an electrical source of natural radiation in the atmosphere, and also explains reports of previously-unidentified luminous phenomena in the air. The new phenomenon is bigger and longer-lived than ball lightning.

INTRODUCTION

In recent years the author has studied a hitherto-unrecognised electromagnetic vortex phenomenon which is caused by a natural electrical effect - possibly a low-heat plasma - which develops within a rapidly-spinning mass of air in the atmosphere. Not only does the assembled evidence point to a vortex of spinning charges and source of natural electromagnetic radiation, with its necessary consequences for radio-communications interference and radar ghost-images, but the work explains numerous reports of previously-unidentified luminous phenoemna seen spinning close to the ground or high in the air. If a plasma is involved, the new electrical phenomenon could be related to ball lightning although immediately by its much larger size and longer lifetimes. Alternatively the electrical effects may be electrostatic in origin. This paper summarises the evidence for the effect, discusses what is known of its origins, and indicates the lines of inquiry which future work is likely to take.

THE CIRCLES EFFECT

The clearest evidence for the development of an electrically-charged atmospheric vortex of unknown type comes from examination of the distinctive circular-symmetric traces left upon the ground and in standing crops by the descent of such vortices of spinning air. The author's investigation of this subject began in August 1980 with his discovery and first scientific analysis of circles formed in Wiltshire in Southern England (1). Since then, information on 850 circles had been gathered by August 1989.

The most common finding is a circular area of damage in which stalks of growing crops have been pushed over by some intense rotating force having axi-symmetrical symmetry. Often the pliable stems are not broken but are bent just above ground level. In most cases a conspicuous spiral trace is discernible, extending from a tight centre up to a diameter of 32 metres or more at which a sharp perimeter appears (Fig. 1, taken by the author from the air over Wiltshire, England, in August 1989, shows a 32-metre ringed circle and some small 2 - 3 metre circles).

Both senses of rotation are found. Sometimes the centre is not flattened, but remains standing as a cone or pyramid showing that it has been struck by a descending ring which ultimately expanded (this is certain proof of a ring vortex). Clearly, the crop-damaging vortices must be descending ones. As this is highly unusual, we need to suppose the occurrence of vortex breakdown of the usual spin-up state.

The simplest indications that electrical effects are involved come from nocturnal eyewitness accounts of luminous balls of light descending into fields and scoring out the circles. At times, sounds typical of electrical charges have been heard. There are other effects too which are discussed later, but first we should explore the principal effects which have been learnt about the circles themselves.

Fig. 1. A huge ringed circle, diameter 32 metres, formed in a wheat crop by the strongly spinning, downward pressure of an ionized-air vortex (on the slopes of a hill at West Overton, Wiltshire, England). Note also the small circles created by smaller vortices. The straight lines are tractor marks.

The most surprising of the circle systems are the five-circle sets comprising a big central circle and four nearly-equidistant satellites of similar, smaller size (Fig. 2) and the various ringed-circle types. However, circles also form in other cluster combinations, some with multiple rings. Sketches or photographs of two dozen principal types are given in the books The circles effect and its mysteries (2) and Circles from the sky (3). Groups of two, three, four and five are relatively common, from which it is clear that there is considerable evidence to indicate that in many cases the parent vortices are interlinked and hence interdependent on one another.

To begin with, consider the simple triplet circle systems composed of threee circles close together in a straight line (see Fig. 2 again). This pattern is found most years in England. In a recent paper (1989) van Heijst and Kloosterziel (4) have shown how a monopolar vortex in a rapidly-rotating fluid medium breaks down naturally into tripolar in-line vortices. That this takes place in laboratory conditions suggests that such spin stability is a natural tendency for vortex phenomena in sheared geophysical fluid environments; hence although the parallel is not exact we infer that our observations in the real atmosphere are witness to this very proposal. It also goes to show how little is known even now about the microscopic behaviour of the atmosphere under conditions of extreme shear and very high spin. These effects, coupled with the electromagnetic ones, go to the very heart of the problem.

Fig. 2. Quintuplet circles formed by the descent of a primary central atmospheric vortex, a dramatic manifestation of the 'circles effect'. The field in which they formed has been harvested but the circles still show up well. The circles in the next field (not harvested) include a set of triple circles in a line. A steep hill, rising more than 100 metres, is immediately to the left of the picture (Bratton, Wiltshire, August 1987).

ACOUSTIC EFFECTS ACCOMPANYING CIRCLE-FORMATION

Several eyewitnesses have reported audible effects at the time of formation of such circles (2). Typical is a humming noise that is often likened to the sound of corona discharge which is a familiar accompaniment to high-tension electrical installations. These sounds imply the presence and motion of electric charges, just as they do for their better-studied atmospheric counerparts the whirlwind and the tornado.

It is well-known to American researchers that the common whirlwind generates its own electric field. The field is electrostatic, arising as it does from the triboelectric effect with the charges carried on dust particles which have been scoured off the surface of the ground by the energy of the rotating winds (5),(6),(7). Electric-field values exceeding $-450\ \mathrm{V\ m^{-1}}$ at a distance of 30 metres have been measured as compared with a pre-existing fair-weather field of $+50\ \mathrm{V\ m^{-1}}$ or more.

The electric field of tornadoes is known to generate acoustic and luminous effects (8). It could be similar with our circles-effect vortices. In fact there is a circle case from France 35 years ago (St Sou plet, Nord, October 1954) in which circle formation was coincident with a red ball of light and acoustic noise, and two similar cases were witnessed this year in England (Wiltshire, 29 June, and Kent, 10 August 1989). Both light and noise imply the existence of electrical discharges. As we shall find later, there is every reason to expect that the naturally-occurring wind vortices which form the circles are also electrical generators.

THE EFFECTS WHICH ARE IMPLIED BY RINGS AROUND THE CIRCLES

When rings surround circles, they are narrow rings concentric with the circle and parallel to the circumference. Moreover, they have steep sides indicative of a certain precision at the time of development. The flattening force evidently exhibited an amazingly fine degree of control. Rings in both directions are found. One needs to explain how a spinning vortex of wind is able to depress the crop so neatly at the edges of the main circle and at the edges of the rings (and satellite circles too).

One suggestion that has been proposed assumes that the intense circulating current in the primary circle induces a secondary flow of ions in the region beyond the main perimeter. Put another way, the central field has an oscillating waveform, the rings developing at peaks in the external field. As many as three concentric rings have been observed, but we have reason to expect that many more remain to be discovered - up to six or eight possibly. In order to understand the steepness of the walls of the circle and the rings one may surmise both the existence of an intra-vortical critical windspeed to knock over the standing crop and an interfacial boundary or tension between the electrically-charged vortex and the extra-vortical surrounding air. Thus one finds the sharp edge at the plasma-pause.

However it looks perhaps to be difficult to set air in motion by electrical forces at N.T.P. Perhaps the much wider rings are formed, rather, by core-wall displacement or ejection, the ionized air in them being propelled by mechanical, i.e. aerodynamic, forces. Either way, remanent effects seem to be left in the soil for some time afterwards, possiblydielectric polarisation of susceptible particles in the soil may be involved. It is to be noted that in Britain all the complex ringed circle systems are found over chalky soils or over limestone (i.e. so far as isknown).

PROPOSED ORIGIN OF THE DESCENDING VORTICES

Our work of the last ten years suggests that the vortices responsible for the circular traces are likely to be created by geophysical columnar- or trailing vortices operating in the lower atmosphere and that they are set into motion by windflow past topographical obstructions. The circles, or perhpas most of them, result from the phenomenon of vortex breakdown. The ideas are detailed elsewhere (2), (3). Here it is opportune to introduce the experimental work of Tamai, Asaeda and Tanaka (9) who showed how arched vortex structures could develop in front of and behind even a regularly-shaped like a hemispherical hump. Figures 3, 4 and 5 of their paper exemplify this impressively, and it is to be hoped that more work of this type will be performed in the future using models of real hills. Already I have been able to show that circle-making vortices do form even as far as 60 to 100 hundred times the hill height in a direction downwind of the hill (e.g. the quintuplet at Aylesbury, Central England, in July 1989) (3).

In Scotland Jenkins et al (10) have studied windflow patterns beyond the quasi-conical island of Ailsa Craig and have found valid evidence for a trailing-vortex structure for certain windspeeds and directions. Similar work has been done for the Rock of Gibraltar (11). The vortex structure may adopt a horseshoe-shaped twin-roll character vertically or horizontally oriented or take the form of a single trailing vortex which resembles in style and behaviour the better-studied aircraft wake vortices. We think that similar effects develop in the lee of hills and escarpments, and that the geophysical vortex which leads to the circular traces or the circles effect in crops has its origin in a true trailing vortex by a mechanism yet to be adequately researched (2), (3).

Nonetheless, we already know that an energetic perturbation intrudes which disturbs the relative tranquillity of the dynamic stability of the spinning column and compels a component of its energy, concentrated into a small volume, to follow a route that leads ultimately to the circles effect (via self-destruction on the ground if directed downwards) or to steady dissipation in the atmosphere if directed upwards or outwards.

An essential pre-requisite for vortex generation seems to be a quiet airmass having good, stable stratification which remains effectively undisturbed even in the presence of a steady, weak airflow. Then another airmass encroaches, marked by a microfrontal boundary, with higher windspeed and often new wind direction and different thermo-hygrometrical and baro-electrical properties. As the windspeed rises vortex tubes develop downwind of critically-shaped obstacles, like round hills and the spurs of scarps, much as suggested by the experiments of Tamai, Asaeda and Tanaka (9), but the final act is a system breakdown which provokes vortex-shedding or ring-vortex generation and causes the circles effect. In practice, the trailing vortex may exist for a finite time ranging from many seconds to many minutes, whereas the breakdown state is short-lived and associated with the final stage of a briskly-rising rising wind. An excellent example of circles-effect formation from Bratton/Westbury (Wiltshire) occurred on 19 July 1989. Nine circles formed in the middle of the night $1\frac{1}{2}$ to 2 km downwind of the hill-spur of Westbury, in exact alignment with the new wind direction which overrode the pre-existing calm-to-light air of a northerly wind. The presence of well-stratified air, as develops in an inversion, is typical, and it is well-known that vorticity is inherent or incipient in the shear zone at the top of the inversion or boundary layer.

Several dated examples can be given, particularly those which are associated with weak fronts of various kinds. In fact, it is partly because of the frequent passage of sea-breeze fronts that hilly (and chalky) Southern England with its numerous cerealfields owes its justified reputation as the home of the circles-effect problem. In the summer months sea-breezes penetrate deep inland, sometimes reaching a hundred kilometres from the coast by late evening or after dark when conditions have otherwise fallen calm. Sometimes the fronts retrograde, pushed back by a rising gradient wind, as at Cheesefoot (Hampshire) after midnight on 18 June 1989 on which occasion a ringed circle was formed. Splendid research on sea-breeze fronts has been carried out in Southern England by J.E. Simpson (12).

Whatever the sequence of critical factors, it is evident that the vortex has acquired a considerable amount of energy compacted into a small volume and whose kinetic origins are ultimately traceable to the energy of the windfield which is diverted and focussed by a hill system or rugged terrain. J.T.Snow has discussed ideas regarding vortex breakdown in private correspondence with the author. A detailed report will be given in 1990 at the first international conference on the circles effect (Circles from the Sky). The airflow into the up-spinning vortex becomes so intense that an unstable ring-vortex develops, which plunges downwards to the ground. Vortex breakdown is well-known from laboratory work, and has been seen to occur in tornadoes (13) and waterspouts, and also in dust-devils.

Foremost is the spin energy of revolution manifest as a fast-spinning axi-symmetrical ring-vortex or spherical vortex. This spin regime leads to direct or indirect excitation of atoms and molecules of air. The spinning charges may be ions carried along on aerosol particles like dust, pollen, sea-salt or water vapour, or they may be metastable assemblies of ions and electrons excited by some other means. In either event it is these excited, charged states which give rise to the observed luminous and acoustic properties of the vortex.

LUMINOUS PROPERTIES AND BALLS OF LIGHT

Sometimes the spinning vortex produces a coloured light, sometimes pulsating. Reds and orange-tinted objects are most frequently seen, colours which correspond to the emission spectrum of atmospheric nitrogen. As an example we cite the Japanese event which happened in the mountain part of Yamagata Province in August 1986 (S. Namiki and N. Yamaguchi). Overnight on 8th-9th August an almost circular area 6-m in diameter was blasted into a pond 6 km from the town of Nishikawa, and identified by a strongly-marked radial pattern in a floating bed of reeds; at the same time 20 tonnes of water vanished from this rice paddy. At 8 p.m. on 8th at a nearby house a period of severe television interference indicative of strong local electromagnetic radiation was possibly associated with this event, as perhaps also a report of the descent of a bright orange light of 'ball lightning' (although the simultaneity of these events is unproven).

Various examples from other countries of the world can be given (cf Meaden 1989) including the French event at St Souplet, mentioned above, in which a red light scoured a three-metre circle into a plot of spinach late one autumn night (2), (3). Humming noises characteristic of corona or other electrical discharge effects are often reported. Illuminated vortex tubes have been sighted on rare occasions. One such case was at Silbury Hill one night in 1988 (2) and another, the same summer, was at Bratton, also in Wiltshire. A seemingly related case is known for Japan (Tomakomai, Hokkaido). At Silbury the tube of light was a kilometre long between cloud and ground. These tubes could possibly be the visible representation of part of the primary columnar vortex along which a concentrated ionized-vortex or its separately-projected constituents may travel.

If the process happens this way, the consequence would be that the aerosol-bearing charges or the ionized atoms responsible for the discharges get gathered from a vast volume of air or cloud-substance and are piped along a lengthy tubular vortex until concentrated into a ball a few metres in diameter by the intense spin known to accompany the circles-effect vortex. This would create a quasi steady-state situation whereby a fresh flow of ions continually replaces the ions lost by discharge and leakage, and so explain the long lifetimes and big diameters of the luminous body. It contrasts with the ball-lightning situation in which if the ball is self-consuming from the moment of its creation under thunderstorm conditions the ball can hardly last for more than a couple of seconds or so (14). Consequently, and because also the circles-effect vortex is not known to have displayed any obvious heating or desiccating effects in the 850 circles we have studied, a low-density plasma state of matter may be indicated, compared with the fully-ionized plasma expected of conventional ball lightning (with its well-known theoretical instability problems) (14). Of course, because of the intense spin the charges get concentrated into the external region of the vortex - the plasma pause - where their density is high enough to generate luminosity and other radio-frequency effects. This would be the result for those ring-vortex or spherical-vortex cases which arise solely from the phenomenon of vortex breakdown.

Bare-earth centres to circles are surprisingly common. It is possible that there are occasions when the vortex aimed for it, or for the pointed heads of the crop adjacent to it, just as lightning aims for the lightning conductor because it is following the path of greatest conductivity. This intimates a liaison with the electric field because the local electron flow was higher, proving of course that the vortex had a field too. Hence, cyclotron-resonance effects occur in which ions and electrons circulate more efficiently if their Larmor radii correspond to the wind-vortex radii (15).

INFERRED ELECTRICAL PROPERTIES

Radio-frequency effects have been inferred for various reasons. There are numerous incidents of radio-communication interference in the vicinity of spinning balls of light (2). It is topical to mention the Osaka aircraft case at this Japanese conference. At the same time that the automatic direction finder failed and radio reception was blotted out by severe interference the pilot noticed an incandescent ball of light behind his port wingtip. This elliptical green vortex, which persisted while the aircraft covered 90 km as it approached Osaka airport, may have been fed by a steady flow of charge into a wimgtip trailing vortex. The time was 1900 on 21 March 1965 (2).

Hundreds of examples of previously-unexplained spinning balls of light are known (2), (3), (14). Sometimes pulsed lights are reported; and of course it is well known that tornado vortices can be tracked by their severe sferics or RF emissions (16).

Tornadoes too generate luminous RF pulsed emissions, the Blackwell (Oklahoma) case being one such well-reported example (17). We also believe that some of the clear-air radar angels observed on plan position radar indicators in the past, especially near hills and in association with fronts, were caused by electrified vortices of the type now under study. Monitoring by conventional radar and by Doppler sodar (a vertically-oriented acoustic radar) is a most promising area for immediate study by those who are able to approach the circles-effect problem from this angle. There also occurred in the Pacific near Hawaii in October 1986 an event involving radar obervations from two Japanes ships. Although nothing could be seen visually, something which Y.H. Ohtsuki suggests was a plasma vortex was seen on both radars. Later, after dark the plasma appeared on both radars again, but this time a light was visible and sound was heard (3).

Of what matter might the electrified vortices be constituted?

A VOLUME OF CHARGES SPINNING RAPIDLY

The evidence of our research points to a vortex of electrically-activated air, probably ionized by externally-fed electric discharges or triboelectric ions, as the origin of the circular-symmetric crop markings which are found worldwide. One possibility is that the charges which are fed into the system originated elsewhere. They could be charges carried along on marine salt in the humid air of the sea-breeze front. Or like the charged dust particles of land devils the charges may be on aerosol particles like pollen, soil-dust, or fertiliser powder etc. On the other hand the electrical phenomenon may be related to the ball lightning problem (2),(14). There is little question nowadays about the reality of ball lightning. The experiments of Powell and Finkelstein at reproducing plasmas similar to short-lived ball lightning in the laboratory also serve to demonstrate the practicability of Kapitsa's theoretical ideas with regard to resonance-focussing of microwave radiation. Ohtsuki too has made ball lightning by electrostatic charging of impurity-laden air (18), and is working on microwave interference in ordinary air with good prospects of success.

Moreover, sightings of luminescent tubes of great length give credence to the hypothesis of long-distance channelling of externally-fed ionized particles along wave-guide type up-spinning vortices as a continuous source of fresh energy. By this means electrical energy is pumped into a specific volume of air, and gives rise to the luminous discharge phenomenon (at the time of vortex breakdown) which is axi-symmetric and self-confining. Its longevity in the atmosphere is problematical as it depends on various local circumstances, but it may be quite long-lived (minutes, or tens of minutes) or short-lived (seconds, or tens of seconds). But of course, if circumstances should direct it towards the ground (as in a vortex breakdown situation) it is normally doomed to extinction because it immediately loses its identity soon after impact. Certainly, either a plasma or an ionized airmass is intimated by the collected evidence - namely, a mixed assembly of ions and electrons giving rise to luminous discharges which might emanate from the energy of recombination or de-excitation of metastable states. For whatever reason, it is suspected that the spinning volume soon acquires a net positive charge, possibly by loss of energetic electrons, and the rotational forces are such that for an appropriately-directed local electric field the imprint that is made at ground level is what we call clockwise.

The intensity of the impact of the ionized vortex with the crop (and indeed at times its precise location) depends on the relative magnitudes and directions of the local electric field. And the net direction of the lying crop is influenced by external aerodynamic factors promoted by the local windflow or electromagnetic influences as via cyclotron-resonance interactions.

The source of the internal radiation is very much debatable at the present time. Molecular dissociation would be thought to be of minor importance. We know that the balls of light under study produce RF noise on the radio-communication bands and television bands and that pulsed luminosity is often reported. In addition, the narrow rings of the circular ground-traces bear witness to possibly oscillatory flow if explicable in terms of RF standing-wave or resonance interactions which in their turn cascade the ionization further. However, the rings and satellites may originate from direct aerodynamical activities - the satellites by ejection from the vortex core at the time of ground impact, the broad rings by displacement of the vortex shell. Either way, it does seem that there are occasions when the circles-effect vortex produces pulsed electromagnetic effects just like tornado and waterspout vortices do.

SUMMARY

As with other known ionization phenomena in the atmosphere (as caused by lightning strokes, meteor tracks, corona discharges, etc), circles-effect vortices are likely to be a source of radio-frequency interference and spurious RF signals affecting radio and aircraft navigation, and so need to be considered in electromagnetic compatibility. The electrically-charged volume seems best described as an ionized vortex, possibly a plasma, which is often ball-shaped and could have some affinities with ball lightning (2), (14).

We may regard the circles effect as a powerful new research tool whereby analyses of the impressions bring us unexpected new methods for studying the behaviour of the lower atmosphere and the electrical effects that develop there. It is indeed a new topic in boundary layer meteorology which is well worthy of high-level seriously-funded research.

REFERENCES

1. Meaden, G.T. (1981). Mystery spirals in a Wiltshire cereal-field. J. Meteorology, 6, 76-80.

2. Meaden, G.T. (1989). The circles effect and its mysteries. Artetech Publishing Co., Bradford-on, Wiltshire, England. £14.00.

3. Meaden, G.T. (1990). Circles from the sky. Proc. First Internat. Conference on the circles effect, Oxford (to be held 23 June 1990). Artetech Publishing Co., Bradford-on-Avon, Wiltshire, England. £18.00 (papers from ten authors).

4. Van Heijst, G.J.F. and Kloosterziel, R.C. Tripolar vortices in a rotating fluid. Nature, 338, 569-571, 1989.

5. Freier, G.D. The electric field of a large dust devil. J. Geophys. Res. 65, 3504.

6. Crozier, W.D. (1964). The electric field of a New Mexico dust devil. J. Geophys. Research 69, 5427-5429.

7. Crozier, W.D. (1970). Dust devil properties. ibid. 75, 4583-4585.

8. Vonnegut, B. (1960). Electrical theory of tornadoes. J. Geophys. Research 65, 203-212.

9. Tamai, N., Asaeda, T. and Tanaka, N. (1987). Vortex structures around a hemispheric hump. Boundary-Layer Meteorology, 39, 301-314.

10. Jenkins, G.J., Mason, P.J., Moores, W.H. and Sykes, R.I. (1981). Measurements of the flow structure around Ailsa Craig, a steep, three-dimensional, isolated hill. Quart. J. Roy. Met. Soc., 107, 833-851.

11. Cook, N.J., Coulson, B.H. and McKay, W. (1978). Wind conditions around the Rock of Gibraltar. J. Ind. Aerodyn., 2, 289-309.

12. Simpson, J.E. (1977). Inland penetration of sea-breeze fronts. Quart. J. Roy. Met. Soc., 103, 47-76.

13. Pauley, R.L. and Snow, J.T. (1988). On the kinematics and dynamics of the 18 July 1986 Minneapolis tornado. Monthly Weather Review, 116, 2731-2736.

14. Meaden, G.T. (Ed.) 1990). Ball lightning studies. Artetech Publishing Co., Frome Road, Bradford-on-Avon, Wiltshire, England.

15. Kikuchi, H. (1989). Private communication.

16. Jones, H.L. (1951). A sferic method of tornado identification and tracking. Bull. Amer. Met. Soc., 32, 380-385.

17. Jones, H.L. (1955). The tornado pulse generator. Weatherwise, 18, 78-79.

18. Ohtsuki, Y.H. (1989). See Science of ball lightning. Proc. First Conf. on Ball Lightning, Tokyo 1988. World Scientific, P.O Box 128, Singapore 9128.

14.3 Electromagnetohydrodynamic Vortices and Corn Circles

H. KIKUCHI
Nihon University, College of Science and Technology, Tokyo, 101 Japan

ABSTRACT

A new type of large-scale vortex formation is indicated in helical turbulence in terms of hydrodynamic, electric, magnetic, and space charge fields in the presence of an external electric (and magnetic) field. It is called "Electromagnetohydrodynamic Vortices" and is generated as a result of self-organization by pumping out some of the energy of helical turbulence into large-scale motions. The theory is based upon a new set of electromagnetohydrodynamic (EMHD) equations and Maxwell's equations and a new equation of EMHD vortices is derived. Possible evidence of EMHD vortices in the lower atmosphere has recently been brought by numerous observations of circular symmetric ground patterns in a corn field in southern England.

1. INTRODUCTION

As is well known, the conventional neutral wind or flow is produced by the gradient of the mechanical pressure and the wind or flow vortex by the rotation of flow, while the conducting or plasma flow and vortex are produced by the total of the mechanical and magnetic forces, the latter being due to the gradient of the magnetic pressure. These hydrodynamic (HD) and magnetohydrodynamic (MHD) vortices have well been investigated for large-scale structures and are presently the subject of active research for small-scale turbulence on the basis of the HD or MHD equations.

The traditional presentation of turbulence is connected with that kind of disordered neutral and/or charged particles or fluid motions in which any large-scale perturbation transfers its energy through random chaotic motions to ever smaller scales. It would seem that such ideas about turbulence are incompatible with the existence of coherent structures in it.

In recent studies of self-organizational processes in nonequilibrium media, it has been shown that hydrodynamic (HD) helical turbulence and/or some additional symmetry-breaking factors can lead to the generation of large-scale vortex structures [1]. Essentially all of the helicity and additional factors are of the nature of a release mechanism which makes it possible to pump some of the energy of helical turbulence into large-scale vortex structures and such a process has naturally been interpreted as a vortex dynamo. Thus, the generation of HD vortices in HD helical turbulence has well been investigated on the basis of the Navier-Stokes equation by using a mathematical procedure, namely functionals and the random-force method [2].

For the meteorologico-electric phenomena, however, involving or accompanying electric discharge, ionization or electrification (such as thunderstorms, cold and polar fronts, jet streams, typhoons, hurricanes, cyclones and tornadoes in the terrestrial environment, and planetary lightnings, rings, cometary tails, dark interstellar clouds in the extraterrestrial environment), the roles of electrical effects such as electric and space charge fields are becoming signif icant. As a result, conventional hydrodynamics (HD) or magnetohydrodynamics (MHD) breaks down, and even conventional plasma physics is also not relevant for these phenomena, because their environments are considered neither fully ionized gases nor collisionless plasmas as often assumed for conventional plasmas. They may be partially ionized and collisional gases, dusty or dirty plasmas containing charged dust grains or aerosols. Consequently, conventional HD or MHD should be extended to a new area of electrohydrodynamics (EHD) and/or electromagnetohydrodynamics (EMHD). In this context, the present author has described a new set of the fundamental equations of electrohydrodynamic (EHD) and/or electromagnetohydrodynamic (EMHD) equations together with Maxwell's equations, taking into consideration the electric or ponderomotive force that is the gradient of the electric or radiation pressure, in another accompanying paper in this chapter [3].

On the basis of new EHD or EMHD equations, the present paper examines how a coherent large-scale of HD vortex structures would be modified by a small scale of additional helical turbulence in terms of the electric, magnetic and space-charge fields in the presence of an external electric (and magnetic) field. With the aid of a mathematical procedure similar to that for HD turbulence, a new equation of large-scale EMHD vortices is obtained, producing two additional terms due to EMHD helical turbulence in the presence of electric and space charge fields. These terms cause additional helicity and facilitates the evolution to large-scale coherent vortex structures. In fact, the growth rate of the

EHD and/or EMHD vortex can be greater than that of the pure HD vortex. Consequently, a new type of large-scale vortex formation is predicted theoretically in EMHD helical turbulence. It may be called "Electromagnetohydrodynamic vortices" and is generated as a result of self-organization by pumping out some of the energy of EMHD helical turbulence into large-scale motions.

Possible evidence of EMHD vortices in the lower atmosphere has recently been brought by a number of observations of circular symmetric ground patterns in a corn field in southern England [4]. Explanations for some of these phenomena, the so-called "Corn Circles" are given, based upon a new theory of EMHD vortices described in this paper.

2. BASIC EQUATIONS FOR THREE COMPONENTS

We consider an EHD and/or EMHD regime in an external electric field constituting the background of helical turbulence, i.e., stationary, homogeneous, isotropic, but without being invariant under reflection, in terms of space charges, electric and magnetic fields in addition to the velocity and density of the fluid considered in the HD regime. When some small large-scale perturbations are created in its turbulent background, the turbulent pulsations acquire additional random fluctuations as a result of their interactions and the total fields can be written in the form for the velocity $\underline{v}$, density ρ, electric field $\underline{E}$, magnetic field $\underline{B}$, and space charge density q, respectively

$$\underline{v} = \langle \underline{v} \rangle + \underline{v}^t + \tilde{\underline{v}}, \qquad \langle \underline{v}^t \rangle = \langle \tilde{\underline{v}} \rangle = 0, \tag{1}$$

$$\rho = \langle \rho \rangle + \rho^t + \tilde{\rho}, \quad \langle \rho^t \rangle = \rho_0 = \text{const}, \langle \tilde{\rho} \rangle = 0, \tag{2}$$

$$\underline{E} = \langle \underline{E} \rangle + \underline{E}^t + \tilde{\underline{E}}, \langle \underline{E}^t \rangle = \underline{E}_0 = \text{const}, \langle \tilde{\underline{E}} \rangle = 0, \tag{3}$$

$$\underline{B} = \langle \underline{B} \rangle + \underline{B}^t + \tilde{\underline{B}}, \qquad \langle \underline{B}^t \rangle = \langle \tilde{\underline{B}} \rangle = 0, \tag{4}$$

$$q = \langle q \rangle + q^t + \tilde{q}, \qquad \langle q^t \rangle = q_0, \langle \tilde{q} \rangle = 0, \quad q = \eta\rho \, (\eta \gtreqless 0) \tag{5}$$

where the angular bracket $\langle \ \rangle$, superscript t and tilde $\sim$ refer to an ensemble average, turbulent pulsation, and additional part of random fluctuations. It has also been shown in a new EHD and/or EMHD regime that if the turbulence is helical, i.e. stationary, homogeneous, isotropic, but not invariant under reflection, the amplification of perturbation seed eddies is possible, leading to the appearance of large-scale EMHD vortex structures by the transfer of energy from small-scale to large-scale sizes, as illustrated typically in Fig. 1 for EHD and/or EMHD cases [5].

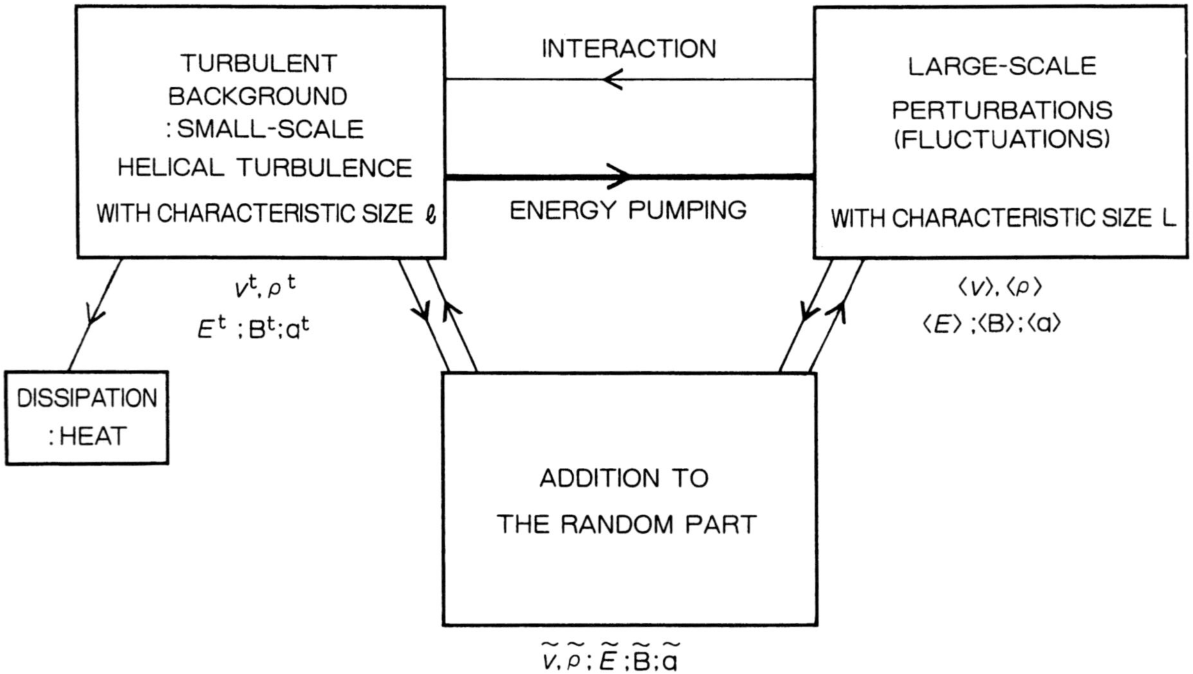

Fig. 1. EHD Scheme with Helical Turbulence

Substituting Eqs. (1) ~ (5) into the EMHD equations together with Maxwell's equations (Eqs. (15) ~ (18) in [3]), we obtain the equations for the turbulent background, averages and additional fields of random fluctuations retaining only the nonlinear terms with respect to turbulent background fields and neglecting other nonlinear terms, as follows:

2.1 Equations for Turbulent Background

$$\frac{\partial v_i^t}{\partial t}+v_k^t\frac{\partial v_i^t}{\partial x_k}=\nu_0\Delta v_i^t-\frac{c_0^2}{\rho_0}\frac{\partial \rho^t}{\partial x_i}+\kappa E^t\frac{\partial E^t}{\partial x_i}+\eta(E_i^t+v_j^tB_k^t-v_k^tB_j^t), \tag{6}$$

$$\frac{\partial \rho^t}{\partial t}+\frac{\partial}{\partial x_k}(\rho^t v_k^t)=0, \tag{7}$$

$$(1+\kappa_s\rho^t)\frac{\partial E_k^t}{\partial x_k}+\kappa_s E_k^t\frac{\partial \rho^t}{\partial x_k}=\frac{\eta}{\epsilon_0}\rho^t , \tag{8}$$

$$\frac{\partial B_i^t}{\partial t}=\frac{\partial E_j^t}{\partial x_k}-\frac{\partial E_k^t}{\partial x_j}, \tag{9}$$

$$\eta\rho^t v_i^t+\epsilon_0\frac{\partial E_i^t}{\partial t}=\frac{1}{\mu_0}\left(\frac{\partial B_k^t}{\partial x_j}-\frac{\partial B_j^t}{\partial x_k}\right) , \tag{10}$$

where

$$\varepsilon_s-1=\kappa_s\rho=\rho\frac{\partial \varepsilon_s}{\partial \rho} \qquad \rho\frac{\partial \varepsilon}{\partial \rho}=\varepsilon_0\kappa_s\rho=\kappa\rho \tag{11}$$

and ν_0 the kinetic viscosity, c_0 the sound velocity, ρ_0 the constant equilibrium density, ε_s the specific dielectric constant, ε the dielectric constant, and μ the permeability.

2.2 Averaged Equations

$$\frac{\partial\langle v_i\rangle}{\partial t}+\langle\bar{v}_k\frac{\partial v_i^t}{\partial x_k}\rangle+\langle v_k^t\frac{\partial \bar{v}_i}{\partial x_k}\rangle=\nu_0\Delta\langle v_i\rangle-\frac{c_0^2}{\rho_0}\frac{\partial\langle\rho\rangle}{\partial x_i}+\kappa E_0\frac{\partial\langle E\rangle}{\partial x_i}$$
$$+\eta[\langle E_i\rangle+\langle\bar{v}_jB_k^t\rangle-\langle\bar{v}_kB_j^t\rangle], \tag{12}$$

$$\frac{\partial\langle\rho\rangle}{\partial t}+\frac{\partial}{\partial x_k}[\langle\tilde{\rho}v_k^t\rangle+\langle\rho^t\bar{v}_k\rangle]=0, \tag{13}$$

$$(1+\kappa_s\rho_0)\frac{\partial}{\partial x_k}\langle E_k\rangle+\kappa_s[\langle\widetilde{E}_k\frac{\partial\rho}{\partial x_k}\rangle+\langle E_k^t\frac{\partial\tilde{\rho}}{\partial x_k}\rangle+\langle\tilde{\rho}\frac{\partial E_k^t}{\partial x_k}\rangle+\langle\rho^t\frac{\partial\widetilde{E}_k}{\partial x_k}\rangle+E_{0k}\frac{\partial\langle\rho\rangle}{\partial x_k}]=\frac{\eta}{\epsilon_0}\langle\rho\rangle, \tag{14}$$

$$\frac{\partial\langle B_i\rangle}{\partial t}=\frac{\partial\langle E_j\rangle}{\partial x_k}-\frac{\partial\langle E_k\rangle}{\partial x_j} , \tag{15}$$

$$\eta[\langle\tilde{\rho}v_i^t\rangle+\langle\rho^t\bar{v}_i\rangle+\rho_0\langle v_i\rangle]+\epsilon_0\frac{\partial\langle E_i\rangle}{\partial t}=\frac{1}{\mu_0}\left(\frac{\partial\langle B_k\rangle}{\partial x_j}-\frac{\partial\langle B_j\rangle}{\partial x_k}\right) . \tag{16}$$

2.3 Equations for additional Fields of Random Fluctuations

$$\frac{\partial \tilde{v}_l}{\partial t}+\tilde{v}_k\frac{\partial v_l^t}{\partial x_k}+v_k^t\frac{\partial \tilde{v}_l}{\partial x_k}+\langle v_k\rangle\frac{\partial v_l^t}{\partial x_k}+v_k^t\frac{\partial}{\partial x_k}\langle v_l\rangle=\nu_0\Delta\tilde{v}_l-\frac{c_0^2}{\rho_0}\cdot\frac{\partial\tilde{\rho}}{\partial x_l}+\eta\tilde{E}_l+\langle\tilde{v}_k\frac{\partial v_l^t}{\partial x_k}\rangle+\langle v_k^t\frac{\partial\tilde{v}_l}{\partial x_k}\rangle$$
$$+\kappa[\tilde{E}\frac{\partial E^t}{\partial x_l}+E^t\frac{\partial\tilde{E}}{\partial x_l}+\langle E\rangle\frac{\partial E^t}{\partial x_l}+E^t\frac{\partial}{\partial x_l}\langle E\rangle-\langle\tilde{E}\frac{\partial E^t}{\partial x_l}\rangle-\langle E^t\frac{\partial\tilde{E}}{\partial x_l}\rangle-E_0\frac{\partial\langle E\rangle}{\partial x_l}]$$
$$+\eta[(\tilde{v}_jB_k^t-\tilde{v}_kB_j^t)+(v_j^t\tilde{B}_k-v_k^t\tilde{B}_j)+(\langle v_j\rangle B_k^t-\langle v_k\rangle B_j^t)+(\langle B_k\rangle v_j^t-\langle B_j\rangle v_k^t)$$
$$+\langle\tilde{v}_kB_j^t\rangle-\langle v_j^t\tilde{B}_k\rangle+\langle v_k^t\tilde{B}_j\rangle-\langle\tilde{v}_jB_k^t\rangle], \tag{17}$$

$$\frac{\partial\tilde{\rho}}{\partial t}+\frac{\partial}{\partial x_k}[\tilde{\rho}v_k^t+\langle\rho\rangle v_k^t+\rho^t\langle v_k\rangle+\rho^t\tilde{v}_k]=\frac{\partial}{\partial x_k}[\langle\tilde{\rho}v_k^t\rangle+\langle\rho^t\tilde{v}_k\rangle], \tag{18}$$

$$\frac{\partial\tilde{E}_k}{\partial x_k}+\kappa_s[\tilde{E}_k\frac{\partial\rho^t}{\partial x_k}+\tilde{\rho}\frac{\partial E_k^t}{\partial x_k}+E_k^t\frac{\partial\tilde{\rho}}{\partial x_k}+\rho^t\frac{\partial\tilde{E}_k}{\partial x_k}+\rho^t\frac{\partial}{\partial x_k}\langle E_k\rangle+\langle\rho\rangle\frac{\partial E_k^t}{\partial x_k}+\langle E_k\rangle\frac{\partial\rho^t}{\partial x_k}+E_k^t\frac{\partial}{\partial x_k}\langle\rho\rangle]$$
$$=\frac{\eta}{\varepsilon_0}\tilde{\rho}+\kappa_s[\rho_0\frac{\partial}{\partial x_k}\langle E_k\rangle+\langle\tilde{E}_k\frac{\partial\rho^t}{\partial x_k}\rangle+\langle E_k^t\frac{\partial\tilde{\rho}}{\partial x_k}\rangle+\langle\tilde{\rho}\frac{\partial E_k^t}{\partial x_k}\rangle+\langle\rho^t\frac{\partial\tilde{E}_k}{\partial x_k}\rangle+E_{0k}\frac{\partial\langle\rho\rangle}{\partial x_k}], \tag{19}$$

$$\frac{\partial\tilde{B}_l}{\partial t}=\frac{\partial\tilde{E}_j}{\partial x_k}-\frac{\partial\tilde{E}_k}{\partial x_j}, \tag{20}$$

$$\eta[\langle\rho\rangle v_l^t+\rho^t\langle v_l\rangle]+\varepsilon_0\frac{\partial\tilde{E}_l}{\partial t}=\frac{1}{\mu_0}(\frac{\partial\tilde{B}_k}{\partial x_j}-\frac{\partial\tilde{B}_j}{\partial x_k}) . \tag{21}$$

3. CORRELATORS OF TURBULENT FIELDS

Assuming that the turbulence is helical, i.e. stationary, homogeneous, isotropic, but not invariant under reflection, the correlators of the turbulent fields, $\underline{v}^t$, ρ^t, $\underline{E}^t$, and $\underline{B}^t$, have the form

$$\langle v_i^t(\underline{x}_1,t_1)\,v_j^t(\underline{x}_2,t_2)\rangle=K_{ij}(\underline{x}_1,\underline{x}_2)\varphi(t_1-t_2), \tag{22}$$

$$\langle\rho^t(\underline{x}_1,t_1)\,v_i^t(\underline{x}_2,t_2)\rangle=R_i(\underline{x}_1,\underline{x}_2)\varphi(t_1-t_2), \tag{23}$$

$$\langle v_i^t(\underline{x}_1,t_1)B_j^t(\underline{x}_2,t_2)\rangle=T_{ij}(\underline{x}_1,\underline{x}_2)\varphi(t_1-t_2), \tag{24}$$

$$\langle\rho^t(\underline{x}_1,t_1)E_i^t(\underline{x}_2,t_2)\rangle=Q_i(\underline{x}_1,\underline{x}_2)\varphi(t_1-t_2), \tag{25}$$

$$K_{ik}(\underline{x}_1,\underline{x}_2)=C(r)\delta_{ik}+B(r)r_ir_k+g(r)\varepsilon_{ikl}r_l, \tag{26}$$

$$R_i(\underline{x}_1,\underline{x}_2)=D(r)r_i, \qquad \underline{r}=\underline{x}_1-\underline{x}_2, \tag{27}$$

$$T_{ik}(\underline{x}_1,\underline{x}_2)=S(r)\delta_{ik}+R(r)r_ir_k+n(r)\varepsilon_{ikl}r_l, \tag{28}$$

$$Q_i(\underline{x}_1,\underline{x}_2)=H(r)r_i, \tag{29}$$

where B, C, D, H, R, S are scalars, and g, n are pseudoscalars with the consequence of helical turbulence.

4. EXPLICIT EXPRESSIONS FOR AVERAGED EQUATIONS

We consider the case when the characteristic scales of large-scale structures, temporal T and spatial L are much larger than the small scales, τ and ℓ, namely

$$T \gg \tau, \qquad L \gg \ell, \tag{30}$$

In that case, we can assume that

$$\varphi(t_1 - t_2) = \delta(t_1 - t_2) \tag{31}$$

and that the random fields of the turbulent fluctuations are to be Gaussian.

Then, we can evaluate the averages in Eqs. (12) ~ (16) by means of a functional technique with variational derivatives. Using Eqs. (22) ~ (29) and (31) and carrying out rather cumbersome calculations, we finally obtain for Eq. (12)

$$\begin{aligned} \frac{\partial \langle v_i \rangle}{\partial t} - \frac{1}{2}[g(0) - 2\eta n(0)]\, \mathrm{rot}_i \langle v \rangle - \frac{\eta}{2} n(0) \langle v_i \rangle - \eta \langle E_i \rangle \\ = \nu \Delta \langle v_i \rangle - \frac{c_0^2}{\rho_0} \frac{\partial \langle \rho \rangle}{\partial x_i} + \kappa E_0 \frac{\partial \langle E \rangle}{\partial x_i} - \eta^2 \delta. \end{aligned} \tag{32}$$

This is an EMHD equation extended to the EMHD regime and a basic equation capable for the meteorologico-electric phenomena.

5. EQUATIONS OF EMHD VORTICES

We restrict the problem to the solenoidal part of the fields, since we are interested in the generation of vortex structures. Taking the rotation of Eq. (32), we obtain, to a first approximation, retaining linear terms with respect to η,

$$\frac{\partial \underline{\Omega}}{\partial t} - \frac{1}{2}\eta n(0)\, \underline{\Omega} - \frac{1}{2}[g(0) - 2\eta n(0)]\, \mathrm{rot}\, \underline{\Omega} = \nu \Delta \underline{\Omega} \tag{33}$$

where

$$\underline{\Omega} = \mathrm{rot} \langle \underline{v} \rangle \tag{34}$$

is the vorticity vector. In Eq. (33), we find that only difference from the HD case is the new inclusion of two terms with respect to η arising from space charges and the EMHD helical turbulence.

For $\eta_l = 0$, Eq. (33) yields

$$\frac{\partial \underline{\Omega}}{\partial t} - \frac{g(0)}{2} \nabla \times \underline{\Omega} = \nu \Delta \underline{\Omega}, \tag{35}$$

and recovers the result obtained by Moiseev et al. [1].

Fourier-transforming Eq. (33) leads to

$$-(i\omega + \frac{1}{2}\eta n)\Omega_j - \frac{1}{2} i(g - 2\eta n) [\underline{k} \times \underline{\Omega}]_j + \nu k^2 \Omega_j = 0. \tag{36}$$

Choosing the system of coordinates such that the z-axis is directed along the vector $\underline{k}$, we obtain the dispersion equation

$$\omega = i\gamma = i[(-\frac{1}{2})(g - 2\eta n)k - \nu k^2 + \frac{1}{2}\eta n], \tag{37}$$

where γ is the growth rate and Re $\omega = 0$, for real k. We easily find from Eq. (37) the largest value of the growth rate γ_{max} and the wave number k_0 for it

$$\gamma_{max} = \gamma(k_0) = \frac{(g - 2\eta n)^2}{16\nu} + \frac{1}{2}\eta n, \tag{38}$$

$$k_0 = \frac{|g - 2\eta n|}{4\nu} \ll \frac{1}{\ell}, g < 0. \tag{39}$$

From Eqs. (38) and (39), we find that the growth rate of the EMHD vortex can be greater than that of the HD vortex for $\eta_l n > 0$, namely in the presence of space charges and EMHD turbulence and that the generation of a large scale of coherent EMHD vortices is possible.

6. EMHD VORTICES AND CORN CIRCLES

While a theoretical study described so far does indicate the appearance of EMHD vortices, possible evidence of it in the atmosphere has recently been brought by numerous observations of circular symmetric ground patterns in a corn field in southern England [4].

According to eyewitness reports of the vortices in action, the vortex phenomenon was accompanied by acoustic, luminous and electromagnetic effects causing radio-communications interference or radar-TV ghost images. This might indicate that the vortex phenomenon is not a pure HD vortex but is possibly related to an EMHD vortex theoretically described above. It might be inferred that the vortices responsible for the ground traces could be created by the EMHD helical turbulence in a dusty plasma with space charges and possibly in the airflow downwind of a hill in the lower atmosphere and that these dusty charged vortices

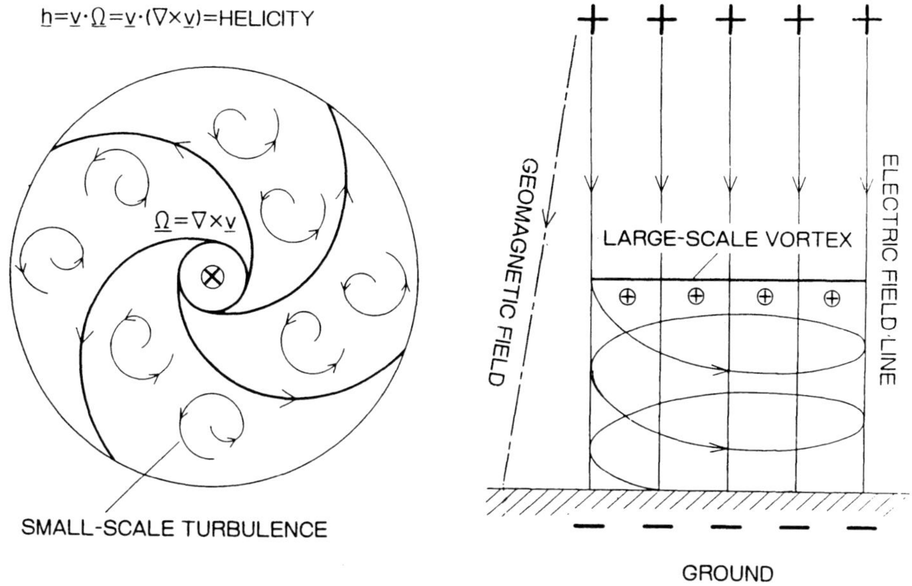

Fig.2. A MODEL FOR FAIR-WEATHER CIRCLE PHENOMENA

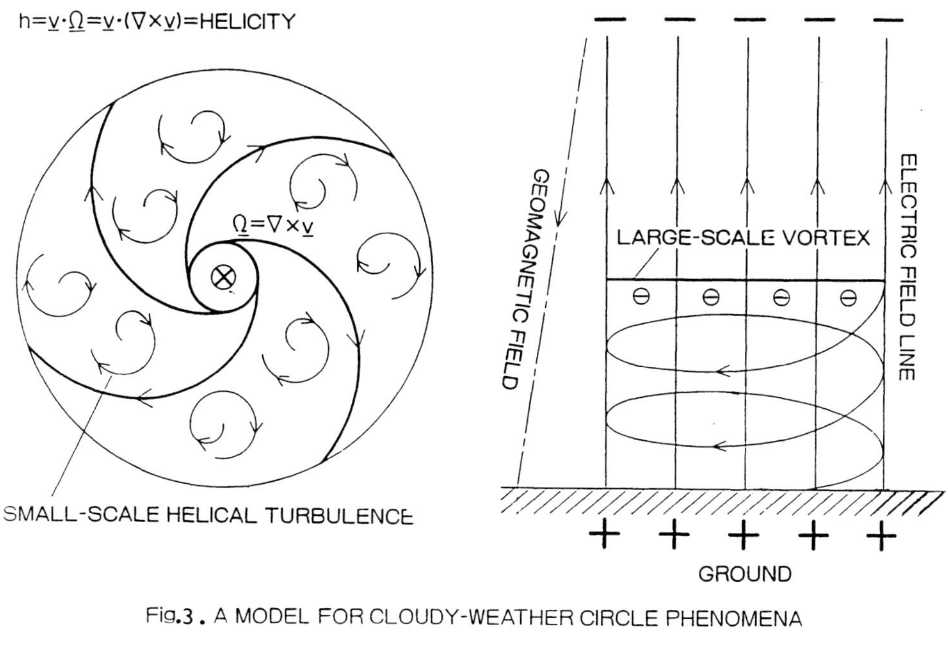

Fig.3. A MODEL FOR CLOUDY-WEATHER CIRCLE PHENOMENA

could be drifting downward in an external vertical electric (and magnetic) field eventually to strike a crop upon the ground, forming a circular trace, often with a helical centre.

Specifically, the circles, which regularly crop up in southern England, are made by dust picked up by winds over the cliffs of Dover. As the dust-laden winds pass over hills, turbulence occurs, creating friction that electrically charges the dust particles. Electromagnetic forces due to natural electric and geomagnetic fields could then combine to produce small-scale helical turbulence and eventually to create a large-scale swirling eddy by the transfer of energy from small-scale to large-scale sizes, which forms a circular trace with an outward-spiraling centre, flattening crops when it moves down by external electric and geomagnetic fields and hits the ground. Accordingly, for fair weather, the circular ground pattern is observed to be a counterclockwise spiral because of a downward electric and geomagnetic field as shown in Fig. 2. In contrast, for cloudy weather, the circular trace is to be a clockwise spiral because of a reversed electric field upwards as shown in Fig. 3.

Using a relation between small and large scale structures, temporal and spatial,

$$\left(\frac{L}{\ell}\right)^2 \sim \left(\frac{T}{\tau}\right) \tag{40}$$

which can be derived from Eqs. (38) and (39), the characteristic horizontal scale and development time of an EMHD vortex structure are estimated as $L \sim 10$ m and $T \sim 10$ s, assuming that the characteristic scale and correlation time of the turbulent fields be $\ell \sim 10$ cm and $\tau \sim 1$ ms. This is in good agreement with relevant data for patterned circle traces.

7. CONCLUSION

A new type of large-scale vortex formation has theoretically been found in helical turbulence in terms of hydrodynamic, electric, magnetic, and space charge fields in an external electric (and magnetic) field. It is called "Electromagnetohydrodynamic Vortices" and is generated as a result of self-organizational processes in nonequilibrium media by the transfer of energy from small-scale to large-scale sizes. Space-charge and electric fields as well as electric and magnetic field turbulence cause additional helicity and facilitate the evolution to large-scale coherent vortex structures. In fact, the growth rate of the EMHD vortex can be greater than that of the pure HD vortex.

The corn circles, which regularly crop up in southern England, have tentatively been identified as a kind of electromagnetohydrodynamic (EMHD) vortices and some of circle-effect phenomena have been explained on the basis of a new theory of EMHD.

REFERENCES

1. Moiseev, S.S., Sagdeev, R.Z., Tur A.V., Khomenko, G.A., Yanovskii, V.V., Sov. Phys. JETP, 58 (6) (1983) 1149.

2. Novikov, E.A., Sov. Phys. JETP, 20 (5) (1965) 1290.

3. Kikuchi, H., in Environmental and Space Electromagnetics, ed. Kikuchi, H. (Springer-Verlag, Tokyo, 1991) this issue.

4. Meaden, G.T., J. Meteorology, 14 (1989) 151.

5. Kikuchi, H., in Nonlinear Dynamics of Structures, ed. Sagdeev, R.Z., Erokhin, N.S. (World Scientific, Singapore, 1991).

15 Assembled Abstracts

15.1 Research at the European Space Agency on Spacecraft/Plasma Interaction and Their Electromagnetic Effects

J. HAMELIN

European Space Agency (ESA), Noordwijk, The Netherlands

Various plasma environments are encountered by spacecraft. Geostationary orbit experiences energetic tenuous plasma 'storms', polar orbits encounter energetic, dense, directed auroral plasma and Low Earth orbits encounter cold, dense plasma. These are in fact very different plasma regimes.

The result of these plasma regimes is to charge the spacecraft to potential levels where electrostatic discharges (ESD) can occur, leading to related EMC problems.

For geostationary spacecraft since 1982 studies have and are still being performed by ESA. These lead to a comprehensive understanding of the phenomena together with prediction and analysis codes and design and test methodology.

For low-earth orbit and in particular for large structures (space-station, shuttle....) similar work has been in progress for about one year.

The object of this paper is to synthesize the present knowledge and to point out in particular the research trends at ESA.

15.2 The Formation of Lightning Upward Discharges

G. Berger

Laboratoire de Physique des Décharges, Centre National de la Recherche Scientifique, Ecole Supérieure d'Electricité, 91192 GIF sur YVETTE Cedex, France

The approach of a negative downward leader induces at the ground level a relatively slow rise of the electric field which may initiate the formation of an upward discharge originated from a protrusion on the earth surface.

These upward discharges have been observed in the nature or experimentally during triggered lightning tests or in a high voltage laboratory.

According to the discharge physics, there are some necessary boundary conditions which have to be satisfied for an upward discharge to be launched. In order to quantify them, a deeper study of the space and time evolution of the electric field above the ground has to be performed. Results show that the electric field intensities are considerably reduced as compared with those measured some tens of meters above the ground due to the injection of corona positive ions in a vicinity of the surface. This leads to a reduction of the initiatory electron source through combination and diffusion processes, thus in turn delays the formation of the upward discharge.

Such results apply to the problem of lightning protection (lightning rod or any other system). Some indications will be given about how to promote the initiation of the ascending discharge with experimental evidences.

At last, we shall show how the problem of the formation of an upward discharge takes part in the build-up of a further modelling of a thunderbolt.

15.3 The Ability of Using Lightning Flash Detector Data for Geophysical Research – A Study Initiated by the ESA Meteosat Program for the Second Generation Meteorological Satellites Instrumentation

V. SCUKA
Uppsala University, Institute of High Voltage Research

Abstract

Thunderstorm thermodynamics, electrification processes in clouds and lightning are interrelated geophysical phenomena not yet well understood. Further research is also needed regarding the fundamental atmospheric electric processes controlling the interaction phenomenology of the thunderstorm clouds chemistry and electrodynamics and the clouds near and distant surrounding space such as the earth surface, the upper atmospheric layers, the ionosphere and the magnetosphere. Today we believe that some of these interaction processes my keep under cover the feed-back mechanisms which accelerate the growth of thunderstrom cells and it has become clear that lightning in all its figures is the central atmospheric electric event in the complexion of the above geophysical phenomena. It is hoped that deeper knowledge regarding the physical characteristics of the above mentioned processes may contribute to better prediction of the long-term effects of e.g. air pollution on the global Earth environment and it is also believed that this knowledge will increase our possibilities to find the means for a better control of the Earth environment. Analyzes of the appearance of lightning on a global scale, which are until now based only on night-time observations which are made from different orbiting satellites do, however, indicate the experimental ability to study the above coupling mechanisms using the lightning discharge as the basic diagnostic tool. Today a second generation of meteorological satellites is under consideration and the space administrations, ESA and NASA, are in progress to develop an advanced optical lightning flash detector to be included into the second generation meteosat instrumentation load. The new lightning flash detectors will register lightning both in the day as well as in the night-time observation period. The real time data from the lightning flash detector shall be used for different warning purposes like warning for severe thunderstorms, hail storms and floods. Referring to the preliminary specifications of these instruments, the lightning data available from the space craft will also be very useful for the above listed scientific research purposes. Using the lightning flash detector data an experimental verification on a direct quantitative base of different existing geophysical models by which we today intend to explain the worldwide geophysical phenomenology will become feasible. Items of preferences for geophysical research for which satellite based lightning flash detector data could be used may be listed as follows: the dynamics and the consistence of the Earth global electrical circuit, the generation process of whistler waves in the magnetosphere and the whistler waves effects on the space plasma particle tripping mechanisms, generation of the atmospheric radio noise and the radio noise time patterns in different geographical locations, the effect of thunderstorms on the ionospheric-winds, on the nature of the horizontal electrical fields in the upper atmosphere and the ionosphere and the variation in time and space of the ionospheric potential.

Using the satellite techniques it is also hoped that it will become possible to find an appropriate atmospheric electric index which will in an efficient way characterize the temporary state of the Earth global atmospheric electric circuit.

We believe that time has come now to consider seriously how to implement the future space based worldwide lightning flash detector data to prove the theories explaining the geophysical phenomenology and to consider the planning of relevant experimental verifications of these theories.

15.4 Detection and Location of Intracloud Lightning Discharges

J.L. BOULAY

Office National d'Etudes et de Recherches Aérospatiales, BP72-92322 Châtillon Cedex, France

Since 1978, the ONERA is involved in basic studies related to the lightning/aircraft interaction problems. One of the first aspect of this research concerns the understanding of the lightning discharge properties and especially the mechanisms associated with the discharges which occurs in altitude.

For example, intracloud lightning processes characterization are essential for the definition of the triggering conditions on a vehicule in altitude (aircraft, missile or launcher) and for the main specifications of the threat (peak current, pulses repetition, electromagnetic fields).

Among several new scientific tools, the ONERA defines and operates an electromagnetic interferometer which permits the detection and the location of intracloud discharges with the following essential requirements:

- frequence range: VHF-UHF,
- detection range: 200 km,
- time resolution: 1 μs,
- location accuracy: 1 at 100 km,
- 2D and 3D reconstruction of the sources propagation.

A 3D system was used first at the Kennedy Space Center and after in the South of France in order to obtain correlations with doppler radar observations.

Two 2D operationnal systems are now used in France (Flight Test Center) and in Frénch Guiana (ARIANE launching site).

In this paper we want to present two categories of situations which can be analysed in the detail with our system:

a - detection and location of intracloud discharges on a very large scale (≃ 200 km) in order to give useful information for operational functions: prevention of lightning accident, air-traffic control, complementary meteorological parameters;

b - detection and location of intracloud discharges on a short range (≃ some 10 km) to permit reconstruction and analysis of different lightning channels; this analysis gives new ideas about basic mechanisms of the lightning initiation and propagation in altitude.

For these two different categories, several examples will be presented and discussed.

15.5 On Triggered Lightnings from Multiple Points

Y. KAWAMATA[1], T. TAKEUTI[1], M. NAKANO[1], and Z-I. KAWASAKI[2]

The Research Institute of Atmospherics, Nagoya University, Toyokawa, 442 Japan[1] and The Department of Electrical Engineering, Osaka University, Osaka, 560 Japan[2]

For several winter seasons, we observed triggered lightnings from multiple points on the coast of the Japan Sea. In winter, the height of thundercloud is lower than that in summer, therefore triggered lightnings, sometimes from multiple points, are easy to generate.

Takagi et al. (1987)(1) reported that 62% of triggered lightnings were those from multiple points, and for triggered lightnings from multiple points, average distance between striking points was 1.2 Km, and average time interval was 12 ms. If some charges in the clouds or charges between ground and clouds move within this short time duration, the corresponding wind velocity should be over 10^5 $m \cdot s^{-1}$. So, it is difficult to explain these phenomena.

Therefore, successive triggered lightnings following the first one did not appear due to the space charge or the cloud charge's field, but due to field changes with cloud, discharge occurred just before the first triggered lightning or due to the field changes of first triggered lightning. So we need to distinguish the first triggered lightning from the successive triggered lightnings.

In this report, based on a past observational result, we discuss the cause of winter multiple triggered lightnings.

(1) Takagi, N., T. Yamamoto, T. Watanabe, I. Arima, T. Takauti, M. Nakano, Z-I. Kawasaki, H. Kinosita, K. Yamamoto, K. Saikawa, T. Higuti, and M. Suzuki. On the simultaneous multipoints lightning flashes, IEE Japan, HV-87-52, (in Japanese), 1987

15.6 Small-Scale Variations of ELF and VLF Magnetic and Electric Field Components in Subauroral Zone

Ja.I. Likhter[1], V.I. Larkina[1], and G.L. Gdalevich[2]
IZMIR AN USSR, Troitsk, Moscow Region, USSR[1] and IKI AN USSR, Moscow, USSR[2]

Abstract

When "Intercosmos 14" and "Intercosmos 19" satellites passed across the region of inhomogeneities of concentration in the topside ionosphere, which are close to the high latitudinal boundary of the main ionospheric trough, it was found out that there were simultaneous and synchronous fluctuations of plasma concentration N_e, plasma temperature T_e and of the amplitudes H_x, E_y of VLF and ELF emissions field components. Probability characteristics of H_x and E_y fluctuations are analyzed. The correlation coefficient $R_{Hx,Ey}$ at frequencies less than 4.65 kHz has turned out to be smaller than 0.5; at higher frequencies it increases up to 0.8 - 0.9. The interpretation of this and similar facts is this: 1) At frequencies less than F_{LHR} a noticeable role belongs to electrostatic waves (drifting or gradient-drifting) which are locally excited. The electrostatic wave have no genetic relation with the electromagnetic ones arrive to the place of registrations from remote regions. 2) At F more than F_{LHR} electromagnetic waves undergoing the influence of the N_e inhomogeneities are recorded. These assumptions explain satisfactorily the phenomena observed.

* IZMIR AN USSR, Troitsk, Moscow region, USSR.
** IKI AN USSR, Moscow, USSR.

15.7 Electromagnetic Wave Environment in the Magnetosphere Observed by Akebono Satellite – Preliminary Results

I. KIMURA[1], K. HASHIMOTO[2], I. NAGANO[3], T. OKADA[4], and M. YAMAMOTO[1]

Department of Electrical Engineering, Kyoto University, Kyoto, 606 Japan[1], Department of Electrical Engineering, Tokyo Denki University, Tokyo, 101 Japan[2], Department of Electrical Engineering, Kanazawa University, Kanazawa, 920 Japan[3], and Research Institute of Atmospherics, Nagoya University, Toyokawa, 442 Japan[4]

A Japanese scientific satellite Akebono (EXOS-D) was launched in an orbit, 10,395 km of apogee, 272km of perigee and 72.5°of inclination, on February 22, 1989 for the exploration of auroral, its energy source and relating electromagnetic environment in the magnetosphere.

Auroral UV and visible imager, energetic particle detectors, electric field detector, plasma probe and magnetic field sensor are on board in addition to VLF, LF and HF wave receivers.

The VLF instrument involves WBA (wideband analyser for a frequency range from 100Hz to 13kHz), PFX (propagation vector analyser by two components of electric field (E) and three components of magnetic field (B) for swept frequency or a fixed frequency in the frequency range from 100Hz to 12.7kHz), MCA (multi-channel analyser in the VLF range for E and B fields), ELF (ELF frequency receivers for E and B fields for frequencies less than 100Hz) and VIP (measurement of vector impedance of crossed dipole antennas of 60m tip-to-tip). Tracking of the PCM S band telemetry and U band analog telemetry is made at Kagoshima Space Center, Syowa station in Antarctica, Prince Albert in Canada and Esrange in Sweden. VLF wide band signals are telemetered by the U band link.

In this report, we introduce preliminary results observed by VLF instruments on board Akebono. All sub-systems of VLF instruments are working successfully. In the VLF range we have found wellknown features of VLF wave phenomena, such as whistlers, VLF emissions, ground-based Omega signals transmitted from Tsushima, Japan and Australia, and funnel shaped VLF hiss emissions especially in the auroral zone. One of our specific characteristics of the VLF instruments is PFX by which we are able to determine wave normal direction of signals of interest for a band width of 50 Hz, the center frequency being swept or fixed by command.

Acknowledgments: We are grateful to ISAS launching team, and NEC and Meisei Electric Company to have assembled the satellite and fabricated our VLF instruments respectively. We also thank other members of VLF team not listed as the authors, T. Yoshino of Electro-communication Univ., H. Matsumoto of RASC, Kyoto University, and M. Ejiri of National Inst. of Polar Res. and thank all staff of scientific instruments on board Akebono for their collaboration.

15.8 Schumann Resonances in Middle Latitudes of Asia

YU.B. BASHKUEV, V.B. HAPTANOV, D.G. BUYANOVA, and E.M. MITKINOV

Nature Science Institute, Siberian Division Academy of Sciences, Ulan-Ude 42, USSR

In the present report the results of the experimental study of the earth-ionosphere cavity global resonances made in middle latitudes of Asia (Transbaikalia) over the years 1979-1988 have been summarized. The integral enegy spectra of the orthogonal electric and magnetic components are received in the frequency band 3-50 Hz. The resonances from 3 to 6 were observed in the most spektra at 7.7,13.7,20.1,26.3, 32.3 Hz (average value), Q-factors were varied from 3 to 8, resonance frequencies were changed from 0.5 to 1 Hz. Maximum spectral intensity levels exceed minimum levels from 1.6 to 4 times. The experimental data of f_n and Q-factors in comparison with the other author's theoretical and experimental ones are presented.

The spectral properties of the global resonances reveal large variability of the level spectral densities, their frequency content and the Q-factors. The main cause of it is the periodical change of the low ionosphere parametres. The propogation parametres ReS=c/v_{ph} and α were determined from the experimental Schumann spectra. The typical value of phase velocity is in the range from 0.71c to 0.79c and increases with the frequency increase. The average values of the attenuation rate α are 0.16,0.24,0.34 dB/Mm at day and 0.22,0.35,0.56 dB/Mm at night at the first, second and third reconance frequencies. Comparison of experimental data of ReS and α with the isotropic and anisotropic daytime and nighttime exponential curves are carried out. Applications of the global reconances to investigate the structure of Earth's crust are considered.

15.9 Type II Interplanetary Radio Bursts

H. Kikuchi
Nihon University, College of Science and Technology, Tokyo, 101 Japan

Conventional ground observations of solar radio bursts are usually limited to a range above 10 MHz in frequency or below some tens of meters in wavelength because of the shielding effect of the ionosphere, and their source origin is thought to be located less than 0.05 AU or 10 solar radio in the solar corona. Recent spacecraft measurements in situ allowed us to cover low-frequency radio bursts down to frequencies as low as 10 kHz, including deca-, hecto-, and kilo-meters in wavelength. Their source positions are extended to interplanetary space not only in the solar corona, thus being referred to as interplanetary radio bursts (IRB). Based upon in situ observations from the HELIOS 1 and 2 Solar Probes, there has been found a new type of interplanetary radio bursts with electron plasma oscillations (EPO) that are indicated as type II in close association with interplanetary shocks, superthermal electrons, and low/medium energy protons, and correspond to ground observations of the type II solar radio burst generated in the coronal region. While these EPO-IRB events are produced at the front of a large scale of interplanetary hydromagnetic shocks, there has been found another kind of EPO-IRB events associated with a small scale of sporadic electrostatic shocks that are characterized by superthermal electrons in a way similar to type II IRB in a highly turbulent downstream region behind a hydromagnetic shock near a HELIOS perihelion. Such a typical event obatined from HELIOS 2 on April 22, 1977 is presented and is tentativety termed as modifed type II. In situ measurements indicate that as close as to 0.3 AU a downstream region behind a large-scale of flare-associated hydromagnetic shocks is often highly variable and turbulent, sometimes producing magnetic clouds as reported and possibly electric clouds as well which could produce local electrostatic shocks with superthermal electrons responsible for EPO-IRB.

15.10 New Results for Lightning Activity on Venus and Jupiter, and Predictions of Lightning Activity on Neptune

W.J. BORUCKI

NASA Ames Research Center, Moffett Field, CA 94035, USA

One goal of planetary lightning research is the identification of the specific conditions that produce lightning on each planet. On Earth, lightning activity is produced by condensates in cumuliform water clouds. On Venus, there are no water clouds, yet there is abundant evidence of lightning. On Jupiter, water-rich air parcels sink rather than rise because water vapor is much denser than the ambient atmosphere. Yet Jupiter shows two, 100,000 km long, lines of thunderstorms. In this paper, a summary is presented of recent analyses of spacecraft data that give us clues about the conditions that produce lightning on Venus and Jupiter.

In particular, the recent papers of Russell and colleagues show that the source of the signals observed by the Pioneer Venus Orbiter electric field detector must be below 200 km and therefore the signals must be due to lightning rather than magnetospheric processes. The observed altitude versus frequency dependence of the signals is similar to that found by Hayakawa for terrestrial whistlers detected by the Ariel 4 satellite. However, preliminary results of a new search for optical pulses from venusian lightning indicate that lightning flashes must be weak, infrequent, or both.

A new analysis of Voyager 2 images of Jupiter confirm the existence of a band of lightning storms at 49 degrees North latitude and show a second band at 17 degrees North latitude. No lightning activity is observed in the Southern Hemisphere. This latitude dependence is very different from that found by Kotarii et. al. for terrestrial lightning. A determination of the latitude and longitude of each storm center and the association of these positions with cloud patterns shows that the northerly line of storms has its origin below the ammonia clouds: possibly in the water clouds.

Based on the observed lightning activity on Venus, Earth, and Jupiter, predictions are presented for the frequency of Neptunian lightning.

15.11 High Power Electromagnetics – A. EMP Hardening

C.E. BAUM[1] and R.L. GARDNER[2]

Weapons Laboratory/NTA (AFSC), Kirtland Air Force Base, NM 87117, USA[1] and Mission Research Corporation, Albuquerque, NM 87106-4245, USA[2]

High Power Electromagnetics (HPE) is a collection of electromagnetic interactions with systems that are strong enough to cause non-linear effects. HPE specifically includes, but is not limited to, EMP, lightning, and high power microwaves. This talk will begin with an overview of HPE and lead into EMP hardening issues. Lightning electromagnetics is covered in a companion talk.

The EMP hardening main section of this talk is divided into Singularity Expansion Method (SEM), Topology, and Transmission line theory. SEM and Topology are considered only for their fundamentals and because of their importance to EMP hardening. SEM divides the system response into a series of decaying exponentials so that response from different paths of a scattering body can be obtained.

Topology provides a practical means of designing a system to be hard to incident electromagnetic fields and currents. It can be approached in a qualitative way which appeals to practicing engineers. Basic rules and applications will be considered.

Finally, most real systems are very complex, and it is difficult to directly calculate or measure their reaction to electromagnetic waves. Methods to estimate the interaction are often based on Transmission line theory. An effective method of formalizing multiconductor transmission line theory is through the use of the BLT equation. The forms of the BLT equation will be shown, followed by new applications and approximations tailored to system like cable configuration are shown.

15.12 High Power Electromagnetics – B. Lightning Electromagnetics

R.L. Gardner

Mission Research Corporation, Albuquerque, NM 87106-4245, USA

This paper will review progress in electromagnetics as it applies to lightning. The review is based on an upcoming book by Hemisphere which, in turn, is an expansion of the special issue on the electromagnetics of lightning in *Electromagnetics*. There are contributions from many of the European and U.S. Lightning Programs. The subject is divided into four categories:

1. Models of Fundamental Lightning Processes;
2. Propagation;
3. Measurement of Lightning Parameters; and
4. Lightning Interaction with Systems.

Models of Fundamental Lightning Processes describes a series of first principles models of the development of lightning currents from initial conditions. There is a concentration of the quantitative physical development of streamers and the transmission line elements that result. These transmission line elements are then used in nonlinear, nonuniform transmission line models which predict the electromagnetic observables from a set of initial conditions.

Propagation is a new section that did not appear in the special issue. It starts with an overview of the propagation of electromagnetic fields over lossy media by J. R. Wait. It is completed by the application of those methods to lightning.

The largest section is called, "Measurement of Lightning Parameters." This section is a compendium of data presentations from a number of the active lightning experimental groups. Of interest are fast measurements of electromagnetic parameters, such as currents and fields.

The final section considers the application of the theory and data presented above and empirical data to particular systems.

15.13 Results of the Scope-Enuwar Project, Which Involved URSI

M.W. Wik

Defence Materiel Administration, Electronics Directorate, S 115 88 Stockholm, Sweden

Abstract

URSI contribution in a broad interdisciplinary scientific program to the SCOPE-ENUWAR project (Scientific Committee On Problems of the Environment, Environmental Consequences of Nuclear War) has previously been presented and published. However, the SCOPE-ENUWAR overall results are less known to the URSI community and are therefore summarized.

The project was initiated by ICSU (International Council of Scientific Unions) in 1982 and involved approximately 300 scientists from more than 30 countries. The scientific results have been published in two volumes by John Wiley & Sons.

The first volume describes the physical and atmospheric effects of a nuclear war and includes the URSI contribution. The second volume describes the ecological and agricultural effects of a nuclear war.

The findings by SCOPE-ENUWAR and URSI have also supported a report to the United Nations. The New Zealand Planning Council was supported by URSI and published a book entitled "New Zealand after Nuclear War" which to a great extent deals with the effects of nuclear electromagnetic pulse effects on society.

15.14 Magnetic Coupling of Precursors at Volcano Eruption

T. YOSHINO
University of Electro-Communications, Tokyo, 182 Japan

At the time of the major volcanic eruptions on the 15th and 21st of November, 1986 at Mt. Mihara in Izu-Ohshima Island, Tokyo territory in Japan, a magnetic field strength detector of a multipoint direction finding network for detection of the seismogenic electromagnetic emissions which was set up at Ohshima Volcanic observatory of Tokyo University had continued the observation. Since October 20th, a lot of magnetic impulsive noise bursts at 82 kHz were recorded, and the occurrence of these anomalous bursts increased after November 3rd. At 10 - 16 JST (Japanese Standard Time, +9 hours from UT) on November 14, a lot of strong noise bursts were observed. The first major eruption occurred at 16 JST on November 15th in the main and old summit crater of this volcano.

At 10 - 12 JST on November 21st, the noise bursts were observed again four hours before the following new violent and sudden eruptions, in which 19 new craters erupted suddenly on the caldera and on the side slope of the mountain body. These records are the first observations of electromagnetic emissions as precursors to volcano eruptions activity correlated with the time of the insertion of dykes inside this mountain body.

The author touches upon briefly on new observations relating to another volcanic activity at Ito which erupted only about one month before this presentation.

15.15 Ball Lightning Properties

A.I. Grigoryev
Yaroslavl State University, USSR

Functions of the density of ball lightning distribution on its existence duration and characteristic linear dimensions have been plotted according to the results of statistically processed by computers 3200 descriptions of ball lightning events (BL) collected from the eyewitnesses. Specific features of BL appearance and disappearance, movement and heat radiation etc. have been analysed. The authors investigated the connection of BL observation frequency with the intensity of thunder activity, which turned out quite rigid. This allowed to draw certain conclusions of the conditions of BL appearance. Pair correlation dependences between BL existence duration, characteristic linear dimensions, brightness and colour have been stated. Specific feature of the dependences of BL existence duration on brightness and colour is nonmonotony. The correlation dependence between BL brightness and its characteristic linear dimensions indicates that radiation proceeds from the BL surface layer the thickness of which is less than the radius. The authors evaluated relative errors in the quantitative data of BL characteristics reported by the eyewitnesses.

The authors also examined the reasons of disagreement in average BL characteristics derived from various independent collections of BL descriptions made in different periods of time by different authors: Mc Nally, Stachanov I.P., 3200 descriptions, prepared for processing by computers in the Yaroslavl State University. The data were processed by two independent investigators, one working with 2I00 and the other with II00 descriptions. The descriptions to be processed were chosen randomly out of $\approx$ 6000 cases available at the University. Statistic analysis of the data showed that average BL characteristics derived from the two collections differed. This fact indicates rather a strong psichological effect of those engaged in processing and analysing the data of BL observation cases on the results obtained.

SPRINGER NATURE

GPSR Compliance

The European Union's (EU) General Product Safety Regulation (GPSR) is a set of rules that requires consumer products to be safe and our obligations to ensure this.

If you have any concerns about our products, you can contact us on ProductSafety@springernature.com

In case Publisher is established outside the EU, the EU authorized representative is:

Springer Nature Customer Service Center GmbH
Europaplatz 3
69115 Heidelberg, Germany

FSC
www.fsc.org
MIX
Papier | Fördert gute Waldnutzung
FSC® C083411

Kolibri 360 GmbH
Ferdinand-Jühlke-Straße 7
99095 Erfurt, Deutschland
produktsicherheit@kolibri360.de